ADVANCED THEORY OF VIBRATION

(Nonlinear Vibration and
One Dimensional Structures)

ADVANCED THEORY OF VIBRATION

(Nonlinear Vibration and One Dimensional Structures)

J S Rao
MTech., PhD, DSc.
Former Professor of Mechanical Engineering
BHEL Chair
Indian Institute of Technology
New Delhi

NEW AGE
REFERENCE

An Imprint of

NEW AGE INTERNATIONAL (P) LIMITED, PUBLISHERS
LONDON • NEW DELHI • NAIROBI
Bangalore • Chennai • Cochin • Guwahati • Hyderabad • Kolkata • Lucknow • Mumbai
Visit us at **www.newagepublishers.com**

Published by New Age International (P) Ltd., Publishers
First Edition: 1992
Reprint: 2017

GLOBAL OFFICES

- **New Delhi** **NEW AGE INTERNATIONAL (P) LIMITED, PUBLISHERS**
7/30 A, Daryaganj, New Delhi-110002, (INDIA)
Tel.: (011) 23253771, 23253472, **Telefax:** 23267437, 43551305
E-mail: contactus@newagepublishers.com • Visit us at www.newagepublishers.com
- **London** **NEW AGE INTERNATIONAL (UK) LTD.**
27 Old Gloucester Street, London, WC1N 3AX, UK
E-mail: info@newacademicscience.co.uk • Visit us at www.newacademicscience.co.uk
- **Nairobi** **NEW AGE GOLDEN (EAST AFRICA) LTD.**
Ground Floor, Westlands Arcade, Chiromo Road (Next to Naivas Supermarket)
Westlands, Nairobi, KENYA, **Tel.:** 00-254-713848772, 00-254-725700286
E-mail: kenya@newagepublishers.com

BRANCHES

- **Bangalore** 37/10, 8th Cross (Near Hanuman Temple), Azad Nagar, Chamarajpet, Bangalore- 560 018
Tel.: (080) 26756823, **Telefax:** 26756820, **E-mail: bangalore@newagepublishers.com**
- **Chennai** 26, Damodaran Street, T. Nagar, Chennai-600 017, **Tel.:** (044) 24353401
Telefax: 24351463, **E-mail: chennai@newagepublishers.com**
- **Cochin** CC-39/1016, Carrier Station Road, Ernakulam South, Cochin-682 016
Tel.: (0484) 2377303, **Telefax:** 4051304, **E-mail: cochin@newagepublishers.com**
- **Guwahati** Hemsen Complex, Mohd. Shah Road, Paltan Bazar, Near Starline Hotel
Guwahati-781 008, **Tel.:** (0361) 2513881, **Telefax:** 2543669
E-mail: guwahati@newagepublishers.com
- **Hyderabad** 105, 1st Floor, Madhiray Kaveri Tower, 3-2-19, Azam Jahi Road, Near Kumar Theater
Nimboliadda Kachiguda, Hyderabad-500 027, **Tel.:** (040) 24652456, **Telefax:** 24652457
E-mail: hyderabad@newagepublishers.com
- **Kolkata** RDB Chambers (Formerly Lotus Cinema) 106A, 1st Floor, S N Banerjee Road
Kolkata-700 014, **Tel.:** (033) 22273773, **Telefax:** 22275247
E-mail: kolkata@newagepublishers.com
- **Lucknow** 16-A, Jopling Road, Lucknow-226 001, **Tel.:** (0522) 2209578, 4045297, **Telefax:** 2204098
E-mail: lucknow@newagepublishers.com
- **Mumbai** 142C, Victor House, Ground Floor, N.M. Joshi Marg, Lower Parel, Mumbai-400 013
Tel.: (022) 24927869, **Telefax:** 24915415, **E-mail: mumbai@newagepublishers.com**
- **New Delhi** 22, Golden House, Daryaganj, New Delhi-110 002, **Tel.:** (011) 23262368, 23262370
Telefax: 43551305, **E-mail: sales@newagepublishers.com**

ISBN: 978-81-224-0442-5
C-16-11-9800

Printed in India at Glorious Printers, Delhi.

NEW AGE INTERNATIONAL (P) LIMITED, PUBLISHERS
7/30 A, Daryaganj, New Delhi-110002
Visit us at **www.newagepublishers.com**
(CIN: U74899DL1966PTC004618)

To

My Wife Indira

and

To the memory of my parents Jammi Chikka Rao and Ramanamma

To

my wife Uma,

and

to the memory of my parents Jajula Chikka Rao and Ramamma

Preface

Vibration analysis is now well recognized as essential for good engineering design. Over the years, the speed, size and complexity to be taken into account in the design of engineering systems, have grown steadily; so also the demand on designers to improve the rigor and sophistication of vibration analysis. The result was development of vibration analysis into a sturdy discipline in its own right, of course, drawing liberally upon from subjects like mechanical and electrical engineering, strength of materials, structural mechanics, analytical methods, nonlinear analysis, variational calculus, and so on, and aided by computer application. Obviously, therefore, what now constitutes the theory of vibration – particularly, advanced theory – is scattered over a large number of publications relating to different disciplines. What has been attempted in the present book is a comprehensive consolidation of them and its presentation in a concise manner for the benefit of those aspiring to specialize in vibration studies at postgraduate and doctoral levels.

The contents of this book have got crystallized over a period of 25 years while teaching and guiding doctoral level research at the Indian Institutes of Technology at Kharagpur and Delhi; and at Institute Nationale des Sciences Appliquee, Lyon, France; Concordia University, Montreal, Canada, Rochester Institute of Technology, Rochester, USA; and Gesamthochschule, Kassel, Germany as visiting professor.

The emphasis in this book is on analysis of continuous systems rather than discrete system models. Starting with single degree of freedom systems with nonlinear stiffness and damping, and variable inertia problems, various analytical, graphical and numerical techniques including stability analysis are discussed. A concise treatment of variational principles and their application to vibration problems is given next. Detailed analysis of one dimensional structural members (longitudinal vibration of bars, torsional vibration of rods with circular and noncircular cross-sections, and lateral vibration of beams) is given. Following Hamilton's principle various theories have been derived in a unified manner giving the involved assumptions. Solutions to problems of free and forced vibrations are presented. Members with variable cross-sections have been considered in detail by using energy method. Vibration theories of viscoelastic materials in longitudinal vibration and lateral vibration have also been considered at length. Various aspects of beams on foundations, moving loads, multilayered beams and combined effects such as bending-torsion, bending-bending and rotating beams are presented in a concise manner. Nonlinear problems of beams in lateral vibration are also discussed. The problem of stress analysis of rods and beams in forced vibration is presented using Reissner's theory. Thus the book seeks to expose the reader to a large variety of problems of one dimensional structures.

The author is indebted to all those who have contributed in one way or other to development of this book. All my students who have interacted with me in the class rooms during the last 25 years have helped me to crystallize my thinking on various topics. Their names are far too many to accomodate in the limited space. However, particular mention needs to be given to those students who worked with me very closely on various problems. They are Dr D.K. Rao, Dr M.K. Khashu, Dr M. Swaminadham, Dr S. Banerjee, Dr E. Raghavacharyulu, Dr R.C. Kar, Dr K.N. Gupta, Dr V.V. Satyanarayana, Dr P.N. Rao, Dr S.S.P. Rao, Dr H.M. Jadvani, Dr A. Cameron, Dr P.V. Reddy, Dr K.B. Subrahmaniam, Dr N.S. Vyas and Dr K.V. Bhaskara Sarma.

As it always happens, the major brunt of diverting my time to writing a book of this nature and volume naturally fell on the members of my family. I am deeply indebted to my wife Indira who has shown immense understanding and extended full cooperation throughout the last 28 years, and particularly while writing this book.

Finally, I am grateful to M/s Wiley Eastern for publishing four of my earlier books and the present one.

New Delhi

J.S. RAO

Contents

1

Introduction

All mechanical systems are inherently prone to vibration problems as their design features themselves generate some pulsating forces. In internal combustion engines, burning of air fuel mixture once in every four or two strokes, generates a highly fluctuating force, which in turn produce several harmonics of lateral forces and torsional moments acting on the crank shaft. These harmonic forces are felt throughout the engine, flywheel and driven unit, such as a compressor or a generator. In a steam or gas turbine, the flow interaction between the stator blades and the moving rotor blades, generates fluctuating aerodynamic forces on the blading and they get transferred through the blades to the disk and the shaft of the system. Similarly, in various other applications, such as textile mill looms, marine drives, mining machinery, etc. periodic loads are generated by their basic work principle itself.

Occurrence of transient loads is another problem with almost all mechanical systems. This again is a phenomenon arising mainly from design features. For example, when the speed of an automobile is changed by increasing or decreasing the fuel supply or by changing gears, the engine develops more power or less power, accordingly. This happens also when the automobile encounters a sudden bump or a ditch on the road or a general unevenness of the road surface. When the load on a generator in a power plant changes depending on the demand, the governor system responds and alters the energy supply, thus changing the excitation forces for a small period of time. While some of these transient loads are predictable, there are others which are not easy to predict. For example, an earthquake can rattle a machine system or a building leading to violent transient vibrations to the point of their total collapse.

There is yet another class of vibration problems which is quite common and generated by the action within the machine itself. Self-excited vibration due to oil film action in a bearing is a typical example of this category. The aerodynamic forces that change with the angle of attack of the upstream flow on an airfoil can cause the vibration of the airfoil to grow without an external source of excitation. The variable mass properties of a system such as a marine diesel engine can cause instabilities. The variable stiffness properties due to a noncircular shaft can cause unstable regimes due to parametric excitations.

The small amplitude vibrations thus produced in a variety of low speed mechanical systems do not cause major problems, and as such vibration study has comparatively less relevance to them. However, vibration study assumes greater importance for complex, high speed, and heavy duty mechanical systems, in view of the critical need to minimize large amplitude vibrations in them, and thereby to minimize occurrence of the resulting dynamic stresses which can cause their failure.

It is sufficient to analyze most mechanical systems by small amplitude vibration theories, by avoiding large amplitude regimes under resonance conditions. It is then possible to apply linear theories. With stiffness and damping forces represented as proportional to displacement and velocity, respectively, the equations of motion become linear and hence the solution becomes easier.

Linear relationship, however, does not always exist, particularly when the deformations are large. The stiffness and damping forces then become nonlinear functions of displacement and velocity, respectively. Sometimes we design systems to be nonlinear to take advantage of their properties under large amplitudes of vibration. A flexible coupling used to connect the drive unit with the driven load of a mechanical system, is a typical nonlinear torsional spring. Such a coupling may be subjected to large angular deformations under steady or transient conditions of operation. In some cases, stiffness may not be independent of time, a typical case being a generator shaft. Variable stiffness can be still linear. However, it is responsible for instabilities due to parametric excitation, and is typically a subject studied under stability of systems in nonlinear vibrations.

The damping present in a mechanical system is very complex and invariably nonlinear. Very often, a theoretical linearized model is used which gives rise to a simple mathematical analysis. This may lead to erroneous results in practice, particularly at or near resonant conditions. Typical cases include the cases of long transmission cables undergoing large amplitudes of vibration, where the damping force is amplitude dependent; and the damping in a rotating blade-disk system due to friction at the interface junction where material deformation is dependent on rotational speed and amplitude of vibration.

The mass of a system is generally constant, except in applications such as space vehicles and missiles. In the case of reciprocating engines, the mass moment of inertia of each cylinder bank varies periodically with several harmonic components of engine speed. In linear theory an average value for this mass moment of inertia is considered in the study of torsional oscillations so as to simplify the problem. However, for large marine diesel engines, this variation may be an important factor in the stability analysis.

In addition to the nonlinear behaviour of inertia force, damping force and stiffness force, we also come across nonlinearities due to geometry. For example, the Coriolis forces of a long rotating blade mounted on a rotating disk are nonlinear in nature. Typically, such nonlinearities in a system give rise to an entirely different dynamic behaviour when compared with linear theories.

There are several books that deal with nonlinear vibration theory, but few that directly deal with mechanical vibration problems for single degree freedom as well as continuous systems. In this book, chapter 2 deals with nonlinear vibration of simple systems by analytical as well as by numerical techniques. Stability considerations are given in chapter 3, by graphical, analytical and numerical techniques. Forced oscillations of nonlinear single degree of freedom systems is dealt in chapter 4. The transient behaviour of forced oscillation systems is also considered in this chapter, by graphical, analytical and numerical methods.

Variational principles form the backbone for an accurate understanding of vibration of continuous systems. Procedures based on strength of materials principles are approximate and they fail to reveal the limitations of such theories, though they give rise to simple theories. Quite often, they are not easy in their application for deriving the theory of a complicated system. Variational principles provide for the most general way of deriving the governing equations of motion of continuous systems. In addition, these variational principles themselves suggest the best way of solving complicated problems of continuous systems, particularly those which are nonlinear. Towards this end, a basic review of tensors, linear elasticity principles, linear viscoelasticity principles and Laplace transform are given in Appendices 1 to 4. With the help of these Appendices, variational principles are dealt with in chapter 5, particularly with an engineering approach. Functions with several discrete independent variables and time, with higher order derivatives, with independent variables of space and time, corresponding to discrete and one dimensional structures, are considered. The principle of virtual work, Hamilton's principle, complementary virtual work and Reissner's principles are presented in detail with examples. Numerical techniques based on variational principles, viz., Lagrange, Ritz and Galerkin methods are presented with examples.

The variational principles are applied to one dimensional structures in the next three chapters. Bars in axial or longitudinal vibration are considered in chapter 6. First the elementary theory is derived using variational principles and the limitations of this basic theory are pointed out. Rayleigh and Bishop's theories are then derived to remove these limitations and derive accurate theories of longitudinal vibration of bars, alongwith the different possible boundary conditions. Both free and forced vibrations are considered. Non-uniform bars are also considered. Then the cases of viscoelastic bars and viscoelastically coated bars are presented.

The elementary theory of torsion of circular rods is first outlined in chapter 7. The St Venant's theory for noncircular rods, Love's theory, Timoshenko-Gere theory are all presented next, by deriving them in an unified manner through variational principles. Prandtl's membrane analogy is also presented for thin cross-sections of rods. The torsion of rectangular cross-section is considered and then applied to vibration of tapered rods. Pre-twist increases torsional rigidity of rods which is then considered according to Carnegie's theory. Torsional vibrations of pre-twisted tapered rods are also considered. The theory of torsional vibration is further improved by deriving the Reissner and Lo-Goulard's theories, Barr's theory and then the most accurate theory of rods are derived in chapter 7. Finally, stress analysis of vibrating rods by classical theory, followed by Lagrange method using potential energy theory and Reissner's variational functional are presented.

In the last chapter, the lateral vibration of beams is considered. First the classical Euler-Bernoulli theory is presented. Rayleigh's modification to this theory by introducing the rotary inertia is next presented. This theory is further refined by including Timoshenko's correction for shear deformation. The limitations of these theories from the point of view of classical linear elasticity are pointed out and Volterra's theory of beams is derived using classical variational principles. Free and forced vibration of various types of beams is then considered. Special cases of beams with axial loads, beams on elastic foundation, and viscoelastic beams on viscoelastic foundation are then presented. Next the cases of laminated beams are presented by first deriving Oberst's theory for two-layered beams, followed by Ditaranto's theory for three-layered beams. Then a general n-layered beam theory is presented with a specific example of three layers with the middle one as a viscoelastic layer. Pre-twisted beams which exhibit coupled bending motion in two different planes are next considered with examples. When the cross-section of a beam is asymmetric, the bending motions are coupled with torsional motion and this coupling is next considered. The influence of rotation on a beam is next presented. In all these cases, variational principles are used to derive the governing equations of motion in a unified manner. The Coriolis forces due to the lateral motion of a rotating beam are next accounted and the solution of the nonlinear problem is presented. The stress analysis of a vibrating beam is then presented by using Reissner's variational functional. Examples of a rotating beam under steady excitation are given using variational solutions.

As outlined above, this book attempts to provide a clear understanding of linear and nonlinear vibration of discrete systems and one dimensional structures, viz., bars, rods and beams of various types. The philosophy of the book is to stress on the basic understanding of the theories and solution procedures, which will be helpful in advanced machine design of high speed machine elements.

2

Nonlinear Vibration of Simple Systems

For small amplitude linear vibration, the equation of motion for a single degree of freedom system is of the form

$$m\ddot{x} + c\dot{x} + kx = F(t) \tag{2.1}$$

In the above, m, c and k represent the mass, damping coefficient and stiffness of the system; and $F(t)$, the excitation function and x, the response of the system due to the excitation force. The parameters m, c and k are taken as constants, representing the proportionality coefficients of the inertia force, damping force and stiffness force, which are taken directly proportional to acceleration, velocity and displacement, respectively. This assumption is valid for many mechanical systems for small amplitude vibration. Since all these forces are linear functions of the independent parameters, the differential equation of motion for the system is linear and the theory of vibrations studied with the help of the above equation is linear theory of vibrations. Such a theory of vibration is very useful to predict the natural frequencies according to linear analysis, which is relatively simple, where large amplitudes of vibration occur. The linear theory is also very useful to predict the dynamic behaviour of a system at frequencies away from resonance. Thus the linear theory serves most of the requirements for design purposes as long as the system is detuned from resonance conditions.

In practice, however, no linear relationship exists, particularly when the deformations are large. An example of nonlinear stiffness is the case of a flexible coupling used to connect the drive unit with the load of a mechanical system. The nonlinearity is characterized by a relationship as shown in Fig. 2.1. Such a coupling may be subjected to large angular deformations under steady or transient conditions of operation. In some cases, the stiffness may not be independent of time, a typical case being that of a generator shaft which may be approximated by a rectangular cross-section as shown in Fig. 2.2. The variable stiffness can be still linear, however, it is responsible for instabilities due to parametric excitation, and is typically a subject studied under stability of systems in nonlinear vibrations.

As an example of nonlinear damping, consider the case of long transmission cables undergoing large amplitudes of vibration; the damping force which is amplitude dependent can be approximated as shown in Fig. 2.3. Similarly, the damping in a blade-disk system due to friction at the interface junction and the material deformation for a given speed of rotation is found to be amplitude dependent and can be approximated as shown in Fig. 2.4. This type of damping significantly alters the dynamic behaviour of the mechanical system, particularly near resonant conditions, which may be responsible for instabilities.

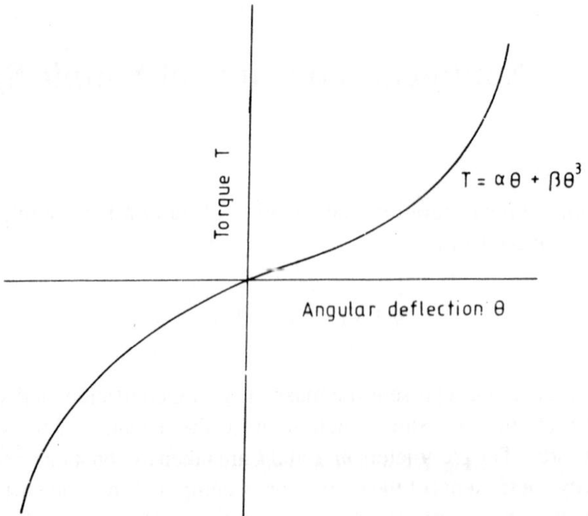

Fig. 2.1 Torgue–deflection characteristics of a coupling

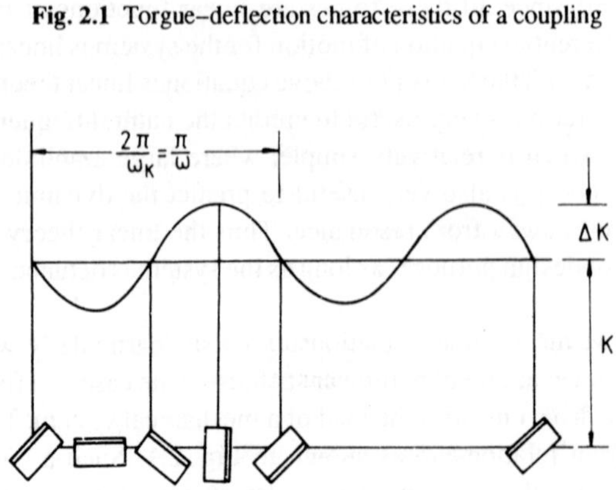

Fig. 2.2 Asymmetric shaft stiffness in one revolution $\omega_k = 2\omega$

The mass of a system is generally constant, except in applications such as space vehicles and missiles as shown in Fig. 2.5. In the case of reciprocating engines, the mass moment of inertia of each cylinder bank varies periodically with several harmonic components of engine speed. In linear theory an average value for this mass moment of inertia is considered in the study of torsional oscillations so as to simplify the problem. However, for large marine diesel engines, this variation as shown in Fig. 2.6 may be an important factor in stability analysis.

In addition to the nonlinearity behaviour of coefficients of inertia force, damping force and stiffness force, we also come across nonlinearities due to geometry. For example, the Coriolis forces of a long rotating blade mounted on a rotating disk are nonlinear in nature. This is illustrated in Fig. 2.7. Typically, such nonlinearities in a

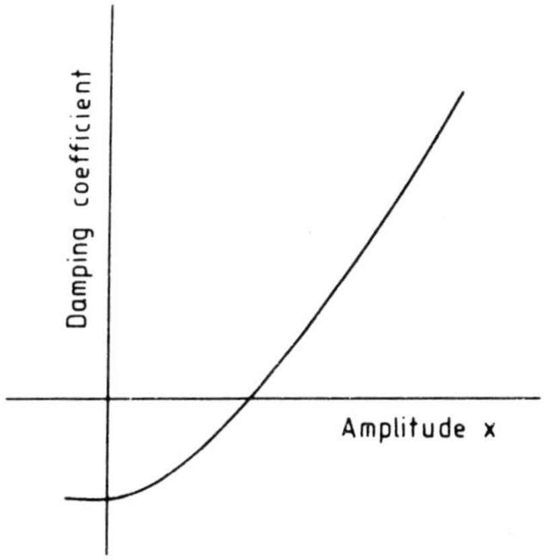

Fig. 2.3 Amplitude dependent damping

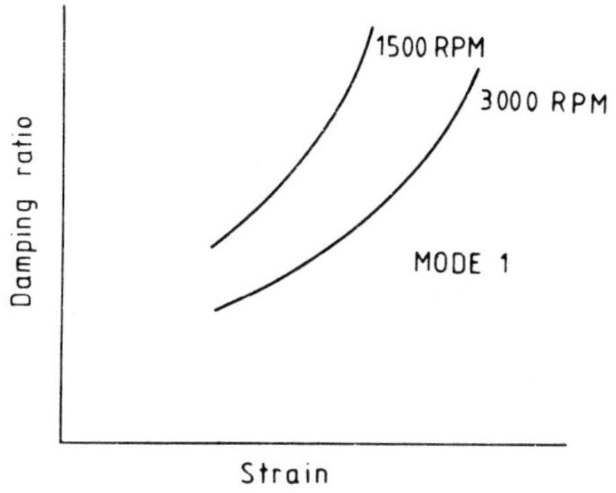

Fig. 2.4 Amplitude dependent damping ratio for different speeds of a blade-disk system

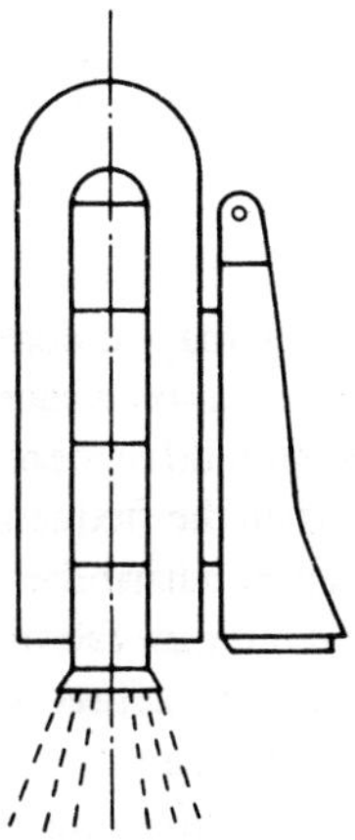

Fig. 2.5 Variable mass system

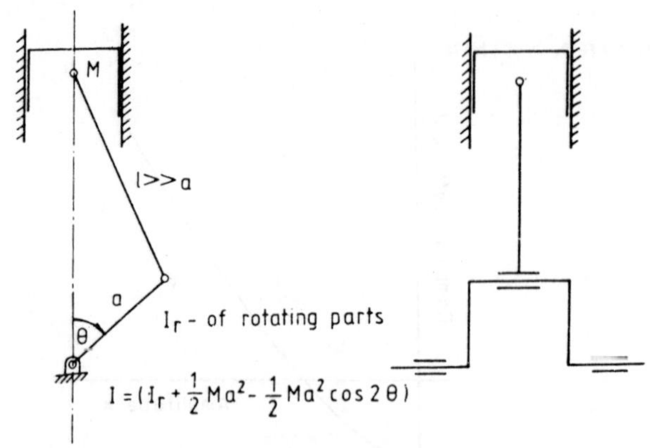

Fig. 2.6 Variable inertia system

system give rise to an entirely different dynamic behaviour when compared with linear theories. The characteristics of dynamic behaviour of a nonlinear system may be summarized as follows.

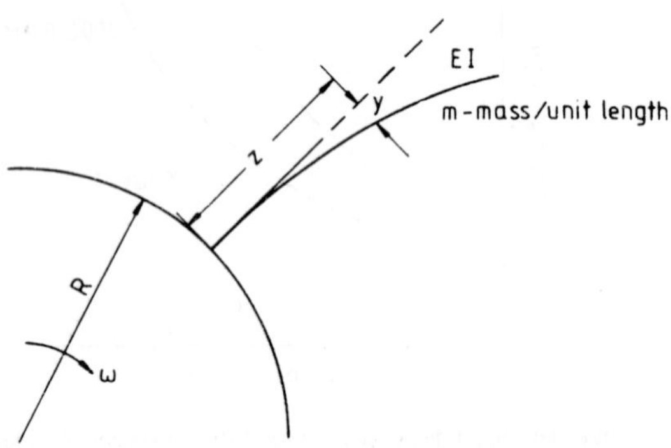

Fig. 2.7 A vibrating beam on a rotating disk causing Coriolis forces

$$-2m\omega\frac{\partial}{\partial z}\left\{y'\int_z^l \dot{y}dz\right\} - m\omega\frac{\partial}{\partial t}\int_0^z y'^2 dz$$

(1) The frequency of vibration may be dependent on the amplitude of vibration.

(2) While the frequency of excitation is increased, the amplitude of vibration may have a significant jump; the amplitude suddenly may fall after reaching a peak value, and suddenly jump upwards when the frequency of excitation is decreased.

(3) When a system is excited by a harmonic force, the response will not only have the basic harmonic component but may consist of higher or subharmonics.

(4) The system can become self excited, and amplitudes of vibration may grow even without any external disturbance.

(5) The system can become unstable under some conditions.

We will study the influence of nonlinearities in stiffness first.

2.1 PERTURBATION METHOD

The equation of motion of an undamped single degree of freedom system vibrating under free conditions, shown in Fig. 2.8, can be written as

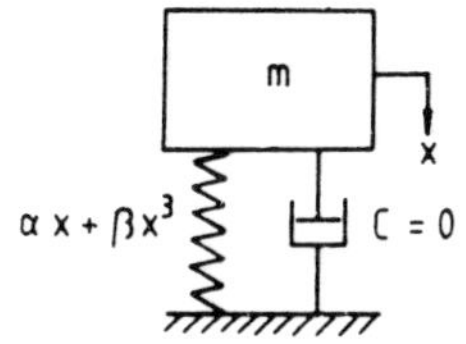

Fig. 2.8 Undamped nonlinear system in free vibration

$$m\frac{d^2x}{dt^2} + \alpha x + \beta x^3 = 0 \tag{2.2}$$

Let

$$p = \sqrt{\frac{\alpha}{m}}$$

$$T = pt \quad dt = \frac{1}{p}dT$$

$$\mu = \frac{\beta}{mp^2} \tag{2.3}$$

then

$$\ddot{x} + x + \mu x^3 = 0 \tag{2.4}$$

$$\cdot = \frac{d}{dT}$$

Equation (2.4) is known as Duffing's equation and it is nonlinear with respect to the restoring force defined by the cubic function. It is not possible to obtain an exact solution to equation (2.4) as in the case of a linear system. We must attempt an approximate solution. Let the initial conditions for the system be given by

$$x(0) = A; \quad \dot{x}(0) = 0 \tag{2.5}$$

A simple approach to obtain the solution of equation (2.4) is to build the solution around the linear case. Let $x_0\ (t)$ be the linear solution ($\beta = 0$). We can then perturb the solution about the linear solution in a series form as

$$x(T) = x_0(T) + \mu x_1(T) + \mu^2 x_2(T) + \ldots \tag{2.6}$$

The coefficient μ in equation (2.4) is a nonlinear parameter and the solution assumed in equation (2.6) converges fast if $\mu << 1$. This is the basic principle of perturbation method. Let us now substitute equation (2.6) in (2.4), and see what happens.

$$(\ddot{x}_0+\mu\ddot{x}_1+\mu^2\ddot{x}_2+\ldots)+(x_0+\mu x_1+\mu^2 x_2+\ldots)+$$

$$\mu(x_0+\mu x_1+\mu^2 x_2+\ldots)^3=0 \tag{2.7}$$

Since μ is a parameter, we can separate out the terms of μ^0, μ^1, μ^2 etc.

$$\ddot{x}_0+x_0=0$$

$$\ddot{x}_1+x_1=-x_0^3$$

...etc. (2.8)

It can be readily seen that the first of equation (2.8) is the linear part and in view of the initial conditions (2.5), has a solution of the form

$$x_0=a_1\cos T+b_1\sin T$$

$$=A\cos T \tag{2.9}$$

We can now substitute (2.9) in the second equation of (2.8) as the exciting term, to get

$$\ddot{x}_1+x_1=-A^3\cos^3 T$$

$$=-\frac{3}{4}A^3\cos^3 T-\frac{1}{4}A^3\cos 3T \tag{2.10}$$

From the above, we observe that there is an exciting term with third harmonic of the natural frequency and thus the solution (2.6) contains higher harmonic terms. This is typical of any nonlinear system. We now split equation (2.10) in two parts, one belonging to the harmonic solution and the second one belonging to the higher harmonic. Thus

$$\ddot{x}_{11}+x_{11}=-\frac{3}{4}A^3\cos T$$

$$\ddot{x}_{12}+x_{12}=-\frac{1}{4}A^3\cos 3T$$

$$x_1=x_{11}+x_{12} \tag{2.11}$$

It may be noted here that equation (2.4) is not a classical forced vibration problem of a system, and the right hand terms in equation (2.11) are due to the perturbation of the nonlinear free vibration problem. Hence, we must consider both the transient and particular integral parts of the equations in (2.11) for the final solution. Noting that the first equation in (2.11) is excited by a forcing term at resonant frequency, we write the following

$$x_{11}=T(a_2\cos T+b_2\sin T)$$

Hence

$$\ddot{x}_{11} = -2a_2 \sin T + 2b_2 \cos T$$

Substituting the above in the first equation of (2.11), we get

$$x_{11} = -\frac{3}{8}A^3 T \sin T \qquad (2.12)$$

Similarly for the higher harmonic, we have

$$x_{12} = a_3 \cos T + b_3 \sin T + \frac{A^3}{32}\cos 3T$$

Since

$$x_{12} = 0;\ \dot{x}_{12} = 0 \text{ at } T = 0$$

we have

$$x_{12} = -\frac{1}{32}A^3(\cos T - \cos 3T) \qquad (2.13)$$

Hence, the complete solution for x_1 is given by

$$x_1 = -\frac{3}{8}A^3 T \sin T - \frac{1}{32}A^3(\cos T - \cos 3T) \qquad (2.14)$$

The first term on the right hand side of the above equation (2.14) grows with time. Such a term is called secular term. It is not physically possible to have a growing solution for the system in Fig. 2.8 as the amplitude of vibration should be limited. This difficulty is overcome by introducing frequency-amplitude interaction by Lindstedt-Poincare method as follows.

For the linear system with $\mu = 0$, in equation (2.4), the solution becomes periodic with period 2 π. For $\mu \neq 0$, we introduce an unspecified function ω as follows

$$\tau = \omega T \qquad (2.15)$$

We choose ω such that the secular term in equation (2.14) can be avoided. With the help of the above equation, we can now write equation (2.4) as

$$\omega^2 \ddot{x} + x + \mu x^3 = 0 \qquad (2.16)$$

$$\cdot = \frac{d}{d\tau}$$

As before, we write the solution to equation (2.16) as

$$x(\tau) = x_0(\tau) + \mu x_1(\tau) + \mu^2 x_2(\tau) + \ldots \qquad (2.17)$$

and in addition choose the function for ω in series as

$$\omega = \omega_0 + \mu\omega_1 + \mu^2\omega_2 + \ldots \qquad (2.18)$$

Therefore

$$(\omega_0+\mu\omega_1+\mu^2\omega_2+\ldots)^2(\ddot{x}_0+\mu\ddot{x}_1+\mu^2\ddot{x}_2+\ldots)+$$

$$+(x_0+\mu x_1+\mu^2 x_2+\ldots)+\mu(x_0+\mu x_1+\mu^2 x_2+\ldots)^3=0 \tag{2.19}$$

As before, we now separate terms of μ^0, μ^1, μ^2, etc. to obtain

$$\omega_0^2\ddot{x}_0+x_0=0$$

$$\omega_0^2\ddot{x}_1+x_1=-2\omega_0\omega_1\ddot{x}_0-x_0^3$$

$$\omega_0^2\ddot{x}_2+x_2=-(2\omega_0\omega_2+\omega_1^2)\ddot{x}_0-2\omega_0\omega_1\ddot{x}_1-3x_0^2x_1$$

... etc. (2.20)

In addition to the initial conditions (2.5), we now prescribe

$$x_i(\tau+2\pi)=x_i(\tau) \tag{2.21}$$

since the solution should be periodic in τ with time period 2π.

The solution of the first equation in (2.20) gives

$$x_0=a_1\cos\frac{\tau}{\omega_0}+b_1\sin\frac{\tau}{\omega_0}$$

In view of the initial conditions (2.5) and the periodicity condition (2.21), the above gives

$$x_0=A\cos\tau \tag{2.22}$$

$$\omega_0=1 \tag{2.23}$$

With the help of the above, the second equation in (2.20) becomes

$$\ddot{x}_1+x_1=2\omega_1 A\cos\tau-A^3\cos^3\tau$$

$$=\left(2\omega_1 A-\frac{3}{4}A^3\right)\cos\tau-\frac{1}{4}A^3\cos 3\tau \tag{2.24}$$

We immediately identify that the bracketed term in the above equation (2.24) gives rise to secular term in the solution. Therefore, to avoid the secular term in the solution, we choose

$$\omega_1=\frac{3}{8}A^2 \tag{2.25}$$

so that there is no term in the solution which grows with time.

Now the solution of equation (2.24) is governed by the third harmonic excitation term and it can be written as

$$x_1 = a_2 \cos\tau + b_2 \sin\tau + \frac{A^3}{32}\cos 3\tau$$

$$= \frac{A^3}{32}(-\cos\tau + \cos 3\tau) \tag{2.26}$$

We thus have successfully avoided the problem of secular term and can now continue the solution further by substituting equation (2.26) in the third equation of (2.20) and determine ω_2 and x_2. Considering these three terms only, the solution of equation (2.16) is given by

$$x(\tau) = \left(A - \frac{1}{32}\mu A^3 + \frac{23}{1024}\mu^2 A^5\right)\cos\tau +$$

$$\left(\frac{1}{32}\mu A^3 - \frac{3}{128}\mu^2 A^5\right)\cos 3\tau + \frac{1}{1024}\mu^2 A^5 \cos 5\tau \tag{2.27}$$

$$\omega = 1 + \frac{3}{8}\mu A^2 - \frac{21}{256}\mu^2 A^4 \tag{2.28}$$

Equation (2.28) clearly demonstrates the dependency of frequency ω on amplitude A given initially. If $\mu > 0$, we have a hardening spring as in Fig. 2.1. If $\mu < 0$, then the spring is softening type as shown in Fig. 2.9. Equation (2.28) is plotted in Fig. 2.10 for both the cases.

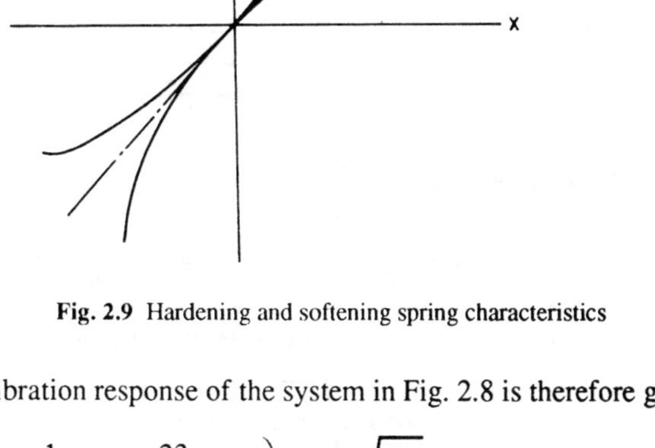

Fig. 2.9 Hardening and softening spring characteristics

The free vibration response of the system in Fig. 2.8 is therefore given by

$$x(t) = \left(A - \frac{1}{32}\mu A^3 + \frac{23}{1024}\mu^2 A^5\right)\cos\omega\sqrt{\frac{\alpha}{m}}\,t +$$

$$\left(\frac{1}{32}\mu A^3 - \frac{3}{128}\mu^2 A^5\right)\cos 3\omega\sqrt{\frac{\alpha}{m}}\,t + \frac{1}{1024}\mu^2 A^5 \cos 5\omega\sqrt{\frac{\alpha}{m}}\,t \tag{2.29}$$

For the linear system, the above reduces to the well known solution.

$$x(t) = A\cos\sqrt{\frac{\alpha}{m}}\,t \tag{2.30}$$

For a simple pendulum shown in Fig. 2.11, we can write the following equation.

$$ml\ddot{\theta} + mg\sin\theta = 0$$

i.e.,

$$\ddot{\theta} + \frac{g}{l}\sin\theta = 0$$

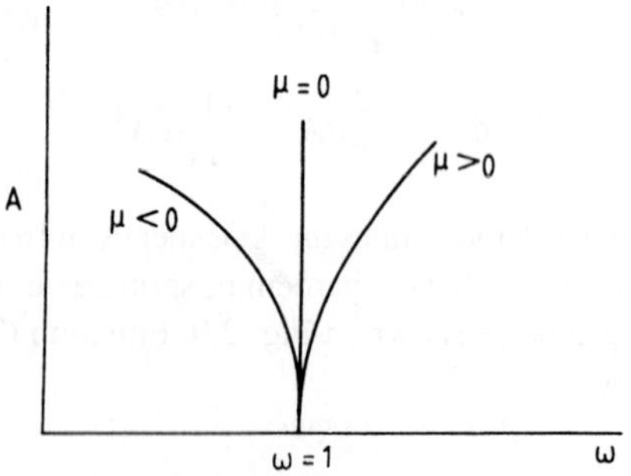

Fig.2.10 Amplitude – frequency relations

For small amplitudes of vibration, we can approximate $\sin\theta = \theta$, resulting in the simple harmonic oscillator. If the amplitudes are not small, the motion is complicated; it is known from observations that the frequency decreases with an increase in the amplitude of vibration. We can obtain this relation by expanding $\sin\theta$ in a power series. Let us consider for simplicity the first two terms in this expansion. Then,

$$\ddot{\theta} + \frac{g}{l}\left(\theta - \frac{\theta^3}{6}\right) = 0$$

Comparing with equation (2.2) and (2.3), we have

$$m = l;\quad \alpha = \frac{g}{l};\quad \beta = -\frac{g}{6l};\quad p^2 = \frac{g}{l}\quad \text{and}\quad \mu = \frac{\beta}{mp^2} = -\frac{1}{6}$$

Hence from equations (2.28) and (2.29) considering only first two terms, we get

$$\theta(t) = \left\{A - \frac{1}{32}\left(-\frac{1}{6}\right)A^3\right\}\cos\omega\sqrt{\frac{g}{l}}\,t + \frac{1}{32}\left(-\frac{1}{6}\right)A^3\cos 3\omega\sqrt{\frac{g}{l}}\,t$$

$$\omega = 1 - \frac{3}{8}\times\frac{1}{6}A^2$$

Let $A = 1$ radian then, the above become

$$\theta(t) = 1.0052083\cos\frac{15}{16}\sqrt{\frac{g}{l}}\,t - 0.0052083\cos\frac{45}{16}\sqrt{\frac{g}{l}}\,t$$

Thus, we see that the simple pendulum acts like a softening spring.

As an example of hardening spring, consider the case of a mass attached to a stretched wire shown in Fig. 2.12. Let the cross-sectional area of the wire be a, E be the Young's modulus and S the axial tension, then the elongation in the wire due to a displacement x is

$$\frac{\sqrt{l^2+x^2}-l}{l} \approx \frac{x^2}{2l^2}$$

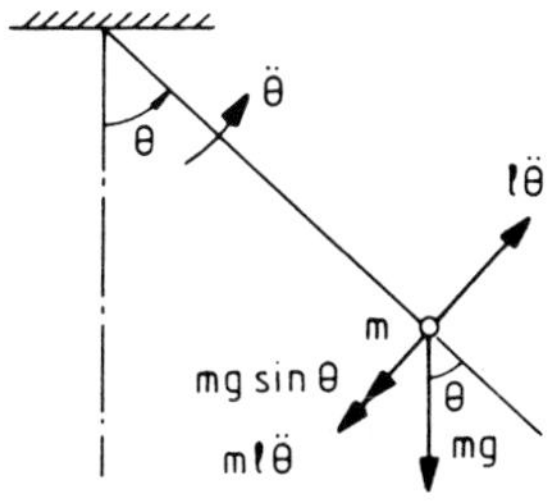

Fig.2.11 A simple pendulum

In the deformed position, the tensile force is therefore

$$S + aE\frac{x^2}{2l^2}$$

The restoring force is therefore

$$\left\{S + aE\frac{x^2}{2l^2}\right\}\frac{2x}{\sqrt{l^2+x^2}} \approx \frac{2Sx}{l} + aE\frac{x^3}{l^3}$$

The differential equation is therefore

$$m\ddot{x} + \frac{2S}{l}x + \frac{aE}{l^3}x^3 = 0$$

Then, we find

$$\alpha = \frac{2S}{l}; \quad \beta = \frac{aE}{l^3}$$

$$p^2 = \frac{2S}{ml}; \quad \mu = \frac{aE}{2Sl^2}$$

Then, the free vibratory response of the system is given by

$$\omega = 1 + \frac{3}{16}\frac{aE}{Sl^2}A^2$$

$$x = \left(A - \frac{1}{64}\frac{aE}{Sl^2}A^3\right)\cos\sqrt{\frac{2S}{ml}}\,\omega t + \frac{1}{64}\frac{aE}{Sl^2}A^3\cos\sqrt{\frac{2S}{ml}}\,3\omega t$$

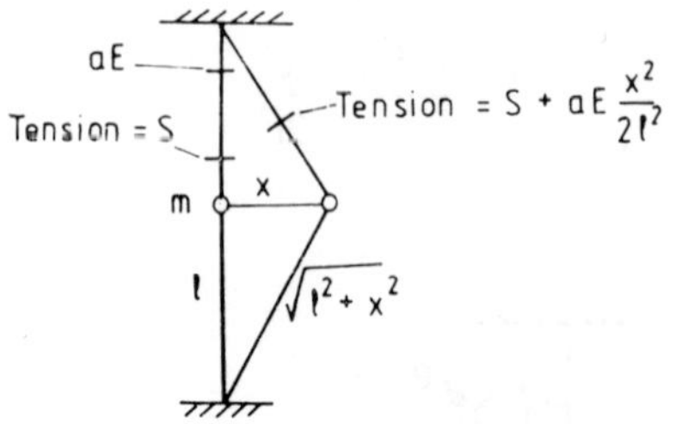

Fig. 2.12 A mass on a wire under tension

2.2 PERIODIC SOLUTIONS OF DAMPED SYSTEMS

The perturbation method used in the previous section 2.1 is useful whenever a periodic solution is to be determined. The moment we introduce some damping in the system, the method becomes complicated. Let us examine equation (2.2) with linear viscous damping.

$$m\frac{d^2x}{dt^2} + c\frac{dx}{dt} + \alpha x + \beta x^3 = 0 \tag{2.2a}$$

$$\omega^2\ddot{x} + 2\xi\omega\dot{x} + x + \mu x^3 = 0 \tag{2.16a}$$

where

$$\cdot = \frac{d}{d\tau}$$

and

$$\xi = \frac{c}{2mp} \tag{2.31}$$

Let us use equations (2.17) and (2.18) and obtain from equation (2.16*a*)

$$(\omega_0 + \mu\omega_1 + \mu^2\omega_2 + \ldots)^2(\ddot{x} + \mu\ddot{x}_1 + \mu^2\ddot{x}_2 + \ldots) +$$

$$+2\xi(\omega_0 + \mu\omega_1 + \mu^2\omega_2 + \ldots)(\dot{x}_0 + \mu\dot{x}_1 + \mu^2\dot{x}_2 + \ldots) +$$

$$+(x_0 + \mu x_1 + \mu^2 x_2 + \ldots) + \mu(x_0 + \mu x_1 + \mu^2 x_2 + \ldots)^3 = 0 \tag{2.19a}$$

As before, we now separate terms of μ^0, μ^1, μ^2, etc. to obtain

$$\omega_0^2\ddot{x}_0 + 2\xi\omega_0\dot{x}_0 + x_0 = 0$$

$$\omega_0^2\ddot{x}_1 + 2\xi\omega_0\dot{x}_1 + x_1 = -2\omega_0\omega_1\ddot{x}_0 - 2\xi\omega_1\dot{x}_0 - x_0^3$$

...etc. (2.20*a*)

In view of the initial conditions (2.5), for small values of damping we can write the generating solution as

$$\omega_0 = 1$$

$$x_0 \approx A e^{-\xi\tau}\cos(\tau - \phi_0)$$

$$\phi_0 = \tan^{-1}\frac{\xi}{\sqrt{1-\xi^2}} \tag{2.32}$$

Using the above equation and its derivatives, we can write the second perturbation as

$$\ddot{x}_1 + 2\xi\dot{x}_1 + x_1 = -2A\,\omega_1 e^{-\xi\tau}\{\xi\sin(\tau-\phi_0) - \cos(\tau-\phi_0)\}$$

$$-e^{-3\xi\tau}A^3\left\{\frac{3}{4}\cos(\tau-\phi_0) + \frac{1}{4}\cos 3(\tau-\phi_0)\right\} \tag{2.33}$$

Since there is damping in the system, we do not find any secular terms as before. The solution of above equation is however lengthy and leads to complex expressions. Also, if the damping is small it has marginal effect on the response. The stiffness nonlinearity does not anyway give rise to classical resonance of the corresponding linear system and hence the solution without damping given in the previous section 2.1 is of more practical usage than the complicated solution that may be obtained from the above. In fact, it may be desirable to obtain graphical or numerical solutions in such cases, as we will discuss them later.

The perturbation method can be effectively used to predict periodic oscillations that arise out of certain specific nonlinear damping characteristics. Such periodic oscillations are observed in the galloping of transmission wires in the northern states of Canada. At subzero temperatures, when sleet is formed on the cables (see Fig. 2.13) between the transmission towers located 100 m apart, they vibrate violently with more than a meter of amplitude at a very low frequency (1 Hz) under transverse wind conditions.

Let us first check the possibility of forced vibration due to transverse wind gusts at a very low frequency. Since the time period is about one second let there be 600 gusts of wind in the first 10 minutes, giving rise to forced vibration. Since these gusts cannot be entirely harmonic or periodic, let us assume that in the next ten minutes, the number of gusts, say, is 601. Then there should be a beating oscillation given by

$$p - \omega = 2\pi\frac{601}{600} - 2\pi\frac{600}{600} = \frac{2\pi}{600} = 2\varepsilon \text{ rad/s}$$

$$T \text{ for beats} = \frac{\pi}{\varepsilon} = 600 \text{ s}$$

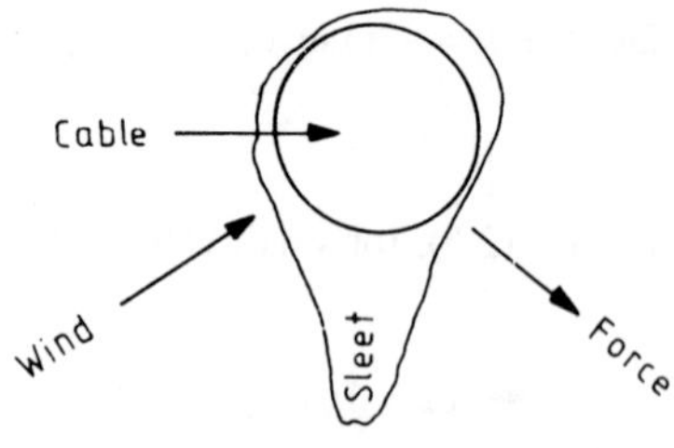

Fig. 2.13 Cross section of cable with sleet around

Therefore, the oscillations should die in the next five minutes and build up again. However, the amplitudes observed are continuous and hence the possibility of forced vibration from the transverse gusts should be discounted. An explanation of the type of phenomena observed is to be found elsewhere.

In Fig. 2.14, a circular and an airfoil cross-section bodies are shown. In the case of the circular body, the direction of force is the same as that of the wind, whereas in the airfoil these directions are different giving an aerodynamic lift for a horizontal direction of wind. Therefore a circular cross-section body remains stable in a horizontal wind, because a displacement, say, in the downward direction produces an upward force, as in Fig. 2.15, thus bringing back the body to equilibrium position.

On the contrary, for a noncircular body like the semicircular cross-section shown in Fig. 2.16, we have a different picture. Here, across a line *a-b*, the stream lines are crowded, giving rise to pressures below atmospheric values at *b*. This results in a further downward force, making the system unstable. Since this force is generated by the motion of the body itself any resulting vibration is called self-excited vibration. A similar phenomenon happens to the cable with sleet in Fig. 2.13. The displacement grows until the elastic restoring forces due to cable deformation take over in reversing downward motion of the cable to an upward motion. The upward motion in turn causes an upward thrust; the cable continues its motion and cross over the equilibrium position until the restoring forces cause a reversal of the upward motion to downward motion. The phenomenon then repeats thus causing large amplitudes of vibration. We will revert to this discussion of stability later.

For the present, we can try and fit the above phenomenon of a periodic motion under self-excitation, into a mathematical model by considering the damping to be amplitude dependent. Since the amplitudes grow to generate a periodic motion, we can assume linear viscous damping in the system, whose strength is dependent on the amplitude of vibration. The viscous damping is assumed to be negative to account for the growing characteristic of vibration, when the amplitudes are small and assumed to be positive for larger amplitudes. Let the amplitude depend on square of the displacement, so that we have a parabolic variation as in Fig. 2.17, giving rise to the following damping force function.

$$F_d = (-c_1 + c_2 x^2)\dot{x} \tag{2.34}$$

Hence, the equation of motion for a system subjected to the type of damping represented by equation (2.34) and no external force, is given by

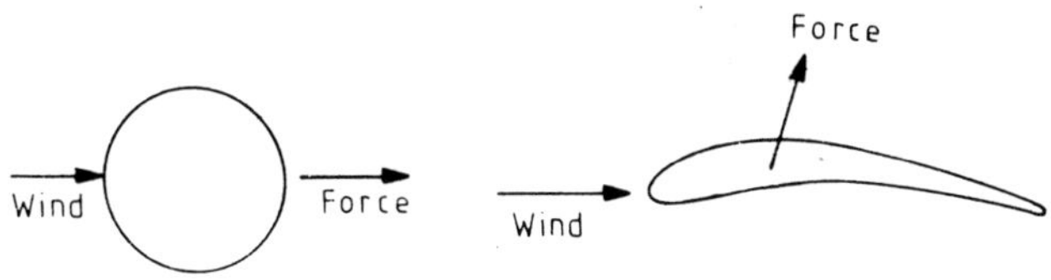

Fig. 2. 14 Directions of wind and force on circular and aerofoil bodies

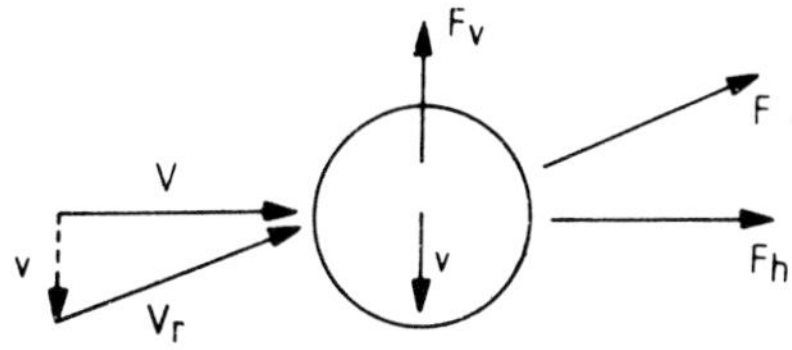

Fig. 2.15 Circular cross section in relative motion-v-downward motion: V – direction of wind velocity; V_r – relative velocity of wind; F - Force; F_v – vertical component of force opposing downward motion.

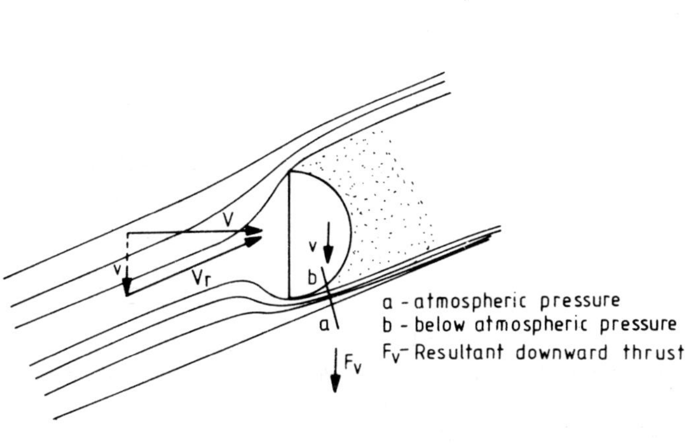

Fig. 2.16 Non-circular section in relative motion

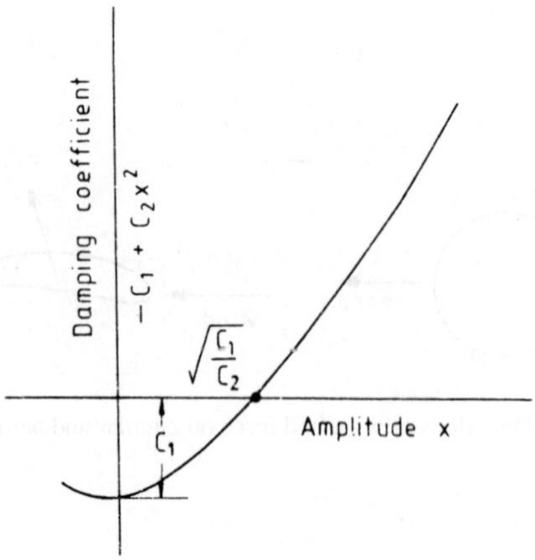

Fig. 2.17 Amplitude dependent damping coefficient

$$m\ddot{x} - (c_1 - c_2x^2)\dot{x} + kx = 0 \tag{2.35}$$

$$\cdot = \frac{d}{dt}$$

Let, $T = pt$, then

$$p^2\ddot{x} - \frac{c_1}{m}\left(1 - \frac{c_2}{c_1}x^2\right)p\dot{x} + p^2x = 0 \tag{2.36}$$

$$\cdot = \frac{d}{dT}$$

Rearranging, we get

$$\ddot{x} - \mu\left(1 - \frac{c_2}{c_1}x^2\right)\dot{x} + x = 0 \tag{2.36a}$$

where

$$\mu = \frac{c_1}{mp} \tag{2.37}$$

Let us further introduce

$$z = \frac{x}{\sqrt{c_1/c_2}} \tag{2.38}$$

then

$$\ddot{z} - \mu(1 - z^2)\dot{z} + z = 0 \tag{2.39}$$

Equation (2.39) is the familiar Van der Pol's equation. We can obtain a periodic solution to this equation by the procedure used in the previous section 2.1. Let

$$\tau = \omega T$$

then

$$\omega^2\ddot{z} - \mu\omega(1-z^2)\dot{z} + z = 0 \tag{2.40}$$

$$\cdot = \frac{d}{d\tau}$$

With the help of equations (2.17) and (2.18), we get

$$\begin{aligned}&(\omega_0+\mu\omega_1+\mu^2\omega_2+\ldots)^2(\ddot{z}_0+\mu\ddot{z}_1+\mu^2\ddot{z}_2+\ldots)\\&-\mu(\omega_0+\mu\omega_1+\mu^2\omega_2+\ldots)\left\{1-(z_0+\mu z_1+\mu^2 z_2+\ldots)^2\right\}\\&\times(\dot{z}_0+\mu\dot{z}_1+\mu^2\dot{z}_2+\ldots)+(z_0+\mu z_1+\mu^2 z_2+\ldots)=0\end{aligned} \tag{2.41}$$

As before, we now separate terms of μ^0, μ^1, μ^2, etc. to obtain

$$\omega_0^2\ddot{z}_0 + z_0 = 0$$

$$\omega_0^2\ddot{z}_1 + z_1 = -2\omega_0\omega_1\ddot{z}_0 + \omega_0(1-z_0^2)\dot{z}_0$$

. . .etc. (2.42)

We seek a periodic solution in τ with 2π as discussed before without specifying any initial amplitude. Thus

$$\begin{aligned}z(\tau+2\pi) &= z(\tau)\\ \dot{z}(0) &= 0\end{aligned} \tag{2.43}$$

From the first equation of (2.42) we get,

$$\begin{aligned}z_0 &= A_0\cos\tau\\ \omega_0 &= 1\end{aligned} \tag{2.44}$$

The value of A_0 is yet to be specified.
The second equation of (2.42) now becomes

$$\begin{aligned}\ddot{z}_1 + z_1 &= 2\omega_1 A_0\cos\tau - (1-A_0^2\cos^2\tau)A_0\sin\tau\\ &= 2\omega_1 A_0\cos\tau - A_0\sin\tau + A_0^3(1-\sin^2\tau)\sin\tau\\ &= 2\omega_1 A_0\cos\tau - A_0\sin\tau + A_0^3\sin\tau - A_0^3\left(\frac{3}{4}\sin\tau - \frac{1}{4}\sin 3\tau\right)\\ &= 2\omega_1 A_0\cos\tau + A_0\left(\frac{1}{4}A_0^2 - 1\right)\sin\tau + \frac{1}{4}A_0^3\sin 3\tau\end{aligned} \tag{2.45}$$

We have two means to avoid secular terms in equation (2.45), viz.,

$$A_0 = 0 \tag{2.46}$$

$$A_0 = 2 \text{ and } \omega_1 = 0 \tag{2.47}$$

We are not interested in (2.46) as this is an equilibrium solution. Equation (2.47), therefore, gives (2.44) as

$$z_0 = 2\cos\tau \tag{2.48}$$

The solution for (2.45) is now possible, which is

$$z_1 = A_1\cos\tau + B_1\sin\tau - \frac{1}{4}\sin 3\tau \tag{2.49}$$

Since

$$\dot{z}_1(0) = 0$$

we get

$$B_1 = \frac{3}{4}$$

Continuing further, we can determine A_1 etc., and obtain

$$x = \sqrt{\frac{c_1}{c_2}}\left\{\left(2 - \frac{1}{8}\mu^2\right)\cos\omega\sqrt{\frac{k}{m}}\,t + \frac{3}{4}\mu\sin\omega\sqrt{\frac{k}{m}}\,t + \frac{3}{16}\mu^2\cos 3\omega\sqrt{\frac{k}{m}}\,t\right\}$$

$$\left\{-\frac{1}{4}\mu\sin 3\omega\sqrt{\frac{k}{m}}\,t - \frac{5}{96}\mu^2\cos 5\omega\sqrt{\frac{k}{m}}\,t + \ldots\right\}$$

$$\omega = 1 - \frac{1}{16}\mu^2 \ldots \tag{2.50}$$

From Fig. 2.17 we find that the damping coefficient is zero when x is $\sqrt{c_1/c_2}$. As we see from the above the major component of response is $2\sqrt{c_1/c_2}$. This is where we can expect the periodic motion to occur. If the instantaneous response is less than half the amplitude $2\sqrt{c_1/c_2}$, the damping is negative and hence will aid building up the motion and if it is more than this half of the amplitude, the damping is positive and therefore begins to damp the motion limiting it to a peak value $2\sqrt{c_1/c_2}$. Such a periodic solution is called limit cycle. The transient response of the system and the stability studies will be considered later.

2.3 MODIFIED EULER METHOD

With the advent of high speed computation facilities, it is now quite common to resort to a numerical solution of nonlinear differential equations, instead of other analytical and graphical methods. We will discuss three numerical methods, first the oldest and simplest due to Euler. To begin with let us consider a first order differential equation in the form

$$\frac{dx}{dt} = f(x, t) \tag{2.51}$$

where x is the dependent variable and t is the independent variable.

The integral of the above gives the desired solution, x as a function of t, which is

$$x = F(t) \tag{2.52}$$

We can represent equation (2.52) in x–t plane as in Fig. 2.18. Euler's method is based on the assumption that the solution curve given by equation (2.52) can be approximated by several linear segments, each one over a very short interval for t. Then, we can write the following relation

$$\Delta x \approx \Delta t \tan\theta$$

$$= \left(\frac{dx}{dt}\right)_0 \Delta t$$

so that

$$x_1 = x_0 + \left(\frac{dx}{dt}\right)_0 \Delta t \tag{2.53}$$

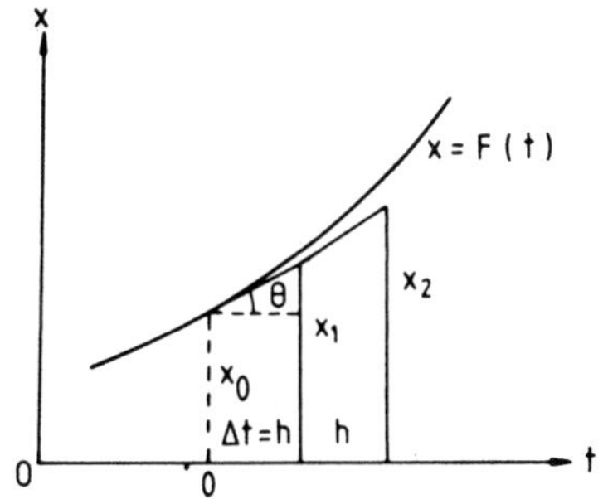

Fig. 2.18 Euler's method

Let us take the interval for Δt to be as small as possible, say, h. Then

$$t_2 = t_1 + h$$

$$t_3 = t_2 + h$$

etc.

Corresponding to the above values of the independent variable t, we can write the following with the help of (2.53).

$$x_2 = x_1 + \left(\frac{dx}{dt}\right)_1 h$$

$$x_3 = x_2 + \left(\frac{dx}{dt}\right)_2 h$$

etc . . . (2.54)

We can thus tabulate the values of x as a function of t. The method, however, is too slow, since one requires a very small value for h to get meaningful results. If h is not sufficiently small, the results will be very inaccurate. Even if h is taken very small for all steps, it is readily seen from Fig. 2.18 that the computed x will deviate farther and farther from the exact solution so long as the curvature of x *versus* t does not change. Thus a modification is made to the above simple procedure. We adopt an iteration procedure for getting the solution with the desired accuracy.

Let us start from the initial value x_0 and compute an approximate value for x_1 from (2.53).

$$x_1^{(1)} = x_0 + \left(\frac{dx}{dt}\right)_0 h \tag{2.55}$$

Substitute this approximate value of x_1 in equation (2.51) to get an approximate value of dx/dt at the end of the first interval.

$$\left(\frac{dx}{dt}\right)_1^{(1)} = f(x_1^{(1)}, t_1) \tag{2.56}$$

Equation (2.56) represents the first approximation of the slope at the end of the first interval. We have the slope of the integral curve at the starting point and therefore an improved value of Δx is found by multiplying h by the average of the values of the slope at the ends of the current interval of t, i.e.,

$$\Delta x = \frac{\left(\frac{dx}{dt}\right)_0 + \left(\frac{dx}{dt}\right)_1^{(1)}}{2} h \tag{2.57}$$

This is obviously a better value for Δx than the value

$$\left(\frac{dx}{dt}\right)_0 h.$$

We are now ready for a second approximation of x at the end of the first interval of t, by using the slope given in equation (2.57) for the integral curve in (2.55).

$$x_1^{(2)} = x_0 + \frac{\left(\frac{dx}{dt}\right)_0 + \left(\frac{dx}{dt}\right)_1^{(1)}}{2} h \tag{2.58}$$

This improved value is now substituted in the given equation (2.51) to get a second approximation for

$$\left(\frac{dx}{dt}\right)_1^{(2)} = f(x_1^{(2)}, t_1) \tag{2.59}$$

The third approximation for x_1, is then obtained from

$$x_1^{(3)} = x_0 + \frac{\left(\frac{dx}{dt}\right)_0 + \left(\frac{dx}{dt}\right)_1^{(2)}}{2} h \tag{2.60}$$

The above two equations (2.59) and (2.60) are applied repeatedly until no change is produced in the value of x_1 to the number of digits desired. This method of averaging the slope in each step and the resulting accuracy compared with the original Euler method are shown in Fig. 2.19. In this figure, *KM* represents the computational process of original Euler method. *PN* is drawn parallel to the tangent at *Q*. Then Δx computed using the slope at *Q* is represented by *KN*. Equation (2.57) represents

$$\Delta x = \frac{\left(\frac{dx}{dt}\right)_0 + \left(\frac{dx}{dt}\right)_1}{2} h$$

$$= \frac{1}{2}\left\{h\left(\frac{dx}{dt}\right)_0 + h\left(\frac{dx}{dt}\right)_1\right\}$$

$$= \frac{1}{2}(KM + KN)$$

$$= KM + \frac{1}{2} MN \tag{2.61}$$

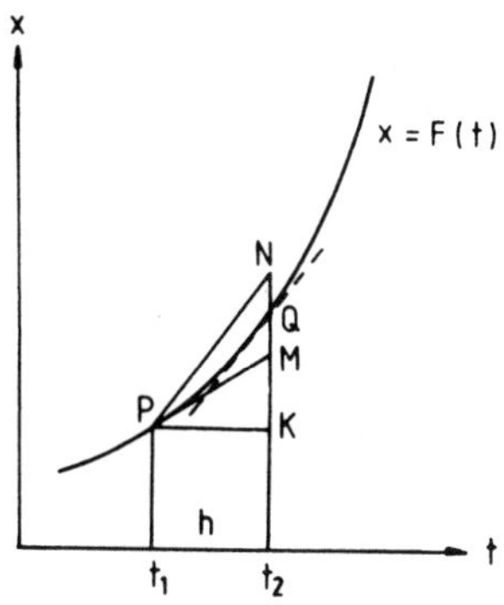

Fig. 2.19 Modified Euler's method

The value of Δx given in equation (2.61) above is very close to the exact value *KQ*. The iterative procedure will converge fast to the desired accuracy if *h* is small.

We can extend now the above procedure to a second order differential equation of the type

$$\frac{d^2x}{dt^2} = f\left(x, \frac{dx}{dt}, t\right) \tag{2.62}$$

Solution of the above is symbolically represented by equation (2.52). Since the governing differential equation is of second order, we have two initial conditions, viz.,

$$x_{(t=0)} = x_0$$

and

$$\left(\frac{dx}{dt}\right)_{(t=0)} = \left(\frac{dx}{dt}\right)_0 \tag{2.63}$$

We can directly write the following iterative equations (j) for the i-th step solution.

$$\left\{\frac{d^2x}{dt^2}\right\}_i^{(j-1)} = f\left(x_i^{(j-1)}, \left\{\frac{dx}{dt}\right\}_i^{(j-1)}, t\right)$$

$$\left\{\frac{dx}{dt}\right\}_i^{(j)} = \left\{\frac{dx}{dt}\right\}_{i-1} + \frac{\left\{\frac{d^2x}{dt^2}\right\}_{i-1} + \left\{\frac{d^2x}{dt^2}\right\}_i^{(j-1)}}{2} h$$

$$x_i^{(j)} = x_{i-1} + \frac{\left\{\frac{dx}{dt}\right\}_{i-1} + \left\{\frac{dx}{dt}\right\}_i^{(j-1)}}{2} h \tag{2.64}$$

The above equations can be easily adopted for fast computing using any modern computer system. It should be remembered that the interval h plays an important role in obtaining the desired convergency.

Let us consider the Van der Pol's equation

$$\ddot{x} - \mu(1 - x^2)\dot{x} + x = 0$$

We rewrite the above suitable to our numerical method as

$$\ddot{x} = f(x, \dot{x}, t)$$

$$= \mu(1 - x^2)\dot{x} - x$$

Let the initial conditions be

$$x(t = 0) = 1$$

$$\dot{x}(t = 0) = 0$$

We can then use the equations in (2.64) and write a small computer program. Let us choose a small interval of time corresponding to

$$\theta = p\tau$$

which is 1.4 degrees. The response obtained is plotted in Fig. 2.20 for $\mu = 1$. It can be readily observed that a periodic solution is obtained corresponding to the natural period with an amplitude equal to 2 as obtained by the Perturbation method. We will discuss this solution in the phase plane while studying the stability in the next chapter.

2.4 RUNGE-KUTTA METHOD

There are several numerical techniques that have been devised based on interpolation of continuous functions. They are better than Euler's method in the sense that one is not required to make a lot of iterations in arriving at a solution at each step with the desired accuracy. However, in methods based on interpolation of polynomial functions, we require to make difference tables and use this data to extrapolate the functions for the next step. Runge-Kutta method eliminates the problems of iteration as well as preparation of the difference data. We, therefore, describe this procedure here without presenting any derivation.

Let the differential equation be represented as before

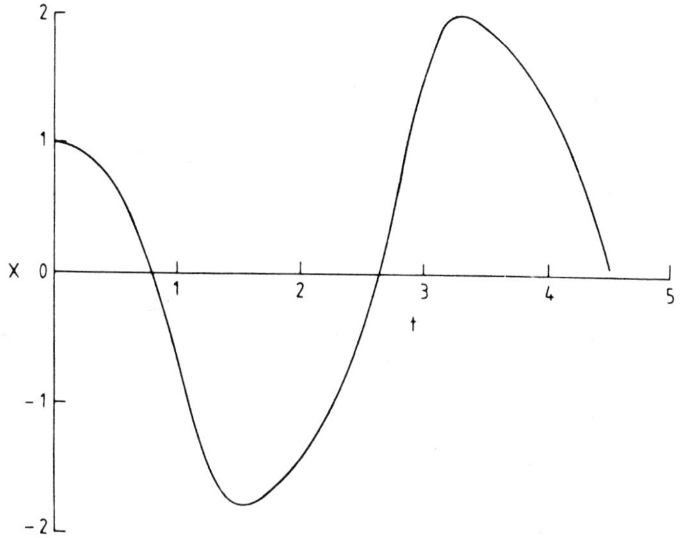

Fig. 2.20 Response of Van der Pol oscillator

$$\frac{d^2x}{dt^2} = f\left(x, \frac{dx}{dt}, t\right) \tag{2.62}$$

i.e.,

$$\ddot{x} = f\left(x, \frac{dx}{dt}, t\right) \tag{2.65}$$

Runge defined the following steps of calculations for the nth step.

$$K_1 = \Delta t f(t_n, x_n, \dot{x}_n)$$

$$K_2 = \Delta t f\left(t_n + \frac{\Delta t}{2}, x_n + \frac{\Delta t}{2}\dot{x}_n + \frac{\Delta t}{8}K_1, \dot{x}_n + \frac{K_1}{2}\right)$$

$$K_3 = \Delta t f\left(t_n + \frac{\Delta t}{2}, x_n + \frac{\Delta t}{2}\dot{x}_n + \frac{\Delta t}{8}K_2, \dot{x}_n + \frac{K_2}{2}\right)$$

$$K_4 = \Delta t f\left(t_n + \Delta t, x_n + \Delta t\dot{x}_n + \frac{\Delta t}{2}K_3, \dot{x}_n + K_3\right) \tag{2.66}$$

The above are used in the increments for x and $\dot{x}$ given by

$$\Delta x = \Delta t\left\{\dot{x}_n + \frac{1}{6}(K_1 + K_2 + K_3)\right\}$$

$$\Delta \dot{x} = \frac{1}{6}(K_1 + 2K_2 + 2K_3 + K_4) \tag{2.67}$$

Hence, the values of x and $\dot{x}$ at step $(n + 1)$ are given by

$$x_{n+1} = x_n + \Delta x$$

$$\dot{x}_{n+1} = \dot{x}_n + \Delta \dot{x} \tag{2.68}$$

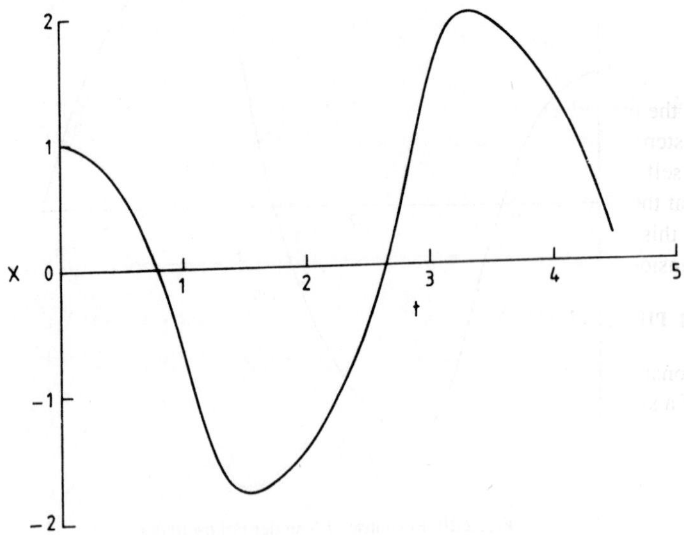

Fig. 2.21 Runge-Kutta solution of Van der Pol oscillator

A simple computer program can be written based on the above. Once again, we consider the Van der Pol's equation with $\mu = 1$ and Runge-Kutta solution is shown in Fig. 2.21.

In the next chapter, we will discuss a graphical procedure based on isoclines in the phase plane for free vibration problems with nonlinear characteristics. For forced oscillations, this procedure is modified in chapter 4, which leads to a numerical method called phase plane-delta method.

3

Stability Considerations

In the previous chapter, we introduced Van der Pol's equation and observed that a system governed by this equation has a periodic solution under no external force due to self excitation. We have also shown from the damping characteristics of the system that the solution either grows or decays depending on the amplitude of the system. In this chapter, we will study the stability considerations in more detail. First, let us consider these graphically in a phase plane.

3.1 PHASE PLANE METHOD

Consider the second order differential equation representing the dynamic behaviour of a system.

$$\frac{d^2x}{dt^2} = f\left(x, \frac{dx}{dt}\right) \tag{3.1}$$

Let

$$\frac{dx}{dt} = y \tag{3.2}$$

Then

$$\frac{dy}{dt} = f(x, y) \tag{3.3}$$

Also

$$\frac{dy}{dx} = \frac{f(x, y)}{y} \tag{3.4}$$

Thus, we can express a second order differential equation by two first order differential equations with the same independent parameter t or by an equation defined by the relation of the form (3.4) without specifically using the independent parameter t. We can generalize the above to express a nonlinear differential equation of second order by

$$\frac{dx}{dt} = X(x, y)$$

$$\frac{dy}{dt} = Y(x, y) \tag{3.5}$$

where $X(x, y)$ and $Y(x, y)$ are polynomials in x and y.

Also

$$\frac{dy}{dx} = \frac{Y(x, y)}{X(x, y)} \tag{3.6}$$

The plane $x - y$, in which the solution is sought is state plane of the system. As a particular case if equation (3.2) is possible, i.e., $X(x, y) = y$, then the state plane becomes a phase plane. The path of $x(t)$ and $y(t)$ in a state plane or phase plane is a trajectory and represents the integral curve of equation (3.6). Before we proceed to discuss characteristics of the integral curves of a general equation (3.6), let us consider the familiar linear equation in phase plane and define certain terms.

Singular Point–Centre

Consider an undamped oscillator

$$\frac{d^2x}{dt^2} + p^2x = 0 \tag{3.7}$$

Then

$$\frac{dx}{dt} = y$$

$$\frac{dy}{dt} = -p^2x$$

$$\frac{dy}{dx} = \frac{-p^2x}{y} \tag{3.8}$$

The solution of equation (3.7) can be written as

$$x = a\cos(pt - \phi)$$

$$y = -ap\sin(pt - \phi) \tag{3.9}$$

From the above we get

$$\frac{x^2}{a^2} + \frac{y^2}{a^2p^2} = 1 \tag{3.10}$$

The above is represented in the phase plane in Fig. 3.1. The trajectories (integral curves) are closed ones and they represent all the possible free vibratory solutions. For this case (equation 3.7), they are ellipses. The specific trajectory is picked up depending on the initial conditions. From equation (3.8), it can be seen that the direction of the tangent is indeterminate at $x = 0$, $y = 0$. Such points are designated as singular points. At these points either no trajectory passes through or several trajectories go through. In the case of centre, none of the integral curves passes through the origin of coordinates. Such a singular point in the neighbourhood of which the integral curves are closed and contain this point, is a centre. Since all the trajectories are closed the motion is periodic and a state of equilibrium exists at the singular point.

Consider any region of permissible deviations from the region η which represents the state of equilibrium in Fig. 3.2. Then the state of equilibrium will be stable, if we can indicate a region δ (η) surrounding the position of equilibrium such that none of the motions initiated within δ (η) penetrates the region η. Obviously the centre is in stable equilibrium since this condition is satisfied.

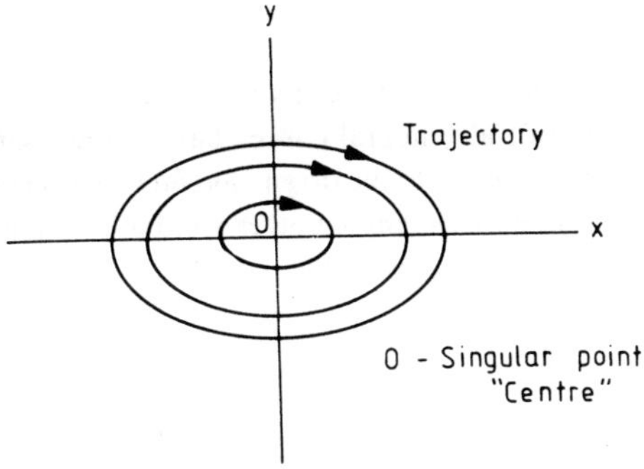

Fig. 3.1 Integral curves of an undamped oscillator, singular point-centre

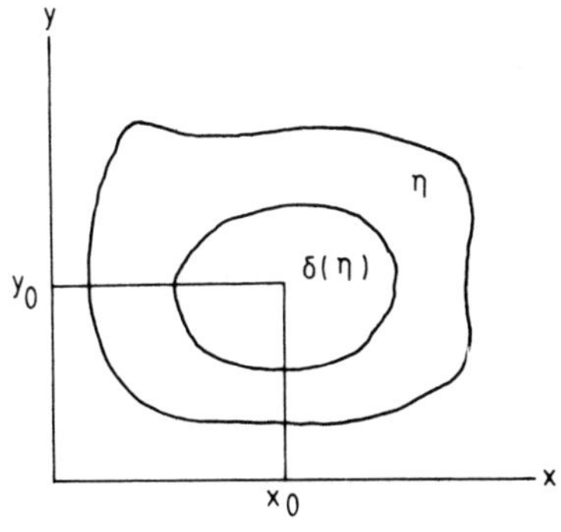

Fig. 3.2 Cocept of stability

Singular Point–Focus

Let us now consider a linear damped oscillator

$$\frac{d^2x}{dt^2} + 2\xi p \frac{dx}{dt} + p^2 x = 0 \tag{3.11}$$

Then

$$\frac{dx}{dt} = y$$

$$\frac{dy}{dt} = -2\xi p\, y - p^2 x$$

$$\frac{dy}{dx} = \frac{-2\xi p\, y - p^2 x}{y} \tag{3.12}$$

The origin is again a singular point and the solution is

$$x = ae^{-\xi pt}\cos(\sqrt{1-\xi^2}\,pt - \phi)$$

$$y = -ape^{-\xi pt}\sin\left\{\sqrt{1-\xi^2}\,pt - \phi + \tan^{-1}\frac{\xi}{\sqrt{1-\xi^2}}\right\} \quad (3.13)$$

The trajectories of equation (3.12) are given in Fig. 3.3 and the singular point, which is asymptotic point of all integral curves having the form of a spiral, is the focus. If $\xi > 0$, the spirals converge to the origin, as can be seen from equation (3.13), and the singularity is a stable focus. On the other hand if $\xi < 0$, the spiral trajectories go out of the origin and the focus is unstable.

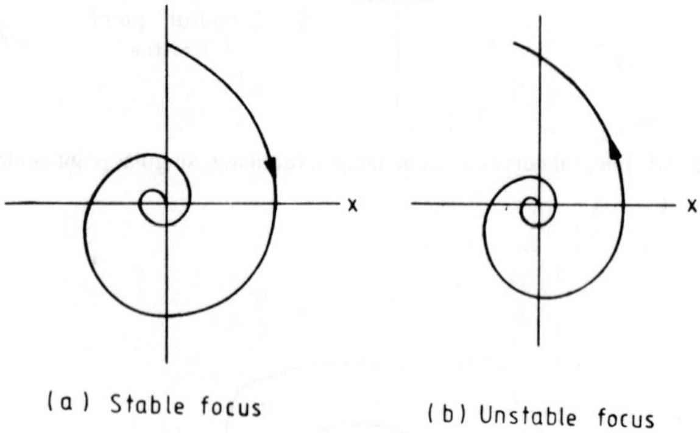

Fig. 3.3 Singular point–Focus

Singular Point–Node

Consider the equation (3.11) with overdamping, i.e., $\xi > 1$. The solution is

$$x = c_1 e^{-s_1 t} + c_2 e^{-s_2 t} \quad (3.14)$$

$$y = -c_1 s_1 e^{-s_1 t} - c_2 s_2 e^{-s_2 t} \quad (3.15)$$

where

$$-s_{1,2} = (-\xi \pm \sqrt{\xi^2 - 1})p$$

Multiply equation (3.14) by s_1 and add to (3.15). Then we get

$$y + s_1 x = c_2(s_1 - s_2)\,e^{-s_2 t} \quad (3.16)$$

Multiply again equation (3.14) by s_2 and add to (3.15) to get

$$y + s_2 x = -c_1(s_1 - s_2)e^{-s_1 t} \quad (3.17)$$

Raise the power on both sides of equation (3.16) by s_1 and equation (3.17) by s_2 and add to give

$$(y + s_1 x)^{s_1} = c(y + s_2 x)^{s_2}$$

i.e.,

$$(y + s_1 x) = c(y + s_2 x)^{s_2/s_1} \tag{3.18}$$

Equation (3.18) is shown in Fig. 3.4, which is a family of parabolas, touching the straight line $y = -s_1 x$ at the origin of the coordinates. When $\xi > 1$, the integral curves are given in Fig. 3.4a and when $\xi \leftarrow 1$, Fig. 3.4b gives the phase plane. This singularity is the node or branch point. Fig. 3.4b depicts an unstable node.

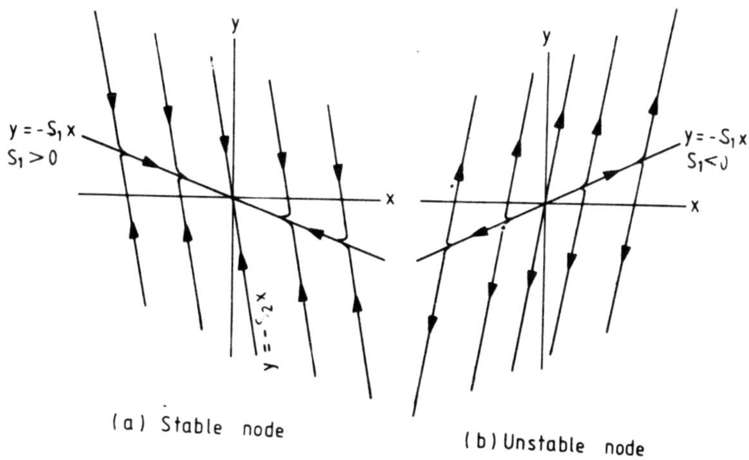

Fig. 3.4 Singular point–Node

Singular Point–Saddle

Now consider a hypothetical case of an undamped oscillator with negative stiffness, i.e.,

$$\frac{d^2x}{dt^2} - kx = 0 \tag{3.19}$$

The solution of the above equation (3.19) is

$$x = c_1 e^{\sqrt{k}\,t} + c_2 e^{-\sqrt{k}\,t}$$

$$y = c_1 \sqrt{k}\, e^{\sqrt{k}\,t} - c_2 \sqrt{k}\, e^{-\sqrt{k}\,t} \tag{3.20}$$

i.e.,

$$x^2 - \frac{y^2}{k} = C \tag{3.21}$$

This is a family of hyperbolas, as shown in Fig. 3.5. For $C = 0$ the two asymptotes of the family of curves are

$$y = \sqrt{k}\, x \text{ and } y = -\sqrt{k}\, x$$

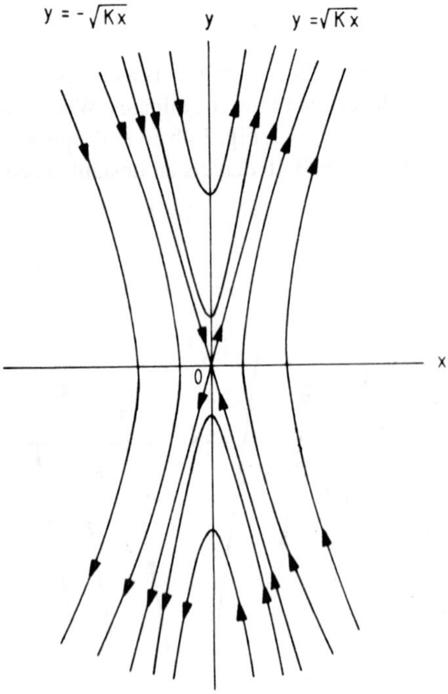

Fig. 3.5 Singular point – Saddle

The origin of the coordinates is again a singular point and the type shown in Fig. 3.5 is saddle. The singular point saddle is inherently unstable as can be readily observed from Fig. 3.5.

With these definitions of singular points in relation to linear systems, we can now consider nonlinear systems.

General Nonlinear Equation–Types of Singularity

Now consider equation (3.5) and express them in the form

$$\frac{dx}{dt} = a_1 x + a_2 y + X_2(x, y)$$

$$\frac{dy}{dt} = b_1 x + b_2 y + Y_2(x, y) \tag{3.22}$$

where $X_2(x, y)$ and $Y_2(x, y)$ are polynomials in x and y, containing the nonlinear terms only, i.e., powers greater than or equal to 2.

Since the origin $x = 0$, $y = 0$ is a singular point, we will use the linear part of the equation (3.22) in its neighbourhood to study the singularities, i.e.,

$$\frac{dy}{dx} = \frac{b_1 x + b_2 y}{a_1 x + a_2 y} \tag{3.23}$$

The solution of the above can be written as

$$x = A e^{\lambda t}$$

$$y = B e^{\lambda t} \tag{3.24}$$

Hence, from the linear part of equation (3.22)

$$\lambda A = a_1 A + a_2 B$$

$$\lambda B = b_1 A + b_2 B$$

i.e.,

$$\begin{bmatrix} (a_1 - \lambda) & a_2 \\ b_1 & (b_2 - \lambda) \end{bmatrix} \begin{Bmatrix} A \\ B \end{Bmatrix} = 0 \tag{3.25}$$

For nontrivial values of A and B, we get

$$\lambda_{1,2} = \frac{1}{2}\left\{a_1 + b_2 \pm \sqrt{(a_1 - b_2)^2 + 4a_2 b_1}\right\} \tag{3.26}$$

and

$$\frac{B_i}{A_i} = \frac{a_1 - \lambda_i}{-a_2} = \frac{-b_1}{b_2 - \lambda_i} \tag{3.27}$$

The general solution is therefore

$$x = A_1 e^{\lambda_1 t} + A_2 e^{\lambda_2 t}$$

$$y = B_1 e^{\lambda_1 t} + B_2 e^{\lambda_2 t} \tag{3.28}$$

From the above, we can study the type of singularity of the origin.

1. Node

When both roots of (3.26) are real and have the same sign, equation (3.28) gives the singularity to be a node or a branch point (*see* Fig. 3.4). The condition for the singularity to be a node is therefore

$$(a_1 - b_2)^2 + 4a_2 b_1 \geq 0$$

$$a_1 b_2 - a_2 b_1 > 0 \tag{3.29}$$

Further the conditions of stability of this node are given by

$$a_1 + b_2 > 0 \quad \text{unstable}$$

$$a_1 + b_2 < 0 \quad \text{stable} \tag{3.30}$$

2. Saddle

When both the roots of (3.26) are real, but of opposite sign, we have a saddle (*see* Fig. 3.5). The condition for the singularity to be saddle is therefore

$$(a_1 - b_2)^2 + 4a_2 b_1 \geq 0$$

$$a_1 b_2 - a_2 b_1 < 0 \tag{3.31}$$

The saddle is inherently unstable, as shown in Fig. 3.5.

3. Focus

If the roots of equation (3.26) are complex, we have a focus (*see* Fig. 3.3) and the condition for this singularity is therefore

$$(a_1 - b_2)^2 + 4a_2b_1 \le 0$$

$$a_1 + b_2 < 0 \quad \text{stable} \tag{3.32}$$

and

$$(a_1 - b_2)^2 + 4a_2b_1 \le 0$$

$$a_1 + b_2 > 0 \quad \text{unstable} \tag{3.33}$$

4. Centre

If the roots are imaginary, we have the singularity as centre (*see* Fig. 3.1). The condition for singularity to be centre is therefore

$$(a_1 - b_2)^2 + 4a_2b_1 \le 0$$

$$a_1 + b_2 + 0 \tag{3.34}$$

We use the above four properties of the linear part of the nonlinear equation to study the stability criterion of a singularity.

Trajectory Construction in Phase Plane

The simplest way of constructing the trajectories in the phase plane is the method of isoclines. Consider the general equation (3.6) and write it as

$$\frac{dy}{dx} = \frac{Y(x, y)}{X(x, y)} = F(x, y) \tag{3.35}$$

Since dy/dx defines the slope of the tangent to a trajectory, we can define a curve

$$F(x, y) = C \tag{3.36}$$

on which the integral curve has the same slope C. i.e., any trajectory intersecting at any point on the above curve (3.36) will go through with a slope equal to C. Such curves (3.36) are isoclines and with their help, the trajectories can be drawn quickly on a phase plane. For illustration, let us consider Van der Pol's equation, for which we found a periodic solution earlier.

$$\ddot{x} - \mu(1 - x^2)\dot{x} + x = 0 \tag{2.39}$$

i.e.,

$$\frac{dx}{dt} = y$$

$$\frac{dy}{dt} = \mu(1 - x^2)y - x$$

$$\frac{dy}{dx} = \frac{\mu(1 - x^2)y - x}{y} \tag{3.37}$$

Therefore, the equation for isoclines is

$$\mu(1-x^2)y - x = Cy \tag{3.38}$$

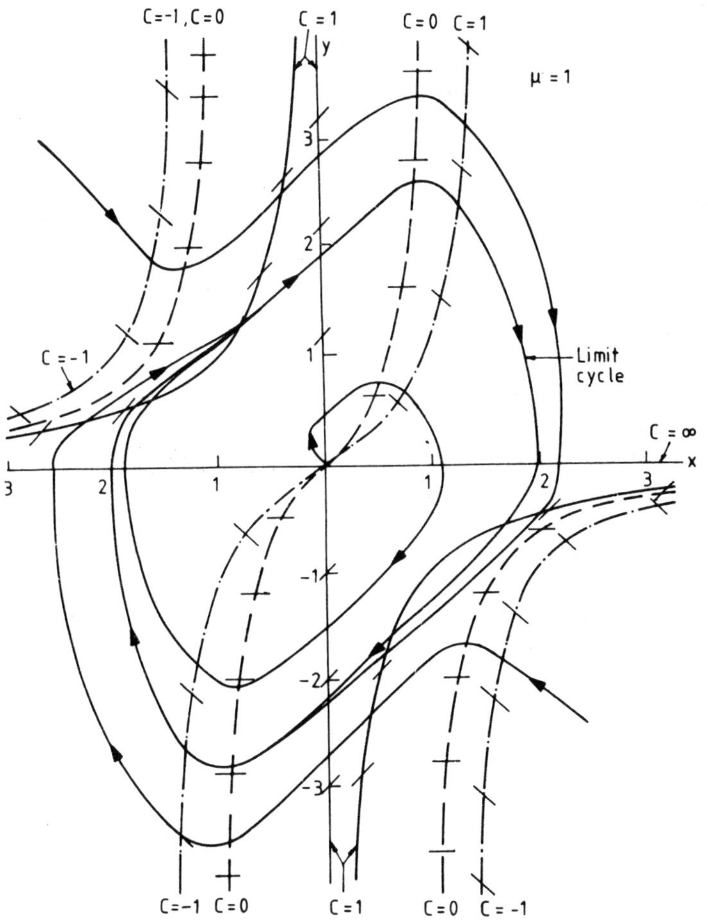

Fig. 3.6 Integral curves of Van der Pol's equation

Let $\mu = 1$. To draw the trajectories, the above equation is first plotted for different values of $C = 0, 1, -1$ and ∞ as shown in Fig. 3.6. The tangents are also marked on these isoclines. The procedure for drawing the integral curves is self-evident. We can choose any initial condition, i.e., starting point on the phase plane and draw the trajectory as indicated in Fig. 3.6, which remains tangential to the slope lines drawn on the isoclines.

From Fig. 3.6, it can be seen that for a Van der Pol oscillator, irrespective of where you start the trajectory, i.e., whatever may be the initial conditions, the integral curve is a closed loop, which is a limit cycle. A very good agreement exists even with one-term approximation of equation (2.39) given in (2.50). Clearly, the origin is an unstable focus. We can establish this from the analysis of singular points discussed earlier.

Comparing equation (3.37) with (3.23) we have

$$a_1 = 0; \quad a_2 = 1$$

$$b_1 = -1; \quad b_2 = 1$$

Also, we have

$$(a_1 - b_2)^2 + 4a_2 b_1 = -3$$

Hence, we conclude from (3.32) that the origin is a focus. Also, since $a_1 + b_2 = 1$, we find that the focus is unstable from equation (3.33).

Existence of a Limit Cycle

From the above discussion, we find that a limit cycle can exist in a nonlinear system. There could be in general several limit cycles for a system in the phase plane. These limit cycles can be stable or unstable depending on whether the integral curves converge to it or diverge away from it. The method of contact curves by Poincare will be useful in determining the possibility of a limit cycle existing for a nonlinear system. The method is as follows.

Consider a set of concentric circles with the singular point, i.e., the origin in the case of Van der Pol oscillator, and find the locus of the points, where these circles are tangent to integral curves. This locus is contact curve. The limit cycles, if any, would exist in the ring domain, with centre as singular point, whose boundaries are the innermost and outermost circles of radii $r_{\min}$ and $r_{\max}$ which touch the contact curve. The family of the circles around the origin are

$$x^2 + y^2 = \text{constant} \tag{3.39}$$

i.e.,

$$\frac{dy}{dx} = -\frac{x}{y} \tag{3.40}$$

For Van der Pol oscillator, with $\mu = 1$, we have

$$\frac{dy}{dx} = \frac{(1 - x^2)y - x}{y} \tag{3.37a}$$

Therefore, the contact curve is given by

$$\frac{\mu(1 - x^2)y - x}{y} = -\frac{x}{y}$$

i.e.,

$$(1 - x^2)y^2 = 0 \tag{3.41}$$

In polar coordinates the above is

$$(1 - r^2\cos^2\theta)\sin^2\theta = 0 \tag{3.42}$$

and

$$r = \frac{2\sin\theta}{\sin 2\theta} \tag{3.43}$$

The minimum and maximum values of r are

$$r_{\min} = \sqrt{2}; \quad r_{\max} = \infty \tag{3.44}$$

Hence, a limit cycle can exist anywhere in the region beyond a circle of radius of $\sqrt{2}$, as can be observed from Fig. 3.6.

When μ is very small, the trajectory will be circular in nature corresponding to the undamped linear system with amplitude equal to 2 units. When μ becomes very large, say 10, the limit cycle gets elongated in the direction of y as shown in Fig. 3.7. Because of this, the oscillatory behaviour is typical like a stick-slip motion. At the peak amplitudes the mass remains at rest or sticks for a considerable time, and suddenly moves or slips from one extreme position to the other. Because of the almost stationary position of the system for a considerable period and sudden jerky motion, this type of oscillation is called relaxation oscillation. A nonlinearly damped system known after Rayleigh is

$$\ddot{x} + x - \mu(\dot{x} - \dot{x}^3) = 0 \tag{3.45}$$

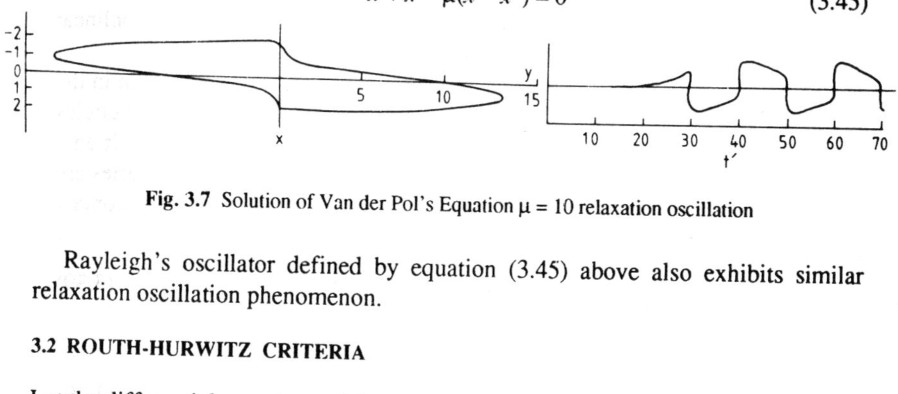

Fig. 3.7 Solution of Van der Pol's Equation $\mu = 10$ relaxation oscillation

Rayleigh's oscillator defined by equation (3.45) above also exhibits similar relaxation oscillation phenomenon.

3.2 ROUTH-HURWITZ CRITERIA

Let the differential equations of the system be

$$\frac{dx}{dt} = X(x, y)$$

$$\frac{dy}{dt} = Y(x, y) \tag{3.5}$$

Now, let (x_0, y_0) be the singular point under consideration for the above equation. We would like to study whether this singular point is stable or not. In order to determine whether the trajectory of the system converges to the singular point in its immediate neighbourhood, let us give a small variation to the point x_0, y_0 and define

$$x = x_0 + \xi$$

$$y = y_0 + \eta \tag{3.46}$$

Equations (3.5) then become

$$\frac{d\xi}{dt} = X(x_0 + \xi, y_0 + \eta)$$

$$\frac{d\eta}{dt} = Y(x_0 + \xi, y_0 + \eta) \tag{3.47}$$

Expanding the right hand side of the above equations in Taylor series, we get

$$\frac{d\xi}{dt} = X(x_0, y_0) + \frac{\partial X}{\partial x}\}_{x_0, y_0}\xi + \frac{\partial X}{\partial y}\}_{x_0, y_0}\eta + \ldots$$

$$\frac{d\eta}{dt} = Y(x_0, y_0) + \frac{\partial Y}{\partial x}\}_{x_0, y_0}\xi + \frac{\partial Y}{\partial y}\}_{x_0, y_0}\eta + \ldots \tag{3.48}$$

Equations (3.48) can be written in a simple form as

$$\frac{d\xi}{dt} = a_{11}\xi + a_{12}\eta$$

$$\frac{d\eta}{dt} = a_{21}\xi + a_{22}\eta \tag{3.49}$$

where a_{11}, a_{12}, a_{21}, a_{22} are the partial derivatives of X and Y at the singular point as indicated in (3.48).

The solution of equations (3.49) can be written as

$$\xi = A_1 e^{\lambda t}$$

$$\eta = A_2 e^{\lambda t} \tag{3.50}$$

Therefore

$$\begin{bmatrix} (a_{11} - \lambda) & a_{12} \\ a_{21} & (a_{22} - \lambda) \end{bmatrix} \begin{Bmatrix} A \\ B \end{Bmatrix} = 0 \tag{3.51}$$

The characteristic equation for λ is then given by

$$\lambda^2 - (a_{11} + a_{22})\lambda + a_{11}a_{22} - a_{12}a_{21} = 0 \tag{3.52}$$

For stability of the system, the real parts of λ should be negative and hence we have

$$-(a_{11} + a_{22}) > 0$$

$$a_{11}a_{22} - a_{12}a_{21} > 0 \tag{3.53}$$

As an example, consider Van der Pol's oscillator with $\mu = 1$

$$X(x,y) = y$$

$$Y(x,y) = (1-x^2)y - x$$

$$\frac{\partial X}{\partial x}\}_{x_0, y_0} = a_{11} = 0$$

$$\frac{\partial X}{\partial y}\}_{x_0, y_0} = a_{12} = 1$$

$$\frac{\partial Y}{\partial x}\}_{x_0, y_0} = a_{21} = -2xy - 1 = -1$$

$$\frac{\partial Y}{\partial y}\}_{x_0, y_0} = a_{22} = 1 - x^2 = 1$$

Hence

$$-(a_{11} + a_{22}) = -1$$

$$a_{11}a_{22} - a_{12}a_{21} = +1 \tag{3.53a}$$

In view of the first condition of (3.53), equations (3.53a) above show that the Van der Pol's oscillator is not stable at the origin. Hence, any disturbance from the rest position will lead to the limit cycle discussed earlier.

3.3 DIRECT STABILITY METHOD

The methods discussed in sections 3.1 and 3.2 typically fail to indicate the stability of the system in the absence of asymptotic stability. The application of previous methods require the linear part of the system to be responsible for stability, which may not be necessarily true in all cases. For example, consider a system governed by

$$\frac{dx}{dt} = X(x,y) = y - x(x^2 + y^2)$$

$$\frac{dy}{dt} = Y(x,y) = -x - y(x^2 + y^2) \tag{3.54}$$

Evidently, the origin $x = 0$, $y = 0$ is a singular point. Let us use the linear part of the above to determine the type of singularity. Writing the above in the form of equation (3.22), we can express equations (3.54) in the form of (3.23) to give

$$\frac{dy}{dx} = \frac{y}{-x} \tag{3.55}$$

Hence

$$a_1 = -1; \quad a_2 = 0$$

$$b_1 = 0; \quad b_2 = 1$$

Therefore

$$(a_1 - b_2)^2 + 4a_2b_1 = 4$$

$$a_1b_2 - a_2b_1 = -1$$

Hence from (3.31) we conclude that the singular point is a saddle and therefore unstable. However, we will soon find out that this is wrong and the prediction of stability behaviour based on the linear part alone is not correct. This is because the linear system given by (3.55) does not possess asymptotic characteristics.

We can also use the Routh-Hurwitz criteria (which also fails to indicate stability in the absence of asymptotic stability) for this problem to check the above deduction. For the system in equation (3.54), the linearity coefficients in (3.49) are

$$\frac{\partial X}{\partial x}\Big\}_{x_0, y_0} = a_{11} = 0$$

$$\frac{\partial X}{\partial y}\Big\}_{x_0, y_0} = a_{12} = 1$$

$$\frac{\partial Y}{\partial x}\Big\}_{x_0, y_0} = a_{21} = -1$$

$$\frac{\partial Y}{\partial y}\Big\}_{x_0, y_0} = a_{22} = 0$$

Therefore

$$-(a_{11} + a_{22}) = 0$$

$$a_{11}a_{22} - a_{12}a_{21} = +1$$

Hence, the two roots of (3.52) are complex conjugate with the real part equal to 0. Hence there is no asymptotic stability of the system. The first equation in (3.53) is not satisfied and therefore the Routh-Hurwitz criterion also makes a wrong prediction regarding the stability.

A second method of Liapunov treats the stability problem directly, in a natural, though somewhat nonconstructive way. Application of the method is not limited to the presence of linearity of the system, but is limited by practical difficulties of constructing what is called as Liapunov function.

The principle behind the method is very simple. Let the integral curve of a system be as shown in Fig. 3.8. We choose a function $V(x, y)$ such that it intersects the integral curve at only one point P. This choice may not be practicable in all cases. If the singularity is stable, then the intersection point P tends to the origin as t tends to ∞.

The simplest choice for $V(x, y)$ is a circle if it is possible. Then

$$\frac{dV}{dt} = 2x\frac{dx}{dt} + 2y\frac{dy}{dt} \tag{3.56}$$

For the system defined by (3.54), the above equation gives

$$\frac{dV}{dt} = 2x\{y - x(x^2 + y^2)\} + 2y\{-x - y(x^2 + y^2)\}$$

$$= -2(x^2 + y^2)^2 \tag{3.57}$$

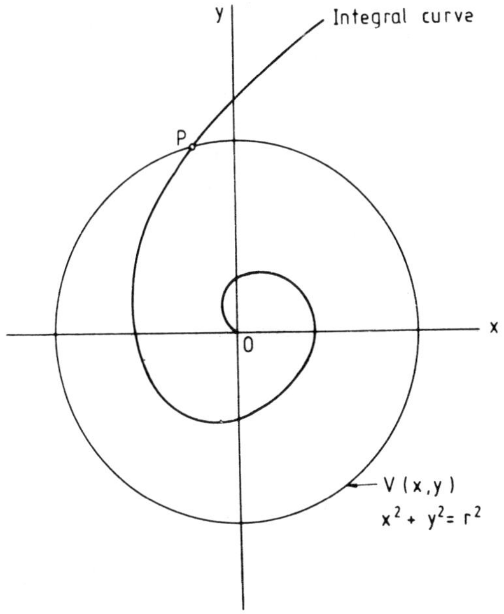

Fig. 3.8 Liapunov direct method

We find that dV/dt is negative for all t and hence the origin is stable. We can also make a plot of the response by using the method of isoclines. This is shown in Fig. 3.9. It can be readily seen that the singular point $x = 0, y = 0$ is stable. Thus the second method of Liapunov is more general. The function V in this method is to be so chosen that there is only one intersection with the trajectory in the neighbourhood of the singularity. This can be a circle or an ellipse or any suitable closed curve. In general

$$\frac{dx_i}{dt} = X_i(x_1, x_2, \ldots x_n) i = 1, 2, \ldots n \tag{3.58}$$

We introduce

$$V(x_1, x_2, \ldots x_n)$$

which is positive definite, i.e., V is continuous as well as $\partial V/\partial x_i$; in a certain domain outside the origin V is positive and zero at the origin. Then

$$\begin{aligned} \frac{dV}{dt} &= \sum_{i=1}^{n} \frac{\partial V}{\partial x_i} \frac{dx_i}{dt} \\ &= \sum_{i=1}^{n} \frac{\partial V}{\partial x_i} X_i \\ &= W(x_i) \end{aligned} \tag{3.59}$$

V is a Liapunov function, provided

$$\frac{dV}{dt} = W(x_i) \le 0$$

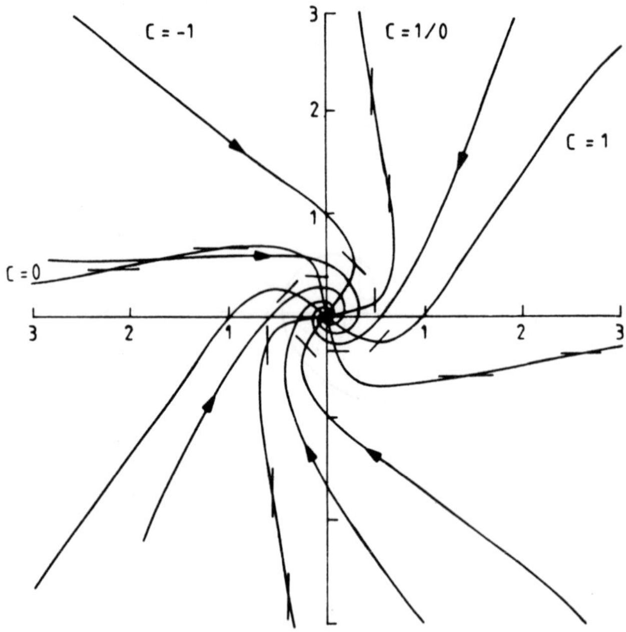

Fig. 3.9 Trajectories of system given by (3.54)

For a general equation of one degree of freedom

$$\frac{d^2x}{dt^2} + f(x)\frac{dx}{dt} + g(x) = 0 \tag{3.60}$$

we can adopt the following procedure. First we convert the above equation into two first order equations by defining

$$F(x) = \int_0^x f(x)dx$$

$$G(x) = \int_0^x g(x)dx \tag{3.61}$$

Then it can be verified that

$$\frac{dx}{dt} = y - F(x)$$

$$\frac{dy}{dt} = -g(x) \tag{3.62}$$

For Liapunov function, we use the total energy of the system not accounting for damping and define

$$V(x, y) = \frac{1}{2}y^2 + G(x) \tag{3.63}$$

Then

$$\frac{dV}{dt} = g(x)\{y - F(x)\} - yg(x)$$

$$= -g(x)F(x) \tag{3.64}$$

Hence, the origin is stable if

$$-g(x)F(x) \leq 0 \tag{3.65}$$

Consider the Van der Pol oscillator

$$\frac{d^2x}{dt^2} - \mu(l - x^2)\frac{dx}{dt} + x = 0$$

Then

$$g(x) = x$$

$$f(x) = -\mu(1 - x^2)$$

$$F(x) = -\mu\left(x - \frac{x^3}{3}\right)$$

$$-g(x)F(x) = \mu x^2\left(1 - \frac{x^2}{3}\right)$$

For $x^2 < 3$, the above quantity is positive and hence the origin is unstable.

3.4 FLOQUET'S THEORY OF STABILITY

Consider a linear differential equation with periodic coefficients

$$\frac{d^2w}{dt^2} + p(t)\frac{dw}{dt} + q(t)w = 0 \tag{3.66}$$

where $p(t)$ and $q(t)$ are functions of time periodic in T. i.e.,

$$p(t + T) = p(t)$$

$$q(t + T) = q(t) \tag{3.67}$$

Since equation (3.66) is a second order differential equation, there must be two independent solutions and let them be represented by $w_1(t)$ and $w_2(t)$. In the present case, we refer to these solutions as fundamental set of solutions or principal solutions for equations with periodic coefficients. The solutions of equations (3.66) can be written as

$$w(t) = c_1w_1(t) + c_2w_2(t) \tag{3.68}$$

Next consider $w(t + T)$ and substitute in equation (3.66), we get

$$\ddot{w}(t + T) = -p(t + T)\dot{w}(t + T) - q(t + T)w(t + T)$$

In view of the periodicity of the functions $p(t)$ and $q(t)$ given in (3.67), the above becomes

$$\ddot{w}(t+T) = -p(t)\dot{w}(t+T) - q(t)w(t+T) \tag{3.69}$$

It is then evident that $w_1(t+T)$ and $w_2\,(t+T)$ are also a fundamental solutions set. Therefore, they can be expressed as a linear combination of the two independent solutions $w_1\,(t)$ and $w_2\,(t)$, as follows

$$w_1(t+T) = a_{11}w_1(t) + a_{12}w_2(t)$$

$$w_2(t+T) = a_{21}w_1(t) + a_{22}w_2(t) \tag{3.70}$$

The above two equations in (3.70) represent the principal solutions at time $t = t + T$ and it is logical to study them in terms of solutions at time t, to understand the stability behaviour of the system. We therefore define a normal solution, which is a certain multiple of the principal solution after the period T.

$$w\,(t+T) = mw(t) \tag{3.71}$$

Our aim should be to determine such a solution as in above equation (3.71) so as to study the stability of equation (3.66). From equations (3.70) and (3.71) we can write

$$w_1(t+T) = mw_1(t) = a_{11}w_1(t) + a_{12}w_2(t)$$

$$w_2(t+T) = mw_2(t) = a_{21}w_1(t) + a_{22}w_2(t) \tag{3.72}$$

Since $w_1\,(t)$ and $w_2\,(t)$ are nonvanishing, we get from (3.72)

$$det\begin{bmatrix} a_{11}-m & a_{12} \\ a_{21} & a_{22}-m \end{bmatrix} = 0 \tag{3.73}$$

There are two roots for m, m_1, and m_2 for the nontrivial solution. Also, in view of (3.71) we have

$$w(t+nT) = m^n w(t) \tag{3.74}$$

Stability would depend on the values of m in equation (3.74). We can also choose a function similar to the damped response of a linear oscillator to study the stability behaviour. For the periodicity of the solution derived in equation (3.72), we consider the following function

$$e^{-\mu_i(t+T)} w_i(t+T) \tag{a}$$

With the help of equation (3.72), the above function (a) becomes

$$e^{-\mu_i(t+T)} m_i w_i(t) \tag{b}$$

Therefore, equating (a) and (b), we get

$$e^{-\mu_i(t+T)} w_i\,(t+T) = m_i e^{-\mu_i T} e^{-\mu_i t} w_i(t) \tag{3.75}$$

Let us choose

$$m_i e^{-\mu_i T} = 1$$

i.e.,

$$m_i = e^{\mu_i T} \tag{3.76}$$

then

$$\phi_i(t) = e^{-\mu_i t} w_i(t) \tag{3.77}$$

will be periodic in T.

The relationship between m and μ is obtained from equation (3.76) as

$$\mu_i T = \ln m_i$$

i.e.,

$$\mu_i = \frac{\ln m_i}{T} \tag{3.76a}$$

Equation (3.77) is known as Floquet's result and we can write from this

$$w_i(t) = e^{\mu_i t} \phi_i(t) \tag{3.78}$$

It can be easily verified on similar analysis that the following is also valid.

$$w_i(t) = e^{-\mu_i t} \psi_i(t) \tag{3.79}$$

Both the functions $\phi_i(t)$ and $\psi_i(t)$ are periodic in T.

The application of the above can be illustrated for a general nonlinear system with an exciting periodic force.

$$\frac{d^2x}{dt^2} + f\left(x, \frac{dx}{dt}\right) = e(t) \tag{3.80}$$

with the external force $e(t)$ to be periodic with t_0, i.e.,

$$e(t + t_0) = e(t) \tag{3.81}$$

Let a periodic solution neglecting the transients be $x_0(t)$, such that

$$x_0(t + t_0) = x_0(t) \tag{3.82}$$

To examine the stability of the above solution, we will consider the following variation

$$x(t) = x_0(t) + \xi \tag{3.83}$$

Substituting (3.83) above in equation (3.80), we have

$$\frac{d^2x_0}{dt^2} + \frac{d^2\xi}{dt^2} + f\left(x_0 + \xi, \frac{dx_0}{dt} + \frac{d\xi}{dt}\right) = e(t) \tag{3.84}$$

Expanding the function f in (3.84) above in Taylor's series and noting that x_0 is a solution, we get

$$\ddot{\xi} + \left\{\frac{\partial f(x,\dot{x})}{\partial \dot{x}}\right\}_0 \dot{\xi} + \left\{\frac{\partial f(x,\dot{x})}{\partial x}\right\}_0 \xi = 0 \tag{3.85}$$

The higher order terms in (3.85) are not considered as we can assume that in the neighbourhood of the singular point, linear terms are sufficient to consider stability. Then, equation (3.85) can be written as

$$\xi + F(t)\xi + G(t)\xi = 0 \tag{3.86}$$

where

$$F(t) = \left\{\frac{\partial f(x,\dot{x})}{\partial \dot{x}}\right\}_0$$

$$G(t) = \left\{\frac{\partial f(x,\dot{x})}{\partial x}\right\}_0$$

We can transform (3.86) to an equation without damping term using

$$\xi = \eta e^{-\frac{1}{2}\int F(t)dt} \tag{3.87}$$

Then

$$\xi = \dot{\eta}\frac{\xi}{\eta} - \frac{1}{2}F\xi$$

$$\xi = \ddot{\eta}\frac{\xi}{\eta} - F\dot{\eta}\frac{\xi}{\eta} + \frac{1}{4}F^2\xi - \frac{1}{2}\dot{F}\xi \tag{3.88}$$

Using the above equations (3.87) and (3.88), equation (3.86) becomes

$$\ddot{\eta} + \left(G - \frac{1}{2}\dot{F} - \frac{1}{4}F^2\right)\eta = 0 \tag{3.89}$$

In view of (3.72)

$$\eta_1(t+T) = a_{11}\eta_1(t) + a_{12}\eta_2(t)$$

$$\dot{\eta}_1(t+T) = a_{11}\dot{\eta}_1(t) + a_{12}\dot{\eta}_2(t)$$

$$\eta_2(t+T) = a_{21}\eta_1(t) + a_{22}\eta_2(t)$$

$$\dot{\eta}_2(t+T) = a_{21}\dot{\eta}_1(t) + a_{22}\dot{\eta}_2(t) \tag{3.90}$$

Therefore

$$\begin{bmatrix} \eta_1(t+T) & \eta_2(t+T) \\ \dot{\eta}_1(t+T) & \dot{\eta}_2(t+T) \end{bmatrix} = \begin{bmatrix} \eta_1(t) & \eta_2(t) \\ \dot{\eta}_1(t) & \dot{\eta}_2(t) \end{bmatrix} = \begin{bmatrix} a_{11} & a_{21} \\ a_{12} & a_{22} \end{bmatrix} \tag{3.91}$$

Let at time $t = 0$

$$\eta_1(0) = 1; \quad \dot{\eta}_1(0) = 0$$

$$\eta_2(0) = 0; \quad \dot{\eta}_2(0) = 1 \tag{3.92}$$

Then, equation (3.91) gives

$$\begin{bmatrix} \eta_1(T) & n_2(T) \\ \dot{\eta}_1(T) & \dot{\eta}_2(T) \end{bmatrix} = \begin{bmatrix} 1 & 0 \\ 0 & 1 \end{bmatrix} \begin{bmatrix} a_{11} & a_{21} \\ a_{12} & a_{22} \end{bmatrix} \tag{3.93}$$

i.e.,

$$\eta_1(T) = a_{11}; \quad \eta_2(T) = a_{21}$$

$$\dot{\eta}_1(T) = a_{12}; \quad \dot{\eta}_2(T) = a_{22} \tag{3.94}$$

From equation (3.73) we get

$$det\begin{bmatrix} \eta_1(T) - m & \dot{\eta}_1(T) \\ \eta_2(t) & \dot{\eta}_2(T) - m \end{bmatrix} = 0 \tag{3.95}$$

Hence

$$m^2 - 2Am + B = 0 \tag{3.96}$$

where

$$A = \frac{1}{2}\{\eta \cdot (T) + \dot{\eta}_2(T)\}$$

$$B = \eta_1(T)\dot{\eta}_2(T) - \eta_2(T)\dot{\eta}_1(T) \tag{3.97}$$

It may be noted that B in equation (3.97) is the Wronskian determinant of η_1 and η_2 and is a constant. This can be also directly verified from equation (3.89), which can be written as

$$\ddot{\eta} + G_1(t)\eta = 0$$

Multiply the above equation first by η_1 and next by η_2 and subtract as follows

$$\ddot{\eta}_1 + G_1(t)\eta_1 = 0 \quad \times \quad \eta_2$$

$$\ddot{\eta}_2 + G_1(t)\eta_2 = 0 \quad \times \quad \eta_1$$

Then

$$\ddot{\eta}_1\eta_2 - \ddot{\eta}_2\eta_1 = 0 \tag{3.98}$$

Integrating the above we get

$$\dot{\eta}_1\eta_2 - \dot{\eta}_2\eta_1 = \text{constant} \tag{3.99}$$

Hence, B is a constant. At $t = 0$, the value of B can be obtained by using (3.92) as equal to 1. Hence

$$m^2 - 2Am + 1 = 0 \tag{3.100}$$

The stability of the system would depend on the value of A. The two roots of m are

$$m_{1,2} = A \pm \sqrt{A^2 - 1} \tag{3.101}$$

The Floquet solution for η can be written as

$$\eta(t) = c_1 e^{\mu t}\phi(t) + c_2 e^{-\mu t}\psi(t) \tag{3.102}$$

where

$$\mu_1 = \frac{\ln m_i}{T}$$

Equation (3.87) now becomes

$$\xi = c_1 e^{\mu t} e^{-\frac{1}{2}\int F(t)dt} \phi(t) + c_2 e^{-\mu t} e^{-\frac{1}{2}\int F(t)dt} \psi(t) \tag{3.103}$$

Since exp $(\mu t)\,\phi(t)$ is a solution of equation (3.89), we can determine μ for a given period T. If $F(t)$ in equation (3.86) is written as

$$F(t) = 2\delta + \text{time dependent terms} \tag{3.104}$$

then

$$\xi = c_1 e^{(\mu-\delta)t} e^{-h(t)} \phi(t) + c_2 e^{(-\mu-\delta)t} e^{-h(t)} \psi(t) \tag{3.105}$$

where $h(t)$ has only higher order terms. Therefore, the stability condition for x_0 to be stable is

$$\mathcal{R}(\pm\mu - \delta) \le 0 \tag{3.106}$$

3.5 MATHIEU'S EQUATION

An important equation of variable parameters in physical systems is Mathieu's equation. This is

$$\frac{d^2x}{d\tau^2} + (a + 16q\cos 2\tau)x = 0 \tag{3.107}$$

where a and q are real numbers. If the shaft of a rotor is asymmetric, like a two-pole generator rotor, its equation of motion for lateral bending can be shown to be of the type above. The solution of the above equation (3.107) is

$$x = c_1 e^{\mu\tau} \phi(\tau) + c_2 e^{-\mu\tau} \phi(-\tau) \tag{3.108}$$

since equation (3.107) remains same if τ is changed to $-\tau$. As discussed earlier, the above solutions in (3.108) are stable, if x remains bounded as $\tau \to \infty$ and unstable if x grows indefinitely as $\tau \to \infty$ and x are neutral, if the solution is periodic. Such neutral solutions are Mathieu functions and a and q should have a definite relationship between them. For example, if $q = 0$, the solutions are

$$a = 1;\ \cos\tau;\ \sin\tau$$

$$a = 4;\ \cos 2\tau;\ \sin 2\tau$$

$$a = 9;\ \cos 3\tau;\ \sin 3\tau \quad \ldots \tag{3.109}$$

For other values of q, we have the Mathieu functions denoted by

$$ce_\eta(\tau, q);\quad se_n(\tau, q)\ n = 1, 2 \ldots \tag{3.110}$$

For periodic solutions, other sets of boundary conditions can be obtained by using the Perturbation method. Let

$$x = x_0 + qx_1 + q^2x_2 + \dots$$

$$a = n^2 + qa_1 + q^2a_2 + \dots$$

Then

$$(\ddot{x}_0 + q\ddot{x}_1 + q^2\ddot{x}_2 + q^3\ddot{x}_3 + \dots)$$

$$+(n^2 + qa_1 + q^2a_2 + q^3a_3 + \dots + 16q\cos 2\tau)(x_0 + qx_1 + q^2x_2 + q^3x_3 + \dots) = 0$$

From the above equation we write

$$\ddot{x}_0 + n^2x_0 = 0$$

$$\ddot{x}_1 + n^2x_1 = -(a_1 + 16\cos 2\tau)x_0$$

$$\ddot{x}_2 + n^2x_2 = -(a_1 + 16\cos 2\tau)x_1 - a_2x_0$$

$$\ddot{x}_3 + n^2x_3 = -(a_1 + 16\cos 2\tau)x_2 - a_2x_1 - a_3x_0$$

$$\ddot{x}_4 + n^2x_4 = -(a_1 + 16\cos 2\tau)x_3 - a_2x_2 - a_3x_1 - a_4x_0 \text{ etc.}$$

From the first equation above, we have

$$x_0 = \cos n\tau$$

$$\sin n\tau, n = 0, 1, 2, 3, \dots$$

Case 1: n = 0

$$x_0 = 1$$

$$\ddot{x}_1 = -a_1 - 16\cos 2\tau$$

$$a_1 = 0; x_1 = 4\cos 2\tau$$

$$\ddot{x}_2 = -(16\cos 2\tau)(4\cos 2\tau) - a_2$$

$$= -(32 + a_2) - 32\cos 4\tau$$

$$a_2 = -32; x_2 = 2\cos 4\tau$$

$$\ddot{x}_3 = -(16\cos 2\tau)(2\cos 4\tau) + 32 \times 4\cos 2\tau - a_3$$

$$= -a_3 + 112\cos 2\tau - 16\cos 6\tau$$

$$a_3 = 0; \quad x_3 = -28\cos 2\tau + \frac{4}{9}\cos 6\tau$$

$$\ddot{x}_4 = -16\cos 2\tau(-28\cos 2\tau + \frac{4}{9}\cos 6\tau + 32\times 2\cos 4\tau - a_4$$

$$= (324 - a_4) + \frac{3460}{9}\cos 4\tau - \frac{32}{9}\cos 8\tau$$

$$a_4 = 324; \quad x_4 = -\frac{865}{4}\cos 4\tau + \frac{1}{18}\cos 8\tau \text{ etc.}$$

Hence, $n = 0$

$$a = -32q^2 + 324q^4 + \ldots$$

$$x = 1 + (4q - 28q^3)\cos 2\tau + \left(2q^2 - \frac{865}{4}q^4\right)\cos 4\tau + \frac{4}{9}q^3\cos 6\tau + \frac{1}{18}q^4\cos 8\tau + \ldots$$

The above represents the case $q = 0$ and $a = 0$, $x = 1$ for $n = 0$. We should study the periodic solutions with $n = 1, 2, 3$, etc. to define the relationship between a and q for obtaining the stability boundaries.

Case 2: n = 1

For

$$x_0 = \cos\tau$$

$$\ddot{x}_1 + x_1 = -(a_1 + 16\cos 2\tau)\cos\tau$$

$$= -(a_1 + 8)\cos\tau - 8\cos 3\tau$$

$$a_1 = -8; \quad x_1 = \cos 3\tau$$

$$\ddot{x}_2 + x_2 = -(-8 + 16\cos 2\tau)\cos 3\tau - a_2\cos\tau$$

$$= -(a_2 + 8)\cos\tau + 8\cos 3\tau - 8\cos 5\tau$$

$$a_2 = -8; \quad x_2 = -\cos 3\tau + \frac{1}{3}\cos 5\tau$$

$$\ddot{x}_3 + x_3 = -(-8 + 16\cos 2\tau)\left(-\cos 3\tau + \frac{1}{3}\cos 5\tau\right) - a_2\cos 3\tau - a_3\cos\tau$$

$$= (-a_3 + 8)\cos\tau - \frac{8}{3}\cos 3\tau + \frac{32}{3}\cos 5\tau - \frac{8}{3}\cos 7\tau$$

$$a_3 = 8 \text{ etc.}$$

From the above

$$a_{c1} = 1 - 8q - 8q^2 + 8q^3 + \ldots \tag{3.111}$$

In the above suffix c denotes $x_0 = \cos\tau$ and suffix 1 denotes $n = 1$.
In a similar manner, for $x_0 = \sin\tau$, we get

$$x_0 = \sin\tau$$

$$a_{s1} = 1 + 8q - 8q^2 - 8q^3 + \dots \tag{3.112}$$

Case 3: n = 2

$$x_0 = \cos 2\tau$$

$$\ddot{x}_1 + 4x_1 = -(a_1 + 16\cos 2\tau)\cos 2\tau$$

$$= -8 - a_1 \cos 2\tau - 8\cos 4\tau$$

$$a_1 = 0; \quad x_1 = -2 + \frac{2}{3}\cos 4\tau$$

$$\ddot{x}_2 + 4x_2 = -16\left(-2 + \frac{2}{3}\cos 4\tau\right)\cos 2\tau - a_2 \cos 2\tau$$

$$= \left(-a_2 + \frac{80}{3}\right)\cos 2\tau - \frac{16}{3}\cos 6\tau$$

$$a_2 = \frac{80}{3}; \; x_2 = \frac{1}{6}\cos 6\tau$$

$$\ddot{x}_3 + 4x_3 = -\frac{16}{6}\cos 2\tau \cos 6\tau - \frac{80}{3}\left(-2 + \frac{2}{3}\cos 4\tau\right) - a_3 \cos 2\tau$$

$$= -a_3 \cos 2\tau + \frac{160}{3} - \frac{172}{9}\cos 4\tau - \frac{4}{3}\cos 8\tau$$

$$a_3 = 0; \; x_3 = \frac{40}{3} + \frac{43}{27}\cos 4\tau + \frac{1}{45}\cos 8\tau$$

$$\ddot{x}_4 + 4x_4 = -16\left(\frac{40}{3} + \frac{43}{27}\cos 4\tau + \frac{1}{45}\cos 8\tau\right)\cos 2\tau - \frac{80}{18}\cos 6\tau - a_4 \cos 2\tau$$

$$= \left(-a_4 - \frac{6104}{27}\right)\cos 2\tau + \frac{2344}{135}\cos 6\tau - \frac{8}{45}\cos 10\tau$$

$$a_4 = -\frac{6104}{27} \quad \text{etc.}$$

From the above

$$a_{c2} = 4 + \frac{80}{3}q^2 - \frac{6104}{27}q^4 + \dots \tag{3.113}$$

In a similar manner, for $x_0 = \sin 2\,\tau$, we get

$$x_0 = \sin 2\tau$$

$$a_{s2} = 4 - \frac{16}{3}q^2 + \frac{40}{27}q^4 + \dots \tag{3.114}$$

Case 4: n = 3

$$x_0 = \cos 3\tau$$

For this case it can be shown in a similar way

$$a_{c3} = 9 + 4q^2 - 8q^3 + \frac{13}{5}q^4 + \ldots \tag{3.115}$$

In a similar manner, for $x_0 = \sin 3\tau$, we get

$$x_0 = \sin 3\tau$$

$$a_{s3} = 9 + 4q^2 + 8q^3 + \frac{13}{5}q^4 + \ldots \tag{3.116}$$

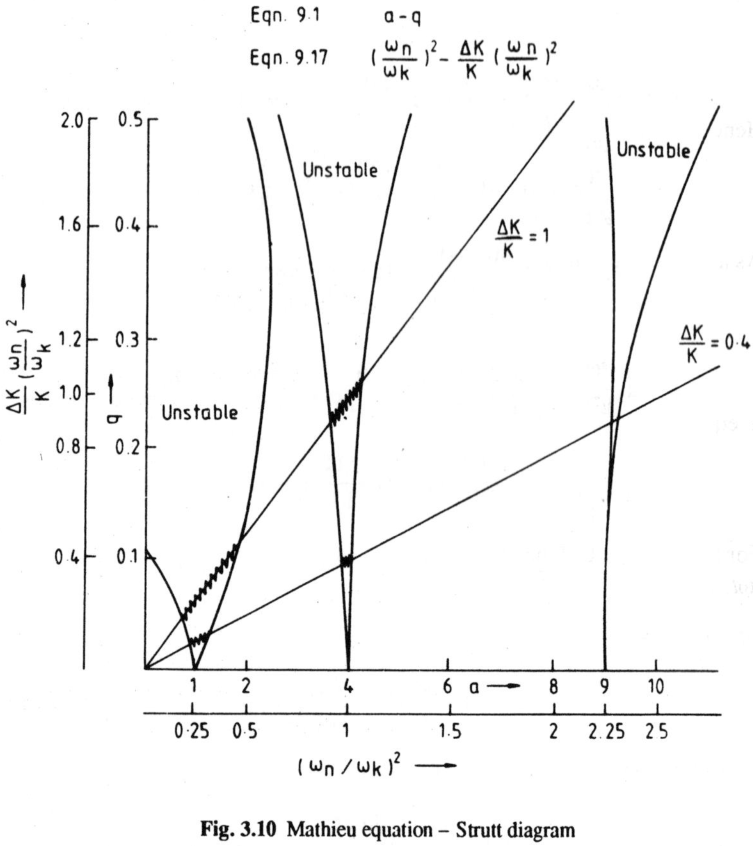

Fig. 3.10 Mathieu equation – Strutt diagram

Equations (3.111) to (3.116) are plotted in Fig. 3.10. The unstable regions occur between the two boundary lines for $a = 1$, $a = 4$, $a = 9$ etc., the regimes when $q = 0$

corresponding to equation (3.109). Therefore, Mathieu equation has unstable solutions for any combination of a and q in the unstable regimes of Fig. 3.10. This type of diagram is also known as Strutt diagram.

Consider a vertical shaft of rectangular cross-section carrying a disk of mass m as shown in Fig. 3.11. The shaft is permitted to have lateral motion in one plane only by means of the guides shown at section A. Let the maximum stiffness of the shaft be $k + \Delta k$ and minimum stiffness $k - \Delta k$. Obviously in one revolution, the stiffness changes in two cycles as shown in Fig. 3.12. The equation of motion for the disk is

$$m\ddot{x} + (k + \Delta k \sin \omega_k t)x = 0 \tag{3.117}$$

where

$$\omega_k = 2\omega$$

If the disk is horizontally supported instead of vertical as shown in Fig. 3.11, the origin gets shifted and we can write

$$y = x + c \tag{3.118}$$

Hence, equation (3.117) for a horizontal shaft becomes

$$m\ddot{y} + (k + \Delta k \sin \omega_k t)y = mg + kc + c\Delta k \sin \omega_k t \tag{3.119}$$

As a first approximation, we will take

$$c = \frac{-mg}{k} (\Delta k \ll k) \tag{3.120}$$

then, equation (3.119) becomes

$$m\ddot{x} + (k + \Delta k \sin \omega_k t)x = mg \tag{3.119a}$$

For free vibration of the above, equation (3.119a) and (3.117) are same. Defining $\tau = \omega t$, we have

$$\ddot{x} + \left(\frac{\omega_n^2}{\omega^2} + \frac{\Delta k}{m\omega^2} \sin 2\tau\right)x = 0 \quad \cdot = \frac{d}{d\tau}$$

i.e.,

$$\ddot{x} + \left\{4\frac{\omega_n^2}{\omega_k^2} + 4\frac{\Delta k\,\omega_n^2}{k\omega_k^2} \sin 2\tau\right\}x = 0 \tag{3.121}$$

Equation (3.121) is a Mathieu equation with

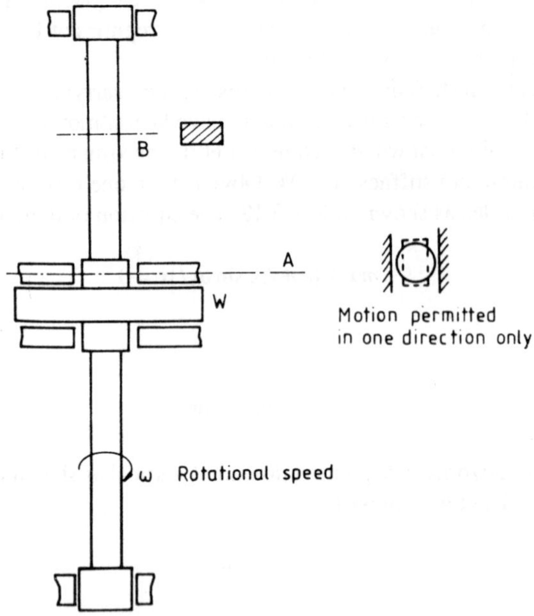

Fig. 3.11 Rectangular shaft carrying a disc *W*

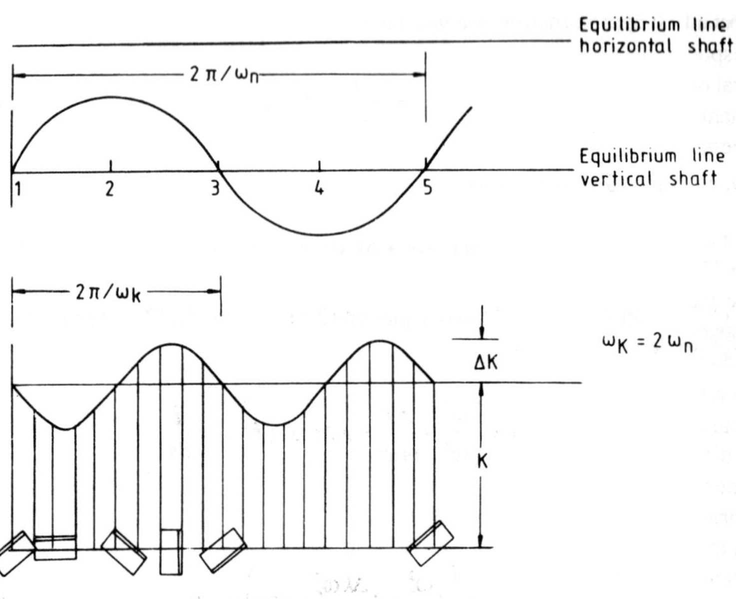

Fig. 3.12 Assymetric shaft running at critical speed ω_n

$$a = 4\frac{\omega_n^2}{\omega_k^2}$$

$$q = \frac{1\Delta k\,\omega_n^2}{4k\,\omega_k^2} \tag{3.122}$$

The scales corresponding to the above equations (3.122) are also indicated in the Strutt diagram of Fig. 3.10. On the y axis

$$\frac{\Delta k\,\omega_n^2}{k\,\omega_k^2}$$

is plotted as a function of

$$\frac{\omega_n^2}{\omega_k^2}$$

given on the x axis. Hence we can draw lines of appropriate slopes on Fig. 3.10 to represent the ratio $\Delta k/k$. Two typical values are shown. The points farther from the origin on the x axis represent low speeds of rotor and those near the origin represent very high speeds of rotation. Thus a rotor, when its speed is increased, passes through several zones of narrow regimes of instability as indicated for the two cases of $\Delta k/k = 0.4$ and 1. In practice all these regimes are not necessarily harmful. There is a substantial zone of instability in the vicinity of $a = 4$ and 1. The regime $a = 4$ corresponds to the shaft rotation at half the critical speed, which is the so called gravity critical observed quite often in practice. For the case $a = 1$, the shaft is running under resonant conditions and we not only have the unstable condition due to variable stiffness, but also large vibrations due to rotor unbalance. This is true for a vertical rotor.

For a horizontal rotor, gravity plays a significant role in the stability of the system. Fig. 3.12 shows a shaft running at the critical speed. Under no gravity conditions, we notice that for the regions 1–2; 3–4, spring force is small when the amplitude is increasing and the spring force is large when the amplitude is decreasing in the regions 2–3; 4–5, and hence the system is unstable. If the shaft is horizontal, however, it can be seen that the total work per cycle is zero and hence there is no instability as in the vertical shaft.

A different conclusion will be reached if we have the horizontal shaft running at half the critical speed as shown in Fig. 3.13. For lateral motion at the natural frequency, the spring force is small when the disk is going away from equilibrium, and is large when the disk is coming towards the equilibrium position. Thus, we have instability at rotational speeds equal to half the lateral natural frequency as in the case of a vertical rotor.

If the variable stiffness is assumed to be of the type given in Fig. 3.14, which is a rectangular ripple rather than a sinusoidal function assumed before, we can use a simple linear theory to study the stability. For such a system, the equation of motion (3.117) can be written as

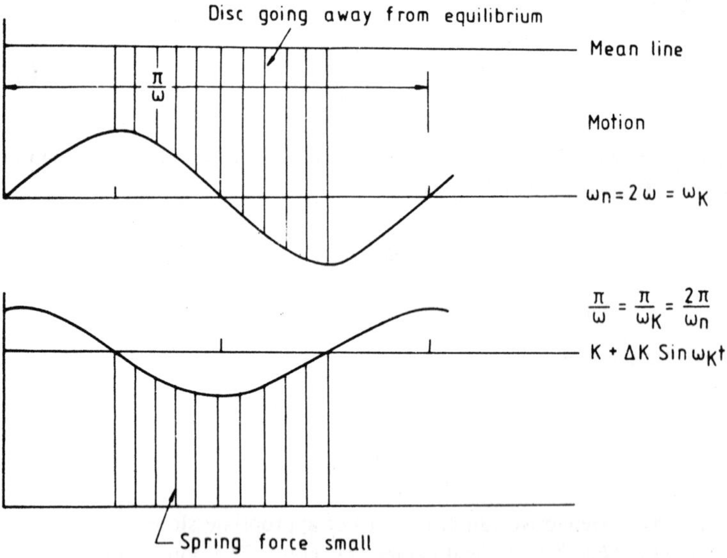

Fig. 3.13 Horizontal shaft running at $\frac{1}{2}$ critical speed

$$\ddot{x} + \left(\omega_n^2 + \frac{\Delta k}{m}\right)x = 0$$

$$0 < \omega_k t < \pi \tag{3.123}$$

$$\ddot{x} + \left(\omega_n^2 - \frac{\Delta k}{m}\right)x = 0$$

$$\pi < \omega_k t < 2\pi \tag{3.124}$$

Let

$$p_1 = \sqrt{\omega_n^2 + \frac{\Delta k}{m}}$$

$$p_2 = \sqrt{\omega_n^2 - \frac{\Delta k}{m}} \tag{3.125}$$

then, the solution of equations (3.123) and (3.124) can be obtained as

$$x_1 = c_1 \sin p_1 t + c_2 \cos p_1 t$$

$$x_2 = c_3 \sin p_2 t + c_4 \cos p_2 t \tag{3.126}$$

We will assume a growing solution of the type shown in Fig. 3.15, from which we can write the following relations.

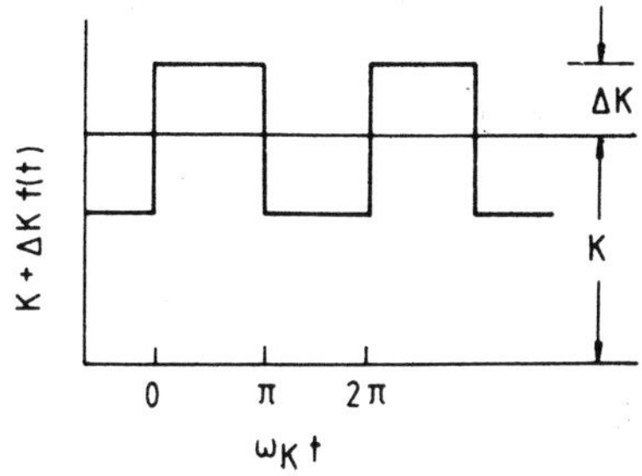

Fig. 3.14 Rectangular ripple variation of stiffness

$$x_1\}_{\omega_k t=0} = x_2\}_{\omega_k t=0}$$

$$\dot{x}_1\}_{\omega_k t=0} = \dot{x}_2\}_{\omega_k t=0}$$

$$x_2\}_{\omega_k t=\pi} = s x_1\}_{\omega_k t=-\pi}$$

$$\dot{x}_2\}_{\omega_k t=\pi} = s \dot{x}_1\}_{\omega_k t=-\pi} \quad (3.127)$$

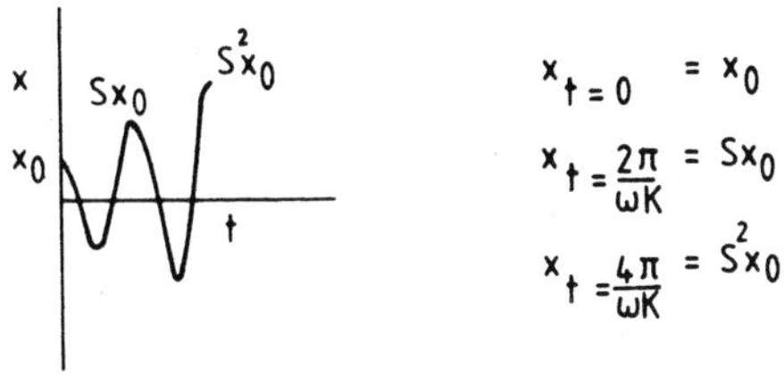

Fig. 3.15 Unstable solution

The first two conditions in (3.127) above, give from (3.126)

$$c_2 = c_4$$

$$c_3 = \frac{p_1}{p_2} c_1 \quad (3.128)$$

The next two conditions in (3.127) give

$$\frac{p_1}{p_2} c_1 \sin\frac{p_2\pi}{\omega_k} + c_2 \cos\frac{p_2\pi}{\omega_k} = -s c_1 \sin\frac{p_1\pi}{\omega_k} + s c_2 \cos\frac{p_1\pi}{\omega_k}$$

$$p_1 c_1 \cos\frac{p_2\pi}{\omega_k} - p_2 c_2 \sin\frac{p_2\pi}{\omega_k} = s p_1 c_1 \cos\frac{p_1\pi}{\omega_k} + s p_1 c_2 \sin\frac{p_1\pi}{\omega_k} \quad (3.129)$$

Eliminating c_1 and c_2, we get

$$s^2 - 2s\left\{\cos\frac{p_1\pi}{\omega_k}\cos\frac{p_2\pi}{\omega_k} - \frac{p_1^2+p_2^2}{2p_1p_2}\sin\frac{p_1\pi}{\omega_k}\sin\frac{p_2\pi}{\omega_k}\right\} + 1 = 0 \qquad (3.130)$$

Writing

$$A = \cos\frac{p_1\pi}{\omega_k}\cos\frac{p_2\pi}{\omega_k} - \frac{p_1^2+p_2^2}{2p_1p_2}\sin\frac{p_1\pi}{\omega_k}\sin\frac{p_2\pi}{\omega_k} \qquad (3.131)$$

Then, the two roots of (3.130) are

$$s_{1,2} = A \pm \sqrt{A^2 - 1} \qquad (3.132)$$

There are three possible cases that arise out of the values of A in (3.131).

1. $A > 1$, clearly the solution is unstable, in view of (3.127).
2. $-1 < A < 1$, the roots of (3.132) are complex. The real component in this case is always < 1. Hence the solution is stable.
3. $A < -1$, the real component is < -1 and hence the system is unstable.

Hence, the stability boundaries are

$$A = \pm 1 \qquad (3.133)$$

Since

$$\frac{p_1}{\omega_k} = \frac{\omega_n}{\omega_k}\sqrt{1+\frac{\Delta k}{k}}$$

$$\frac{p_2}{w_k} = \frac{\omega_n}{\omega_k}\sqrt{1-\frac{\Delta k}{k}}$$

$$\frac{p_1^2+p_2^2}{2p_1p_2} = \frac{1}{\sqrt{1+\frac{\Delta k}{k}}\sqrt{1-\frac{\Delta k}{k}}} \qquad (3.134)$$

we can write (3.133) as

$$\cos\pi\frac{\omega_n}{\omega_k}\sqrt{1+\frac{\Delta k}{k}}\cos\pi\frac{\omega_n}{\omega_k}\sqrt{1-\frac{\Delta k}{k}}$$

$$-\frac{1}{\sqrt{1+\frac{\Delta k}{k}}\sqrt{1-\frac{\Delta k}{k}}}\sin\pi\frac{\omega_n}{\omega_k}\sqrt{1-\frac{\Delta k}{k}}\sin\pi\frac{\omega_n}{\omega_k}\sqrt{1-\frac{\Delta k}{k}} = \pm 1 \qquad (3.135)$$

With the help of (3.135) above, we can make a Strutt diagram as shown in Fig. 3.16.

3.6 HILL'S EQUATION

Consider equation (3.89) and express it in series form

$$\ddot{x} + \left(\theta_0 + 2\sum_{k=1}^{\infty}\theta_k\cos 2kt\right)x = 0 \qquad (3.136)$$

This equation is Hill's equation which was used to discuss the motion of lunar perigee. We obtain a similar equation while considering the variation over one of the singular points (i.e., equilibrium positions) of a forced nonlinear oscillator. This will be demonstrated later in the next chapter. For the present, we will discuss the procedure for studying the stability of a system governed by Hill's equation (3.136). As discussed before, a particular solution of this equation is

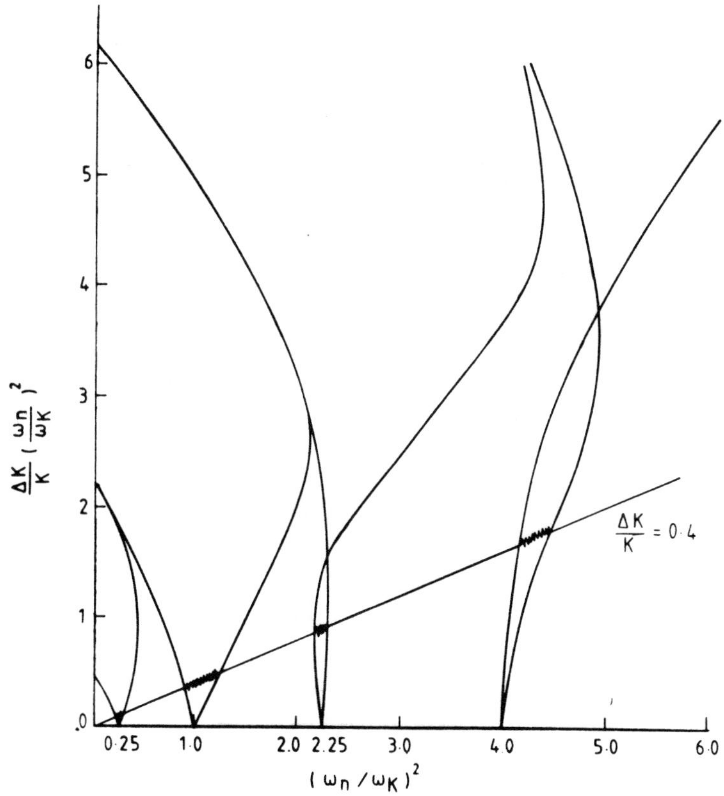

Fig. 3.16 Stability diagram – variable elasticity shaft

$$x = e^{\mu\tau}\phi(\tau)$$

$$\dot{x} = \mu e^{\mu\tau}\phi(\tau) + e^{\mu\tau}\dot{\phi}(\tau)$$

$$\ddot{x} = \mu^2 e^{\mu\tau}\phi(\tau) + 2\mu e^{\mu\tau}\dot{\phi}(\tau) + e^{\mu\tau}\ddot{\phi}(\tau) \qquad (3.137)$$

With the help of above, Hill's equation becomes

$$\ddot{\phi} + 2\mu\dot{\phi} + \left(\theta_0 + \mu^2 + 2\sum_{k=1}^{\infty}\theta_k \cos 2kt\right)\phi = 0 \qquad (3.138)$$

From equation (3.137), we see that x remains stable if it is bounded as $t \to \infty$ and unstable if it grows unboundedly as $t \to \infty$. The solution is neutral if it is periodic in π or 2π. There will also be several unstable regimes and we can write the solution for the nth unstable regime as

$$\phi(t) = \sin(nt - \sigma)$$

$$n = 1, 2, \ldots \qquad (3.139)$$

Using the above equation (3.138) becomes

$$-n^2 \sin nt \cos\sigma + n^2 \cos nt \sin\sigma + 2\mu n \cos nt \cos\sigma + 2\mu n \sin nt \sin\sigma$$

$$+\left(\theta_0+\mu^2+2\sum_{k=1}^{\infty}\theta_k \cos 2kt\right)(\sin nt\cos\sigma-\cos nt\sin\sigma)=0 \qquad (3.140)$$

Considering the value of $k = n$ only in the series term of above equation, we get

$$-n^2 \sin nt \cos\sigma + n^2 \cos nt \sin\sigma + 2\mu n \cos nt \cos\sigma + 2\mu n \sin nt \sin\sigma$$

$$+(\theta_0+\mu^2+2\theta_n \cos 2nt)(\sin nt\cos\sigma-\cos nt\sin\sigma)=0 \qquad (3.140a)$$

Now separating out sin $n\tau$ and cos $n\tau$, we get

$$2\mu n \sin\sigma + (\theta_0+\mu^2-n^2)\cos\sigma - \theta_n\cos\sigma = 0$$

$$2\mu n \cos\sigma - (\theta_0+\mu^2-n^2)\sin\sigma - \theta_n\sin\sigma = 0 \qquad (3.141)$$

By multiplying the first equation in the above by sin σ and the second one by cos σ and adding we get

$$2\mu n = \theta_n \sin 2\sigma \qquad (3.142)$$

Therefore

$$\sin 2\sigma = \frac{2\mu n}{\theta_n}$$

$$\cos 2\sigma = \sqrt{1-\frac{4n^2\mu^2}{\theta_n^2}} \qquad (3.143)$$

Similarly, by multiplying the first equation in (3.141) by cos σ and the second one by sin σ and subtracting we get

$$\theta_0+\mu^2-n^2 = \theta_n\cos 2\sigma \qquad (3.144)$$

Using (3.143), equation (3.144) becomes

$$\theta_0+\mu^2-n^2 = \sqrt{\theta_n^2-4n^2\mu^2} \qquad (3.145)$$

Squaring both sides of the above equation, we get

$$(\theta_0-n^2)^2+\mu^4+2(\theta_0-n^2)\mu^2 = \theta_n^2-4n^2\mu^2$$

i.e.,

$$(\theta_0+n^2)^2-4n^2\theta_0+\mu^4+2(\theta_0+n^2)\mu^2 = \theta_n^2$$

Rearranging the above terms

$$(\theta_0+\mu^2+n^2)^2 = \theta_n^2+4n^2\theta_0$$

Rewriting the above we can get the roots for μ^2 as

$$\mu^2 = -(\theta_0+n^2)\pm\sqrt{\theta_n^2+4n^2\theta_0} \qquad (3.146)$$

If μ is imaginary the nth regime will be stable and hence we consider the positive sign only in the above equation for further analysis. For unstable condition $\mu^2 > 0$ and therefore equation (3.146) gives

$$-(\theta_0 + n^2) + \sqrt{\theta_n^2 + 4n^2\theta_0} > 0 \tag{3.147}$$

i.e.,

$$\theta_n^2 + 4n^2\theta_0 > (\theta_0 + n^2)^2$$

$$\theta_n^2 > (\theta_0 - n^2)^2$$

Hence

$$|\theta_n| > |\theta_0 - n^2| \tag{3.148}$$

On the boundaries of nth unstable regime, $\mu = 0$ and they are defined by

$$\theta_0 = n^2 \pm \theta_n \tag{3.149}$$

It can be shown that the above relations are valid for a more general Hill's equation given by

$$\ddot{x} + \left(\theta_0 + 2\sum_{k=1}^{\infty} \theta_{kc} \cos 2kt + 2\sum_{k=1}^{\infty} \theta_{ks} \sin 2kt\right)x = 0 \tag{3.136a}$$

For equation (3.80), we have shown in (3.106) that

$$\mathcal{R}(\pm\mu - \delta) \le 0 \tag{3.106}$$

for stability, where δ is half constant term in

$$F(t) = \left(\frac{\partial f}{\partial \dot{x}}\right)_0$$

Equation (3.89) can be expressed in the form of (3.136a) and we can use (3.146) to determine the values of μ^2.

If μ is imaginary, from (3.106) we get

$$\delta > 0$$

If μ is real, then

$$\delta > 0$$

$$\delta^2 > \mu^2$$

i.e.,

$$\delta^2 > -(\theta_0 + n^2) + \sqrt{\theta_n^2 + 4n^2\theta_0}$$

$$\delta^2 + (\theta_0 + n^2) > \sqrt{\theta_n^2 + 4n^2\theta_0}$$

or

$$\delta^4 + 2\delta^2(\theta_0 + n^2) + (\theta_0 + n^2)^2 > 4n^2\theta_0 + \theta_n^2$$

Rearranging the above, we get

$$\delta^4 + 2\delta^2(\theta_0 + n^2) + (\theta_0 - n^2)^2 > \theta_n^2 \quad n = 1, 2, \ldots \tag{3.150}$$

The above approach for stability consideration will be more clear when we consider the forced oscillations of nonlinear systems in the next chapter.

3.7 SYSTEM WITH VARIABLE INERTIA

We will now consider a single cylinder reciprocating engine connected to a flywheel as shown in Fig. 3.17. The system executes torsional oscillations as shown. The crankshaft of radius a is a purely rotating body and its mass moment of inertia can be easily found and let this be equal to I. Let the reciprocating parts of the piston, piston pin, rings, etc. be of mass M and for simplicity of analysis, we will assume that the connecting rod BC is sufficiently large in comparison to a, so that the reciprocating mass has pure harmonic motion. The torsional stiffness of the shaft from the cylinder centre line to the flywheel A be K. The flywheel being a rotating body, its mass moment of inertia can be estimated and let this be I_A.

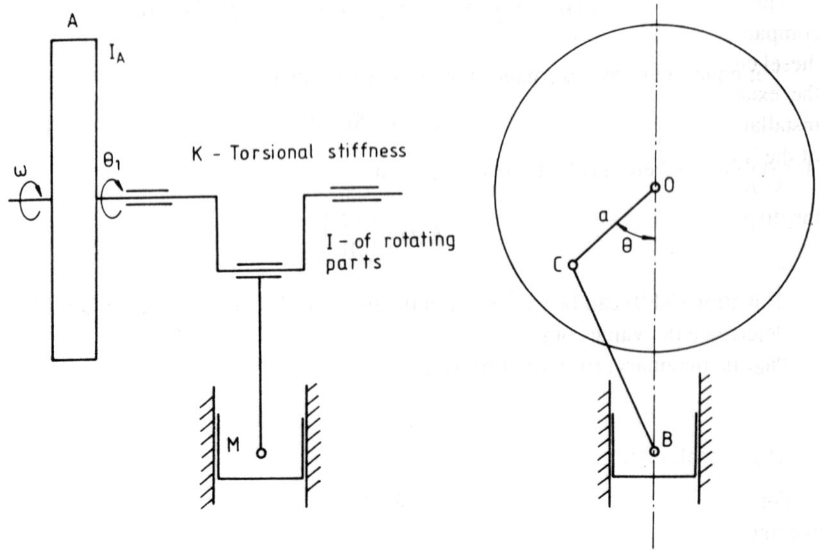

Fig. 3.17 Variable inertia system

The cylinder of the engine has rotating as well as reciprocating parts and therefore the determination of mass moment of inertia is not a straight forward matter. Consider the piston shown in Fig. 3.18 in two different positions and imagine the crankshaft with a mass moment of inertia I, to be not rotating but executing torsional oscillations. In Fig. 3.18a, there is no motion for the piston, with small oscillations of the crank and hence the equivalent inertia of the piston is zero. However, in Fig. 3.18b, the piston has practically the same acceleration as that of the crank pin and the equivalent mass moment of inertia is Ma^2. Hence, when the crank shaft is rotating, the mass

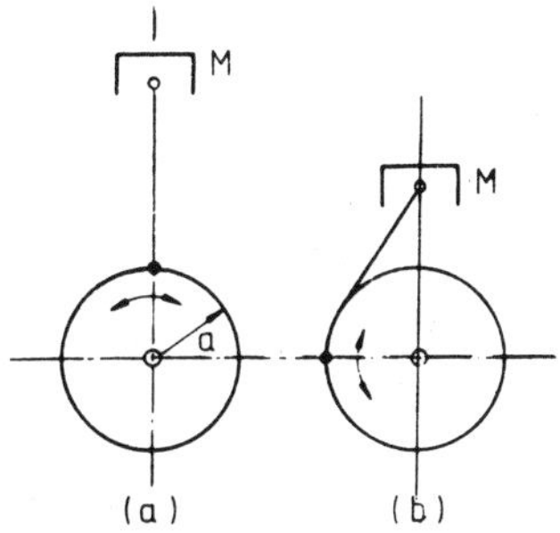

Fig. 3.18 Equivalent mass moment of inertia of a piston

moment of inertia of the cylinder as a whole varies from I to $I + Ma^2$. We recall that in linear theory, we neglect this variation and assume a constant mass moment of inertia for the cylinder given by the average value

$$I_c = I + \frac{1}{2}Ma^2 \tag{3.151}$$

The above expression is somewhat valid, when the change in inertia is small compared with the average inertia. For engines with heavy pistons as in some marine diesel engines, this assumption is not valid and simple linear theory may not predict the exact behaviour of the system. In practice, for such reciprocating machine installations, large amplitude vibrations were observed at speeds which are about half of the linear system critical speed.

As mentioned before, let us assume simple harmonic motion to the piston. Then the displacement of the piston is

$$x = a\cos\theta$$

$$x = -a\sin\theta\dot{\theta} \tag{3.152}$$

The square of the linear velocity of the piston is therefore

$$\dot{x}^2 = \frac{1}{2}a^2\dot{\theta}^2(1 - \cos 2\theta) \tag{3.153}$$

For simplicity, let us assume that inertia of flywheel is very large compared to the average inertia of the cylinder and hence consider that there is a node at the flywheel. Thus we can ignore the kinetic energy of the flywheel in the oscillatory motion and write

$$T = \frac{1}{2}I\dot{\theta}^2 + \frac{1}{2}Ma^2\frac{\dot{\theta}^2}{2}(1 - \cos 2\theta) + \frac{1}{2}I_A\omega^2 \tag{3.154}$$

The strain energy in the shaft is

$$V = \frac{1}{2}K(\theta - \theta_1^2) \tag{3.155}$$

where

$$\theta_1 = \omega t$$

Since $T + V$ is constant, we have

$$\frac{d}{dt}(T+V) = 0$$

Upon substitution of T and V given in equations (3.154) and (3.155), the above gives

$$\ddot{\theta}\left(I + \frac{1}{2}Ma^2 - \frac{1}{2}Ma^2\cos 2\theta\right) +$$

$$\frac{1}{2}Ma^2\dot{\theta}^2\sin 2\theta + K(\theta - \theta_1) = 0 \tag{3.156}$$

Writing

$$\theta = \omega t + \gamma$$

$$\tau = \omega t$$

$$\mu = \frac{\frac{1}{2}Ma^2}{I + \frac{1}{2}Ma^2}$$

$$\kappa^2 = \frac{K}{\omega^2\left(I + \frac{1}{2}Ma^2\right)} \tag{3.157}$$

and neglecting terms like

$$\mu\gamma\ddot{\gamma}, \mu\dot{\gamma}^2 \ldots$$

we get

$$\ddot{\gamma}(1 - \mu\cos 2\tau) + 2\mu\dot{\gamma}\sin 2\tau$$

$$+\gamma(\kappa^2 + 2\mu\cos 2\tau) = -\mu\sin 2\tau \tag{3.158}$$

$$\cdot = \frac{d}{d\tau}$$

Equation (3.158) above governs the vibratory motion of the system in Fig. 3.17. We note that equation (3.158) is linear with all the coefficients variable in time. First let us consider the complementary function of the above. Then

$$\ddot{\gamma}(1 - \mu\cos 2\tau) + 2\mu\dot{\gamma}\sin 2\tau + \gamma(\kappa^2 + 2\mu\cos 2\tau) = 0 \tag{3.159}$$

Goldsbrough assumed the solution to the above as similar to the solution of Floquet and we write

$$\gamma = a_1 e^{c\tau}\phi(\tau) + a_2 e^{-c\tau}\phi(-\tau) \tag{3.160}$$

On similar lines discussed before for Mathieu and Hill equations, we can write

$$c = \mu c_1 + \mu^2 c_2 + \ldots$$

$$\phi = \cos \kappa\tau + A \sin \kappa\tau + \mu x_1 + \mu^2 x_2 + \ldots \tag{3.161}$$

Substituting (3.161) in (3.160) and then in (3.159) and separating the terms of μ, we get

$$\ddot{x}_1 + \kappa^2 x_1 + 2c_1(-\kappa \sin \kappa\tau + \kappa A \cos \kappa\tau)$$

$$+\kappa^2 \cos 2\tau(\cos \kappa\tau + A \sin \kappa\tau) + 2 \sin 2\tau(-\kappa \sin \kappa\tau + \kappa A \cos \kappa\tau) +$$

$$+2 \cos 2\tau(\cos \kappa\tau + A \sin \kappa\tau) = 0 \tag{3.162}$$

To avoid the secular terms, we make

$$c_1 = 0 \tag{3.163}$$

and then we have from the solution of (3.162)

$$x_1 = \frac{\frac{1}{2}\kappa^2 + \kappa + 1}{4(\kappa + 1)}\{\cos(\kappa + 2)\tau + A \sin(\kappa + 2)\tau\}$$

$$-\frac{\frac{1}{2}\kappa^2 - \kappa + 1}{4(\kappa - 1)}\{\cos(\kappa - 2)\tau + A \sin(\kappa - 2)\tau\} \tag{3.164}$$

In the above A is still to be determined. On separating the terms of μ^2 and using the above, we get the following

$$\ddot{x}_2 + \kappa^2 x_2 + 2c_2\kappa(-\sin \kappa\tau + A \cos \kappa\tau)$$

$$+\{\cos(\kappa + 2)\tau + A \sin(\kappa + 2)\tau\}\left\{\frac{\frac{1}{2}\kappa^2 + \kappa + 1}{4(\kappa + 1)}\right\}\{(\kappa + 2)^2 \cos 2\tau + 2\}$$

$$-\frac{\frac{1}{2}\kappa^2 - \kappa + 1}{4(\kappa - 1)}\{\cos(\kappa - 2)\tau + A \sin(\kappa - 2)\tau\}\{(\kappa - 2)^2 \cos 2\tau + 2\}$$

$$+\frac{1}{2}\sin 2\tau \frac{(\kappa + 2)\left(\frac{1}{2}\kappa^2 + \kappa + 1\right)}{(\kappa + 1)}\{-\sin(\kappa + 2)\tau + A \cos(\kappa + 2)\tau\}$$

$$-\frac{1}{2}\sin 2\tau \frac{(\kappa - 2)\left(\frac{1}{2}\kappa^2 - \kappa + 1\right)}{(\kappa - 1)}\{-\sin(\kappa - 2)\tau + A \cos(\kappa - 2)\tau\} = 0 \tag{3.165}$$

We avoid the secular terms in the above by making the coefficients of $\cos \kappa\tau$ and $\sin \kappa\tau$ equal to zero. i.e.,

$$2c_2 A + P = 0$$

$$PA - 2c_2 = 0 \tag{3.166}$$

where

$$P = \frac{1}{8}\frac{\kappa(\kappa+2)\left(\frac{1}{2}\kappa^2+\kappa+1\right)}{(\kappa+1)} - \frac{1}{8}\frac{\kappa(\kappa-2)\left(\frac{1}{2}\kappa^2-\kappa+1\right)}{(\kappa-1)}$$

From (3.166), eliminating A, we get

$$4c_2^2 + P^2 = 0 \tag{3.167}$$

i.e.,

$$c_2 = \pm\frac{1}{2}Pi \tag{3.167a}$$

Also, from (3.166), we have

$$A = -\frac{P}{2c_2} \tag{3.168}$$

Hence, from (3.167a) we get

$$A = \pm i \tag{3.169}$$

Then, the solution of equation (3.165) gives terms in the expression for x_2 of the type

$$\frac{\cos(\kappa\pm 4)\tau}{8(\kappa\pm 2)}$$

$$\frac{\sin(\kappa\pm 4)\tau}{8(\kappa\pm 2)} \tag{3.170}$$

So we can continue the solution and note that no real terms arise in the value of the exponent c, and therefore ultimately γ consists of a series of circular functions. Also, we note from (3.170), that the solution fails, owing to the presence of the denominators of the form $\kappa\pm r$, r integral, whenever κ has a value approximately $+r$ or $-r$.

In addition to the complementary function just discussed above, the complete solution of (3.158) requires the determination of the particular integral due to $-\mu \sin 2\tau$ term. Hence, again, it is necessary to construct special solutions for the cases where $\kappa = \pm 1, \pm 2, \pm 3\ldots$ Therefore, as in the case of Mathieu equation, we seek special solutions for the complementary function of equation (3.158), when κ is a positive or negative integer. In order to cover not only the case when κ is an integer, but also the values in the vicinity, we write

$$\kappa^2 = r^2 + \mu\alpha + \mu^2\alpha_2 + \ldots \tag{3.171}$$

In the above α is an arbitrary parameter, and by expressing ultimately $\alpha_2, \alpha_3\ldots$ in terms of α, we can obtain κ in terms of r and α. In place of equation (3.159), we now have

$$\ddot{\gamma}(1-\mu\cos 2\tau) + 2\mu\gamma\sin 2\tau + \gamma(r^2 + \mu\alpha + \mu^2\alpha_2 + \ldots + 2\mu\cos 2\tau) = 0 \tag{3.172}$$

Using equations (3.160) and (3.161), the terms factored by μ give

$$\ddot{x}_1 + x_1 + (3\cos 2\tau + \alpha)(\cos\tau + A\sin\tau)$$

$$+ (2\sin 2\tau + 2c_1)(-\sin\tau + A\cos\tau) = 0 \tag{3.173}$$

Proceeding as before, we get

$$\alpha + \frac{1}{2} + 2c_1 A = 0$$

$$-2c_1 + \left(\alpha - \frac{1}{2}\right) A = 0 \tag{3.174}$$

Equation (3.174) above gives

$$c_1 = \pm \frac{1}{2}\sqrt{\frac{1}{4} - \alpha^2}$$

$$A = \mp \sqrt{\frac{\frac{1}{2} + \alpha}{\frac{1}{2} - \alpha}} \tag{3.175}$$

and

$$x_1 = \frac{5}{16}\cos 3\tau + \frac{5}{16} A \sin 3\tau \tag{3.176}$$

We can continue the solution above and then proceed for cases of $r = 2, 3$ etc. The solutions for these cases are summarized below.

Case 1: r = 1

$$\kappa^2 = 1 + \mu\alpha + \mu^2\alpha_2 + \mu^3\alpha_3 + \ldots$$

$$x_0 = \cos\tau + A\sin\tau$$

$$x_1 = \frac{5}{16}\cos 3\tau + \frac{5}{16} A \sin 3\tau$$

$$A = \mp \sqrt{\frac{\frac{1}{2} + \alpha}{\frac{1}{2} - \alpha}}$$

$$\alpha_2 = -\frac{27}{32} + \frac{1}{4}\alpha^2$$

$$\alpha_3 = -\frac{1}{32} - \frac{9}{256}\alpha + \frac{1}{8}\alpha^2 - \frac{1}{4}\alpha^3$$

$$c_1 = \pm \frac{1}{2}\sqrt{\frac{1}{4} - \alpha^2}$$

$$c_2 = 0$$

$$c_3 = \pm \sqrt{\frac{1}{4} - \alpha^2}\left(\frac{17}{512} + \frac{1}{4}\alpha^2\right)$$

Case 2: r = 2

$$\kappa^2 = 4 + \mu^2\alpha + \mu^3\alpha_3 + \ldots$$

$$x_0 = \cos 2\tau + A \sin 2\tau$$

$$x_1 = -\frac{1}{4} + \frac{5}{12}\cos 4\tau + \frac{5}{12} A \sin 4\tau$$

$$A = -\frac{\alpha + \frac{25}{12}}{2c_2}$$

$$\alpha_3 = 0$$

$$c_1 = 0$$

$$c_2 = \pm\sqrt{-\left(\frac{25}{12} + \alpha\right)\left(\frac{19}{12} + \alpha\right)}$$

$$c_3 = 0$$

Case 3: r = 3

$$\kappa^2 = 9 - \frac{157}{32}\mu^2 + \mu^3\alpha + \ldots$$

$$x_0 = \cos 3\tau + A \sin 3\tau$$

$$c = \pm\frac{1}{6}\mu^3\sqrt{\left(\frac{25}{128}\right)^2 - \alpha^2} \tag{3.177}$$

For unstable conditions, we note that c should be real. For case 1, from equation (3.177), we have

$$c = \frac{1}{2}\sqrt{\frac{1}{4} - \alpha^2}\;\mu + \sqrt{\frac{1}{4} - \alpha^2}\left(\frac{17}{512} + \frac{1}{4}\alpha^2\right)\mu^3 \tag{3.178}$$

Hence, c is real for values of α in the range

$$-\frac{1}{2} < \alpha < \frac{1}{2} \tag{3.179}$$

We have from (3.157) and

$$\kappa^2 = \frac{K}{\omega^2\left(I + \frac{1}{2}Ma^2\right)} = 1 + \mu\alpha + \mu^2\left(-\frac{27}{32} + \frac{1}{4}\alpha^2\right) + \mu^3\left(-\frac{1}{32} - \frac{9}{256}\alpha + \frac{1}{8}\alpha^2 - \frac{1}{4}\alpha^3\right) + \ldots$$

The vibrations will be therefore unstable in the range of speeds ω_1 to ω_2, where

$$\omega_1^2 = \frac{K}{\left(I + \frac{1}{2}Ma^2\right)}\left(1 + \frac{1}{2}\mu - \frac{25}{32}\mu^2 - \frac{25}{512}\mu^3 + \ldots\right)^{-1}$$

$$\omega_2^2 = \frac{K}{\left(I + \frac{1}{2}Ma^2\right)}\left(1 - \frac{1}{2}\mu - \frac{25}{32}\mu^2 + \frac{25}{512}\mu^3 + \ldots\right)^{-1} \qquad (3.180)$$

The magnitudes of both c and the range of instability in equations (3.178) and (3.179) depend on μ. We find from (3.178) that c is a function of μ and its higher powers. Hence, this range is important for instability investigation.

For case 2, we have

$$\kappa^2 = \frac{K}{\omega^2\left(I + \frac{1}{2}Ma^2\right)} = 4 + \mu^2\alpha$$

$$c = \frac{1}{2}\sqrt{-\left(\frac{25}{12} + \alpha\right)\left(\frac{19}{12} + \alpha\right)}\;\mu^2 \qquad (3.181)$$

As in case 1, we have similar results. The vibrations are unstable for values of the parameter α between $-19/12$ and $-25/12$. Outside this range, the motion is stable. The range of speeds are then given by

$$\omega_1^2 = \frac{K}{I + \frac{1}{2}Ma^2}\left(4 - \frac{19}{12}\mu^2 + \ldots\right)^{-1}$$

$$\omega_2^2 = \frac{K}{I + \frac{1}{2}Ma^2}\left(4 - \frac{25}{12}\mu^2 + \ldots\right)^{-1} \qquad (3.182)$$

The range of instability and c in this case depend on μ^2 term and therefore are both smaller than in case 1.

For the case 3, κ^2 is approximately 9; the range of instability and c are not dependent on μ^3 term; and therefore are much diminished and of not practical importance. The instability regimes for the case $r = 1$ and $r = 2$ are both plotted in Figs. 3.19 and 3.20.

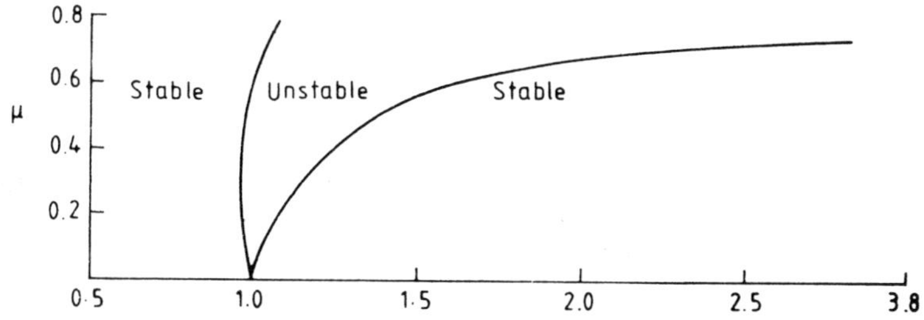

Fig. 3.19 Unstable region for $1/\kappa = 1$

We can obtain the above instability regions by using a numerical method to solve the equation (3.158). We can follow the modified Euler method and express equation (3.158) as

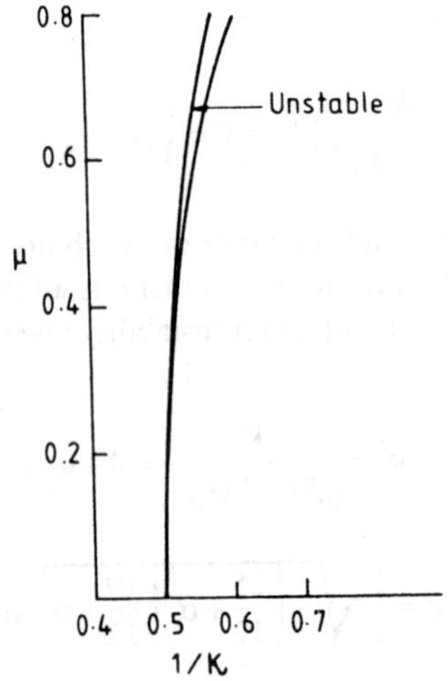

Fig. 3.20 Unstable region for $1/\kappa = 0.5$

$$\ddot{\gamma} = f(\tau, \gamma, \dot{\gamma}) \tag{3.183}$$

The solution of (3.183) above is written as

$$\gamma = F(\tau) \tag{3.184}$$

Since we are interested in the stability analysis of the system we would like to know whethe. the system has sustained steady state oscillations when it is disturbed with any arbitrary initial conditions. Let at time $t = 0$, the initial conditions be

$$\gamma_0 = 1$$

$$\dot{\gamma}_0 = 0 \tag{3.185}$$

We can adopt equations (2.64) to write the following:

$$\left\{\frac{d^2\gamma}{d\tau^2}\right\}_i^{(j-1)} = f\left(\gamma_i^{(j-1)}, \left\{\frac{d\gamma}{d\tau}\right\}_i^{(j-1)}, \tau\right)$$

$$\left\{\frac{d\gamma}{d\tau}\right\}_i^{(j)} = \left\{\frac{d\gamma}{d\tau}\right\}_{i-1} + \frac{\left\{\frac{d^2\gamma}{d\tau^2}\right\}_{i-1} + \left\{\frac{d^2\gamma}{d\tau^2}\right\}_i^{(j-1)}}{2}\Delta\tau$$

$$\gamma_i^{(j)} = \gamma_{i-1} + \frac{\left\{\frac{d\gamma}{d\tau}\right\}_{i-1} + \left\{\frac{d\gamma}{d\tau}\right\}_i^{(j-1)}}{2}\Delta\tau \tag{3.186}$$

The above equations (3.186) can be programmed to determine the solution $\gamma = F(\tau)$ for a given μ and κ. Typical response curves for $1/\kappa = 0.06, 0.5, 0.55$ and 1 are shown in Figs. 3.21 to 3.24. The growing solutions in Figs. 3.23 and 3.24 show the instability in the system.

The stability boundaries can be obtained by varying κ for a given μ and noting the values where the solution is unstable. The stability charts thus obtained are shown in Figs. 3.25a and 3.25b for the regions shown in Figs. 3.19 and 3.20.

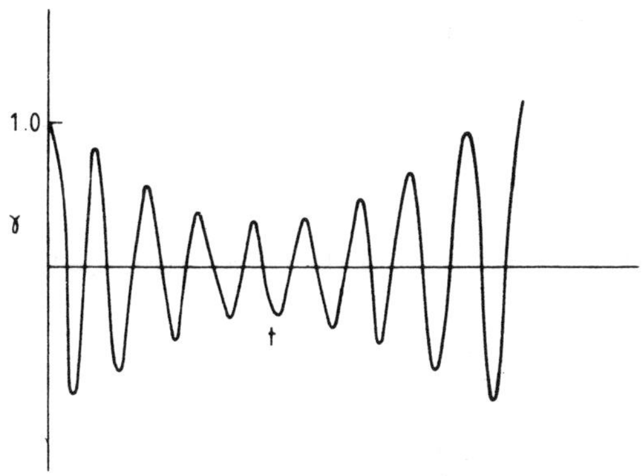

Fig. 3.21 Response of variable inertia system $\omega/\omega_n = 0.06$

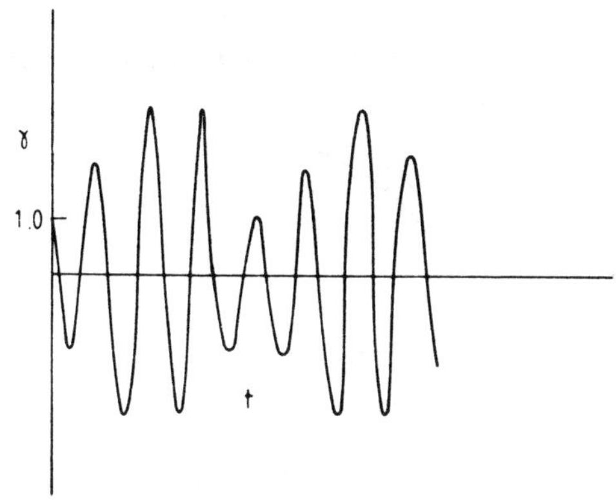

Fig. 3.22 Response of variable inertia system $\omega/\omega_n = 0.5$

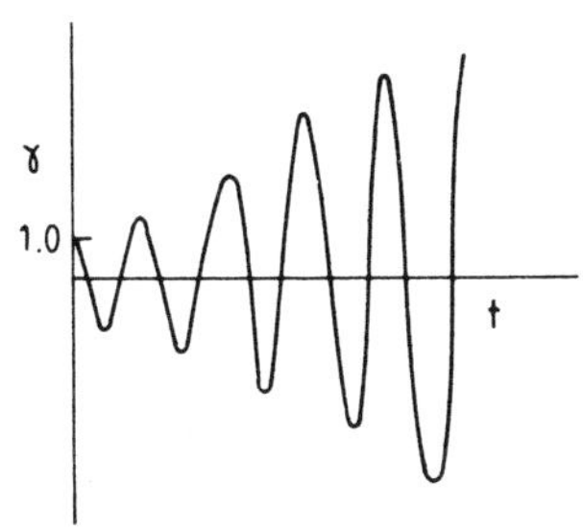

Fig. 3.23 Response of variable inertia system $\omega/\omega_n = 0.55$

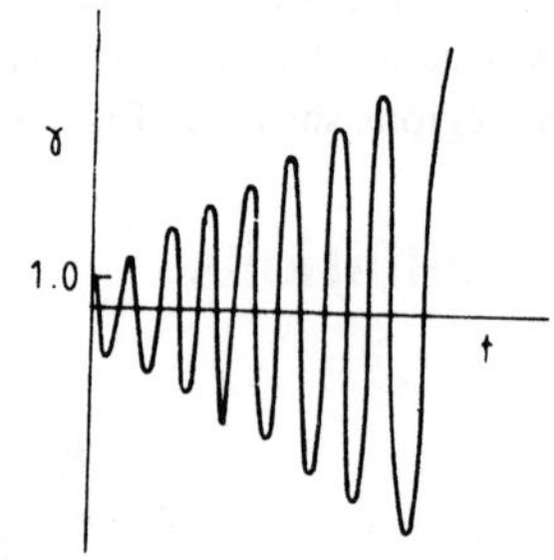

Fig. 3.24 Response of variable inertia system $\omega/\omega_n = 1.0$

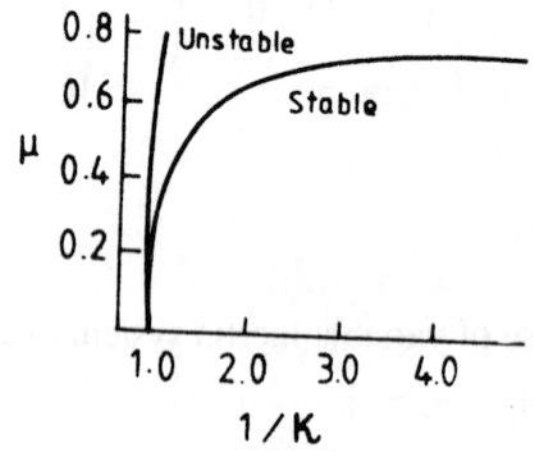

Fig. 3.25 (a) Stability diagram of variable inertia system $1/\kappa = 1$ region

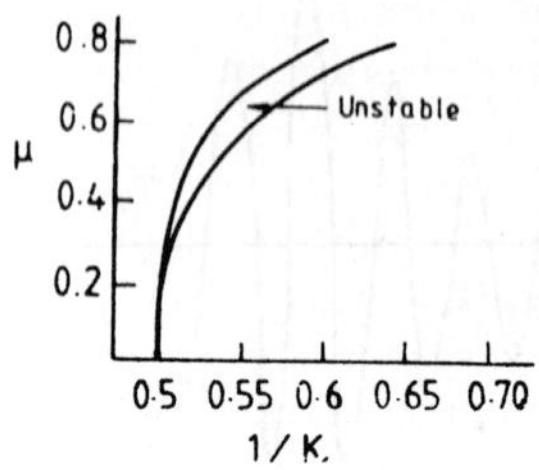

Fig. 3.25 (b) Stability of variable inertia system $1/\kappa = 0.5$ region

4

Forced Oscillations of Nonlinear Systems

In chapters 2 and 3, we studied the characteristics of free vibration of systems with nonlinear stiffness and nonlinear damping. With the application of an external harmonic force, the system experiences large amplitudes of vibration when the frequency of excitation is near the resonant frequency of the linear system. However, the response does not become infinity at resonance as in the case of linear system. Also, in the linear system, the response in the post-resonant zone decreases gradually depending on the damping present in the system. In the case of the nonlinear systems, the response keeps increasing as the frequency increases beyond resonance and then has a sudden jump. In linear systems, we would expect the same behaviour at a given frequency whether the frequency of excitation is increased gradually or decreased gradually at that location. Typically the behaviour of a nonlinear system may have a different response at a given frequency depending on whether the system has reached this excitation point with increasing or decreasing frequency. Some of these solutions may not be stable and, in fact, we may never reach a mathematically possible solution because of some of the typical characteristics of a nonlinear system. Another significant departure from the linear system is that there could be higher or subharmonic response even when the system is excited by a pure harmonic force. Some of these points are discussed in this chapter.

4.1 HARMONIC OSCILLATIONS OF DUFFING'S EQUATION

To begin with, we will consider a simple nonlinear system with one degree of freedom and linear damping. The nonlinearity in the spring is considered to be symmetric and cubic in nature. Let us assume the excitation force to be harmonic. Then, the equation of motion of such a system is given by

$$m\ddot{x} + c\dot{x} + ax + \beta x^3 = F \cos \omega t \tag{4.1}$$

Dividing throughout by m we obtain

$$\ddot{x} + 2\xi p\dot{x} + p^2 x + p^2 \mu x^3 = \frac{F}{m} \cos \omega t \tag{4.2}$$

where

$$p^2 = \frac{\alpha}{m} \quad \mu = \frac{\beta}{\alpha} \quad \xi = \frac{c}{2mp}$$

Let

$$\tau = pt$$

then, equation (4.2) becomes

$$\ddot{x}+2\xi\dot{x}+x+\mu x^3=X_{st}\cos r\tau \tag{4.3}$$

where

$$X_{st}=\frac{F}{k}$$

is the static deflection of the linear system when treating the force F as steady force, and $r=\omega/p$ is the frequency ratio.

Let us first consider a harmonic steady state solution, with no transient part.

$$x_0=a\sin r\tau+b\cos r\tau \tag{4.4}$$

Substituting (4.4) above and its derivatives in (4.3), we get

$$-r^2a\sin r\tau-r^2b\cos r\tau+2\xi ar\cos r\tau-2\xi br\sin r\tau$$

$$+a\sin r\tau+b\cos r\tau+\mu a^3\left(\frac{3}{4}\sin r\tau-\frac{1}{4}\sin 3r\tau\right)$$

$$+\mu b^3\left(\frac{3}{4}\cos r\tau+\frac{1}{4}\cos 3r\tau\right)+3\mu a^2b\left(\frac{1}{4}\cos r\tau-\frac{1}{4}\cos\cos 3r\tau\right)$$

$$+3\mu b^2a\left(\frac{1}{4}\sin r\tau+\frac{1}{4}\sin 3r\tau\right)=X_{st}\cos r\tau \tag{4.5}$$

We now separate $\sin r\tau$ and $\cos r\tau$ terms to obtain

$$-r^2a-2\xi br+a+\frac{3}{4}\mu a^3+\frac{3}{4}\mu ab^2=0$$

$$-r^2b+2\xi ar+b+\frac{3}{4}\mu b^3+\frac{3}{4}\mu a^2b=X_{st} \tag{4.6}$$

Let us define

$$A=r^2-1-\frac{3}{4}\mu X^2$$

$$X^2=a^2+b^2 \tag{4.7}$$

Then, equation (4.6) becomes

$$Aa+2r\xi b=0$$

$$2r\xi a-Ab=X_{st} \tag{4.8}$$

Squaring and adding the above in (4.8), we get

$$(A^2+4r^2\xi^2)X^2=X_{st}^2 \tag{4.9}$$

It should be noted that A is a function of X. In obtaining the solution for equation (4.3), we balanced the harmonic terms and ignored the higher harmonics. Hence, this method is called Harmonic Balance procedure. For free vibrations, $X_{st}=0$ and hence

$$X^2=0$$

$$A^2+4\xi^2r^2=0 \tag{4.10}$$

For undamped vibrations, $\xi = 0$ hence $A^2 = 0$ and we have

$$X^2 = 0$$

$$r^2 - 1 - \frac{3}{4}\mu X^2 = 0 \tag{4.11}$$

Let us consider the following data

$$r^2 = \frac{\omega^2}{p^2} = 10, \mu = 9 \text{ and } \quad X_{st} = 0.1$$

Then, for the undamped free vibration, we have

$$X^2 = 0$$

$$X^2 = \frac{4}{3} \tag{4.12}$$

For the case of undamped forced vibration, we get from (4.9) and (4.7)

$$\frac{9}{16}\mu^2 X^6 + \frac{3}{2}\mu(1 - r^2)X^4 + \{r^4 - 2r^2 + 1\}X^2 - X_{st}^2 = 0 \tag{4.13}$$

For the case of forced damped vibration, we have from (4.9) and (4.7)

$$\frac{9}{16}\mu^2 X^6 + \frac{3}{2}\mu(1 - r^2)X^4 + \{r^4 - 2r^2(1 - 2\xi^2) + 1\}X^2 - X_{st}^2 = 0 \tag{4.14}$$

The above equation (4.14) gives the solution for Duffing oscillator and we notice that there are three roots for X^2 for a given value of the nonlinear parameter μ, frequency ratio $r = \omega/p$, damping ratio = ξ and the excitation force defined by X_{st}. All the three roots may be real or at least one root will be real for specific conditions. In the first instance let us consider a constant frequency of excitation, but excitation force varying. Let

$$\mu = 9; \quad r^2 = 10$$

Then, equation (4.14) becomes

$$X^6 - 2.67X^4 + (1.78 + 0.8779\xi^2)X^2 - 0.02195X_{st}^2 = 0 \tag{4.15a}$$

The roots of the above equation (4.15a) are given in Table 4.1, for ξ = 0,0.1,0.4,1.0.

In Fig. 4.1, the above X^2 are plotted from equation (4.14) as a function of static deflection of the linear system due to the dynamic force treated as stationary load for $r^2 = 10$ and $\mu = 9$. for $\xi = 0$ at $X_{st} = 0$, $X^2 = 1.33$ as given by equation (4.12). For linear systems, the response is directly proportional to the magnitude of the harmonic force applied. For nonlinear systems, it is typical to observe a sudden jump in the amplitude as the magnitude of the force is increased. Also, typically a downward jump in the amplitude is observed as the magnitude of the force is decreased and consequently a part of the solution depicted in Fig. 4.1 is never realized. We will subsequently show that solutions in this region are unstable. For $\xi = 1$ the solution is nonoscillatory and hence does not show any jump characteristic.

Consider now a constant force of excitation, but of varying frequency. Let

Table 4.1

Roots of Duffing oscillator as a function of exciting force

Damping ratio = 0

X_{st}	X_1^2	X_2^2	X_3^2
0.0	0.000	1.333	1.333
0.5	0.003	1.415	1.253
1.0	0.012	1.467	1.191
2.0	0.053	1.576	1.040
3.0	0.138	1.681	0.851
4.0	0.419	1.781	0.474
4.05	–	1.786	–
4.5	–	1.830	–
5.0	–	1.878	–
6.0	–	1.970	–
Damping ratio = 0.1			
0.0	0.0	–	–
0.6	0.0	1.339	1.324
1.0	0.012	1.431	1.228
2.0	0.053	1.559	1.059
3.0	0.137	1.670	0.864
4.0	0.390	1.773	0.508
4.02	0.414	1.790	0.479
5.0	–	1.871	–
6.0	–	1.964	–
Damping ratio = 0.4			
0.0	0.000	–	–
1.0	0.012	–	–
2.0	0.049	–	–
2.5	0.080	–	–
2.9	0.120	1.338	1.219
3.0	0.123	1.395	1.153
4.0	0.280	1.615	0.772
4.35	0.452	1.665	0.548
5.0	–	1.753	–
6.0	–	1.868	–
Damping ratio = 1.0			
0.0	0.000	–	–
1.0	0.008	–	–
2.0	0.034	–	–
3.0	0.081	–	–
4.0	0.155	–	–
5.0	0.274	–	–
6.0	0.505	–	–

$$\mu = 9; \quad X_{st} = 10$$

Then, equation (4.14) becomes

$$X^6 - 0.2963(1 - r^2)X^4 + \{0.02195r^4 - 0.0439r^2(1 - 2\xi^2) + 1\}X^2 - 100 = 0 \qquad (4.15b)$$

The roots of the above equation (4.15b) are given in Table 4.2 for $\xi = 0.1, 0.4$.

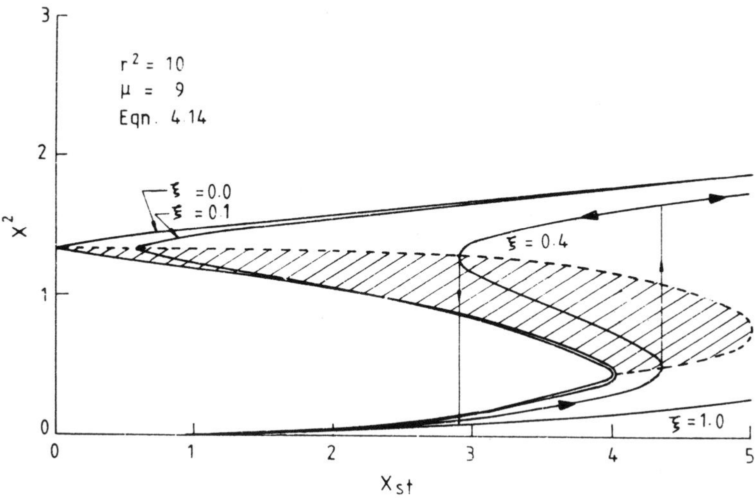

Fig. 4.1 Harmonic solution of Duffing equation

Table 4.2

Roots of Duffing oscillator as a function of exciting frequency

Damping ratio = 0.1

r	X_1^2	X_2^2	X_3^2
0.0	1.203	–	–
1.0	1.229	–	–
2.0	1.610	–	–
3.0	2.184	–	–
4.0	3.061	–	–
4.2	3.274	0.952	0.704
5.0	4.258	2.659	0.194
6.0	5.775	4.511	0.084
7.0	7.607	6.572	0.044
8.0	9.744	8.897	0.025
9.0	12.182	11.506	0.016
10.0	14.910	14.413	0.010
11.0	17.906	17.643	0.007
11.4	19.130	19.074	0.006
11.5	–	–	0.006
Damping ratio = 0.4			
0.0	1.203	–	–
1.0	1.296	–	–
2.0	1.594	–	–
3.0	2.135	–	–
4.0	2.944	–	–
4.2	3.137	1.219	0.574
4.5	3.446	1.928	0.330
5.0	4.000	2.923	0.188
5.5	4.569	3.977	0.121
5.75	4.708	4.693	0.099
6.0	–	–	0.083
7.0	–	–	0.043

The values of X^2 given by equation (4.15b) in Table 4.2 are plotted in Fig. 4.2 now as a function of ω/p, for $X_{st} = 10$, $\mu = 9$ and two different values of $\xi = 0.1$ and 0.4. The free vibration solution given by equation (4.11) is also plotted in the same figure. As discussed above the influence of nonlinearity of the spring is clearly demonstrated in Fig. 4.2 by the amplitude frequency relationship given by equation (4.11). For the present case this is given by

$$X^2 = \frac{r^2 - 1}{6.75} \tag{4.11a}$$

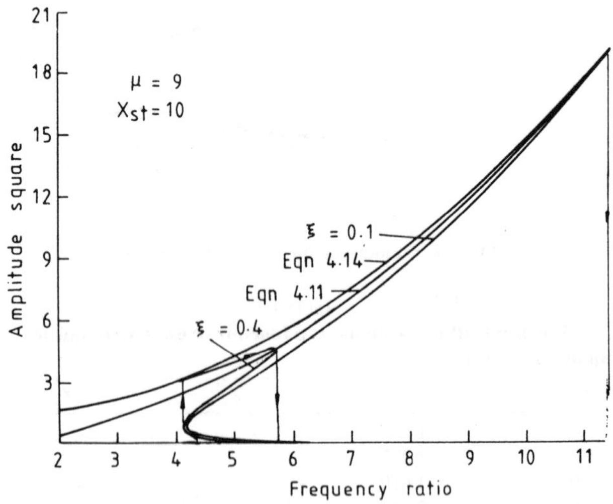

Fig. 4.2 Harmonic solution of Duffing equation-Jump characteristic

Also, the typical upward jump characteristic of the forced oscillation as the frequency of excitation is increased can be observed in Fig. 4.2. When the frequency of excitation is gradually decreased from a higher value, a sudden upward jump occurs and once again a part of the solution is not realizable in practice. This part of the solution will be shown to be unstable later.

The influence of decreasing the magnitude of the force is shown in Fig. 4.3 for $\xi = 0.1$ and two different values of $X_{st} = 10$ and 1. The values corresponding to $X_{st} = 10$ are given in Table 4.2. For $X_{st} = 1$, Table 4.3 gives the roots of the following equation.

$$X^6 - 0.2963(1 - r^2)X^4 + \{0.02195r^4 - 0.0439r^2(1 - 2\xi^2) + 1\}X^2 - 1 = 0 \tag{4.15c}$$

For a larger magnitude of the force, the amplitudes are larger as in Table 4.2 and the nonlinearity of the system becomes pronounced. For small magnitude of the force, the nonlinearity is less pronounced as can be seen in Table 4.3 and in Fig. 4.3.

4.2 STABILITY OF DUFFING SOLUTION

To determine the stability of solutions x_0, discussed above, let

$$x = x_0 + \zeta \tag{4.16}$$

From (4.3), we get

Table 4.3
Roots of Duffing oscillator as function of exciting frequency
Damping ratio = 0.1

r	X_1^2	X_2^2	X_3^2
0.0	0.191	–	–
1.0	0.279	–	–
2.0	0.623	–	–
2.1	0.675	–	–
2.2	0.730	0.312	0.096
2.5	0.914	0.602	0.040
3.0	1.282	1.073	0.016
3.6	1.806	1.736	0.007
3.7	–	–	0.006
4.0	–	–	0.004

$$\ddot{x}_0 + \ddot{\zeta} + 2\xi\dot{x}_0 + 2\xi\dot{\zeta} + x_0 + \zeta + \mu(x_0+\zeta)^3 = X_{st}\cos r\tau \tag{4.16a}$$

Since x_0 is a singular point or a steady solution of the system, we get

$$\ddot{\zeta} + 2\xi\dot{\zeta} + \zeta + 3\mu x_0^2\zeta = 0 \tag{4.17}$$

In arriving at the above equation, the ζ^2 and ζ^3 terms are neglected. Introducing the following transformation (*see* equation 3.87),

$$\zeta = e^{-\xi\tau}\eta \tag{4.18}$$

we get

$$\ddot{\eta} + (1 + 3\mu x_0^2 - \xi^2)\eta = 0 \tag{4.19}$$

For the solution given in equation (4.4), equation (4.19) is

$$\ddot{\eta} + \{1 + 3\mu(a\sin r\tau + b\cos r\tau)^2 - \xi^2\}\eta = 0 \tag{4.20}$$

Equation (4.20) above can be expressed as

$$\ddot{\eta} + \{\theta_0 + 2\theta_1\cos(2r\tau - \epsilon)\}\eta = 0 \tag{4.21}$$

where

$$\theta_0 = 1 - \xi^2 + \frac{3}{2}\mu X^2$$

$$\theta_1 = \frac{3}{4}\mu X^2$$

$$\epsilon = \tan^{-1}\frac{2ab}{b^2 - a^2} \tag{4.22}$$

In view of (3.87) and (3.104)

$$\delta = \xi \tag{4.23}$$

Hence, the stability condition of equation (3.150) becomes

$$\xi^4 + 2\xi^2(\theta_0 + r^2) + (\theta_0 - r^2)^2 > \theta_1^2 \tag{4.24}$$

Using equation (4.22), we have

$$\xi^4 + 2\xi^2\left(1 - \xi^2 + \frac{3}{2}\mu X^2 + r^2\right) + \left(1 - \xi^2 + \frac{3}{2}\mu X^2 - r^2\right)^2 > \frac{9}{16}\mu^2 X^4 \tag{4.25}$$

Using equation (4.7), (4.25) can be simplified as

$$A^2 - \frac{3}{2}\mu A X^2 + 4\xi^2 r^2 > 0 \tag{4.26}$$

From (4.7) and (4.9), the condition in (4.26) above reduces to

$$\frac{dX_{st}^2}{dX^2} > 0 \tag{4.27}$$

Hence, the hatched region in Fig. 4.1 is unstable. We came to this conclusion from a discussion on the jump characteristics of the system.

In a similar way, we can show that the dotted line of the solution in Fig. 4.2 and Fig. 4.3 is unstable.

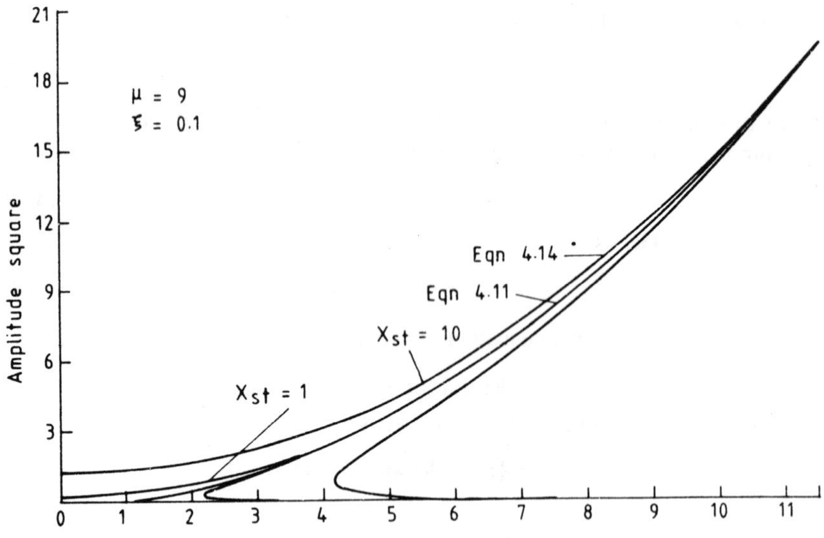

Fig. 4.3 Harmonic solution of Duffing equation

In view of the jump characteristic of a nonlinear system, the system response in time domain will typically differ from the corresponding linear system. For example, if we keep the forcing frequency constant and gradually increase the magnitude of the force, as seen from Fig. 4.1, a sudden jump in the amplitude occurs, which in time domain will look as in Fig. 4.4a. In linear system, the increase in time response will be gradual proportional to the applied force and without any jump for any force. Fig. 4.4b similarly shows the time response when the force is decreased from a given value for a nonlinear system.

Similarly, for a constant value of the amplitude of the force with its frequency increasing, the typical time response of a nonlinear system is as shown in Fig. 4.5a.

The amplitude of the response increases initially in a gradual manner and then suddenly falls from the peak and decreases further in a gradual manner. When the frequency is decreased, initially the amplitude increases with a sudden jump to a larger value and then it decreases gradually as shown in Fig. 4.5b.

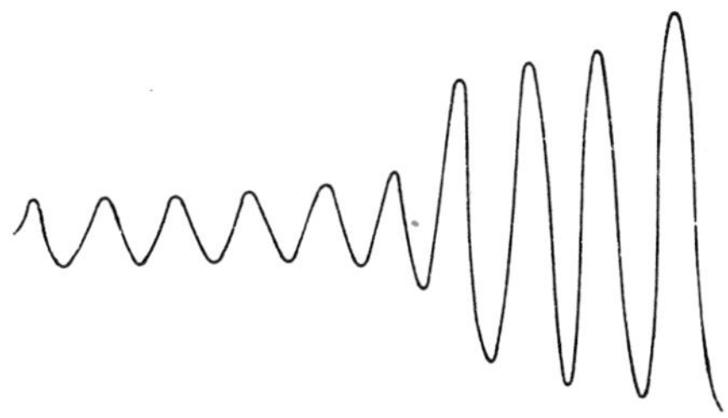

(a) Force increasing

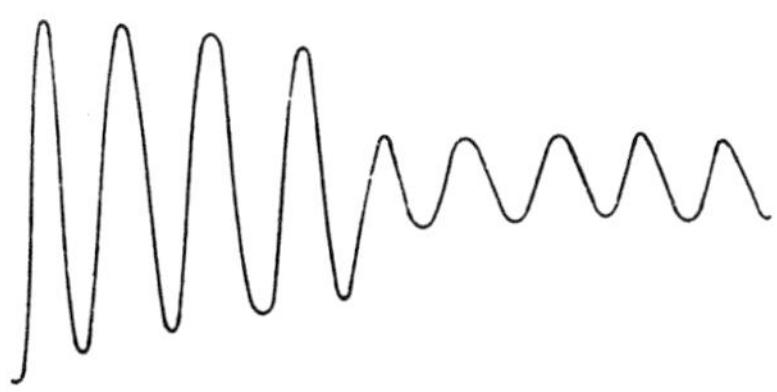

(b) Force decreasing

Fig. 4.4 Response with varying force – constant frequency

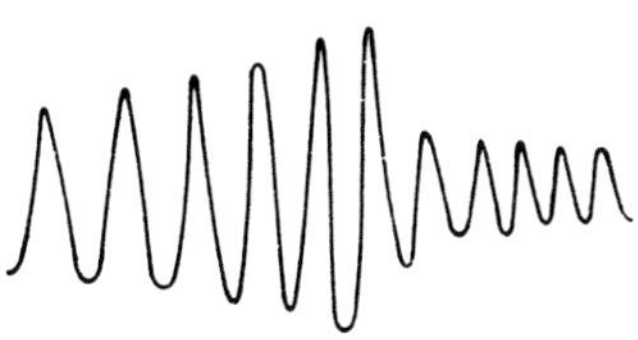

(a) Frequency increasing

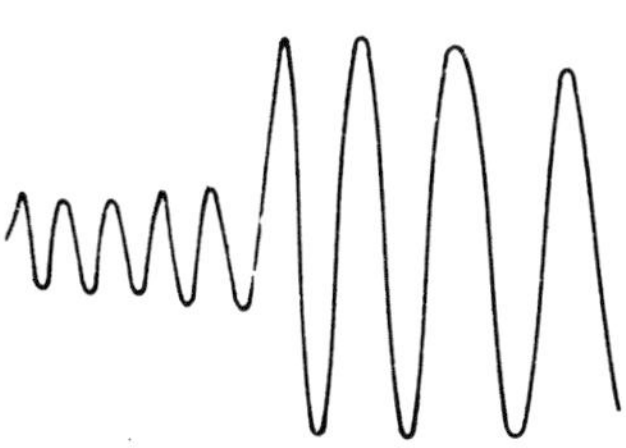

(b) Frequency decreasing

Fig. 4.5 Response with varying frequency – constant force

4.3 TRANSIENT BEHAVIOUR OF FORCED OSCILLATIONS

For a Duffing oscillator operating under a specific excitation force, the steady state solution is given by equation (4.4). If this system is disturbed by a transient condition from the dynamic equilibrium, it will asymptotically reach the steady position again provided the system is stable. If the system is not stable, any transient disturbance then will produce large amplitudes vibration. The condition of stability of a specific singular point can be studied as follows. For this purpose, we write equation (4.4) as

$$x(\tau) = a(\tau)\sin r\tau + b(\tau)\cos r\tau \tag{4.28}$$

Therefore

$$\dot{x}(\tau) = \dot{a}\sin r(\tau) + ar\cos r\tau + \dot{b}\cos r\tau - br\sin r\tau$$

$$\ddot{x}(\tau) = \ddot{a}\sin r(\tau) + 2\dot{a}r\cos r\tau - ar^2\sin r\tau$$

$$+\ddot{b}\cos r\tau - 2\dot{b}r\sin r\tau - br^2\cos r\tau \tag{4.29}$$

Substituting the above in (4.3) and separating the sin $r\tau$ and cos $r\tau$ terms, we get

$$\ddot{a} - ar^2 - 2\dot{b}r + 2\xi\dot{a} - 2\xi br + a + \frac{3}{4}\mu a^3 + \frac{3}{4}\mu ab^2 = 0$$

$$\ddot{b} - br^2 + 2\dot{a}r + 2\xi\dot{b} + 2\xi ar + b + \frac{3}{4}\mu b^3 + \frac{3}{4}\mu ba^2 = X_{st} \tag{4.30}$$

We can assume that a and b are slowly varying and that the damping is sufficiently small to ignore the damping terms, i.e.,

$$\ddot{a} \sim 0 \qquad \ddot{b} \sim 0$$

$$\xi\dot{a} \sim 0 \qquad \xi\dot{b} \sim 0$$

then equation (4.30) simplifies to

$$-ar^2 - 2\dot{b}r - 2\xi br + a + \frac{3}{4}\mu a^3 + \frac{3}{4}\mu ab^2 = 0$$

$$-br^2 + 2\dot{a}r + 2\xi ar + b + \frac{3}{4}\mu b^3 + \frac{3}{4}\mu ba^2 = X_{st} \tag{4.31}$$

$$\frac{da}{d\tau} = \frac{1}{2}\left(\frac{A}{r}b + \frac{X_{st}}{r} - 2\xi a\right) = P(a,b)$$

$$\frac{db}{d\tau} = \frac{1}{2}\left(-\frac{A}{r}a - 2\xi b\right) = Q(a,b) \tag{4.32}$$

Singular points for the system are obtained from

$$\frac{da}{d\tau} = 0 \quad \frac{db}{d\tau} = 0$$

and equation (4.32) lead to equation (4.9) again for the steady state solution. Further

$$a_0^2 = \frac{4X_{st}^2 r^2 \xi^2}{(A^2 + 4\xi^2 r^2)^2} = \frac{X^2}{1 + \frac{A^2}{4\xi^2 r^2}}$$

$$b_0^2 = \frac{A^2 X_{st}^2}{(A^2 + 4\xi^2 r^2)^2} = \frac{X^2}{1 + \frac{4\xi^2 r^2}{A^2}} \tag{4.33}$$

For stability of the singular point, we consider a small variation

$$a = a_0 + \zeta \quad b = b_0 + \eta$$

From (3.48) and (3.49), we get

$$\frac{d\zeta}{dt} = a_{21}\zeta + a_{12}\eta$$

$$\frac{d\eta}{dt} = a_{21}\zeta + a_{22}\eta$$

where

$$a_{11} = \frac{\partial P}{\partial a}\}_{\begin{Bmatrix} a \\ b \end{Bmatrix} = \begin{Bmatrix} a_0 \\ b_0 \end{Bmatrix}} = -\xi - \frac{3}{4r}\mu a_0 b_0$$

$$a_{12} = \frac{\partial P}{\partial b}\}_{\begin{Bmatrix} a \\ b \end{Bmatrix} = \begin{Bmatrix} a_0 \\ b_0 \end{Bmatrix}} = \frac{A}{2r} - \frac{3}{4r}\mu b_0^2$$

$$a_{21} = \frac{\partial Q}{\partial a}\}_{\begin{Bmatrix} a \\ b \end{Bmatrix} = \begin{Bmatrix} a_0 \\ b_0 \end{Bmatrix}} = -\frac{A}{2r} + \frac{3}{4r}\mu a_0^2$$

$$a_{22} = \frac{\partial Q}{\partial b}\}_{\begin{Bmatrix} a \\ b \end{Bmatrix} = \begin{Bmatrix} a_0 \\ b_0 \end{Bmatrix}} = -\xi + \frac{3}{4r}\mu a_0 b_0 \tag{4.34}$$

Using (3.53), the first condition leads to

$$\xi > 0 \tag{4.35}$$

and the second condition for stability gives

$$-\frac{3\mu}{2r^2} A X_0^2 + \frac{A^2}{r^2} + 4\xi^2 > 0 \tag{4.36}$$

which is same as equation (4.26).

4.4 TRANSIENT SOLUTIONS IN PHASE PLANE

In section 4.3, we studied the transient behaviour of the Duffing oscillator for the three different possible singular points under a given set of conditions of operation. We also noted that there is only one real root for the system governed by the cubic for the solution for certain conditions of operation. The stability condition for the system singularities was also discussed in the previous sections. We now will consider the solution in the phase plane. The isoclines are defined from (4.32) as

$$\frac{db}{da} = \frac{-\frac{A}{r}a - 2\xi b}{\frac{A}{r}b + \frac{X_{st}}{r} - 2\xi a} = C \tag{4.37}$$

Substituting for A from equation (4.7), and rearranging we get

$$\frac{3}{4}\mu C b^3 + \frac{3}{4}\mu a b^2 + \left\{C\left(1 - r^2 + \frac{3}{4}\mu a^2\right) - 2\xi r\right\} b$$

$$+\frac{3}{4}\mu a^3 + (1 + 2\xi C r - r^2)a - CX_{st} = 0 \tag{4.38}$$

For $C = 0$, the above is

$$\frac{3}{4}\mu a b^2 - 2\xi r b + \frac{3}{4}\mu a^3 + (1 - r^2)a = 0 \tag{4.39}$$

For $C = \infty$, equation (4.37) leads to

$$\frac{3}{4}\mu b^3 + \left(1 - r^2 + \frac{3}{4}\mu a^2\right)b + 2\xi r a - X_{st} = 0 \tag{4.40}$$

To make phase plane plots, we have to first generate the isoclines using the above three equations for different values of C. Let us draw the phase plane diagram corresponding to

$$\mu = 9 \quad r = \frac{\omega}{p} = 5$$

$$\xi = 0.4 \quad X_{st} = 10$$

Equations (4.38) to (4.40) are to be solved for three possible roots of, say, b, for a given value of a. For $C = 0$, there are two roots only. Repeating this for various values of a, we can plot the isoclines. The roots are given in Table 4.4.

In Fig. 4.6, these isoclines are plotted along with the trajectories. A magnified diagram of the isoclines is shown in Fig. 4.7. From this figure, all the three singular points can be quickly identified from the fact that many isoclines pass through the singular point. The singular points are

1.85, 0.75
1.15, −1.275
0.15, − 0.4

Thus, the values of X^2 are

3.985
2.948
0.183

We can compare these values in Table 4.2 and also in Fig. 4.2. In Fig. 4.6, the trajectories are drawn following the slopes of the isoclines. We observe that the two stable roots determined before are represented by stable foci and the unstable singularity is a saddle. It can be readily observed that this singularity can never be achieved, no matter what the initial disturbance is given.

Table 4.4
Roots of equation (4.38)

C = 0

a	*b1*	*b2*	*b3*
−1.89	−0.067	−0.246	–
−1.7	0.660	−1.009	–
−1.5	0.962	−1.357	–
−1.0	1.330	−1.922	–
−0.5	1.320	−2.505	–
0.0	0.000	0.000	–
0.5	−1.320	2.505	–
1.0	−1.330	1.922	–
1.5	−0.962	1.357	–
1.7	−0.660	1.009	–
1.89	0.067	0.246	–
C = 1			
−5.0	5.156	–	–
−4.0	4.210	–	–
−3.0	3.318	–	–
−2.0	2.588	–	–
−1.6	2.405	–	–
−1.5	2.373	−0.306	−0.567
−1.0	2.285	0.147	−1.432
−0.5	2.254	−0.032	−1.722
0.0	2.196	−0.369	−1.827
0.5	2.060	−0.769	−1.792
1.0	1.804	−1.166	−1.638
1.5	1.339	−1.753	−1.087
1.9	0.460	−2.099	−0.261
2.0	–	−2.191	–
3.0	–	−3.141	–
4.0	–	−4.113	–
5.0	–	−5.095	–
C = −1			
−5.0	–	–	−4.837
−4.0	–	–	−3.772
−3.0	–	–	−2.608
−2.5	–	–	−1.830
−2.0	–	–	0.647
−1.5	–	–	1.361
−1.0	–	–	1.716
−0.5	–	–	1.892
−0.1	−0.784	−1.249	1.934
0.0	−0.559	−1.372	1.931
0.5	0.169	−1.507	1.838
1.0	0.781	−1.355	1.574
1.2	1.055	−1.224	1.369
1.5	–	−0.914	–
2.0	–	1.791	–
2.5	–	2.330	–
3.0	–	2.854	–
4.0	–	3.884	–
5.0	–	4.904	–

C = ∞			
−5.0	0.207	–	–
−4.0	0.307	–	–
−3.0	0.565	–	–
−2.0	1.280	–	–
−1.0	1.908	–	–
−0.7	2.003	−1.239	−0.764
−0.5	2.043	−1.439	−0.605
0.0	2.067	−1.626	−0.441
0.5	1.976	−1.602	−0.374
1.0	1.750	−1.383	−0.367
1.4	1.409	−0.956	−0.476
1.5	1.324	–	–
2.0	0.455	–	–
3.0	−0.054	–	–
4.0	−0.071	–	–
5.0	−0.069	–	–

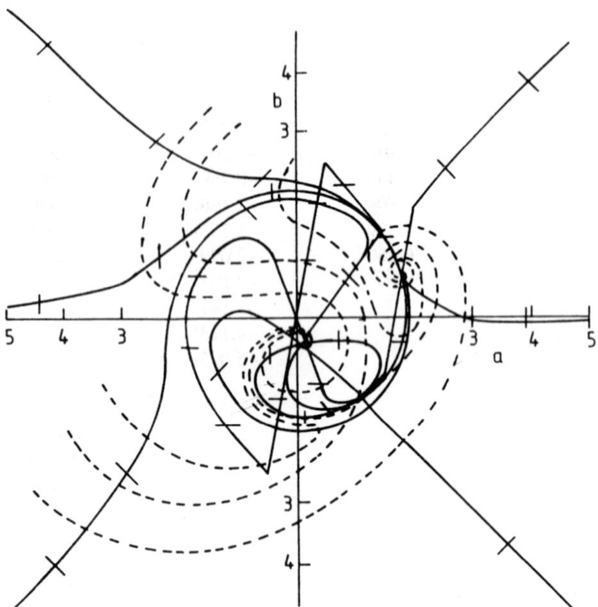

Fig. 4.6 Transient solution of forced Duffing system

For values of r less than 4.2 and greater than 5.75, there is only one root for the system as given in Table 4.2. Such a singularity is normally a stable node, like the one depicted in Fig. 4.8.

If the nonlinear spring is soft, i.e., μ is negative, then the curves in Fig. 4.2 and 4.3 lean to the left, instead of to the right from the corresponding linear case. The reader may like to try solving such a case.

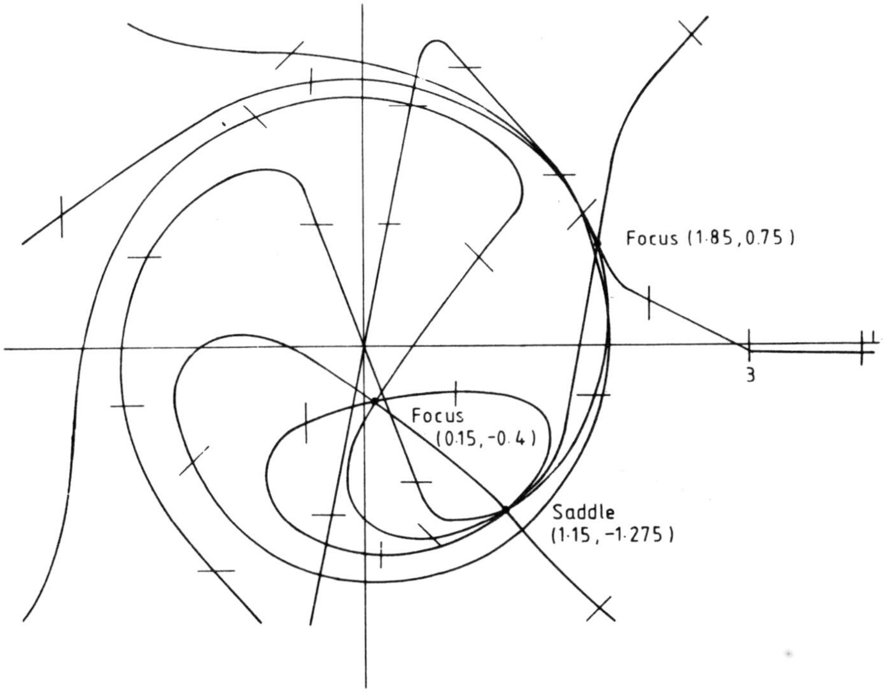

Fig. 4.7 Singularities of Duffing equation

4.5 HIGHER HARMONIC OSCILLATIONS

For the system described in equation (4.1), we obtained the steady state harmonic solution given by (4.4). When we used the harmonic balancing procedure, we deliberately ignored third harmonic terms in equation (4.5) and went ahead to study the dynamic behaviour of harmonic solution. As we discussed before, nonlinear systems exhibit a typical property of harmonic response at higher and submultiples of the frequency of excitation. Let us consider the system in equation (4.3).

$$\ddot{x} + 2\xi\dot{x} + x + \mu x^3 = X_{st}\cos r\tau \tag{4.3}$$

For simplicity we assume that the linear stiffness coefficient is very small in comparison with μ so that the nonlinear behaviour predominates and the algebra involved is simplified. We will also consider only one case of excitation with frequency constant at $r = 1$. Then equation (4.3) becomes

$$\ddot{x} + 2\xi\dot{x} + \mu x^3 = X_{st}\cos\tau \tag{4.41}$$

We now consider a higher harmonic response in addition to the harmonic solution and write

$$x_0 = a_1\sin\tau + b_1\cos\tau + a_3\sin 3\tau + b_3\cos 3\tau \tag{4.42}$$

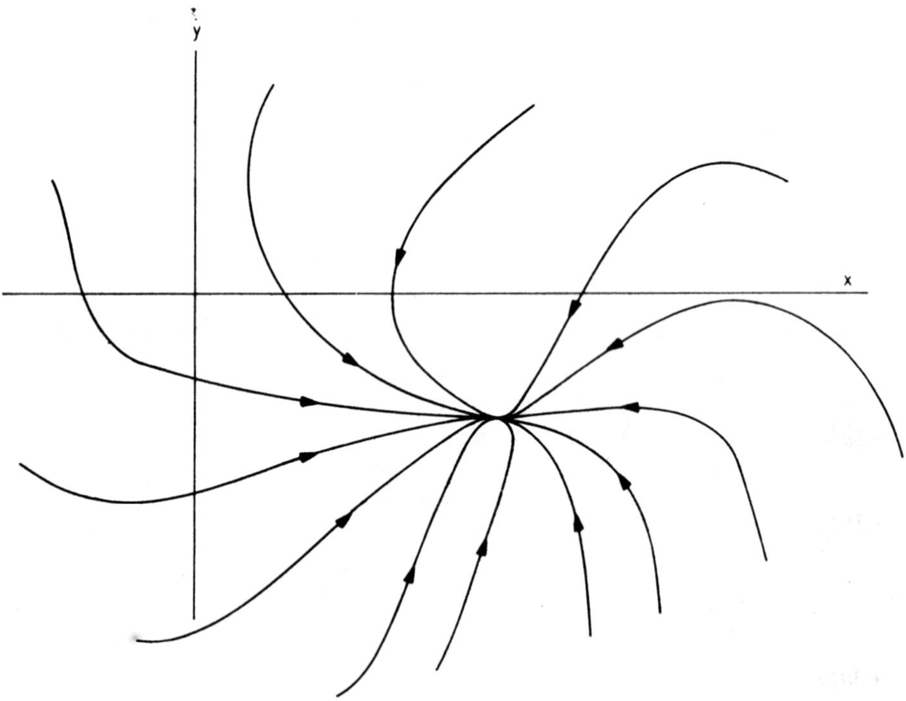

Fig. 4.8 Stable Node-Singular point of Duffing's equation–Harmonic solution

As can be seen shortly, the algebra involved will be very lengthy and we consider only the third harmonic. Since the stiffness nonlinearity is symmetric, we find the second harmonic will not be present as can be seen from equation (4.5). There will be subharmonic response also and we will consider this separately. Substituting equation (4.42) and its derivatives in equation (4.41), we get

$$
\begin{aligned}
&-a_1\sin\tau - b_1\cos\tau - 9a_3\sin 3\tau - 9b_3\cos 3\tau \\
&+2\xi(a_1\cos\tau - b_1\sin\tau + 3a_3\cos 3\tau - 3b_3\sin 3\tau) \\
&+a_1^3\sin^3\tau + b_1^3\cos^3\tau + 3a_1^2b_1\sin^2\tau\cos\tau + 3b_1^2a_1\cos^2\tau\sin\tau \\
&+a_3^3\sin^3 3\tau + b_3^3\cos^3 3\tau + 3a_3^2b_3\sin^2 3\tau\cos 3\tau + 3b_3^2a_3\cos^2 3\tau\sin 3\tau \\
&+3(a_1^2\sin^2\tau + b_1^2\cos^2\tau + 2a_1b_1\sin\tau\cos\tau)(a_3\sin 3\tau + b_3\cos 3\tau) \\
&+3(a_3^2\sin^2 3\tau + b_3^2\cos^2 3\tau + 2a_3b_3\sin 3\tau\cos 3\tau)(a_1\sin\tau + b_1\cos\tau) = X_{st}\cos\tau
\end{aligned}
$$

Expanding the above

$$-a_1\sin\tau - b_1\cos\tau - 9a_3\sin 3\tau - 9b_3\cos 3\tau$$

$$+2\xi(a_1\cos\tau - b_1\sin\tau + 3a_3\cos 3\tau - 3b_3\sin 3\tau)$$

$$+\frac{3}{4}a_1^3\sin\tau - \frac{1}{4}a_1^3\sin 3\tau + \frac{3}{4}b_1^3\cos\tau + \frac{1}{4}b_1^3\cos 3\tau$$

$$+\frac{3}{4}a_1^2b_1\cos\tau - \frac{3}{4}a_1^2b_1\cos 3\tau + \frac{3}{4}a_1b_1^2\sin\tau + \frac{3}{4}a_1b_1^2\sin 3\tau$$

$$+\frac{3}{4}a_0^3\sin 3\tau - \frac{1}{4}a_0^3\sin 9\tau + \frac{3}{4}b_0^3\cos 3\tau + \frac{1}{4}b_0^3\cos 9\tau$$

$$+\frac{3}{4}a_3^2b_3\cos 3\tau - \frac{3}{4}a_3^2b_3\cos 9\tau + \frac{3}{4}a_3b_3^2\sin 3\tau + \frac{3}{4}a_3b_3^2\sin 9\tau$$

$$+3a_1^2a_3\left(-\frac{1}{4}\sin\tau + \frac{1}{2}\sin 3\tau - \frac{1}{4}\sin 5\tau\right) + 3a_1^2b_3\left(-\frac{1}{4}\cos\tau + \frac{1}{2}\cos 3\tau - \frac{1}{4}\cos 5\tau\right)$$

$$+3b_1^2a_3\left(\frac{1}{4}\sin\tau + \frac{1}{2}\sin 3\tau + \frac{1}{4}\sin 5\tau\right) + 3b_1^2b_3\left(\frac{1}{4}\cos\tau + \frac{1}{2}\cos 3\tau + \frac{1}{4}\cos 5\tau\right)$$

$$+\frac{3}{2}a_1b_1a_3(\cos\tau - \cos 5\tau) + \frac{3}{2}a_1b_1b_3(-\sin\tau + \sin 5\tau)$$

$$+3a_3^2a_1\left(\frac{1}{2}\sin\tau + \frac{1}{4}\sin 5\tau - \frac{1}{4}\sin 7\tau\right) + 3a_3^2b_1\left(\frac{1}{2}\cos\tau - \frac{1}{4}\cos 5\tau - \frac{1}{4}\cos 7\tau\right)$$

$$+3b_3^2a_1\left(\frac{1}{2}\sin\tau - \frac{1}{4}\sin 5\tau + \frac{1}{4}\sin 7\tau\right) + 3b_3^2b_1\left(\frac{1}{2}\cos\tau + \frac{1}{4}\cos 5\tau + \frac{1}{4}\cos 7\tau\right)$$

$$+\frac{3}{2}a_3b_3a_1(\cos 5\tau - \cos 7\tau) + \frac{3}{2}a_3b_3b_1(\sin 5\tau + \sin 7\tau) = X_{st}\cos\tau$$

Following harmonic balance procedure, we have
sin τ terms

$$A_1a_1 + 2\xi b_1 + \frac{3}{4}(a_1^2 - b_1^2)a_3 + \frac{3}{2}a_1b_1b_3 = 0$$

cos τ terms

$$-A_1b_1 + 2\xi a_1 - \frac{3}{4}(a_1^2 - b_1^2)b_3 + \frac{3}{2}a_1b_1a_3 = X_{st} \tag{4.43}$$

sin 3τ terms

$$A_3a_3 + 6\xi b_3 + \frac{1}{4}a_1(a_1^2 - 3b_1^2) = 0$$

cos 3τ terms

$$-A_3b_3 + 6\xi a_3 - \frac{1}{4}b_1(3a_1^2 - b_1^2) = 0 \tag{4.44}$$

In (4.44) above

$$A_1 = 1 - \frac{3}{4}(X_1^2 + 2X_3^2)$$

$$A_3 = 9 - \frac{3}{4}(2X_1^2 + X_3^2)$$

$$X_1^2 = a_1^2 + b_1^2$$

$$X_3^2 = a_3^2 + b_3^2 \tag{4.45}$$

By squaring and adding, we can eliminate a and b components in equations (4.43) and (4.44) and after rearranging the terms, we get

$$\left[\left\{A_1 - \frac{3A_3X_3^2}{X_1^2}\right\}^2 + 4\xi^2\left\{1 + \frac{9X_3^2}{X_1^2}\right\}^2\right]X_1^2 = X_{st}^2$$

$$(A_3^2 + 36\xi^2)X_3^2 = \frac{1}{16}X_1^6 \tag{4.46}$$

In the above, X_1 and X_3 are coupled and we can have three roots for each one as before. The real roots of equation (4.46) are plotted in Fig. 4.9, for $\xi = 0.1$.

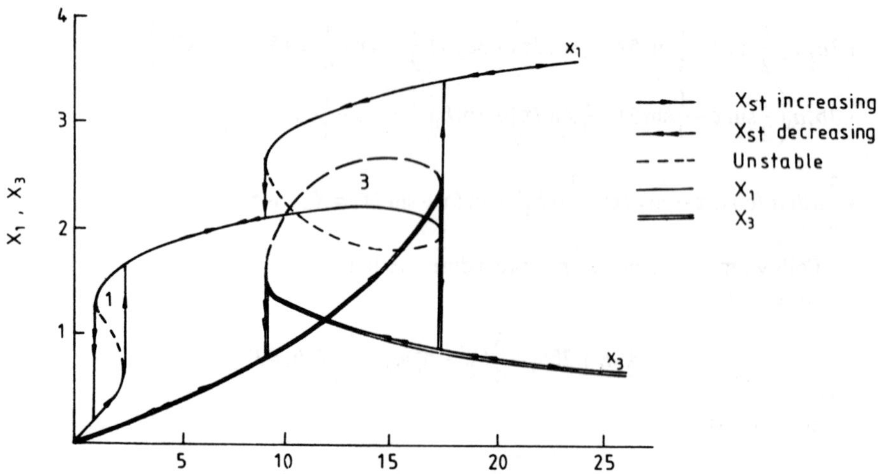

Fig. 4.9 Higher harmonic oscillations–Duffing equation

To study the stability of the response thus obtained, we give a small variation to the singular point

$$x = x_0 + \zeta \tag{4.47}$$

Substituting equation (4.47) in equation (4.41) and noting that X_0 is a solution of (4.41), we get the following variational equation

$$\ddot{\zeta} + 2\xi\dot{\zeta} + 3\mu x_0^2\zeta = 0 \tag{4.48}$$

In arriving at the above equation, the ζ^2 and ζ^3 terms are neglected. Introducing (*see* equation 3.87),

$$\zeta = e^{-\xi\tau}\eta$$

we get

$$\ddot{\eta} + \left\{\theta_0 + 2\sum_{n=1}^{3}\theta_n \cos(2n\tau - \epsilon_n)\right\}\eta = 0 \tag{4.49}$$

where

$$\theta_0 = -\xi^2 + \frac{3}{2}\mu(X_1^2 + X_3^2)$$

$$\theta_n^2 = \theta_{nc}^2 + \theta_{ns}^2$$

$$\epsilon_n = \tan^{-1}\frac{\theta_{ns}}{\theta_{nc}}$$

$$\theta_{1c} = -\frac{3}{4}(a_1^2 - b_1^2) + \frac{3}{2}(a_1 a_3 + b_1 b_3)$$

$$\theta_{1s} = \frac{3}{2}(a_1 b_1 - a_1 b_3 + a_3 b_1)$$

$$\theta_{2c} = \frac{3}{2}(-a_1 a_3 + b_1 b_3)$$

$$\theta_{2s} = \frac{3}{2}(a_1 b_3 + b_1 a_3)$$

$$\theta_{3c} = -\frac{3}{4}(a_3^2 - b_3^2)$$

$$\theta_{3s} = \frac{3}{2}a_3 b_3 \tag{4.50}$$

The stability criterion is given by

$$\delta^4 + 2\delta^2(\theta_0 + n^2) + (\theta_0 - n^2)^2 > \theta_n^2 \quad n = 1, 2 \ldots \tag{3.150}$$

For $\delta = \xi = 0.1$, we get for $n = 1, 2, 3$

$$\theta_0^2 - 1.98\theta_0 + 1.0201 = \theta_1^2 \tag{a}$$

$$\theta_0^2 - 7.98\theta_0 + 16.0801 = \theta_2^2 \tag{b}$$

$$\theta_0^2 - 17.98\theta_0 + 81.1801 = \theta_3^2 \tag{c}$$

The above three relations are plotted in Fig. 4.10 and denoted as (1), (2) and (3) regions of stability.

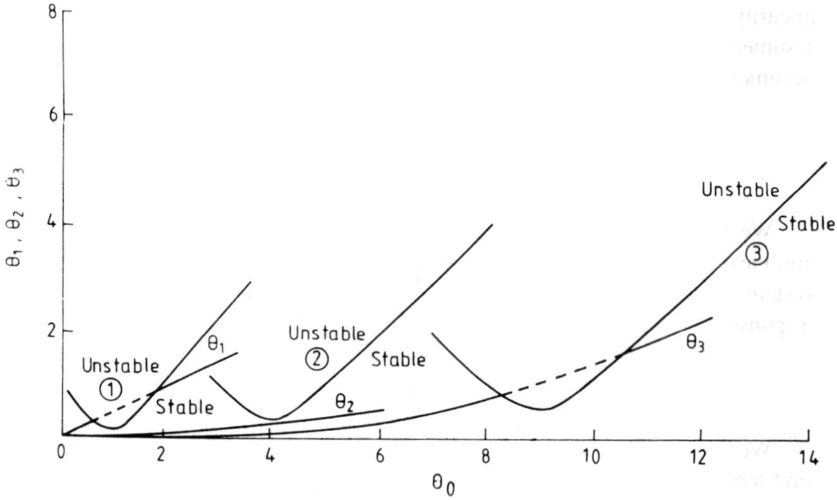

Fig. 4.10 Stability regimes–Duffing equation higher harmonics

For a given excitation force X_{st}, X_1 and X_3 are known from equation (4.46) or Fig. 4.9 and using equation (4.50), the variation of θ_1 with θ_0 as X_{st} increases according to the first stability relation (a) above, is given in Fig. 4.10. We have instability in the first region when the function (a) is shown by discontinuous line in Figs. 4.9 and 4.10. The corresponding solutions are unstable and cannot be realized in practice.

Similar arguments can be given for the second and third unstable regions with the help of relations (b) and (c) and they are shown in Figs. 4.9 and 4.10. From Fig. 4.9, it can be observed that, in addition to jump occurring for X_1 in the first region, jump also occurs for both X_1 and X_3 as X_{st} is further increased. We could have considered the frequency response, as we did in the case of harmonic solutions in the previous section 4.1, which would then demonstrate the jump characteristics for a given X_{st} with increasing or decreasing frequency.

Further, an analysis of transients can be made in a similar way as made in previous section 4.3, and the stability studied in the phase plane.

4.6 SUBHARMONIC OSCILLATIONS

In the previous section, we studied the higher harmonic response of a nonlinear system besides the harmonic response due to a harmonic excitation force. In addition, a nonlinear system response can also contain subharmonic oscillations. This response is usually at a fraction of the excitation frequency. Usually subharmonic response exists only for certain conditions of operation of the system. To simplify the algebra involved, we consider a simple undamped system defined by

$$\ddot{x} + x^3 = X_{st} \cos 3\tau \tag{4.51}$$

The solution of the above is written for the harmonic component and subharmonic component as follows

$$x_0 = a \sin \tau + b \cos \tau + X_3 \cos 3\tau \tag{4.52}$$

In (4.52) above the first two terms represent the subharmonic of the solution, which we assume to be only one third of the excitation frequency because the nonlinearity is symmetric. Also, since there is no damping, the harmonic component is assumed without any phase difference with the excitation force. Equation (4.51) then becomes

$$-a\sin\tau - b\cos\tau - 9X_{st}\cos 3\tau$$

$$+(a\sin\tau + b\cos\tau + X_3\cos 3\tau)^3 = X_{st}\cos 3\tau \tag{4.53}$$

We will make another simple assumption that the harmonic solution for the nonlinear system can be approximated by the corresponding solution of the linear system. This will simplify the algebra involved to determine the subharmonic response of the system. Then,

$$X_3 = -\frac{1}{8}X_{st} \tag{4.54}$$

We follow the harmonic balance now to obtain
$\sin\tau$ terms

$$\left[1 - \frac{3}{4}(a^2 + b^2) - \frac{3}{2}X_3^2\right]a = -\frac{3}{4}X_3 2ab$$

$\cos\tau$ terms

$$\left[1 - \frac{3}{4}(a^2 + b^2) - \frac{3}{2}X_3^2\right]b = -\frac{3}{4}X_3(a^2 - b^2) \tag{4.55}$$

Multiply the first equation by b and the second by a in the above and subtract to obtain

$$a(3b^2 - a^2) = 0 \tag{4.56}$$

Hence

$$a = 0 \tag{4.57}$$

or

$$a = \pm\sqrt{3}\,b \tag{4.58}$$

For $a = 0$ as in equation (4.57), the second equation in (4.55) gives

$$\left(1 - \frac{3}{4}b^2 - \frac{3}{2}X_3^2\right)b = \frac{3}{4}X_3 b^2$$

i.e.,

$$b^2 + X_3 b + 2X_3^2 - \frac{4}{3} = 0 \tag{4.59}$$

In other words, there are two values of b possible if $a = 0$. Similarly, for a given a in equation (4.58), we get from (4.55)

$$a = +\sqrt{3b}$$

$$b^2 - \frac{1}{2}X_3 b + \frac{1}{2}X_3^2 - \frac{1}{3} = 0 \tag{4.60}$$

Hence, there are again two possible solutions for b. Similarly, we also find the same roots as in (4.60) for

$$a = -\sqrt{3}b$$

Therefore, for a given value of X_{st}, we have six possible subharmonic solutions. Let us consider the roots corresponding to $a = 0$, given in (4.57). Equation (4.59) is plotted for different values of X_3 in Fig. 4.11. If X_3 is more than -0.8615, the subharmonic with $a = 0$, does not exist. In the region where the subharmonic exists, its stability can be determined by taking the variation

$$x = x_0 + \zeta \tag{4.61}$$

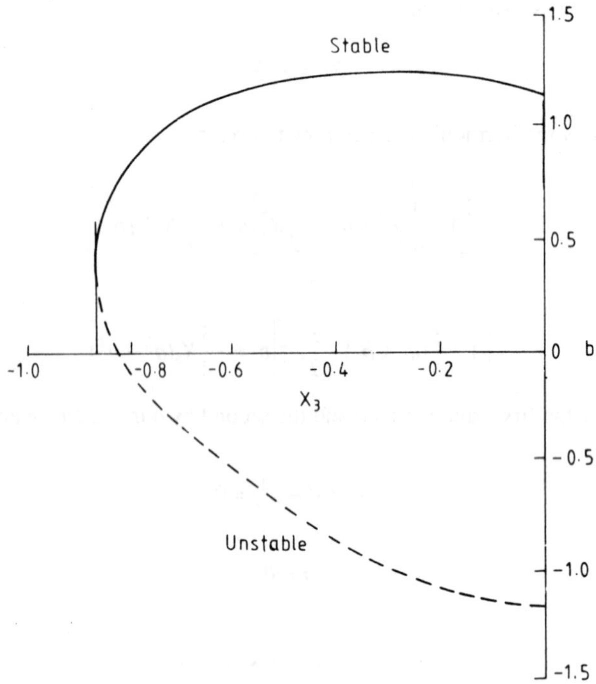

Fig. 4.11 Subharmonic solution of Duffing equation ($a = 0$)

Substituting the above in (4.51) and neglecting higher powers of ζ, we get

$$\zeta + 3x_0^2\zeta = 0 \tag{4.62}$$

It should be noted that x_0 in (4.62) above is

$$x_0 = b\cos\tau + X_3\cos 3\tau \tag{4.63}$$

Hence, (4.62) becomes

$$\zeta + \left(\theta_0 + 2\sum_{n=1}^{3}\theta_n \cos 2n\tau\right)\zeta = 0 \tag{4.64}$$

where

$$\theta_0 = \frac{3}{2}(b^2 + X_3^2)$$

$$\theta_1 = \frac{3}{4}(b^2 + 2X_3 b)$$

$$\theta_2 = \frac{3}{2} b X_3$$

$$\theta_3 = \frac{3}{4} X_3^2 \tag{4.65}$$

For no damping in the system, i.e., $\delta = 0$, the stability condition (3.150) becomes

$$\text{mod}\,[\theta_0 - n^2] > \text{mod}\,[\theta_n] \quad n = 1, 2, 3 \tag{4.66}$$

For $n = 1$, with the help of equation (4.65), (4.66) becomes

$$\text{mod}\left[\frac{3}{2}(b^2 + X_3^2) - 1\right] > \text{mod}\left[\frac{3}{4}(b^2 + 2X_3 b)\right]$$

i.e.,

$$\text{mod}\left[2b^2 + 2X_3^2 - \frac{4}{3}\right] > \text{mod}\,[b^2 + 2X_3 b] \tag{4.67}$$

With the help of equation (4.59), the above equation (4.67) can be written as

$$\text{mod}\,[b^2 - X_3 b] > \text{mod}\,[b^2 + 2X_3 b]$$

i.e.,

$$\text{mod}\,[b - X_3] > \text{mod}\,[b + 2X_3] \tag{4.68}$$

The condition in (4.68) above gives the curve in discontinuous lines in Fig. 4.11 as unstable.

For other roots of subharmonic response in equation (4.58), the analysis above can be repeated. Also, the transients can be studied as followed for higher harmonics and the singularities with their stability criteria determined in a phase plane.

4.7 PERTURBATION METHOD FOR FORCED OSCILLATIONS

Consider an undamped nonlinear Duffing system given by

$$m\frac{d^2x}{dt^2} + \alpha x + \beta x^3 = F\cos\omega t \tag{4.69}$$

Let

$$p = \sqrt{\frac{\alpha}{m}}$$

$$T = pt \qquad dt = \frac{1}{p}dT$$

$$X_{st} = \frac{F}{mp^2}$$

$$\mu = \frac{\beta}{mp^2} \tag{4.70}$$

then

$$\ddot{x} + x + \mu x^3 = X_{st} \cos rT \tag{4.71}$$

$$\cdot = \frac{d}{dT}$$

Further, let

$$\tau = rT$$

and let X_{st} be small with μ so that

$$X_{st} = \mu X_{st0} \tag{4.72}$$

Then, equation (4.71) becomes

$$r^2\ddot{x} + x + \mu x^3 = \mu X_{st0} \cos \tau \tag{4.73}$$

$$\cdot = \frac{d}{d\tau}$$

We can write the solution to the above as

$$x(\tau) = x_0(\tau) + \mu x_1(\tau) + \mu^2 x_2(\tau) + \ldots \tag{4.74}$$

and in addition choose the function for r in series as

$$r = r_0 + \mu r_1 + \mu^2 r_2 + \ldots \tag{4.75}$$

Therefore

$$(r_0 + \mu r_1 + \mu^2 r_2 + \ldots)^2 (\ddot{x}_0 + \mu \ddot{x}_1 + \mu^2 \ddot{x}_2 + \ldots)$$

$$+ (x_0 + \mu x_1 + \mu^2 x_2 + \ldots) + \mu(x_0 + \mu x_1 + \mu^2 x_2 + \ldots)^3 = \mu X_{st0} \cos \tau \tag{4.76}$$

We now separate terms of μ^0, μ^1, μ^2 etc to obtain

$$r_0^2 \ddot{x}_0 + x_0 = 0$$

$$r_0^2 \ddot{x}_1 + x_1 = -2r_0 r_1 \ddot{x}_0 - x_0^3 + X_{st0} \cos \tau$$

$$r_0^2 \ddot{x}_2 + x_2 = -(2r_0 r_2 + r_1^2)\ddot{x}_0 - 2r_0 r_1 \ddot{x}_1 - 3x_0^2 x_1 \quad \ldots \text{etc.} \tag{4.77}$$

Since the solution should be periodic in τ with time period 2π, we seek a periodic solution such that

$$x_i(\tau + 2\pi) = x_i(\tau) \tag{4.78}$$

Further, let the initial conditions be

$$x(0) = A;\ \dot{x}(0) = 0 \tag{4.79}$$

The solution of the first equation in (4.77) is

$$x_0 = a_0 \cos\frac{\tau}{r_0} + b_0 \sin\frac{\tau}{r_0} \tag{4.80}$$

In view of the initial conditions (4.79) and the periodicity condition (4.78), equation (4.80) gives

$$x_0 = A\cos\tau$$

$$r_0 = 1 \tag{4.81}$$

With the help of equation (4.81) above, the second equation in (4.77) becomes

$$\ddot{x}_1 + x_1 = 2r_1 A\cos\tau - A^3\cos^3\tau + X_{st0}\cos\tau$$

$$= \left(2r_1 A - \frac{3}{4}A^3 + X_{st0}\right)\cos\tau - \frac{1}{4}A^3\cos 3\tau \tag{4.82}$$

We immediately identify that the bracketed term in the above gives rise to secular term in the solution. Therefore, to avoid the secular term in the solution, we choose

$$r_1 = \frac{3}{8}A^2 - \frac{X_{st0}}{2A} \tag{4.83}$$

Now the solution of equation (4.82) is governed by the third harmonic excitation term and it can be written as

$$x_1 = a_1\cos\tau + b_1\sin\tau + \frac{A^3}{32}\cos 3\tau$$

$$= \frac{A^3}{32}(-\cos\tau + \cos 3\tau) \tag{4.84}$$

At this stage, we can write the harmonic part of the solution as

$$x = \left(A - \mu\frac{A^3}{32}\right)\cos\tau \tag{4.85}$$

$$r = 1 + \left(\frac{3}{8}A^2 - \frac{X_{st0}}{2A}\right)\mu \tag{4.86}$$

Hence

$$x = \left(A - \mu\frac{A^3}{32}\right)\cos\left\{1 + \left(\frac{3}{8}A^2 - \frac{X_{st0}}{2A}\right)\mu\right\}t \tag{4.87}$$

We can proceed further to refine the solution. However, this solution converges for very small value of μ and much less than 1. For obtaining the harmonic, higher harmonic and subharmonic solutions, it is therefore more convenient to use harmonic balance method than the perturbation method. The phase plane method, however, may be useful to develop a graphical solution without depending much on lot of mathematics. This is described in the next section.

4.8 PHASE PLANE – DELTA METHOD

In chapter 3, we considered the phase plane method for free vibration problems. The method was simple and straight forward, in the absence of forcing terms since the slope at any point in the phase plane is constant (excepting at singularities). The isoclines could be easily constructed and the trajectories drawn to obtain the integral curves. The problem becomes complicated when we have the forcing terms since the slope at any point is no more independent of time and it depends upon the current value of the forcing function when the trajectory passes through the point under consideration. Let us consider the following equation representing the dynamic behaviour of a system.

$$\ddot{x} + G(x, \dot{x}, t) = 0 \tag{4.88}$$

As in the phase plane analysis, let us write

$$\dot{x} = y \tag{4.89}$$

Then

$$\dot{y} = -G(x, \dot{x}, t) \tag{4.90}$$

$$\frac{dy}{dx} = \frac{-G(x, \dot{x}, t)}{y} \tag{4.91}$$

Hence, if we choose to develop the solution in the conventional phase plane, we find that there are no fixed isoclines as before. The isoclines are time-dependent since we cannot define

$$\frac{-G(x, \dot{x}, t)}{y} = C$$

To circumvent this problem, let us rewrite equation (4.88) as follows:

$$\ddot{x} + \{G(x, \dot{x}, t) - p^2 x\} + p^2 x = 0 \tag{4.92}$$

and define

$$\delta = \frac{1}{p^2}\{G(x, \dot{x}, t) - p^2 x\} \tag{4.93}$$

If we imagine δ to remain constant for a small interval of time, then we can construct the trajectory of equation (4.92) for this interval of time. We then have to develop a process of going through several small intervals of time to construct the total trajectory in a given period of time. In view of (4.93), equation (4.92) is written as

$$\ddot{x} + p^2(x + \delta) = 0 \tag{4.94}$$

As in the previous phase plane analysis, let us choose one coordinate of the plane to be x itself. It will be convenient to choose the second coordinate as

$$v = \frac{\dot{x}}{p} \tag{4.95}$$

Then

$$\frac{dx}{dt} = \dot{x} = vp \tag{4.96}$$

and

$$\frac{dv}{dt} = \frac{\ddot{x}}{p} = -p(x + \delta) \tag{4.97}$$

From the above two equations (4.96) and (4.97), we get

$$\frac{dv}{dx} = -\frac{x+\delta}{v} \tag{4.98}$$

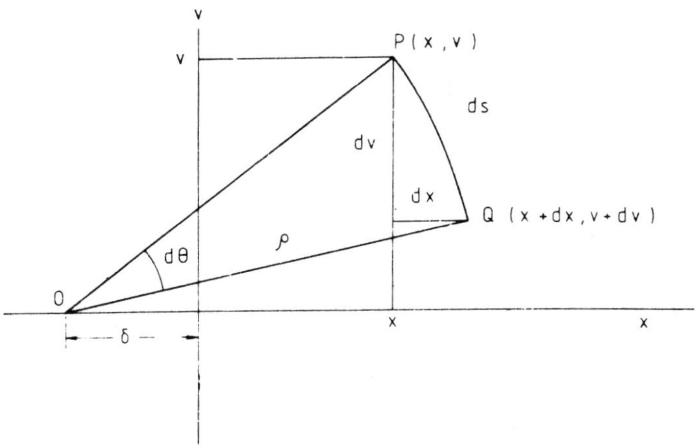

Fig. 4.12 Phase plane–method for forced oscillations

As mentioned before, let us hold δ to be constant for a small interval of time and consider the trajectory for this interval in $x - v$ plane as shown in Fig. 4.12. For a positive value of δ at any given step, locate the centre O of the trajectory curve (approximated as circular arc) at a distance $-\delta$ from the origin on the x axis. From the starting point P (x, v) in this step we can draw a circular arc of an angle $d\theta$ to obtain the end point of the trajectory Q $(x + dx, v + dv)$. If this angle $d\theta$ is small, then we see that equation (4.98) is satisfied. The angle θ and t are connected as follows from (4.96)

$$dt = \frac{dx}{pv} \tag{4.99}$$

Also

$$d\theta = \frac{ds}{\rho}$$

$$= \frac{\sqrt{dx^2 + dv^2}}{\sqrt{v^2 + (x+\delta)^2}}$$

$$= \frac{\sqrt{1 + \left(\frac{dv}{dx}\right)^2}}{\sqrt{1 + \left\{\frac{x+\delta}{v}\right\}^2}} \quad \frac{dx}{v}$$

$$= \frac{dx}{v} \tag{4.100}$$

Using the above in (4.99), we get

$$dt = \frac{d\theta}{p} \tag{4.101}$$

The solution at Q is then given by

$$x_Q = x_P + v_p d\theta$$

$$v_Q = v_P - (x_p + \delta)d\theta \tag{4.102}$$

The procedure then, is as follows:

(1) Use x–v plane as defined in equation (4.95).

(2) Start the trajectory construction at a point with the initial conditions.

$$t_0 = 0$$

$$x_0 = x(t = 0)$$

$$v_0 = \frac{\dot{x}(t = 0)}{p}$$

(3) Express the differential equation in the form (4.94) and determine δ according to (4.93).

(4) Locate the point O on the x axis at $-\delta$ and draw an arc from the starting point through a small angle $\Delta\theta$, corresponding to a small time Δt as defined by (4.101).

(5) At the end point of the arc thus constructed we have a solution given by

$$t_1 = t_0 + \Delta t$$

$$x_1 = x(t = t_1)$$

$$v_1 = v(t = t_1)$$

(6) Repeat the steps 3 to 5, starting from the end point obtained in step 5 at time $t = t_1$.

In the above δ is determined based on the function in (4.93) at the beginning of the interval in any step. Obviously, δ is not a constant in this entire period, however small it may be in each step. Therefore, such a solution as above is very similar to the one obtained by the basic Euler's method and it requires extremely small interval of time to minimize the cumulative error involved.

To circumvent this problem, we would like to develop a procedure where the value of δ is taken as an average value in each step, rather than that based on the initial conditions at the beginning of this step. Thus we use the following iteration scheme for the $(i + 1)$ step, given the values at the end of ith step.

$$t_{i+1} = t_i + \Delta t$$

$$\Delta\theta = p\Delta t \tag{4.103}$$

For the first iteration, we use the following starting values taken from the ith step.

$$x_{(i+1)}^{(1)} = x_i$$

$$\dot{x}_{(i+1)}^{(1)} = \dot{x}_i \tag{4.104}$$

Let the average values of x and $\dot{x}$ be represented by χ and $\dot{\chi}$ respectively, for the $(i+1)$ step in ith iteration. They are

$$\chi^{(j)} = \frac{x_i + x_{(i+1)}^{(j)}}{2}$$

$$\dot{\chi}^{(j)} = \frac{\dot{x}_i + \dot{x}_{(i+1)}^{(j)}}{2} \tag{4.105}$$

The average value of time during this step is

$$T = t_i + \frac{1}{2}\Delta t \tag{4.106}$$

Then the value of δ in this iteration is given by

$$\delta_i^{(j)} = \frac{1}{p^2}[G(\chi^{(j)}, \dot{\chi}^{(j)}, T) - p^2\chi^{(j)}] \tag{4.107}$$

With the help of (4.102), we then write

$$x_{i+1}^{(j+1)} = x_i + \dot{x}_i \Delta\theta$$

$$\dot{x}_{i+1}^{(j+1)} = \dot{x}_i - (x_i + \delta_i^{(j)})\Delta\theta \tag{4.108}$$

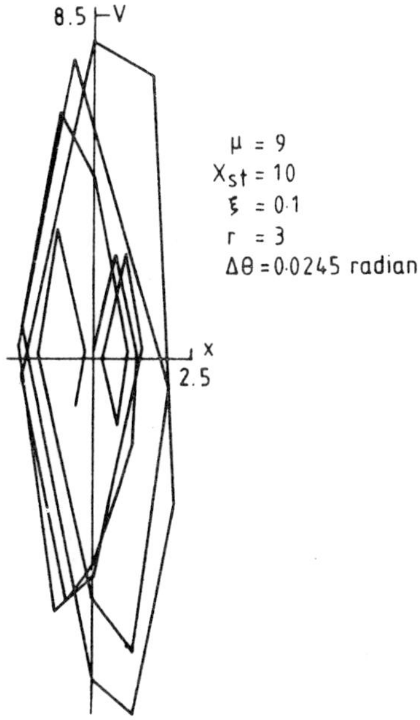

Fig. 4.13 Phase plane solution of Duffing equation

At this stage the $(j+1)$ iteration values are compared with j iteration values and if we are satisfied with the accuracy, then the iterations can be terminated here and proceed to (i + 2) step. Otherwise, the iteration is repeated beginning from (4.105) with j incremented by 1. It should be emphasized here once again that the time step Δt is very important in these calculations.

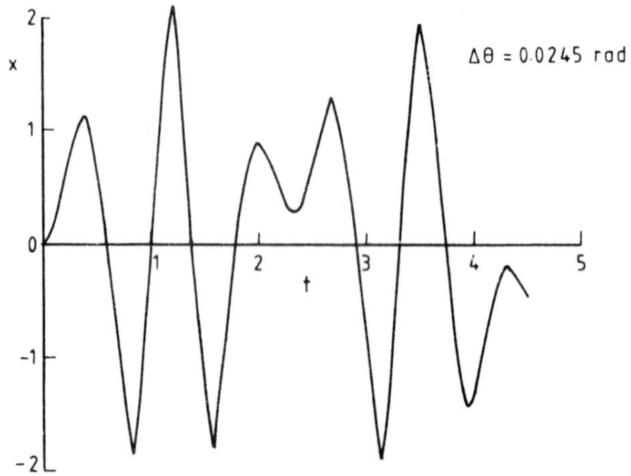

Fig. 4.14 (a) Response of Duffing oscillator

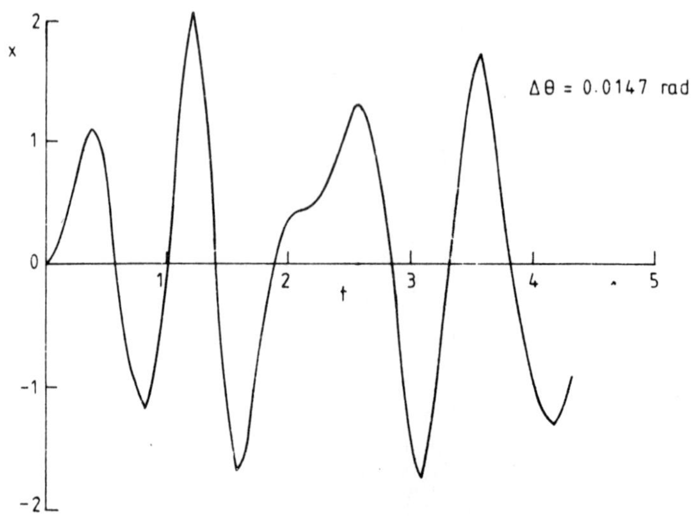

Fig. 4.14 (b) Response of Duffing oscillator

Based on the above method, the solution obtained for the Duffing oscillator is given in Fig. 4.13 with initial conditions for x and $v = 0$. The time step used in this solution $\Delta\theta$ is 0.0245 radian. Actually, as this angle is so small, it is practically difficult to do the graphical construction by hand. One has to depend on a computer. The plot in Fig. 4.13 is made using points every 0.245 radians. The corresponding displacement response x is given in Fig. 4.14a.

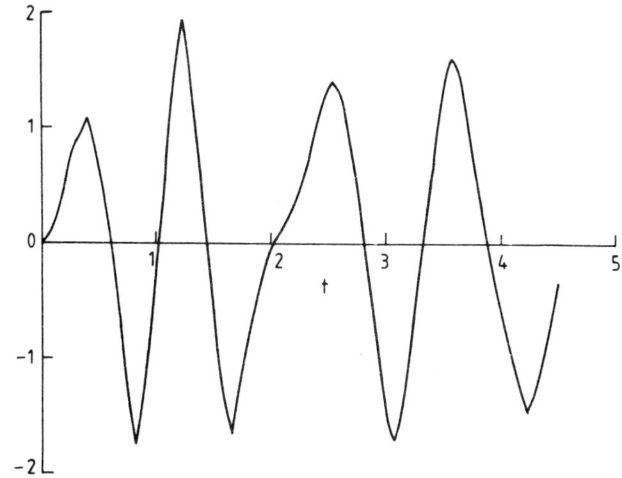

Fig. 4.15 Euler solution of Duffing oscillator

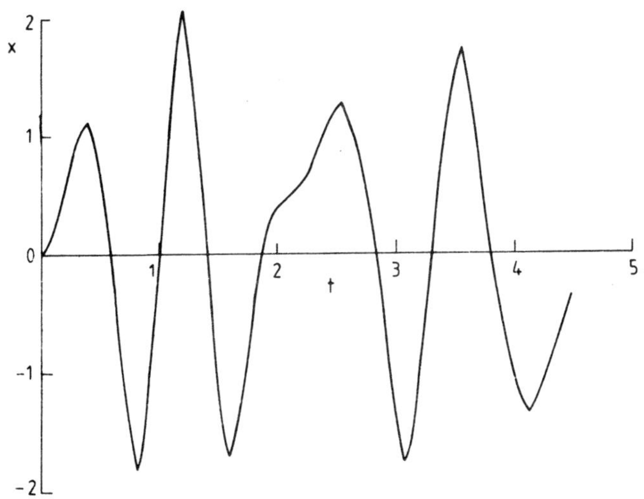

Fig. 4.16 Runge–Kutta solution of Duffing oscillator

One finds from Fig. 4.14a, a kink in the middle which may be due to higher harmonics present in the system or it may be due to an error being carried cumulatively if the time step is not small enough for the solution. To verify this, we reduce the time step further and check if there is a significant difference between these results. With the time step given by $\Delta\theta = 0.0147$ radians, the result obtained is given in Fig. 4.14b. By decreasing the step further not much improvement is found and we may take this as final solution.

We could use the modified Euler's method of section 2.3 or Runge-Kutta method of section 2.4 to obtain the solution of the above problem. The results obtained from these programs are plotted in Figs. 4.15 and 4.16 for $\Delta\theta = 0.0245$ radians.

5

Variational Principles

During the introductory course on vibrations we become familiar with two degrees of freedom systems, then with multi degree of freedom and continuous systems with infinite degrees of freedom, after the study of single degree of freedom systems. Energy principles are also studied to derive the equations of motion for such systems where it is not so apparent to distinguish the inertia forces and restoring forces from Newtonian and D'Alembert principles. Variational principles offer the most powerful tool to formulate the equations of motion of multi degree and continuous systems with a clear understanding on any approximations made during the derivation process. We have also seen that the solution of the problem becomes complicated with an increase in degrees of freedom and departure from linearity of the system forces. Variational principles again offer simple but powerful tools for the solution of such equations of motion, thus enabling us to understand the dynamic behaviour of complex systems. Before we take up the vibration problems, we will first study the basics of calculus of variations without going into the mathematical principles in depth, but only to the extent we need the information to derive the equations of motion and set up approximate methods of solution of the same. To set the tone for an understanding of the subject of calculus of variations, we will briefly recapitulate the calculus of a single variable.

5.1 CALCULUS OF A SINGLE VARIABLE

We studied in the very first course of calculus the behaviour of a function of a single variable e.g., $x = f(t)$. In this, t is the independent variable and x is a function that depends on the value of t. If in a given range of t, say t_1 to t_2, x is continuous, we derived conditions for this function to have a maximum or a minimum value. For a local maxima, say, at $t = a$, we set the condition that in the neighbourhood of $t = a$, x has a value less than $f(a)$. Similarly, for a local minimum at $t = b$, the value of x is more than $f(b)$, in the neighbourhood of $t = b$. This is illustrated in Fig. 5.1. We may call the values of t in the neighbourhood of $t = a$ or $t = b$ as admissible values for checking the local maxima and minima conditions.

To establish the conditions for local maximum or minimum, we expand the function $f(t)$ in Taylor's series about the position $t = a$ or $t = b$. For the region around $t = a$, we get

$$f(t) = f(a) + \left\{\frac{df}{dt}\right\}_{t=a}(t-a) + \frac{1}{2!}\left\{\frac{d^2f}{dt^2}\right\}_{t=a}(t-a)^2 + \frac{1}{3!}\left\{\frac{d^3f}{dt^3}\right\}_{t=a}(t-a)^3 + \dots$$

which can be written as

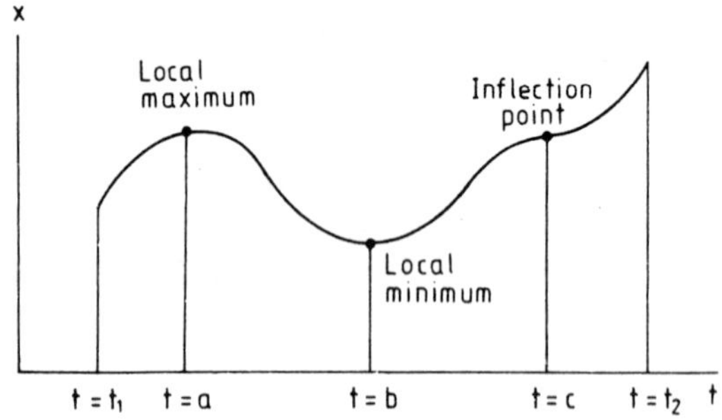

Fig. 5.1 Extremal points of $x = f(t)$

$$f(t) - f(a) = \{\dot{f}(a)\}\ (t-a) + \frac{1}{2!}\{\ddot{f}(a)\}\ (t-a)^2 + \ldots \tag{5.1}$$

For $f(a)$ to be a maximum, it is necessary that the above is a negative quantity for all values of t in the neighbourhood of a. The quantity $(t-a)$ can be positive or negative. Therefore, it is necessary that $\dot{f}(a) = 0$. A similar condition is valid for $f(b)$ to be minimum, i.e., $\dot{f}(b) = 0$, from the following

$$f(t) - f(b) = \{\dot{f}(b)\}\ (t-b) + \frac{1}{2!}\{\ddot{f}(b)\}\ (t-b)^2 + \ldots \tag{5.2}$$

Therefore, for $f(a)$ to be a maximum, i.e., $\{f(t) - f(a)\}$ in equation (5.1) to be negative, we get the condition that

$$\dot{f}(a) = 0$$

$$\ddot{f}(a) < 0 \tag{5.3a}$$

Similarly from (5.2), $f(b)$ is minimum if

$$\dot{f}(b) = 0$$

$$\ddot{f}(b) > 0 \tag{5.3b}$$

We also have a possibility that

$$\dot{f}(c) = 0$$

$$\ddot{f}(c) = 0 \tag{5.3c}$$

In the above case (5.3c) we have neither a maximum nor a minimum for $t = c$, but instead an inflection point.

The points satisfying the conditions (5.3a) to (5.3c) are called extremal points of the function, which can be a maximum, a minimum or an inflection point depending on the conditions in (5.3). Calculus of variations deals with functionals which are functions of a function rather than functions. Both the subjects are invented by Newton, who also discovered the laws of motion. It is worth noting the way in which these are invented by Newton. It is said that Newton in 1663, when he was just 20 years old, bought a book on astrology out of a curiosity to see what is there in it. When he could not understand an illustration in this book, he bought a book on

Trigonometry, and to follow the geometry in this book in turn, he bought the book of Euclid's elements of geometry. Armed with these books, he discovered in 1665, the subject of differential calculus, while he was still an undergraduate student at Cambridge. Almost thirty years later in 1696 another incident happened. Johann Bernoulli challenged fellow mathematicians at that time to solve an unresolved issue called Brachistochrone problem, specifying the curve connecting two points displaced from each other laterally, along which a body, acted upon only by gravity, would fall in the shortest time. He set a dead line of six months originally and extended it to a year and a half at the request of another fellow mathematician. It is said that the challenge was delivered to Newton at 4.00 pm on January 29, 1697. Newton before leaving for work next morning invented calculus of variations to solve this problem and sent off the solution for anonymous publication. When Bernoulli saw this solution, he commented: ''We recognize the lion by his claw''. Be that as it may, the subject of calculus of variations plays an important role in the theory of vibrations that we are currently interested in. An understanding of the subject to the level we need is given in the following.

5.2 FIRST VARIATION

If calculus dealt with functions as we recapitulated in the section above, the calculus of variations deals with functionals. A functional is a function of function or functions. We talked about the Brachistochrone problem before, specifying the curve connecting two points displaced from each other laterally, along which a body, acted upon only by gravity, would fall in the shortest time. Let the two points be denoted by (1) and (2). Let the speed of the body be V and s is the total distance along the curve $y = f(x)$. The time for descent from point (1), x_1, y_1 to (2) x_2, y_2 is given by

$$t = \int_{(1)}^{(2)} \frac{ds}{V}$$

$$= \int_{(1)}^{(2)} \frac{\sqrt{dx^2 + dy^2}}{V}$$

$$= \int_{x_1}^{x_2} \frac{\sqrt{1 + \left\{\frac{dy}{dx}\right\}^2}}{V} dx \tag{5.4}$$

If V_1 is the speed at point (1), x_1, y_1, then at any position x, y during the travel, we have

$$\frac{1}{2} m V_1^2 + m g y_1 = \frac{1}{2} m V^2 + m g y$$

Therefore, the speed at any position x, y is

$$V = \sqrt{V_1^2 - 2g(y - y_1)} \tag{5.5}$$

Substituting the above equation (5.5) in (5.4), we get

$$t = \int_{x_1}^{x_2} \sqrt{\frac{1 + \left\{\frac{dy}{dx}\right\}^2}{\sqrt{V_1^2 - 2g(y - y_1)}}} dx \tag{5.6}$$

We can rewrite equation (5.6) symbolically as

$$I = \int_{x_1}^{x_2} F(x, y, y')dx \tag{5.7}$$

which is a typical functional, a simple one, that we will often come across in the discussion of vibration problems. This functional has only one independent variable x, the dependent variable y and the first order derivative of the dependent variable y. We will begin our discussion with such simple functionals and extend this later to more complex functionals.

The value of I in (5.7) depends on the function $y = f(x)$ between the points (x_1, y_1) and (x_2, y_2). Brachistochrone problem was finding the path which gives the least time for travel between (x_1, y_1) and (x_2, y_2). We assume that such a path exists and try to find the one amongst several such possible paths in the neighbourhood which satisfies the conditions we desire. This is similar to calculus of functions we discussed in section 5.1. Here, we call the function $y = f(x)$ an admissible path and the paths in its neighbourhood as varied paths, $\tilde{y}(x)$. They can be expressed as

$$\tilde{y}(x) = y(x) + \in \eta(x) \tag{5.8}$$

In the above expression (5.8) $\in$ is a small parameter, since we are in the neighbourhood of the admissible path and $\eta(x)$ is a differentiable function at least two times with the condition

$$\eta(x_1) = \eta(x_2) = 0 \tag{5.9}$$

There are several possibilities for the varied paths $\tilde{y}(x)$. All these paths however go through the points (x_1, y_1) and (x_2, y_2). The value of the integral taken over the varied paths is then

$$\tilde{I} = \int_{x_1}^{x_2} F(x, \tilde{y}, \tilde{y}')dx$$

$$= \int_{x_1}^{x_2} F(x, y + \in \eta, y' + \in \eta')dx \tag{5.10}$$

We adopt the same strategy as in calculus, and expand the above functional as

$$\tilde{I} = \{\tilde{I}\}_{\in=0} + \left\{\frac{d\tilde{I}}{d\in}\right\}_{\in=0} \in + \left\{\frac{d^2\tilde{I}}{d\in^2}\right\}_{\in=0} \frac{\in^2}{2!} + \ldots$$

i.e.,

$$\tilde{I} - (\tilde{I}\}_{\in=0} =$$

$$\tilde{I} - I = \left\{\frac{d\tilde{I}}{d\in}\right\}_{\in=0} \in + \left\{\frac{d^2\tilde{I}}{d\in^2}\right\}_{\in=0} \frac{\in^2}{2!} + \ldots \tag{5.11}$$

We readily recognize that

$$\left\{\frac{d\tilde{I}}{d\in}\right\}_{\in=0} = 0 \tag{5.12}$$

for $I = \bar{I}(\epsilon = 0)$ to be an extremum. From (5.10) equation (5.12) becomes

$$\int_{x_1}^{x_2} \left\{ \frac{\partial F}{\partial \bar{y}} \frac{\partial \bar{y}}{\partial \epsilon} + \frac{\partial F}{\partial \bar{y}'} \frac{\partial \bar{y}'}{\partial \epsilon} \right\} dx \Bigg|_{\epsilon = 0} = 0 \tag{5.13}$$

In view of (5.8) and that

$$\left\{ \frac{\partial F}{\partial \bar{y}} \right\}_{\epsilon = 0} = \frac{\partial F}{\partial y}$$

equation (5.13) becomes

$$\int_{x_1}^{x_2} \left\{ \frac{\partial F}{\partial y} \eta + \frac{\partial F}{\partial y'} \eta' \right\} dx = 0 \tag{5.14}$$

The second term in the above is

$$\int_{x_1}^{x_2} \frac{\partial F}{\partial y'} \eta' dx = \frac{\partial F}{\partial y'} \eta \Big|_{x_1}^{x_2} - \int_{x_1}^{x_2} \left\{ \frac{d}{dx} \left(\frac{\partial F}{\partial y'} \right) \right\} \eta dx$$

$$= - \int_{x_1}^{x_2} \left\{ \frac{d}{dx} \left(\frac{\partial F}{\partial y'} \right) \right\} \eta dx$$

since η is zero at the end points. Equation (5.14) now becomes

$$\int_{x_1}^{x_2} \left\{ \frac{\partial F}{\partial y} - \frac{d}{dx} \left(\frac{\partial F}{\partial y'} \right) \right\} \eta dx = 0 \tag{5.15}$$

In the above equation (5.15) η (x) is arbitrary between the end points and therefore

$$\frac{\partial F}{\partial y} - \frac{d}{dx} \frac{\partial F}{\partial y'} = 0 \tag{5.16}$$

This is Euler-Lagrange equation that makes I an extremum. The solution of the above equation for y will give us the function that extremizes the integral I. To just complete our discussion of the Brachistochrone problem, equation (5.6) for the special case $y_1 = 0$ and $V_1 = 0$, we get

$$t = \frac{1}{\sqrt{2g}} \int_{x_1}^{x_2} \sqrt{\frac{1 + \left\{ \frac{dy}{dx} \right\}^2}{y}} dx$$

Hence, the function F for this problem is

$$F(x, y, y') = \left[\frac{1 + \left\{ \frac{dy}{dx} \right\}^2}{y} \right]^{\frac{1}{2}}$$

Using the above in Euler-Lagrange equation (5.16), we get

$$y'' = -\frac{1 + y'^2}{2y} \tag{a}$$

Let us make a substitution

$$\frac{dy}{dx} = \phi \tag{b}$$

then

$$\frac{d^2y}{dx^2} = \frac{d\phi}{dx}$$

$$= \frac{d\phi}{dy}\frac{dy}{dx}$$

$$= \phi\frac{d\phi}{dy} \tag{c}$$

Using (b) and (c) in (a), we get

$$\phi\frac{d\phi}{dy} = -\frac{1+\phi^2}{2y}$$

i.e.,

$$\frac{\phi}{1+\phi^2}d\phi + \frac{1}{2y}dy = 0 \tag{d}$$

Integrating

$$y(1+\phi^2) = C_1$$

or

$$y(1+y'^2) = C_1 \tag{e}$$

We can write (e) as

$$dx = \frac{\sqrt{y}}{\sqrt{C_1 - y}}dy$$

Hence

$$x = \int \frac{\sqrt{y}}{\sqrt{C_1 - y}}dy + x_0 \tag{f}$$

Now, let

$$y = C_1 \sin^2\frac{t}{2}$$

$$= \frac{1}{2}C_1(1-\cos t) \tag{g}$$

Then

$$x = \frac{1}{2}C_1(t - \sin t) + x_0 \tag{h}$$

In view of $x = 0$, at $t = 0$, we get

$$x_0 = 0$$

and the solution is given by

$$x = \frac{1}{2}C_1(t - \sin t)$$

$$y = \frac{1}{2}C_1(1 - \cos t)$$

These two equations represent the desired curve for Brachistochrone's problem, for the least travel of time between the two points. The curve is a cycloid.

5.3 THE DELTA OPERATOR

We considered in the previous section the first variation which defined the condition for an extremum of the functional of the form in equation (5.7). This has been done by considering a family of varied paths in the neighbourhood of the admissible path. We then introduced the functional with a varied path to determine the conditions for the extremizing function $y(x)$ to give an extremum to the functional. This whole operation can be carried out in a very convenient way by defining a variational operator, called δ operator, just as we define a differential operator dy or dx. In Fig. 5.2, we have the admissible path $y(x)$ with some possible varied paths in the interval of x from x_1 to x_2. Note that the varied paths and the admissible path have the same value at both the end points, according to (5.8).

Let us define the increments of the varied paths over the extremizing path as δy, corresponding to *ab* or *ac* for two typical curves shown in Fig. 5.2 at the position x. The increment of varied path δy is variation of y. It should be noted here that δy is not associated with a δx as one would associate a dy with a dx in differential calculus. This can be observed for the varied path shown in Fig. 5.3, where a dy is associated with a dx for $y(x)$, but there is no such definition for δy which is simply the vertical distance between the extremized path and the varied path at a given position x.

We can take a variation for any other function, e.g., dy/dx, remembering that we should fix x while taking such a variation. Therefore

$$\begin{aligned} \delta\left\{\frac{dy}{dx}\right\} &= \frac{d\bar{y}}{dx} - \frac{dy}{dx} \\ &= \frac{d}{dx}(\bar{y} - y) \\ &= \frac{d(\delta y)}{dx} \end{aligned} \qquad (5.17)$$

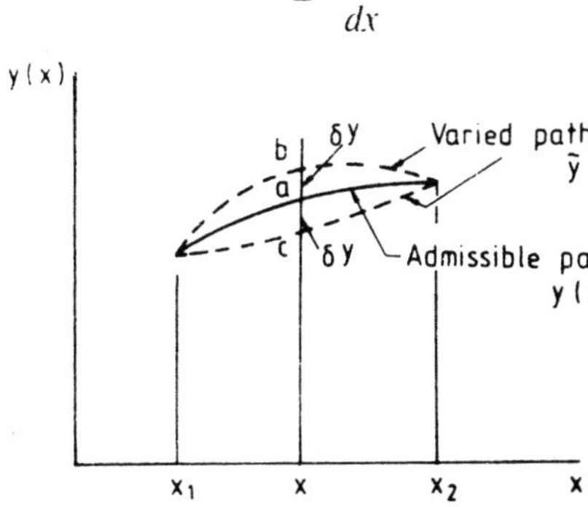

Fig. 5.2 Admissible and varied paths

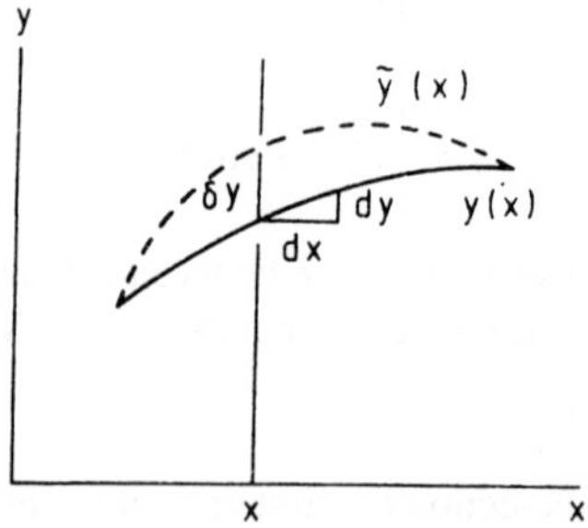

Fig. 5.3 Variational and differential operators

From the above, we note that δ operator is commutative with the differential operator. We can also show that δ operator is commutative with the integral operator as well. This will enable us to use δ operator for taking variations of functions or functionals, just as we use a differential operator. Let us retrace our steps of the previous section by using the δ operator. From (5.8) we write

$$\delta y = \tilde{y} - y = \epsilon\eta \tag{5.18a}$$

$$\delta y' = \tilde{y}' - y' = \epsilon\eta' \tag{5.18b}$$

The functional over the varied path can be written as follows

$$\tilde{F} = F(x, y + \delta y, y' + \delta y') \tag{5.19}$$

Using Taylor's series, the above is

$$\tilde{F} = F(x, y + \delta y, y' + \delta y')$$

$$= F(x, y, y') + \left\{\frac{\partial F}{\partial y}\delta y + \frac{\partial F}{\partial y'}\delta y'\right\} + O(\delta^2) \tag{5.20}$$

where $O(\delta^2)$ refers to the terms containing $(\delta y)^2$, $(\delta y')^2$, $(\delta y)^3$, etc. which are of negligibly higher order.

In view of several terms in the Taylor's series, let us define the total variation of F as

$$\delta^T F = F(x, y + \delta y, y' + \delta y') - F(x, y, y') = \left\{\frac{\partial F}{\partial y}\delta y + \frac{\partial F}{\partial y'}\delta y'\right\} + O(\delta^2) \tag{5.21}$$

We can write the above as

$$\delta^T F = F(x, y + \delta y, y' + \delta y') - F(x, y, y') = \delta^1 F + O(\delta^2) \tag{5.21a}$$

where

$$\delta^1 F = \left\{\frac{\partial F}{\partial y}\delta y + \frac{\partial F}{\partial y'}\delta y'\right\} \tag{5.21b}$$

is the first variation.

Equation (5.11) through variational operation can be obtained by integrating (5.21) as

$$\delta^T I = \tilde{I} - I$$

$$= \int_{x_1}^{x_2} F(x, y + \delta y, y' + \delta y')dx - \int_{x_1}^{x_2} F(x, y, y')dx$$

$$= \int_{x_1}^{x_2} \left\{ \frac{\partial F}{\partial y} \delta y + \frac{\partial F}{\partial y'} \delta y' \right\} dx + O(\delta^2) \tag{5.22}$$

In (5.22) above, we recognize

$$\delta^1 I = \int_{x_1} \left\{ \frac{\partial F}{\partial y} \delta y + \frac{\partial F}{\partial y'} \delta y' \right\} dx \tag{5.23}$$

Similar to the operation in (5.15), integrating by parts the second term in the above, we get

$$\delta^1 I = \int_{x_1}^{x_2} \left\{ \frac{\partial F}{\partial y} - \frac{d}{dx} \frac{\partial F}{\partial y'} \right\} \delta y dx + \left[\frac{\partial F}{\partial y'} \delta y \right]_{x_1}^{x_2} \tag{5.24}$$

Since $\delta y = 0$ at x_1 and x_2 the above equation (5.24) becomes

$$\delta^1 I = \int_{x_1}^{x_2} \left\{ \frac{\partial F}{\partial y} - \frac{d}{dx} \frac{\partial F}{\partial y'} \right\} \delta y dx \tag{5.25}$$

Equation (5.22) can now be written as

$$\delta^T I = \delta^1 I + O(\delta^2)$$

$$= \int_{x_1}^{x_2} \left\{ \frac{\partial F}{\partial y} - \frac{d}{dx} \frac{\partial F}{\partial y'} \right\} \delta y dx + O(\delta^2) \tag{5.22a}$$

As before we recognize that

$$\delta^1 I = \int_{x_1}^{x_2} \left\{ \frac{\partial F}{\partial y} - \frac{d}{dx} \frac{\partial F}{\partial y'} \right\} \delta y dx = 0 \tag{5.25a}$$

as the condition for $y(x)$ to be the extremizing path for the functional in (5.7) to be an extremum. For convenience, we here afterwards drop the superscript 1 for the first variation, as this is the only variation that we are interested in our present analysis.

Hence, for

$$I = \int_{x_1}^{x_2} F(x, y, y')dx \tag{5.7}$$

to have an extremum value, the extremizing path $y(x)$ can be obtained by simply taking the following variation and making it to zero.

$$\delta I = \delta \int_{x_1}^{x_2} F(x, y, y')dx = 0 \tag{5.26}$$

In view of (5.22a), we have once again the Euler-Lagrange equations to determine the extremizing path.

$$\frac{\partial F}{\partial y} - \frac{d}{dx}\frac{\partial F}{\partial y'} = 0 \tag{5.16}$$

Consider the simple undamped Duffing oscillator with mass *m* and spring force, F_s for a displacement x given by

$$F_s = \alpha x + \beta x^3$$

The kinetic energy at any instant for a displacement x, is

$$T = \frac{1}{2} m \dot{x}^2$$

The potential energy stored in the spring is

$$U = \int_0^x (\alpha x + \beta x^3) dx$$

$$= \frac{1}{2}\alpha x^2 + \frac{1}{4}\beta x^4$$

Hamilton's principle (*see* section 5.10 for a formal derivation) for this problem can be stated as

$$\delta \int_{t_1}^{t_2} (T - U) dt = 0$$

Therefore

$$\delta \int_{t_1}^{t_2} \frac{1}{2} m \dot{x}^2 - \left(\frac{1}{2}\alpha x^2 + \frac{1}{4}\beta x^4 \right) dt = 0 \tag{a}$$

The first term in (a) above is

$$\delta \int_{t_1}^{t_2} \frac{1}{2} m \dot{x}^2 dt = \int_{t_1}^{t_2} m \dot{x} \delta \dot{x} \, dt$$

$$= m \dot{x} \delta x \, \Big|_{t_1}^{t_2} - \int_{t_1}^{t_2} m \ddot{x} \delta x \, dt$$

Therefore, equation (a) is

$$m \dot{x} \delta x \Big|_{t_1}^{t_2} - \int_{t_1}^{t_2} m \ddot{x} \delta x \, dt - \int_{t_1}^{t_2} (\alpha x \delta x + \beta x^3 \delta x) dt = 0 \tag{b}$$

Since

$$\delta x = 0 \quad \text{at} \quad t_1 \quad \text{and} \quad t_2$$

equation (b) becomes

$$\int_{t_1}^{t_2} (m \ddot{x} + \alpha x + \beta x^3) \delta x \, dt = 0 \tag{c}$$

For (c) to be true, the equation of motion is given by

$$m\ddot{x} + \alpha x + \beta x^3 = 0$$

We could have arrived at the same equation by using Euler Lagrange equation (5.16), rather than employing the delta operator from the first principles. From (a) we write the functional F as

$$F(x,\dot{x}) = \frac{1}{2}m\dot{x}^2 - \left(\frac{1}{2}\alpha x^2 + \frac{1}{4}\beta x^4\right)$$

The Euler-Lagrangian equation for this case is

$$\frac{\partial F}{\partial x} - \frac{d}{dt}\frac{\partial F}{\partial \dot{x}} = 0$$

Therefore

$$\frac{\partial F}{\partial x} = -(\alpha x + \beta x^3)$$

$$\frac{d}{dt}\frac{\partial F}{\partial \dot{x}} = m\ddot{x} \qquad (d)$$

Substituting equation (d) above in Euler-Lagrangian equation, we get

$$m\ddot{x} + \alpha x + \beta x^3 = 0$$

To get more acquaintance with the δ operator, consider another example of a Porter governor, shown in Fig. 5.4. Each arm of this governor is of length l and can be considered massless when compared with the two heavy masses m, they carry. The mass of the sleeve is M. Mass moment of inertia of all parts excepting that of the two balls is I. For a rotational speed Ω rad/s, the arms make an angle α from the vertical. From this equilibrium position, the governor balls can oscillate due to transient excitations from the sleeve. We see that writing the differential equation of motion for this system is not such a straight forward matter using the conventional Newtonian principles. We find that Hamilton's principle and use of δ operator becomes very handy in exactly obtaining the equation of motion. This is illustrated below.

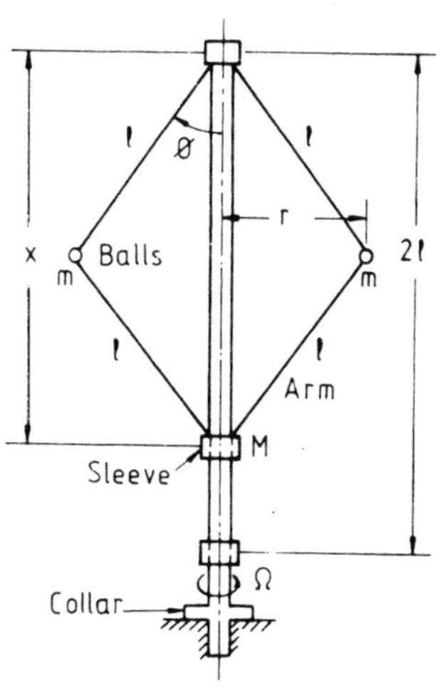

Fig. 5.4 Porter governor

Let ϕ represents the displaced position from the mean equilibrium position α of the arms. The angular oscillation of the arm at the instant shown is θ. Thus

$$\phi = \alpha + \theta$$

Let the sleeve, for the position shown, occupy a position at a distance x from the top of the arm joint. Then the kinetic energy T of the moving parts is

$$T = \frac{1}{2}I\Omega^2 + 2\times\frac{1}{2}m\Omega^2 r^2 + 2\times\frac{1}{2}ml^2\dot{\phi}^2 + \frac{1}{2}M\dot{x}^2 \qquad (e)$$

In equation (e) above

$$r = l\sin\phi$$

$$x = 2l\cos\phi$$

$$\dot{x} = -2l\sin\phi\dot{\phi}$$

and therefore

$$T = \frac{1}{2}\Omega^2(I + 2ml^2\sin^2\phi) + \dot{\phi}^2 l^2(m + 2M\sin^2\phi) \qquad (f)$$

The potential energy of the system with reference to zero speed position is

$$V = Mg(2l - x) + 2mg\left(l - \frac{x}{2}\right)$$

$$= 2lg(M + m)(1 - \cos\phi) \qquad (g)$$

From Hamilton's principle, we get

$$\delta\int_{t_1}^{t_2}(T - V)dt = \delta\int_{t_1}^{t_2}\Bigg[\frac{1}{2}\Omega^2(I + 2ml^2\sin^2\phi)$$

$$+ \dot{\phi}^2 l^2(m + 2M\sin^2\phi) - 2lg(M + m)(1 - \cos\phi)\Bigg]dt = 0 \qquad (h)$$

Taking the variation of (h) above, the following can be obtained.

$$\int_{t_1}^{t_2}\Bigg[\frac{1}{2}\Omega^2 2ml^2 2\sin\phi\cos\phi\delta\phi$$

$$+ \dot{\phi}l^2(2m + 4M\sin^2\phi)\delta\dot{\phi} + 2Ml^2\dot{\phi}^2 2\sin\phi\delta\phi - 2lg(M + m)\sin\phi\delta\phi\Bigg]dt = 0 \qquad (i)$$

We use integration by parts for the $\delta\dot{\phi}$ term so that all the terms can be grouped together with $\delta\phi$ term in the above equation.

$$\int_{t_1}^{t_2}\dot{\phi}l^2(2m + 4M\sin^2\phi)\delta\dot{\phi}dt = \dot{\phi}l^2(2m + 4M\sin^2\phi)\delta\phi\Big|_{t_1}^{t_2}$$

$$-\int_{t_1}^{t_2}\{2l^2(m + 2M\sin^2\phi)\ddot{\phi} + 4Ml^2\sin 2\phi\dot{\phi}^2\}\delta\phi dt \qquad (j)$$

Since

$$\delta\phi = 0 \quad \text{at} \quad t_1 \quad \text{and} \quad t_2$$

equation (j) becomes

$$\int_{t_1}^{t_2} [2\Omega^2 m l^2 \sin\phi\cos\phi - 2l^2(m + 2M\sin^2\phi)\ddot{\phi}$$

$$- 4Ml^2 \sin 2\phi\dot{\phi}^2 + 4Ml^2\dot{\phi}^2 \sin\phi - 2lg(M + m)\sin\phi]\,\delta\phi\, dt = 0$$

Therefore, the equation of motion is

$$(m + 2M\sin^2\phi)\ddot{\phi} - \sin\phi\cos\phi(m\Omega^2 - 4M\dot{\phi}^2) + (M + m)\frac{g}{l}\sin\phi = 0$$

It is left to the reader to obtain the above expression by Euler-Lagrangian equation.

5.4 CASE OF SEVERAL DEPENDENT VARIABLES

In the previous section we studied the first variation with only one dependent variable, as in the case of single degree freedom systems, with the displacement dependent on a single independent variable, time. We will now consider the case where the functional may have any number of dependent variables, with still one independent variable. Such a case corresponds to the study of multi-degree of freedom systems. Let these functions be denoted by $x_1, x_2, x_3, \ldots x_n$ and their derivatives with respect to the independent variable, viz., time. Let the functional to be extremized be given by

$$I = \int_{t_1}^{t_2} F(x_1, x_2, x_3, \ldots x_n, \dot{x}_1, \dot{x}_2, \dot{x}_3, \ldots \dot{x}_n, t)\,dt \tag{5.27}$$

The problem is to find $x_1, x_2, x_3, \ldots x_n$, which are at least twice differentiable, extremizing the functional in (5.27) above.

As before, let us assume that the admissible paths for the extremized path are given by

$$\tilde{x}_1(t) = x_1(t) + \epsilon\eta_1(t)$$

$$\tilde{x}_2(t) = x_2(t) + \epsilon\eta_2(t)$$

$$\ldots$$

$$\tilde{x}_n(t) = x_n(t) + \epsilon\eta_n(t) \tag{5.28}$$

where $\eta_1(t), \eta_2(t), \ldots \eta_n(t)$ are arbitrary functions, which are twice differentiable and that they are zero at the end points t_1 and t_2. We can now write the functional over the varied paths as

$$\tilde{I} = \int_{t_1}^{t_2} F(\tilde{x}_1, \tilde{x}_2, \tilde{x}_3, \ldots \tilde{x}_n, \dot{\tilde{x}}_1, \dot{\tilde{x}}_2, \dot{\tilde{x}}_3, \ldots \dot{\tilde{x}}_n, t)\,dt \tag{5.29}$$

Let us recall equation (5.11) here

$$\tilde{I} - (\tilde{I}\}_{\epsilon=0} =$$

$$\tilde{I} - I = \left\{\frac{d\tilde{I}}{d\epsilon}\right\}_{\epsilon=0} \epsilon + \left\{\frac{d^2\tilde{I}}{d\epsilon^2}\right\}_{\epsilon=0} \frac{\epsilon^2}{2!} + \ldots \tag{5.11}$$

and note that

$$\left\{\frac{d\bar{I}}{d\epsilon}\right\}_{\epsilon=0} = 0$$

for $x_i\ (t)$ to be the extremal paths. Hence

$$\left[\int_{t_1}^{t_2}\left\{\frac{\partial F}{\partial \bar{x}_1}\eta_1 + \frac{\partial F}{\partial \bar{x}_2}\eta_2 + \ldots + \frac{\partial F}{\partial \bar{x}_n}\eta_n\right.\right.$$

$$\left.\left. + \frac{\partial F}{\partial \dot{\bar{x}}_1}\dot{\eta}_1 + \frac{\partial F}{\partial \dot{\bar{x}}_2}\dot{\eta}_2 + \ldots + \frac{\partial F}{\partial \dot{\bar{x}}_n}\dot{\eta}_n\right\}dt\right]_{\epsilon=0} = 0$$

i.e.,

$$\int_{t_1}^{t_2}\left\{\frac{\partial F}{\partial x_1}\eta_1 + \frac{\partial F}{\partial x_2}\eta_2 + \ldots + \frac{\partial F}{\partial x_n}\eta_n\right.$$

$$\left. + \frac{\partial F}{\partial \dot{x}_1}\dot{\eta}_1 + \frac{\partial F}{\partial \dot{x}_2}\dot{\eta}_2 + \ldots + \frac{\partial F}{\partial \dot{x}_n}\dot{\eta}_n\right\}dt = 0 \tag{5.30}$$

Let us choose $\eta_1\ (t)$ to be arbitrary and let all the rest $\eta\ (t)$ be equal to zero. Then, integrating by parts the $\dot{\eta}(t)$ terms in (5.30), we get

$$\int_{t_1}^{t_2}\left\{\frac{\partial F}{\partial x_1} - \frac{d}{dt}\frac{\partial F}{\partial \dot{x}_1}\right\}\eta_1 dt = 0 \tag{5.31a}$$

i.e.,

$$\frac{\partial F}{\partial x_1} - \frac{d}{dt}\frac{\partial F}{\partial \dot{x}_1} = 0 \tag{5.32a}$$

Now, let us choose $\eta_2\ (t)$ to be arbitrary and let all the rest $\eta\ (t)$ be equal to zero. Then, we obtain

$$\int_{t_1}^{t_2}\left\{\frac{\partial F}{\partial x_2} - \frac{d}{dt}\frac{\partial F}{\partial \dot{x}_2}\right\}\eta_2 dt = 0 \tag{5.31b}$$

i.e.,

$$\frac{\partial F}{\partial x_2} - \frac{d}{dt}\frac{\partial F}{\partial \dot{x}_2} = 0 \tag{5.32b}$$

etc. We can combine all equations (5.32) as

$$\frac{\partial F}{\partial x_i} - \frac{d}{dt}\frac{\partial F}{\partial \dot{x}_i} = 0 \quad i = 1, 2 \ldots n \tag{5.32}$$

Equations (5.32) are Euler-Lagrangian equations for the case of several dependent variables. We can use them to derive the equations of motion or directly employ the δ operator. Let us consider the example of an n-degree freedom system shown in Fig. 5.5.

The total kinetic energy of the system is

$$T = \frac{1}{2}(M\dot{x}_1^2 + M\dot{x}_2^2 + \ldots + M\dot{x}_n^2)$$

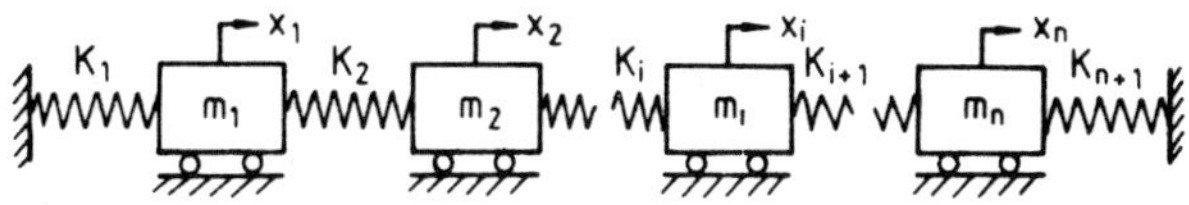

Fig. 5.5 Multi degree system

The potential energy in all the springs is

$$V = \frac{1}{2}\{k_1x_1^2 + k_2(x_2 - x_1)^2 + \ldots + k_n(x_n - x_{n-1})^2 + k_{n+1}x_n^2\}$$

From Hamilton's principle, we get

$$\delta \int_{t_1}^{t_2} \left[\frac{1}{2}(M\dot{x}_1^2 + M\dot{x}_2^2 + \ldots + M\dot{x}_n^2) \right.$$

$$\left. -\frac{1}{2}\{k_1x_1^2 + k_2(x_2 - x_1)^2 + \ldots + k_n(x_n - x_{n-1})^2 + k_{n+1}x_n^2\} \right] dt = 0$$

Taking the variation

$$\int_{t_1}^{t_2} [M\dot{x}_1\delta\dot{x}_1 + M\dot{x}_2\delta\dot{x}_2 + \ldots + M\dot{x}_n\delta\dot{x}_n$$

$$- \{k_1x_1\delta x_1 + k_2(x_2 - x_1)\delta(x_2 - x_1) + \ldots$$

$$+ k_n(x_n - x_{n-1})\delta(x_n - x_{n-1}) + k_{n+1}x_n\delta x_n\}]\, dt = 0$$

Integrating by parts

$$\Sigma M\dot{x}_i\delta x_i|_1^2 - \int_{t_1}^{t_2} \Sigma M\ddot{x}_i\delta x_i$$

$$+ \{k_1x_1 + k_2(x_1 - x_2)\}\,\delta x_1 + \{k_2(x_2 - x_1) + k_3(x_2 - x_3)\}\,\delta x_2 + \ldots$$

$$+ \{k_n(x_n - x_{n-1}) + k_{n+1}x_n\}\,\delta x_n = 0$$

Noting that $\delta x_i = 0$ at t_1 and t_2 and separating $\delta x_1, \delta x_2 \ldots$ terms, we get the following differential equations.

$$M\ddot{x}_1 + k_1x_1 + k_2(x_1 - x_2) = 0$$

$$M\ddot{x}_2 + k_2(x_2 - x_1) + k_3(x_2 - x_3) = 0$$

$$M\ddot{x}_n + k_n(x_n - x_{n-1}) + k_{n+1}x_n = 0$$

One can apply the Euler-Lagrangian equations in equation (5.32) to the functional

$$L = \left[\frac{1}{2}(M\dot{x}_1^2 + M\dot{x}_2^2 + \ldots + M\dot{x}_n^2) \right.$$

$$\left. -\frac{1}{2}\{k_1 x_1^2 + k_2(x_2 - x_1)^2 + \ldots + k_n(x_n - x_{n-1})^2 + k_{n+1}x_n^2\} \right]$$

and obtain the same differential equations as obtained by the δ operator method.

5.5 FUNCTIONALS WITH HIGHER ORDER DERIVATIVES

So far we considered functionals with first order derivatives. We often come across functionals with second order derivatives and higher order while dealing with continuous systems. Let us take the case of the following functional with second order derivative.

$$I = \int_{x_1}^{x_2} F(x, y, y', y'')dx \tag{5.33}$$

Let

$$\tilde{y} = y + \epsilon\eta$$

then

$$\tilde{I} = \int_{x_1}^{x_2} F(x, \tilde{y}, \tilde{y}', \tilde{y}'')dx \tag{5.34}$$

For $y(x)$ to be an extremizing path for the functional in (5.33) to be an extremum, we have

$$\left\{\frac{d\tilde{I}}{d\epsilon}\right\}_{\epsilon=0} = \left\{\frac{d}{d\epsilon}\int_{x_1}^{x_2} F(x, \tilde{y}, \tilde{y}', \tilde{y}'')dx\right\}_{\epsilon=0} = 0 \tag{5.35}$$

Equation (5.35) above gives

$$\left\{\int_{x_1}^{x_2}\left(\frac{\partial F}{\partial \tilde{y}}\eta + \frac{\partial F}{\partial \tilde{y}'}\eta' + \frac{\partial F}{\partial \tilde{y}''}\eta''\right)dx\right\}_{\epsilon=0} = 0$$

i.e.,

$$\int_{x_1}^{x_2}\left(\frac{\partial F}{\partial y}\eta + \frac{\partial F}{\partial y'}\eta' + \frac{\partial F}{\partial y''}\eta''\right)dx = 0 \tag{5.36}$$

Using integration by parts

$$\int_{x_1}^{x_2}\frac{\partial F}{\partial y'}\eta' dx = \frac{\partial F}{\partial y'}\eta \Big|_{x_1}^{x_2} - \int_{x_1}^{x_2}\frac{d}{dx}\left(\frac{\partial F}{\partial y'}\right)\eta dx$$

$$\int_{x_1}^{x_2}\frac{\partial F}{\partial y''}\eta'' dx = \frac{\partial F}{\partial y''}\eta' \Big|_{x_1}^{x_2} - \int_{x_1}^{x_2}\frac{d}{dx}\left(\frac{\partial F}{\partial y''}\right)\eta' dx$$

$$= \frac{\partial F}{\partial y''}\eta' \Big|_{x_1}^{x_2} - \frac{d}{dx}\left(\frac{\partial F}{\partial y''}\right)\eta \Big|_{x_1}^{x_2} + \int_{x_1}^{x_2}\frac{d^2}{dx^2}\left(\frac{\partial F}{\partial y''}\right)\eta dx \tag{a}$$

Using (*a*) above in (5.36), we get

$$\int_{x_1}^{x_2}\left[\frac{d^2}{dx^2}\left(\frac{\partial F}{\partial y''}\right)-\frac{d}{dx}\left(\frac{\partial F}{\partial y'}\right)+\frac{\partial F}{\partial y}\right]\eta dx$$

$$+\frac{\partial F}{\partial y''}\eta']_{x_1}^{x_2}+\left\{-\frac{d}{dx}\left(\frac{\partial F}{\partial y''}\right)+\frac{\partial F}{\partial y'}\right\}\eta]_{x_1}^{x_2}=0 \tag{5.37}$$

For

$$\eta \text{ and } \eta'=0 \text{ at } x_1 \text{ and } x_2 \tag{5.38}$$

equation (5.37) becomes

$$\int_{x_1}^{x_2}\left[\frac{d^2}{dx^2}\left(\frac{\partial F}{\partial y''}\right)-\frac{d}{dx}\left(\frac{\partial F}{\partial y'}\right)+\frac{\partial F}{\partial y}\right]\eta dx=0$$

i.e., the Euler-Lagrangian equation for this case is

$$\frac{d^2}{dx^2}\left(\frac{\partial F}{\partial y''}\right)-\frac{d}{dx}\left(\frac{\partial F}{\partial y'}\right)+\frac{\partial F}{\partial y}=0 \tag{5.39}$$

5.6 BOUNDARY CONDITIONS

In the sections above, we so far considered the extremizing process of the functional like I, by specifying the function $y\ (x)$ to go through the points x_1, y_1 and x_2, y_2. All the varied paths over the admissible path are also made to go through these points x_1, y_1 and x_2, y_2. These conditions specifying the value of the dependent variable y at x_1 and x_2, are called kinematic boundary conditions. To obtain an extremum of the functional like I, it is strictly not necessary for us to go through the specification of such kinematic boundary conditions; it is possible to achieve the extremization by alternate paths. Let us consider the functional in equation (5.33) again,

$$I=\int_{x_1}^{x_2}F(x,y,y',y'')dx \tag{5.33}$$

We determined the conditions for the first variation of the above to vanish, which are given by the equation (5.37)

$$\int_{x_1}^{x_2}\left[\frac{d^2}{dx^2}\left(\frac{\partial F}{\partial y''}\right)-\frac{d}{dx}\left(\frac{\partial F}{\partial y'}\right)+\frac{\partial F}{\partial y}\right]\eta dx$$

$$+\frac{\partial F}{\partial y''}\eta'\Big|_{x_1}^{x_2}+\left\{-\frac{d}{dx}\left(\frac{\partial F}{\partial y''}\right)+\frac{\partial F}{\partial y'}\right\}\eta\Big|_{x_1}^{x_2}=0 \tag{5.37}$$

It is not necessary for us to specify the values in equation (5.9), which leads to the Euler-Lagrangian equation. We can achieve the same result by specifying alternately, the following

$$\frac{\partial F}{\partial y''}\Big|_{x_1}^{x_2}=0$$

and

$$\left\{-\frac{d}{dx}\left(\frac{\partial F}{\partial y''}\right)+\frac{\partial F}{\partial y'}\right\}\Big|_{x_1}^{x_2}=0 \tag{5.40}$$

Therefore, we have multiple choices of specifying the end conditions to make equation (5.37) reduce to Euler-Lagrangian equation. At $x = x_1$ and $x = x_2$, we can satisfy the following

$$\text{Specify } y' \quad \text{i.e.,} \quad \eta' = 0 \quad \text{or} \quad \frac{\partial F}{\partial y''} = 0$$

$$\text{Specify } y \quad \text{i.e.,} \quad \eta = 0 \quad \text{or} \quad -\frac{d}{dx}\left(\frac{\partial F}{\partial y''}\right) + \frac{\partial F}{\partial y'} = 0 \tag{5.41}$$

The first set of conditions in the above are the kinematic conditions. It is generally possible to determine them from simple inspection of a system under investigation. The second set of equations are not always easy to determine from other than variational principles. These conditions which come out naturally from the extremizing process are called natural conditions. We therefore can extremize the functional in (5.33) by the Euler-Lagrangian equation (5.39) subject to the boundary conditions, kinematic or natural as the case may be. An added advantage of the variational principles is to derive the boundary conditions of a system besides setting up of its equations of motion.

5.7 FUNCTIONAL WITH TWO INDEPENDENT VARIABLES

Let us now consider a functional with two independent variables, one that of time and the other a space coordinate, as we come across in vibrating beams. Let the functional be defined by

$$I = \int_{t_1}^{t_2} \int_{x_1}^{x_2} F(x, t, y, \dot{y}, y', y'') dx\, dt \tag{5.42}$$

where x and t are the independent variables and y is the dependent variable.

Let

$$\bar{y}(x, t) = y(x, t) + \in \eta(x, t) \tag{5.43}$$

be the varied path over the admissible path $y\,(x, t)$ for extremizing the functional in (5.42).

We make the function η satisfy the following conditions.

$$\eta(x, t_1) = \eta(x, t_2) = 0 \tag{5.44}$$

For an extremum of the functional, we know that

$$\left\{\frac{d\bar{I}}{d\in}\right\}_{\in=0} = \left\{\frac{d}{d\in}\int_{t_1}^{t_2}\int_{x_1}^{x_2} F(x, t, \bar{y}, \dot{\bar{y}}, \bar{y}', \bar{y}'') dx\, dt\right\}_{\in=0} = 0 \tag{5.45}$$

The above equation (5.45) gives

$$\left\{\int_{x_1}^{x_2}\int_{t_1}^{t_2}\left(\frac{\partial F}{\partial \dot{\bar{y}}}\dot{\eta} + \frac{\partial F}{\partial \bar{y}}\eta\right.\right.$$

$$\left.\left. + \frac{\partial F}{\partial \bar{y}'}\eta' + \frac{\partial F}{\partial \bar{y}''}\eta''\right) dx\, dt\right\}_{\in=0} = 0$$

i.e.,

$$\int_{x_1}^{x_2}\int_{t_1}^{t_2}\left(\frac{\partial F}{\partial \dot{y}}\dot{\eta}+\frac{\partial F}{\partial y}\eta+\frac{\partial F}{\partial y'}\eta'+\frac{\partial F}{\partial y''}\eta''\right)dx\,dt=0 \tag{5.46}$$

The last three terms have been considered before (*see* equation 5.37). The first term in the above can be integrated by parts with respect to t to give

$$\int_{x_1}^{x_2}\int_{t_1}^{t_2}\frac{\partial F}{\partial \dot{y}}\dot{\eta}\,dx\,dt=\int_{x_1}^{x_2}\eta\frac{\partial F}{\partial \dot{y}}\Big|_{t_1}^{t_2}dx$$

$$-\int_{x_1}^{x_2}\int_{t_1}^{t_2}\frac{d}{dt}\left(\frac{\partial F}{\partial \dot{y}}\right)\eta\,dx\,dt \tag{5.47}$$

In view of (5.44), equation (5.47) simplifies to

$$\int_{x_1}^{x_2}\int_{t_1}^{t_2}\frac{\partial F}{\partial \dot{y}}\dot{\eta}\,dx\,dt=-\int_{x_1}^{x_2}\int_{t_1}^{t_2}\frac{d}{dt}\left(\frac{\partial F}{\partial \dot{y}}\right)\eta\,dx\,dt \tag{5.47a}$$

With the help of (5.37) and (5.47*a*), equation (5.46) gives

$$\int_{t_1}^{t_2}\Big[\int_{x_1}^{x_2}\left[\frac{d^2}{dx^2}\left(\frac{\partial F}{\partial y''}\right)-\frac{d}{dx}\left(\frac{\partial F}{\partial y'}\right)+\frac{\partial F}{\partial y}-\frac{d}{dt}\left(\frac{\partial F}{\partial \dot{y}}\right)\right]\eta\,dx$$

$$+\frac{\partial F}{\partial y''}\eta'\Big|_{x_1}^{x_2}+\left\{-\frac{d}{dx}\left(\frac{\partial F}{\partial y''}\right)+\frac{\partial F}{\partial y'}\right\}\eta\Big|_{x_1}^{x_2}\Big]dt=0 \tag{5.48}$$

From the above equation (5.48), we obtain the Euler-Lagrangian equation for two independent variables.

$$\frac{d^2}{dx^2}\left(\frac{\partial F}{\partial y''}\right)-\frac{d}{dx}\left(\frac{\partial F}{\partial y'}\right)+\frac{\partial F}{\partial y}-\frac{d}{dt}\left(\frac{\partial F}{\partial \dot{y}}\right)=0 \tag{5.49}$$

The boundary conditions are

$$\frac{\partial F}{\partial y''}\eta'\Big|_{x_1}^{x_2}+\left\{-\frac{d}{dx}\left(\frac{\partial F}{\partial y''}\right)+\frac{\partial F}{\partial y'}\right\}\eta\Big|_{x_1}^{x_2}=0 \tag{5.50}$$

i.e.,

$$y' \text{ prescribed or } \frac{\partial F}{\partial y''}=0$$

$$y \text{ prescribed or } -\frac{d}{dx}\left(\frac{\partial F}{\partial y''}\right)+\frac{\partial F}{\partial y'}=0 \tag{5.51}$$

Let us consider the case of a simple beam according to Euler-Bernoulli theory. The kinetic energy is given by

$$T=\int_0^L\frac{1}{2}m\dot{y}^2dx$$

The strain energy is

$$V=\int_0^L\frac{1}{2}EIy''^2dx$$

where m is the mass per unit length of the beam, EI, the flexural rigidity and L is the length of the beam.

From Hamilton's principle,

$$\delta \int_{t_1}^{t_2} \int_0^L \left(\frac{1}{2} m \dot{y}^2 - \frac{1}{2} EI y''^2 \right) dx\, dt = 0$$

Hence, the functional F is given by

$$F = \frac{1}{2} m \dot{y}^2 - \frac{1}{2} EI y''^2$$

We now determine

$$\frac{\partial F}{\partial y''} = -EI y''$$

$$\frac{d^2}{dx^2}\left(\frac{\partial F}{\partial y''} \right) = -[EI y'']''$$

$$\frac{\partial F}{\partial \dot{y}} = m \dot{y}$$

$$-\frac{d}{dt}\left(\frac{\partial F}{\partial \dot{y}} \right) = -m \ddot{y}$$

Therefore, the equation of motion is

$$[EI y'']'' + m \ddot{y} = 0 \tag{a}$$

The boundary conditions from (5.51) are

$$y' \text{ prescribed} \quad \text{or} \quad -EI y'' = 0$$

$$y \text{ prescribed} \quad \text{or} \quad (EI y'')' = 0 \tag{b}$$

Instead of using the Euler-Lagrangian equation, we could employ the δ operator.

$$\delta \int_{t_1}^{t_2} \int_0^L \left(\frac{1}{2} m \dot{y}^2 - \frac{1}{2} EI y''^2 \right) dx\, dt = 0$$

i.e.,

$$\int_{t_1}^{t_2} \int_0^L (m \dot{y} \delta \dot{y} - EI y'' \delta y'') dx\, dt = 0 \tag{c}$$

Integrating by parts with respect to time, the first term in (c) above is

$$\int_0^L \left[m \dot{y} \delta y \, \Big|_{t_1}^{t_2} \right] dx - \int_{t_1}^{t_2} \int_0^L m \ddot{y} \delta y \, dx\, dt$$

$$= -\int_{t_1}^{t_2} \int_0^L m \ddot{y} \delta y \, dx\, dt \tag{d}$$

For the second term, we integrate by parts with respect to x, to obtain

$$\int_{t_1}^{t_2} -EIy''\delta y'\Big|_0^L dt + \int_{t_1}^{t_2}\int_0^L \{EIy''\}'\delta y' dx\,dt$$

i.e.,

$$\int_{t_1}^{t_2} [-EIy''\delta y'\Big|_0^L + (EIy'')'\delta y\Big|_0^L]dt - \int_{t_1}^{t_2}\int_0^L \{EIy''\}''\delta y\,dx\,dt \qquad (e)$$

It can be readily seen that equation (e) above leads to the same differential equation (a) and boundary conditions (b).

5.8 FUNCTIONALS CONTAINING INTEGRALS

So far, we considered functionals which contained only derivatives of functions with respect to one or two independent variables. There are examples where we come across functionals which contain integrals, a typical case being of a rotating beam. We will only state here the functional without deriving this for the time being to illustrate the process of extremizing such a functional. In the example of beam of previous section, let it be mounted on a disk of radius R and rotate at an angular velocity ω rad/s. The Hamilton's principle gives for this case

$$\delta\int_{t_1}^{t_2}\int_0^L\left(\frac{1}{2}m\dot{y}^2 - \frac{1}{2}EIy''^2 + \frac{1}{2}m\omega^2y^2 - \frac{1}{2}m\omega^2(R+x)\int_0^x y'^2dx\right)dx\,dt = 0 \qquad (5.52)$$

We know how to handle the first three terms in the functional above. For the fourth term containing the integral, we will bring the integrand containing the derivative of y outside the integral so that the variational process can be carried out in the usual way. To facilitate this let us consider the following.

$$\frac{d}{dx}\left\{\int_x^L f_1(x)dx\int_0^x f_2(x)dx\right\} = f_2(x)\int_x^L f_1(x)dx + \int_0^x f_2(x)dx\frac{d}{dx}\left\{\int_x^L f_1(x)dx\right\} \qquad (5.53)$$

with the condition that

$$\int_x^L f_1(x)dx = 0 \quad \text{at} \quad x = L$$

$$\int_0^x f_2(x)dx = 0 \quad \text{at} \quad x = 0 \qquad (5.54)$$

In the equation (5.53) above, we can rewrite the last term using the following relationship:

$$\frac{d}{dx}\int_x^L \psi(x)dx = -\psi(x) \qquad (5.55)$$

Hence, equation (5.53) becomes

$$\frac{d}{dx}\left\{\int_x^L f_1(x)dx \int_0^x f_2(x)dx\right\} = f_2(x)\int_x^L f_1(x)dx$$

$$-f_1(x)\int_0^x f_2(x)dx \tag{5.56}$$

Integrating both sides of the above equation (5.56), we get

$$\int_x^L f_1(x)dx \int_0^x f_2(x)dx\}_0^L = \int_0^L\left[f_2(x)\int_x^L f_1(x)dx\right]dx$$

$$-\int_0^L\left[f_1(x)\int_0^x f_2(x)dx\right]dx \tag{5.57}$$

The left hand side of the above equation (5.57) is zero in view of the conditions (5.54) and therefore

$$\int_0^L\left[f_1(x)\int_0^x f_2(x)dx\right]dx = \int_0^L\left[f_2(x)\int_x^L f_1(x)dx\right]dx \tag{5.58}$$

Using the above transformation, we can write the functional for the rotating beam now in a form suitable for taking the variation

$$\delta\int_{t_1}^{t_2}\int_0^L\left(\frac{1}{2}m\dot{y}^2 - \frac{1}{2}EIy''^2 + \frac{1}{2}m\omega^2 y^2\right.$$

$$\left.-y'^2\int_x^L \frac{1}{2}m\omega^2(R+x)dx\right)dx\,dt = 0 \tag{5.59}$$

Taking the variation in the above

$$\int_{t_1}^{t_2}\int_0^L\left[m\dot{y}\delta\dot{y} - EIy''\delta y'' + m\omega^2 y\delta y\right.$$

$$\left.-y'\delta y'\int_x^L m\omega^2(R+x)dx\right]dx\,dt = 0 \tag{5.60}$$

Integrating by parts with respect to time, the first term in (5.60) gives

$$\int_0^L m\dot{y}\delta y\,|_{t_1}^{t_2}\,dx - \int_{t_1}^{t_2}\int_0^L m\ddot{y}\delta y\,dx\,dt \tag{5.61a}$$

Integrating by parts with respect to x, the second term of (5.60) gives

$$\int_{t_1}^{t_2} -EIy''\delta y'\,|_0^L\,dt + \int_{t_1}^{t_2}\int_0^L [EIy'']'\delta y'\,dx\,dt$$

i.e.,

$$\int_{t_1}^{t_2}\{-EIy''\delta y'\,|_0^L + [EIy'']'\delta y\,|_0^L\}\,dt - \int_{t_1}^{t_2}\int_0^L [EIy'']''\delta y\,dx\,dt \tag{5.61b}$$

There is no need to change the third term of (5.60) as it is already reduced to δy. The fourth term of (5.60) is

$$-\int_{t_1}^{t_2}\left(y'\delta y \int_x^L m\omega^2(R+x)dx\Big|_0^L\right)dt$$

$$+\int_{t_1}^{t_2}\int_0^L \frac{d}{dx}\left[y'\int_x^L m\omega^2(R+x)dx\right]\delta y\,dx\,dt \tag{5.61c}$$

Taking the third term from equation (5.60) directly and using (5.61a), (5.61b) and (5.61c) the equation of motion is

$$[EIy'']'' - m\omega^2 y - \frac{d}{dx}\left[y'\int_x^L m\omega^2(R+x)dx\right] + m\ddot{y} = 0 \tag{5.62}$$

In (5.61c), the first term is zero for a rotating beam with either y or $y' = 0$ at $x = 0$ and the integral between limits x to L is also zero at $x = L$. Therefore, the boundary conditions for this case from (5.61b) are

$$y' \text{ prescribed or } -EIy'' = 0$$

$$y \text{ prescribed or } (EIy'')' = 0 \tag{5.63}$$

Thus, we see that the variational calculus is very useful in deriving the equations of motion and the boundary conditions for a given vibrating system. We will illustrate several cases in later chapters. We will consider in the remaining part of this chapter some very useful procedures for solving vibration problems, that emanate directly from variational calculus.

5.9 PRINCIPLE OF VIRTUAL WORK

During the first course of mechanics, we are introduced to the concept of virtual work while dealing with problems of particles. We gave a hypothetical displacement, called virtual displacement consistent with any constraints on the system and determined the work done by all the forces acting on the system with this displacement. This work was called virtual work and while determining this, we maintained the forces on the system to be constant. This is a very useful principle and plays a major role in the study of dynamic behaviour of continuous systems as well.

We can use the variational operator δ to denote the virtual displacement, virtual work, etc., thus, as δw, δW, corresponding to a displacement w and work W respectively. Any virtual displacement that we choose should be consistent with system constraints, e.g., a simply supported beam in Fig. 5.6, should have the virtual displacement satisfy the conditions at $x = 0$ and $x = L$. These are w and w'' both equal to zero at $x = 0$ and $x = L$. We can conveniently choose the following trigonometric function for this virtual displacement

$$\delta w = A \sin \pi \frac{x}{L} \tag{5.64}$$

Let us consider the case of a stationary beam under steady forces defined in Fig. 5.6. The forces consist of a distributed force on the neutral axis, and two concentrated forces for which the virtual work is given by

$$\delta W = \int_0^L p(x) A \sin \pi \frac{x}{L} dx + F_1 A \sin \pi \frac{x_1}{L} + F_2 A \sin \pi \frac{x_2}{L} \tag{5.65}$$

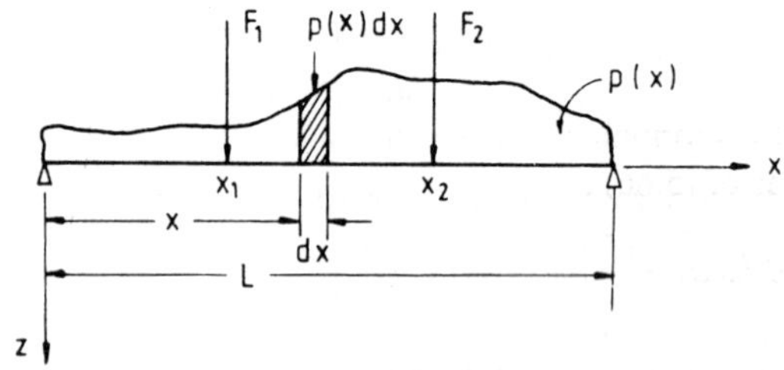

Fig. 5.6 Simply supported beam with external forces

We can generalize the above equation for any three dimensional body shown in Fig. 5.7. We first identify the force system. The forces can be considered in general of two types, one the body forces and the other the traction forces. The body force is a distributed force which acts directly on the distributed matter of the body and is represented in index notation by $B_i\,(x_1, x_2, x_3, t)$. This is generally expressed as a force per unit volume of the material of the body. On the contrary the traction forces are those applied on the surface of the body and therefore are also denoted as surface tractions. They are expressed as force per unit area and represented by

$$T_i^{(v)}(x_1, x_2, x_3, t)$$

where $\tilde{v}$ is the unit outward normal vector to the surface of the body.

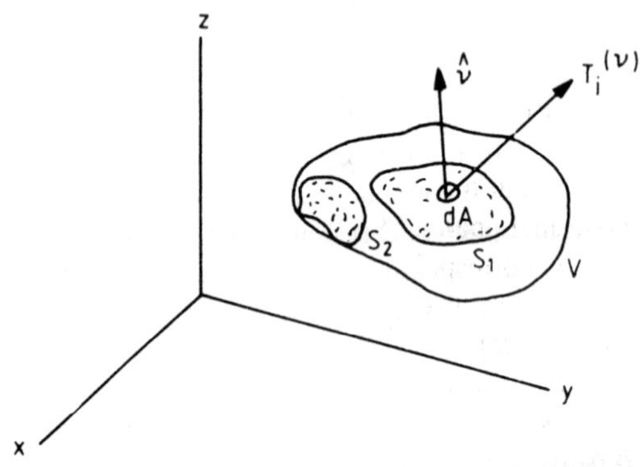

Fig. 5.7 A body of volume V with traction forces on S_1 and displacement field constraints on S_2

Let the body forces in Fig. 5.7 be $B_i\ (x_1, x_2, x_3)$. The traction forces are assumed to act on the surface S_1 and the displacement field is prescribed on the surface S_2. Since we are restricted to choose the virtual displacements to be consistent with the constraints on the surface S_2, the virtual work is done by the traction forces on the surface S_1 only. In general if we denote the whole surface area of the body as S, the virtual work of the traction forces can be evaluated over the whole area, noting that the work is zero on those areas of the surface, where there is no force acting. Then the virtual work is given by the following general expression.

$$\delta W = \iiint_V B_i \delta u_i dv + \iint_S T_i^{(v)} \delta u_i dA \tag{5.66}$$

In the above u_i is the displacement field of the body, consistent with the constraints on the surface S_2.

We use Cauchy's formula to replace the traction forces by the stresses and write the right hand side of (5.66) as

$$\iiint_V B_i \delta u_i dv + \iint_S T_i^{(v)} \delta u_i dA = \iiint_V B_i \delta u_i dv + \iint_S \tau_{ij} \nu_j \delta u_i dA$$

We can then use the divergence theorem to express the work of traction forces over the entire body volume so that we can combine the two integrals. Then

$$\iiint_V B_i \delta u_i dv + \iint_S T_i^{(v)} \delta u_i dA = \iiint_V B_i \delta u_i dv + \iiint_V (\tau_{ij} \delta u_i)_{,j} dv$$

Writing the second integral on the right hand side as

$$\iiint_V (\tau_{ij} \delta u_i)_{,j} dv = \iiint_V \tau_{ij,j} \delta u_i dv + \iiint_V \tau_{ij} (\delta u_i)_{,j} dv$$

we get

$$\iiint_V B_i \delta u_i dv + \iint_S T_i^{(v)} \delta u_i dA = \iiint_V (B_i + \tau_{ij,j}) \delta u_i dv$$

$$+ \iiint_V \tau_{ij} (\delta u_i)_{,j} dv \tag{5.67}$$

From the strain displacement relations, we have

$$\begin{aligned} (\delta u_i)_{,j} &= \delta(u_{i,j}) \\ &= \delta(\varepsilon_{ij} + \omega_{ij}) \\ &= \delta\varepsilon_{ij} + \delta\omega_{ij} \end{aligned} \tag{5.68}$$

Substituting the above in equation (5.67) and noting that rotation tensor ω_{ij} is skew-symmetric, i.e.,

$$\tau_{ij} \delta\omega_{ij} = 0$$

we get

$$\iiint_V B_i \delta u_i dv + \iint_S T_i^{(v)} \delta u_i dA = \iiint_V (B_i + \tau_{ij,j}) \delta u_i dv$$

$$+ \iiint_V \tau_{ij} \delta\varepsilon_{ij} dv \tag{5.69}$$

If there is static equilibrium, the following is true at any point in the body.

$$B_i + \tau_{ij,j} = 0$$

Hence, equation (5.69) is

$$\iiint_V B_i \delta u_i dv + \iint_S T_i^{(v)} \delta u_i dA = \iiint_V \tau_{ij} \delta\varepsilon_{ij} dv \tag{5.70}$$

The volume integral in the left hand side of the above equation (5.70) is the work done by the body forces on the virtual displacements. The surface integral is the work done by the traction forces on the surface of the body with the virtual displacement field. Together, these two integrals represent the external virtual work in the system. The right hand side of equation (5.70) represents the internal virtual work which is the work done by the internal stress distribution on the strain field due to virtual displacements. Thus equation (5.70) is a statement of virtual work principle that is for a deformable body in static equilibrium, the external work of all the body and traction forces on the virtual displacement is equal to the internal work of the stress field on the strains resulting from the same virtual displacements.

5.10 HAMILTON'S PRINCIPLE

We can derive Hamilton's principle directly from Newton's law as follows. For a continuous deformable body, the inertia force distribution from D'Alembert principle is given by

$$-\rho\ddot{\bar{u}}$$

where ρ is the mass density, and $\bar{u}$ is the vector of displacements at any point in the body. This force is obviously a body force and is given by

$$\bar{B}_{\text{inertia}} = -\iiint_V \rho \frac{d^2 u_i}{dt^2} dv \tag{5.71}$$

The virtual work principle in the previous section can now be generalized as follows

$$-\iiint_V \rho \frac{d^2 u_i}{dt^2} \delta u_i dv + \iiint_V B_i \delta u_i dv + \iint_S T_i^{(v)} \delta u_i dS$$

$$= \iiint_V \tau_{ij} \delta\varepsilon_{ij} dv \tag{5.72}$$

Let

$$V = -\iiint_V B_i u_i dv - \iint_S T_i^{(v)} u_i dS \tag{5.73}$$

Then, the second and third terms of equation (5.72) can be written as

$$\iiint_V B_i \delta u_i dv + \iint_S T_i^{(v)} \delta u_i dS = -\delta V \tag{5.74}$$

If we represent the strain energy density function for the body as U_0

$$U_0 = \int_0^{\varepsilon_{ij}} \tau_{ij}\, d\varepsilon_{ij}$$

so that

$$\tau_{ij} = \frac{\partial U_0}{\partial \varepsilon_{ij}}$$

then, equation (5.72) becomes

$$-\iiint_V \rho \frac{d^2u_i}{dt^2}\delta u_i dv - \delta V - \delta U = 0 \tag{5.75}$$

where

$$U = \iiint_V U_0 dv$$

Let the motion of the system begin at time t_1, then integrating the above from t_1 to t_2 we have

$$-\int_{t_1}^{t_2}\iiint_V \rho \frac{d^2u_i}{dt^2}\delta u_i dv\,dt - \int_{t_1}^{t_2}\delta V dt - \int_{t_1}^{t_2}\delta U dt = 0 \tag{5.76}$$

We rewrite (5.76) as

$$-\iiint_V\left[\rho\int_{t_1}^{t_2}\frac{d^2u_i}{dt^2}\delta u_i dt\right]dv - \delta\int_{t_1}^{t_2} V dt - \delta\int_{t_1}^{t_2} U dt = 0 \tag{5.77}$$

The time integral in the first term above can be integrated by parts to give

$$\int_{t_1}^{t_2}\frac{d^2u_i}{dt^2}\delta u_i dt = \frac{du_i}{dt}\delta u_i \Big|_{t_1}^{t_2} - \int_{t_1}^{t_2}\frac{du_i}{dt}\frac{d}{dt}(\delta u_i)\, dt$$

Let us make the virtual displacements at the time limits t_1 and t_2 zero, then the above becomes

$$\int_{t_1}^{t_2}\frac{d^2u_i}{dt^2}\delta u_i dt = -\int_{t_1}^{t_2}\frac{du_i}{dt}\frac{d}{dt}(\delta u_i)dt$$

i.e.,

$$\int_{t_1}^{t_2}\frac{d^2u_i}{dt^2}\delta u_i dt = -\int_{t_1}^{t_2}\frac{1}{2}\delta\left\{\frac{du_i}{dt}\right\}^2 dt$$

Hence, equation (5.77) can be written as

$$\iiint_V \frac{1}{2}\rho\int_{t_1}^{t_2}\delta\left\{\frac{du_i}{dt}\right\}^2 dt\,dv - \delta\int_{t_1}^{t_2} V dt - \delta\int_{t_1}^{t_2} U dt = 0 \tag{5.78}$$

The first integral in the above equation (5.78) can be rearranged as

$$\iiint_V \frac{1}{2}\rho\int_{t_1}^{t_2}\delta\left\{\frac{du_i}{dt}\right\}^2 dt\,dv = \int_{t_1}^{t_2}\delta\left[\iiint_V \frac{1}{2}\rho\left\{\frac{du_i}{dt}\right\}^2 dv\right]dt$$

i.e.,

$$\iiint_V \frac{1}{2}\rho\int_{t_1}^{t_2}\delta\left\{\frac{du_i}{dt}\right\}^2 dt\,dv = \delta\int_{t_1}^{t_2} T dt$$

where T is the kinetic energy of the body at any instant of time.
Therefore, equation (5.78) becomes

$$\delta\int_{t_1}^{t_2}(T - V - U)dt = 0 \tag{5.79}$$

Denoting the total potential energy $V + U$ of the system as π, and writing the Lagrangian expression

$$L = T - \pi$$

we finally arrive at the following

$$\delta \int_{t_1}^{t_2} L dt = 0 \tag{5.80}$$

This is the Hamilton's principle which states that the integral taken between any two arbitrary instants of time of the Lagrangian expression of a system is stationary. A significant fact to be noticed in this principle is that the displacement and stress fields are not related by any constitutive law. Thus, Hamilton's principle is applicable to any material within the limitation of small deformations. Though it is not shown here, the stationary condition for the Lagrangian in equation (5.80) is a local minimum. Hamilton's principle offers a very powerful tool to derive the governing equations of motion and the boundary conditions of a vibrating system. The minimization principle also offers powerful techniques of solution of these equations of motion.

Historically, advanced vibration theories have been applied to determine the free and forced vibration characteristics, but the emphasis has been laid on finding the natural frequencies and displacement fields, i.e., the response of the system. Approximate methods like Ritz and Galerkin procedures which are directly derivable from Hamilton's principle give very good results for such calculations of complex machine members and structures. However, the corresponding stress field may be in considerable error. This is important when we are interested in the stress field for fatigue calculations. We therefore discuss, the complementary virtual work and Reissner's principle in the next sections.

5.11 COMPLEMENTARY VIRTUAL WORK

In section 5.9, we varied the displacement field with the external forces kept fixed to determine the virtual work. The principle based on virtual work was directly extended to obtain the Hamilton's principle in section 5.10. We pointed out that Hamilton's principle derived from the virtual work principle suffers from the shortcoming that the stress field determined may be in error. This is because of the rapid deterioration in the accuracy of an approximate solution for the displacement field, when differentiations are required for the determination of stress field. The complementary energy principle removes this problem as we see here.

In the complementary virtual work principle, we vary the stress field and the external forces keeping the displacements fixed. We will of course restrict ourselves for the variations of the stress field and the external forces which satisfy the equations of equilibrium and the boundary conditions, i.e.,

$$\{\delta\tau_{ij}\}_{,j} + \delta B_i = 0 \tag{5.81a}$$

with

$$\{\delta\tau_{ij}\} \nu_j = \delta T_i^{(\nu)} \tag{5.81b}$$

on the boundary S_2 and

$$\delta T_i^{(\nu)} = 0 \tag{5.81c}$$

on the boundary S_1.

The complementary virtual work is defined as

$$\delta W^* = \iiint_V u_i \delta B_i dv + \iint_S u_i \delta T_i^{(v)} dA \tag{5.82}$$

Following in a similar manner as in section 5.9, the right hand side of the above equation is

$$\iiint_V u_i \delta B_i dv + \iint_S u_i \delta T_i^{(v)} dA = \iiint_V u_i \delta B_i dv + \iint_S u_i (\delta\tau_{ij}) v_j dA$$

$$= \iiint_V u_i \delta B_i dv + \iiint_V [u_i \delta\tau_{ij}]_{,j} dv$$

i.e.,

$$\iiint_V u_i \delta B_i dv + \iint_S u_i \delta T_i^{(v)} dA = \iiint_V u_i \{\delta B_i + (\delta\tau_{ij})_{,j}\} dv$$

$$+ \iiint_V u_{i,j} \delta\tau_{ij} dv \tag{5.83}$$

In view of (5.81a), the above is

$$\iiint_V u_i \delta B_i dv + \iint_s u_i \delta T_i^{(v)} dA = \iiint_V u_{i,j} \delta\tau_{ij} dv \tag{5.84}$$

Replacing

$$u_{i,j} = \varepsilon_{ij} + \omega_{ij}$$

and noting that ω_{ij} is skew-symmetric, the complementary virtual work principle is

$$\iiint_v u_i \delta B_i dv + \iint_S u_i \delta T_i^{(v)} dA = \iiint_V \varepsilon_{ij} \delta\tau_{ij} dv \tag{5.85}$$

If we represent the complementary energy density function for the body as U_0^*

$$U_0^* = \int_0^\tau \varepsilon d\tau$$

so that

$$\varepsilon_{ij} = \frac{\partial U_0^*}{\partial \tau_{ij}}$$

then equation (5.85) becomes

$$\iiint_V u_i \delta B_i dv + \iint_S u_i \delta T_i^{(v)} dA = \iiint_V \frac{\partial U_0^*}{\partial \tau_{ij}} \delta\tau_{ij} dv \tag{5.86}$$

Let us define a potential function

$$V^* = -\iiint_V u_i B_i dv - \iint_S u_i T_i^{(v)} dA$$

so that

$$\delta V^* = -\iiint_v u_i \delta B_i dv - \iint_S u_i \delta T_i^{(v)} dA$$

Then, equation (5.86) becomes

$$\delta(U^* + V^*) = 0 \tag{5.87}$$

In any approximation technique employing the above principle, we will get good approximation for the stress field. However, the displacement field could be in error. It is however necessary that our displacement field is not in error to obtain good approximations for the natural frequencies of the system. Therefore, the method of complementary virtual work is not very suitable for dynamic analysis. To get over the complication, where we want good approximations for both the displacement and stress fields, we find that Reissner's principle is most appropriate to determine the natural frequencies as also forced vibration displacement and stress response. This principle is discussed in the next section.

5.12 REISSNER'S PRINCIPLE

The principles of virtual work and Hamilton's principle are all minimum principles. Reissner's principle is only a stationary principle in the sense that the nature of the extremum of this functional is not known. Reissner's functional is

$$I_R = \iiint_V \{\tau_{ij}\varepsilon_{ij} - U_0^*(\tau_{ij})\}\,dv$$

$$-\iiint_V B_i u_i dv - \iint_s T_i^{(\nu)} u_i dA \tag{5.88}$$

In the equation (5.88) above the body forces and traction forces are both prescribed and we will vary the stresses, strains and displacements, subject to

$$\delta\varepsilon_{ij} = \frac{1}{2}(\delta u_{i,j} + \delta u_{j,i}) \tag{5.89a}$$

and

$$\tau_{ij} = \tau_{ji} \tag{5.89b}$$

Taking the variation of (5.88), we get

$$\delta I_R = \iiint_V \left\{\delta\tau_{ij}\varepsilon_{ij} + \tau_{ij}\delta\varepsilon_{ij} - \frac{\partial U_0^*}{\partial \tau_{ij}}\delta\tau_{ij}\right\} dv$$

$$-\iiint_V B_i \delta u_i dv - \iint_{S1} T_i^{(\nu)} \delta u_i dA \tag{5.90}$$

In view of (5.89), we express the second term in the right hand side of the above as

$$\iiint_V \tau_{ij}\delta\varepsilon_{ij} dv = \iiint_V \tau_{ij}\delta u_{i,j} dv$$

$$= \iiint_V (\tau_{ij}\delta u_i)_{,j} dv - \iiint_V \tau_{ij,j}\delta u_i dv \tag{5.91}$$

Using the divergence theorem, equation (5.91) can be written as

$$\iiint_V \tau_{ij}\delta\varepsilon_{ij} dv = \iint_S \tau_{ij}\nu_j \delta u_i dA - \iiint_V \tau_{ij,j}\delta u_i dv \tag{5.92}$$

Substituting (5.92) in equation (5.90), we get

$$\delta I_R = \iiint_V \delta\tau_{ij}\varepsilon_{ij}dv + \iint_S \tau_{ij}\nu_j\delta u_i dA - \iiint_V \tau_{ij,j}\delta u_i dv$$

$$-\iiint_V \frac{\partial U_0^*}{\partial \tau_{ij}}\delta\tau_{ij}dv - \iiint_V B_i\delta u_i dv - \iint_S T_i^{(\nu)}\delta u_i dA \tag{5.93}$$

We can regroup the volume and surface integrals in the above and obtain

$$\delta I_R = \iiint_V \left\{\left(\varepsilon_{ij} - \frac{\partial U_0^*}{\partial \tau_{ij}}\right)\delta\tau_{ij} - (\tau_{ij,j} + B_i)\delta u_i\right\} dv$$

$$-\iint_{S1} (\tau_{ij}\nu_j - T_i^{(\nu)}\delta u_i dA - \iint_{S2} \tau_{ij}\nu_j\delta u_i dA \tag{5.94}$$

For

$$\delta I_R = 0 \tag{5.95}$$

Equation (5.95) above gives in V

$$\varepsilon_{ij} = \frac{\partial U_0^*}{\partial \tau_{ij}} \tag{5.96}$$

$$\tau_{ij,j} + B_i = 0 \tag{5.97}$$

and on the surface

$$\tau_{ij}\nu_j = T_i^{(\nu)} \quad \text{for} \quad S_1 \tag{5.98}$$

$$\delta u_i = 0 \quad \text{on} \quad S_2 \tag{5.99}$$

So we see that equation (5.95) yields both the stress-strain law (5.96) and the equations of equilibrium (5.97) at the same time satisfying stress and displacement boundary conditions. The strain displacement relations are already included in the process through (5.89a). Therefore, we expect Reissner's principle to give better approximations for both the stress and displacement fields unlike the virtual work principles which either give a good approximation for displacements or stresses, as the case may be. We will illustrate the advantage of the Reissner's principle over virtual work principles later while calculating the stresses in certain problems.

5.13 LAGRANGE METHOD

The variational principles offer us some of the most powerful techniques for the solution of dynamic problems. The first method is a direct consequence of Euler-Lagrange equations. The approximation that is made is assuming the mode shapes in series form and applying the equations of motion directly. To illustrate the method let us consider a simple example of a cantilever beam undergoing free vibratory motion. The kinetic energy is

$$T = \frac{1}{2}\int_0^l m\dot{y}^2 dz$$

where m is the mass per unit length, l is the total length of the beam and y is displacement of the beam at any distance z from the fixed end.

Let the beam have a uniform cross-section to simplify the illustration. Writing $Z = z/l$, the kinetic energy becomes

$$T = \frac{1}{2} ml \int_0^1 \dot{y}^2 dZ \tag{5.100}$$

The strain energy of the beam is

$$U = \frac{1}{2} \int_0^l EI \left\{ \frac{d^2 y}{dz^2} \right\}^2 dz$$

i.e.,

$$U = \frac{1}{2} \frac{EI}{l^3} \int_0^1 y''^2 dZ \tag{5.101}$$

$$' = \frac{d}{dZ}$$

Now we introduce a mode shape that satisfies the boundary conditions. This is essential since the variational process yields the Euler-Lagrange equations and the boundary conditions. Both must be satisfied for an extremum of the Hamilton's integral. We have four boundary conditions for the cantilever viz.,

$$y = 0 \quad y' = 0 \quad Z = 0 \tag{5.102a}$$
$$y'' = 0 \quad y''' = 0 \quad Z = 1 \tag{5.102b}$$

We would like to generate a series function with n terms in Z for the mode shapes that will satisfy the above boundary conditions. Let this be represented by

$$y = \sum_{i=1}^{n} f_i(Z) q_i(t) \tag{5.103}$$

We can choose any type of functions for f_i (Z), e.g., polynomial functions, which are easy for numerical calculations. Since, there are four boundary conditions and we like to retain one arbitrary parameter in the mode shape, we can assume the following for the ith mode shape.

$$f_i(Z) = A_{i-l} Z^{i-l} + A_i Z^i + A_{i+1} Z^{i+1} + A_{i+2} Z^{i+2} + A_{i+3} Z^{i+3} \tag{5.104}$$

In view of (5.102a), equation (5.104) reduces to

$$f_i(Z) = A_{i+1} Z^{i+1} + A_{i+2} Z^{i+2} + A_{i+3} Z^{i+3} \tag{5.105}$$

Taking the derivatives of the above with respect to Z, we have

$$f_i''(Z) = A_{i+1} i(i+1) Z^{i-1} + A_{i+2}(i+1)(i+2) Z^i + A_{i+3}(i+2)(i+3) Z^{i+1} \tag{5.106a}$$

and

$$f_i'''(Z) = A_{i+1}(i-1) i(i+1) Z^{i-2}$$
$$+ A_{i+2} i(i+1)(i+2) Z^{i-1} + A_{i+3}(i+1)(i+2)(i+3) Z^i \tag{5.106b}$$

Using (5.102b), equations (5.106a) and (5.106b) give

$$\begin{bmatrix} i(i+1) & (i+1)(i+2) \\ (i-1)i(i+1) & i(i+1)(i+2) \end{bmatrix} \begin{Bmatrix} A_{i+1} \\ A_{i+2} \end{Bmatrix} = -A_{i+3} \begin{Bmatrix} (i+2)(i+3) \\ (i+1)(i+2)(i+3) \end{Bmatrix} \tag{5.107}$$

From equation (5.107) above we have

$$A_{i+1} = A_{i+3} \frac{(i+2)(i+3)}{i(i+1)}$$

$$A_{i+2} = A_{i+3} \frac{-2(i+3)}{(i+1)} \tag{5.108}$$

Using above equation (5.108) in (5.105) and writing A_{i+3} as A_i we get the assumed mode shape functions as

$$f_i(Z) = \Sigma A_i[(i+2)(i+3)Z^{i+1} - 2i(i+3)Z^{i+2} + i(i+1)Z^{i+3}] \tag{5.109}$$

For simplicity, we will account for only one term in the above series to illustrate the Lagrange method for the cantilever problem. Then

$$y(Z,t) = (6 - 4Z + Z^2)Z^2 q(t) \tag{5.110}$$

We need the following two derivatives

$$\dot{y}(Z,t) = (6 - 4Z + Z^2)Z^2 \dot{q}(t) \tag{5.111a}$$

$$y''(Z,t) = (12 - 24Z + 12Z^2)q(t) \tag{5.111b}$$

Substituting equation (5.111a) in (5.100), the kinetic energy is

$$T = \frac{1}{2} ml \int_0^1 \{(6 - 4Z + Z^2)Z^2\}^2 \dot{q}^2(t) dZ \tag{5.112}$$

Similarly, substituting (5.111b) in (5.101), the potential energy is

$$U = \frac{1}{2} \frac{EI}{l^3} \int_0^1 (12 - 24Z + 12Z^2)^2 q^2(t) dZ \tag{5.113}$$

The Euler-Lagrangian equation (5.49) for the present case is

$$\frac{d}{dt}\left(\frac{\partial T}{\partial \dot{q}}\right) - \frac{\partial T}{\partial q} + \frac{\partial U}{\partial q} = 0 \tag{5.114}$$

The following terms can now be evaluated.

$$\frac{\partial T}{\partial \dot{q}} = ml\dot{q} \int_0^1 \{(6 - 4Z + Z^2)Z^2\}^2 dZ$$

$$\frac{d}{dt}\frac{\partial T}{\partial \dot{q}} = ml\ddot{q} \int_0^1 \{36Z^4 - 48Z^5 + 28Z^6 - 8Z^7 + Z^8\} dZ$$

i.e.,

$$\frac{d}{dt}\frac{\partial T}{\partial \dot{q}} = \frac{104}{45} ml\ddot{q} \tag{5.115}$$

$$\frac{\partial T}{\partial q} = 0 \tag{5.116}$$

$$\frac{\partial U}{\partial q} = \frac{EI}{l^3} q \int_0^1 (12 - 24Z + 12Z^2)^2 dZ$$

i.e.,

$$\frac{\partial U}{\partial q} = \frac{EI}{l^3} q \int_0^1 \{144 - 576Z + 864Z^2 - 576Z^3 + 144Z^4\} dZ$$

Hence

$$\frac{\partial U}{\partial q} = \frac{144}{5} \frac{EI}{l^3} q \tag{5.117}$$

Using equations (5.115) to (5.117) in (5.114), we get the equation of motion as

$$\frac{104}{45} ml\ddot{q} + \frac{144}{5} \frac{EI}{l^3} q = 0 \tag{5.118}$$

Thus the Lagrange method allowed us to reduce the distributed mass system to a single degree of freedom system. The natural frequency of the system is given by

$$p^2 = \frac{144}{5} \frac{EI}{l^3} \frac{45}{104} \frac{1}{ml}$$

which gives

$$p = 3.53\sqrt{\frac{EI}{ml^4}} \text{ rad/s} \tag{5.119}$$

The exact value for the natural frequency of the cantilever is given by

$$p_{\text{exact}} = 3.5156\sqrt{\frac{EI}{ml^4}} \text{ rad/s} \tag{5.120}$$

Therefore, we notice that the Lagrange method gives a remarkably close value to the exact natural frequency even with one-term approximation.

To determine the higher modes of vibration, we should consider more number of terms in our approximate solution (5.109). We can then write

$$T = \frac{1}{2} m \int_0^l \dot{q}_i^2 f_i^2 dz \tag{5.121}$$

$$U = \frac{1}{2} EI \int_0^l q_i^2 f_i''^2 dz \tag{5.122}$$

$$\frac{\partial T}{\partial \dot{q}_i} = m \int_0^l \dot{q}_i f_i^2 dz$$

$$\frac{d}{dt}\left(\frac{\partial T}{\partial \dot{q}_i}\right) = m\ddot{q}_i \int_0^l \Sigma f_i f_j dz$$

$$\frac{\partial V}{\partial q_i} = EIq_i \int_0^l f_i''^2 dz = EIq_i \int_0^l \Sigma f_i'' f_i'' dz$$

With the help of Euler-Lagrangian equations for several independent variables q_i,

$$\frac{d}{dt}\left(\frac{\partial T}{\partial \dot{q}_i}\right) - \frac{\partial T}{\partial q_i} + \frac{\partial U}{\partial q_i} = 0 \tag{5.123}$$

we get the following matrix equations

$$M\{\ddot{q}\} + K\{q\} = 0 \tag{5.124}$$

where the mass matrix $[M]$ is given by

$$[M] = m \int_0^l \Sigma f_i f_j dz \tag{5.125}$$

and the stiffness matrix $[K]$ is

$$[K] = EI \int_0^l \Sigma f_i'' f_j'' dz \tag{5.126}$$

We require a computer algorithm to solve the above eigen value problem after evaluating the elements of mass and stiffness matrices by a numerical integration process for a chosen number of terms in the solution. This book will not discuss these techniques as all scientific computer centres have such routines built into their library. The reader should develop a computer program according to the above steps to determine the natural frequencies. Let us consider the following data of a simple cantilever.

Length of the blade $l = 10$ cm
Second moment of area $I = 0.00675$ cm^4
Young's modulus $E = 2.11 \times 10^{11}$ Pa
Area of cross-section of the beam $A = 0.9$ cm^2
Density of the material $\rho = 0.008$ kg/cm^3

The results from a computer program for 1-, 2-, 3-, 4-, and 5-term solutions are given in Table 5.1 along with the exact values to study the convergency of the solution by Lagrange method.

From Table 5.1 we can conclude that there is a fast convergency in obtaining the natural frequencies of the problem. The first mode has converged with a two-term solution itself for a two-decimal-digit accuracy. We might say that the Lagrangian method with five-term solution gives first three modes to a very high accuracy. Of course, no one would like to determine the natural frequencies of a uniform cantilever by an approximate method when there is a closed form solution available directly. However, this calculation illustrates the advantage of the method to determine the natural frequencies of a complex system for which no exact solution can be determined. We will consider more examples in subsequent chapters where the approximate methods are used to determine the free and forced vibration
different types of systems.

Table 5.1
Natural frequencies of a cantilever by Lagrange method–Hz

Exact	No. of terms	Lagrange method
248.65	1	249.67
1558.36		
4363.89	2	248.68
8551.67		1606.40
14135.01		
	3	248.68
		1558.50
		4686.10
	4	248.68
		1558.50
		4368.60
		9638.70
	5	248.68
		1558.40
		4366.60
		8599.70
		16901.00

5.14 RITZ METHOD

Like in the Lagrange method, in the Ritz method also we begin with an assumed solution for the shape functions with the equations (5.103).

$$y = \sum_{i=1}^{n} f_i(Z)q_i(t) \tag{5.103}$$

We begin using the formulation based on Rayleigh's average energy or maximum energy principle. For example, if p is the frequency of motion, we can average out the kinetic energy as

$$\overline{T} = \int_0^{\frac{2\pi}{p}} T\,dt \tag{5.127}$$

Similarly, the potential energy is also averaged as

$$\overline{U} = \int_0^{\frac{2\pi}{p}} U\,dt \tag{5.128}$$

This enables us to get rid of one of the independent parameters, time. Then the Lagrangian will be a function of the arbitrary parameters in the assumed mode shapes.

$$L = L(A_i) \tag{5.129}$$

In Ritz method we minimize the above expression in (5.129) with respect to A_i parameters.

$$\frac{\partial L}{\partial A_i} = 0 \tag{5.130}$$

The Ritz method begins with an averaging process (*see* equations 5.127 and 5.128 as in Rayleigh's method, the procedure is also called Rayleigh-Ritz method or Ritz averaging method. A major advantage of the Ritz method is in the assumption of shape functions. Since the method is not derived after the extremizing process in the variational principle, the question of natural boundary conditions is not a part of this procedure. If one uses the Euler-Lagrangian equations as in the previous section, it is necessary that we satisfy the natural boundary conditions that are derived in the extremizing process, (*see* equation 5.37). However, we should always satisfy the kinematic conditions at the boundaries.

For the example of previous section 5.13 with the kinetic and potential energies given by (5.100) and (5.101), respectively, we can write for free vibrations

$$y = \hat{y}\sin pt \tag{5.131}$$

then

$$\dot{y} = \hat{y}p\cos pt \tag{5.131a}$$

$$y'' = \hat{y}''\sin pt \tag{5.131b}$$

If we use maximum energy values in our formulation, the kinetic energy is

$$T = \frac{1}{2}ml\int_0^1 \hat{y}^2 p^2 dZ \tag{5.132}$$

Similarly, the expression for the potential energy is

$$\hat{U} = \frac{1}{2}\frac{EI}{l^3}\int_0^1 \hat{y}''^2 dZ \tag{5.133}$$

$$' = \frac{d}{dZ}$$

Using equations (5.132) and (5.133), we can write the Lagrangian function as

$$L = \frac{1}{2}ml\int_0^1 \hat{y}^2 p^2 dZ - \frac{1}{2}\frac{EI}{l^3}\int_0^1 \hat{y}''^2 dZ \tag{5.134}$$

We will consider a one term approximation to illustrate the procedure. Let us assume

$$\hat{y}(Z) = A(6 - 4Z + Z^2)Z^2 \tag{5.135a}$$

then

$$\hat{y}''(Z) = A(12 - 24Z + 12Z^2) \tag{5.135b}$$

The Lagrangian expression now becomes

$$L = \frac{1}{2}ml\int_0^1 A^2\{(6 - 4Z + Z^2)Z^2\}^2 p^2 dZ$$

$$-\frac{1}{2}\frac{EI}{l^3}\int_0^1 A^2(12 - 24Z + 12Z^2)^2 dZ \tag{5.136}$$

With the help of (5.130), equation (5.136) above becomes

$$ml\int_0^1 \{(6-4Z+Z^2)Z^2\}^2 p^2 dZ = \frac{EI}{l^3}\int_0^1 (12-24Z+12Z^2)^2 dZ \qquad (5.137)$$

The evaluation of integrals in the above leads to the same answer in (5.119).

The major advantage of the Ritz averaging procedure is in solving nonlinear vibration problems. Let us consider the problem of a rotating cantilever beam with the effects of Coriolis forces included. For the moment we will not derive the following energy terms, but accept them, so that the Ritz averaging process is illustrated for nonlinear systems.

Let the cantilever of length l, with mass per unit length, m, be mounted on a disk of radius R which is rotating at ω rad/s. The vibration is assumed to take place in the plane of rotation represented by $y\ (z, t)$. Then the kinetic energy is given by

$$T = \int_0^l \frac{1}{2}\left\{m\dot{y}^2 + m\omega^2 y^2 - m\omega^2(R+z)\int_0^z y'^2 dz\right.$$

$$\left. + m\omega y\frac{\partial}{\partial t}\int_0^z y'^2 dz - m\omega\dot{y}\int_0^z y'^2 dz\right\} dz \qquad (5.138)$$

Let EI be the flexural rigidity for the beam in the plane of vibration. Then, the potential energy is

$$U = \frac{1}{2}EI\int_0^l y''^2 dz \qquad (5.139)$$

Let the beam be acted on by a harmonic exciting force $F \sin pt$ so that the response can be assumed as

$$y(Z,t) = A_1 f_1(Z)\sin pt + B_1 f_2(Z)\sin 2pt \qquad (5.140)$$

We choose the shape functions in (5.140) above to be the first two terms from (5.109) as

$$f_1(Z) = 6Z^2 - 4Z^3 + Z^4 \qquad (5.141a)$$

$$f_2(Z) = 10Z^3 - 10Z^4 + 3Z^5 \qquad (5.141b)$$

Equations (5.141) above satisfy the boundary conditions of a cantilever. The average kinetic and potential energies are obtained from (5.127) and (5.128).

$$\overline{T} = \int_0^{\frac{2\pi}{p}} T\,dt \qquad (5.127)$$

$$\overline{U} = \int_0^{\frac{2\pi}{p}} U\,dt \qquad (5.128)$$

The average work of the exciting force is

$$\overline{W} = \int_0^{\frac{2\pi}{p}} F\sin pt(A_1 f_{1,z=l}\sin pt + B_1 f_{2,z=l}\sin 2pt)dt \qquad (5.142)$$

The Lagrangian is now

$$L = \frac{1}{2} m\pi l p \left\{ \frac{104}{45} A_1^2 + \frac{5216}{693} B_1^2 \right\} + \frac{1}{2} m \omega^2 \frac{\pi l}{p} \left\{ \frac{104}{45} A_1^2 + \frac{1304}{693} B_1^2 \right\}$$

$$-\frac{1}{2} m\omega^2 \frac{\pi}{p} \left\{ A_1^2 \left(\frac{18}{5} R + \frac{122}{45} l \right) + B_1^2 \left(\frac{10}{3} R + \frac{1850}{693} l \right) \right\} - \frac{32}{65} m\omega\pi A_1^2 B_1$$

$$-\frac{EI}{2} \frac{\pi}{pl^3} \left(\frac{144}{5} A_1^2 + \frac{240}{7} B_1^2 \right) - \frac{3FA_1\pi}{p} \tag{5.143}$$

The Ritz process requires

$$\frac{\partial L}{\partial A_1} = 0$$

$$\frac{\partial L}{\partial B_1} = 0 \tag{5.144}$$

Applying the above to (5.143), we get

$$\frac{104}{45} l A_1 (\omega^2 + p^2) - \omega^2 A_1 \left(\frac{18}{5} R + \frac{122}{45} l \right) - \frac{64}{65} \omega p A_1 B_1$$

$$-\frac{144}{5} \frac{EI}{ml^3} A_1 - 3\frac{F}{m} = 0 \tag{5.145a}$$

$$\frac{1304}{693} l B_1 (\omega^2 + 4p^2) - \omega^2 B_1 \left(\frac{10}{3} R + \frac{1850}{693} l \right) - \frac{32}{65} \omega p A_1^2$$

$$-\frac{240}{7} \frac{EI}{ml^3} B_1 = 0 \tag{5.145b}$$

As one would expect, we get three possible roots for A_1 and B_1 from the above equations (5.145) and they are

$$C_3 A_1^3 + C_2 A_1^2 + C_1 A_1 + C_0 = 0 \tag{5.146}$$

$$B_1 = \frac{\frac{32}{65} A_1^2 p\omega}{\frac{1304}{693} l(\omega^2 + 4p^2) - \omega^2 \left(\frac{10}{3} R + \frac{1850}{693} l \right) - \frac{240}{7} \frac{EI}{ml^3}} \tag{5.147}$$

where

$$C_3 = \frac{-\frac{2048}{4225} \omega^2 p^2}{\frac{1304}{693} l(\omega^2 + 4p^2) - \omega^2 \left(\frac{10}{3} R + \frac{1850}{693} l \right) - \frac{240}{7} \frac{EI}{ml^3}}$$

$$C_2 = 0$$

$$C_1 = \frac{104}{45} l(\omega^2 + p^2) - \omega^2 \left(\frac{18}{5} R + \frac{122}{45} l \right) - \frac{144}{5} \frac{EI}{ml^3}$$

and

$$C_0 = -3\frac{F}{m}$$

For free vibrations, $F = 0$, and, therefore, equations (5.145) reduce to

$$\frac{104}{45}l(\omega^2 + p^2) - \omega^2\left(\frac{18}{5}R + \frac{122}{45}l\right) - \frac{64}{65}\omega p B_1 - \frac{144}{5}\frac{EI}{ml^3} = 0 \qquad (5.148a)$$

$$\frac{1304}{693}lB_1(\omega^2 + 4p^2) - \omega^2 B_1\left(\frac{10}{3}R + \frac{1850}{693}l\right) - \frac{32}{65}\omega p A_l^2$$

$$-\frac{240}{7}\frac{EI}{ml^3}B_1 = 0 \qquad (5.148b)$$

If $\omega = 0$, for a nonrotating beam, the Coriolis forces disappear and above equations become linear, giving

$$\frac{104}{45}lp^2 = \frac{144}{5}\frac{EI}{ml^3}$$

and as before we get the same value for the first natural frequency p_1 of the cantilever.

Dividing equation (5.148a) by $104l/45$, we get

$$p^2 - \frac{72}{169}\frac{\omega p}{l}B_1 = p_1^2 + \frac{9}{52}\omega^2\left(1 + 9\frac{R}{l}\right) \qquad (5.149)$$

If we consider only the linear terms in (5.149), i.e., $B_1 = 0$ then we get a relationship between the natural frequency of a rotating blade and the nonrotating blade.

$$p_{1r}^2 = p_1^2 + \omega^2\left(0.173 + 1.5577\frac{R}{l}\right) \qquad (5.150)$$

To understand the dynamic behaviour of the nonlinear system, let us consider the following data.

Length of the blade = 9.1948 cm
Disk radius = 26.3652 cm
Area of cross-section of the beam = 0.8258 cm^2
Density = 0.00784 kg/cm^3
Young's modulus = 2.068×10^{11} Pa
Second moment of area = 0.05777 cm^4
Magnitude of force at the tip = 2224 N

Defining the amplitude ratio as A_1/A_{st} and frequency ratio p/p_{1r}, where A_{st} is the static deflection of the cantilever obtained by setting $p = 0$ in equation (5.145a), the first harmonic response determined from a computer for four different speeds of rotation 60, 120, 180 and 240 Hz are plotted in Figs. 5.8 to 5.11. The corresponding results for the second harmonic B_1/A_{st} are given in Figs. 5.12 to 5.15. The nonlinearity in the system is clear from the hardening characteristic of the first harmonic and the softening characteristic of the higher harmonic.

5.15 GALERKIN METHOD

Consider the case of a functional with two independent variables that we considered in section 5.7. For an extremum of the functional in (5.42), we obtained

$$\int_{t_1}^{t_2}\left[\int_{x_1}^{x_2}\left[\frac{d^2}{dx^2}\left(\frac{\partial F}{\partial y''}\right)-\frac{d}{dx}\left(\frac{\partial F}{\partial y'}\right)+\frac{\partial F}{\partial y}-\frac{d}{dt}\left(\frac{\partial F}{\partial \dot{y}}\right)\right]\eta\, dx\right.$$

$$\left.+\frac{\partial F}{\partial y''}\eta'\Big|_{x_1}^{x_2}+\left\{-\frac{d}{dx}\left(\frac{\partial F}{\partial y''}\right)+\frac{\partial F}{\partial y'}\right\}\eta\Big|_{x_1}^{x_2}\right]dt=0 \tag{5.48}$$

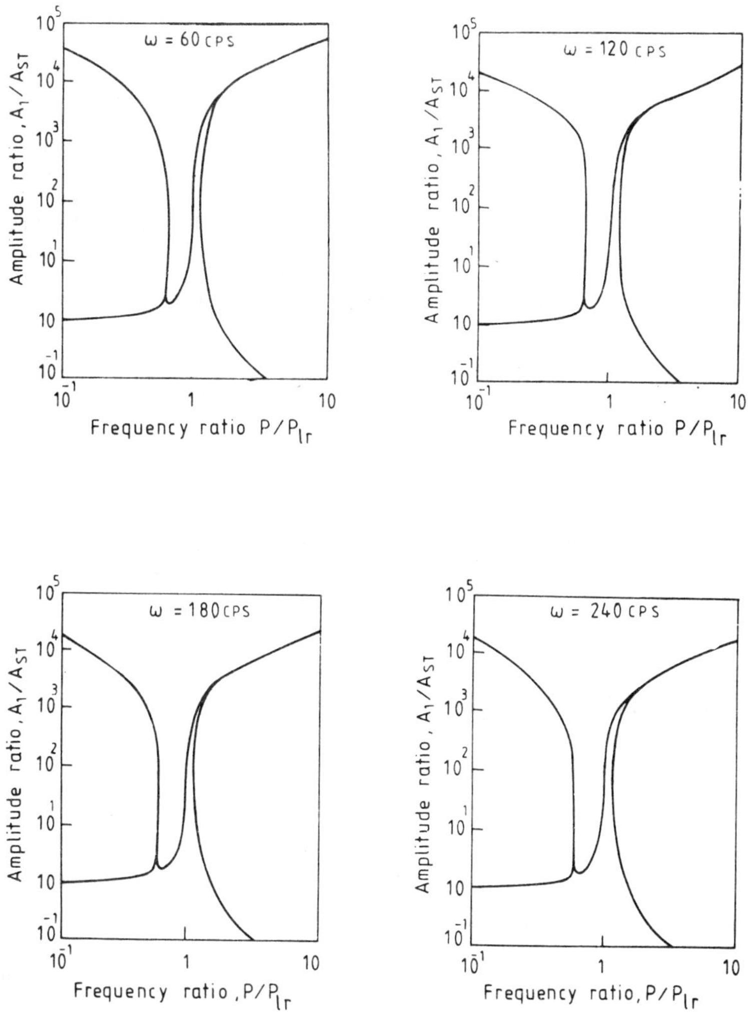

Fig. 5.8 to 5.11 Amplitude ratio A_1 / A_{st} against frequency ratio p / p_{lr} for four different speeds of rotation

which resulted in the Euler-Lagrangian equation

$$\frac{d^2}{dx^2}\left(\frac{\partial F}{\partial y''}\right)-\frac{d}{dx}\left(\frac{\partial F}{\partial y'}\right)+\frac{\partial F}{\partial y}-\frac{d}{dt}\left(\frac{\partial F}{\partial \dot{y}}\right)=0 \tag{5.49}$$

with the boundary conditions

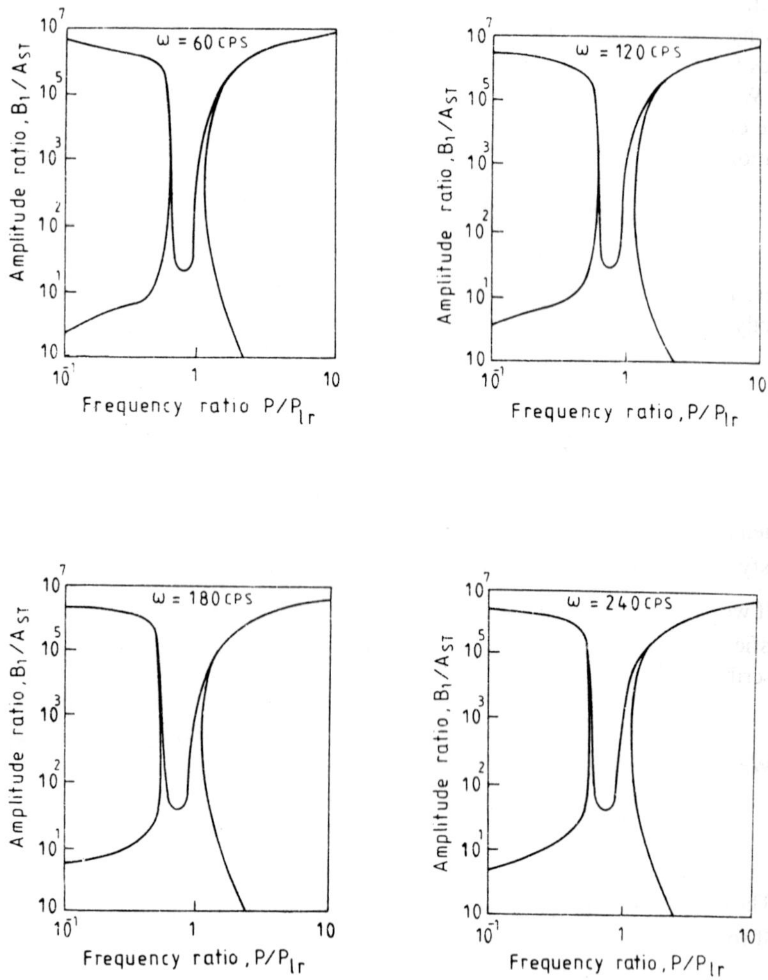

Fig. 5.12 to 5.15 Amplitude ratio B_1 / A_{st} against frequency ratio p / p_{1r} for four different speeds for rotation

$$\frac{\partial F}{\partial y''}\eta'\Big|_{x_1}^{x_2} + \left\{-\frac{d}{dx}\left(\frac{\partial F}{\partial y''}\right)+\frac{\partial F}{\partial y'}\right\}\eta\Big|_{x_1}^{x_2}=0 \tag{5.50}$$

Equation (5.49) gives the differential equation of the system. For a simple beam this is obtained as

$$[EIy'']'' + m\ddot{y} = 0 \tag{a}$$

with the boundary conditions

$$y' \text{ prescribed or } -EIy'' = 0$$

$$y \text{ prescribed or } (EIy'')' = 0 \tag{b}$$

Under normal procedures, we determine the solution of (a) subject to the boundary conditions in (b). Galerkin method is used where the solution of the differential equation (a) is not possible, e.g., both I and m are functions of the distance coordinate varying along the length of the beam. We consider a solution again in the form of a series function in which each term satisfies the appropriate boundary conditions in (b). When we substitute the assumed solution in the differential equation, we find quite evidently it is not satisfied, because the assumed shape functions are not the solutions of the differential equation. In general, we can represent equation (5.48) as

$$\int_{t_1}^{t_2}\left[\int_{x_1}^{x_2}[Ly]\eta dx+[gy]\eta'\Big|_{x_1}^{x_2}+[hy]\eta\Big|_{x_1}^{x_2}\right]dt=0 \tag{5.151}$$

where L, g and h are differential operators. To satisfy the above equation (5.151) exactly so that y is an extremizing path, we make

$$Ly=0 \tag{5.152}$$

and

$$gy=0\Big|_{x_1}^{x_2} \quad \text{and} \quad hy=0\Big|_{x_1}^{x_2} \tag{5.153}$$

In lieu of (5.153) above, we can also prescribe η^1 and η at $x=x_1$ or x_2 as required to satisfy equation (5.151).

If we find it difficult to solve (5.152), we assume an approximate solution $\bar{y}$ that satisfies the natural boundary conditions in (5.153) or the kinematic conditions by prescribing η or η' and substitute this in the differential equation to find the error

$$\in=L\bar{y} \tag{5.154}$$

We can still satisfy equation (5.151) by using (5.154) above, provided we make

$$\int_{x_1}^{x_2}[L\bar{y}]\eta dx=0 \tag{5.155}$$

The difficulty is, what η should we choose. In the Galerkin method, we choose $\bar{y}$ itself for η, i.e.,

$$\int_{x_1}^{x_2}\in\bar{y}dx=0 \tag{5.156}$$

To illustrate the procedure, let us consider the cantilever beam once again. Its differential equation of motion is

$$[EIy'']''=mp^2y \tag{5.157}$$

Let the cross-section of the beam be rectangular with breadth b and thickness t and tapered along the length as defined by the following

$$b=b_0(1-\alpha Z)$$

$$t=t_0(1-\beta Z) \tag{5.158}$$

Then, the second moment of area, I and the mass per unit length m, are given by

$$I = I_0(1 - \alpha Z)(1 - \beta Z)^3$$

$$m = m_0(1 - \alpha Z)(1 - \beta Z) \tag{5.159}$$

Substituting equations (5.158) and (5.159) in (5.157), we get

$$[(1 - \alpha Z)(1 - \beta Z)^3 y'']'' = \frac{m_0 p^2}{EI_0}(1 - \alpha Z)(1 - \beta Z)y \tag{5.160}$$

Let us take a one-term approximation given by

$$y = AZ^2(6 - 4Z + Z^2) \tag{5.161}$$

then

$$y'' = \frac{A}{l^2}(12 - 24Z + 12Z^2) \tag{5.162}$$

Substituting equations (5.161) and (5.162) in (5.160), the error in the differential equation is

$$\in = \left[\frac{A}{l^2}(1 - \alpha Z)(1 - \beta Z)^3(12 - 24Z + 12Z^2)\right]''$$

$$- A\frac{m_0 p^2}{EI_0}(1 - \alpha Z)(1 - \beta Z)Z^2(6 - 4Z + Z^2) \tag{5.163}$$

i.e.,

$$\in = \frac{12A}{l^4}[2(1 - \alpha Z)(1 - \beta Z)^3 + 12\beta(1 - Z)(1 - \alpha Z)(1 - \beta Z)^2$$

$$+ 4\alpha(1 - Z)(1 - \beta Z)^3 + 6\beta^2(1 - \alpha Z)(1 - \beta Z)(1 - 2Z + Z^2)$$

$$+ 6\alpha\beta(1 - \beta Z)^2(1 - 2Z + Z^2)]$$

$$- \frac{m_0 p^2}{EI_0}(1 - \alpha Z)(1 - \beta Z)AZ^2(6 - 4Z + Z^2) \tag{5.164}$$

Galerkin's procedure then gives

$$\int_0^1 \left\{\frac{12A}{l^4}[2(1 - \alpha Z)(1 - \beta Z)^3 + 12\beta(1 - Z)(1 - \alpha Z)(1 - \beta Z)^2]\right.$$

$$+ 4\alpha(1 - Z)(1 - \beta Z)^3 + 6\beta^2(1 - \alpha Z)(1 - \beta Z)(1 - 2Z + Z^2)$$

$$+ 6\alpha\beta(1 - \beta Z)^2(1 - 2Z + Z^2)]$$

$$\left.- \frac{m_0 p^2}{EI_0}(1 - \alpha Z)(1 - \beta Z)AZ^2(6 - 4Z + Z^2)\right\} Z^2(6 - 4Z + Z^2)\, dZ = 0 \tag{5.165}$$

The integrals in the above can be evaluated as follows.

$$I_1 = \int_0^1 \frac{12}{l^4} 2(1-\alpha Z)(1-\beta Z)^3 Z^2 (6-4Z+Z^2)dZ$$

$$= \frac{12}{l^4}\left\{\frac{12}{5} - \frac{26}{15}\alpha - \frac{26}{5}\beta + \frac{142}{35}\alpha\beta + \frac{142}{35}\beta^2\right.$$

$$\left. - \frac{93}{28}\alpha\beta^2 - \frac{31}{28}\beta^3 + \frac{59}{63}\alpha\beta^3\right\}$$

$$I_2 = \int_0^1 \frac{12}{l^4} 12\beta(1-\alpha Z)(1-Z)(1-\beta Z)^2 Z^2 (6-4Z+Z^2)dZ$$

$$= \frac{12}{l^4}\left\{4\beta - \frac{16}{7}\alpha\beta - \frac{32}{7}\beta^2 + \frac{103}{35}\alpha\beta^2 + \frac{103}{70}\beta^3 - \frac{43}{42}\alpha\beta^3\right\}$$

$$I_3 = \int_0^1 \frac{12}{l^4} 4\alpha(1-\alpha Z)(1-\beta Z)^3 Z^2 (6-4Z+Z^2)dZ$$

$$= \frac{12}{l^4}\left\{\frac{4}{3}\alpha - \frac{16}{7}\alpha\beta + \frac{103}{70}\alpha\beta^2 - \frac{43}{126}\alpha\beta^3\right\}$$

$$I_4 = \int_0^1 \frac{12}{l^4} 6\beta^2(1-\alpha Z)(1-\beta Z)(1-2Z+Z^2)Z^2(6-4Z+Z^2)dZ$$

$$= \frac{12}{l^4}\left\{\frac{6}{7}\beta^2 - \frac{57}{140}\alpha\beta^2 - \frac{57}{140}\beta^3 + \frac{47}{210}\alpha\beta^3\right\}$$

$$I_5 = \int_0^1 \frac{12}{l^4} 6\alpha\beta(1-\beta Z)^2(1-2Z+Z^2)Z^2(6-4Z+Z^2)dZ$$

$$= \frac{12}{l^4}\left\{\frac{6}{7}\alpha\beta - \frac{57}{70}\alpha\beta^2 + \frac{47}{210}\alpha\beta^3\right\}$$

$$I_6 = \frac{m_0 p^2}{EI_0}\int_0^1 (1-\alpha Z)(1-\beta Z)Z^4(6-4Z+Z^2)^2 dZ$$

$$= \frac{m_0 p^2}{EI_0}\left\{\frac{104}{45} - \frac{584}{315}\alpha - \frac{584}{315}\beta + \frac{5353}{3465}\alpha\beta\right\}$$

Substituting the above integral values in equation (5.165) and rearranging the terms we get

$$p^2 = \frac{12EI_0}{m_0 l^4}\,\frac{\frac{12}{5} - \frac{2}{5}\alpha - \frac{6}{5}\beta + \frac{12}{35}\alpha\beta + \frac{12}{35}\beta^2 - \frac{9}{70}\alpha\beta^2 - \frac{3}{70}\beta^3 + \frac{2}{105}\alpha\beta^3}{\frac{104}{45} - \frac{584}{315}\alpha - \frac{584}{315}\beta + \frac{5353}{3465}\alpha\beta} \tag{5.166}$$

The above expression gives pretty good results in determining the natural frequency of a doubly tapered beam. However, we would not expect good accuracy for large value of tapers, say, in the region of $\alpha = \beta = 1$. We can then use an N-term solution and write a computer program to determine the natural frequencies as in section 5.13 with the Lagrange method.

Let us use the mode shapes in equation (5.109)

$$y = \sum_i A_i f_i(Z)$$

$$= \sum_i A_i[(i+2)(i+3)Z^{i+1} - 2i(i+3)Z^{i+2} + i(i+1)Z^{i+3}] \quad (5.109)$$

The error in the differential equation now becomes

$$\epsilon = (1-\alpha Z)(1-\beta Z)^3 \Sigma A_i f_i''''$$
$$-2\{\alpha(1-\beta Z)^3 + 3\beta(1-\alpha Z)(1-\beta Z)^2\} \Sigma A_i f_i''''$$
$$+\{6\alpha\beta(1-\beta Z)^2 + 6\beta^2(1-\alpha Z)(1-\beta Z)\} \Sigma A_i f_i''$$
$$-\lambda^2(1-\alpha Z)(1-\beta Z)\} \Sigma A_i f_i \quad (5.167)$$

where

$$\lambda^2 = \frac{m_0 l^4}{EI_0} p^2 \quad (5.168)$$

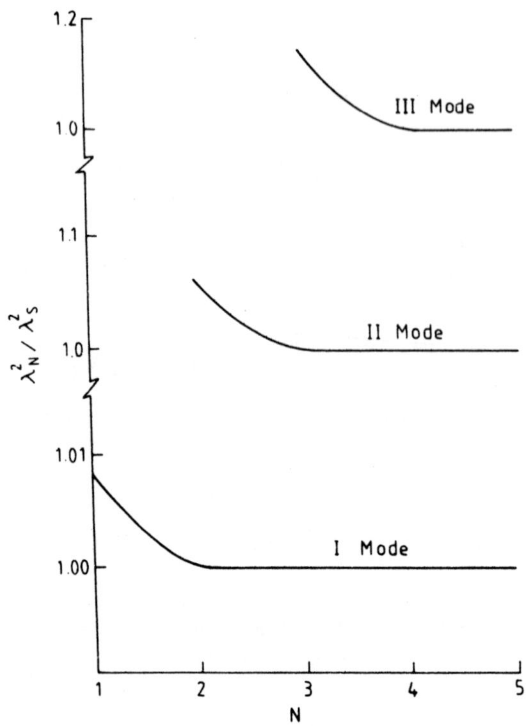

Fig. 5.16 Effect of number of terms in the assumed solution on the frequency parameter ratio λ_N^2/λ_S^2

We now use the Galerkin method and write

$$\int_0^1 \in f_j dZ = 0$$

$$j = 1, 2 \ldots \tag{5.169}$$

For an N-term solution the above can be written as the following eigen value problem.

$$[P] - \lambda^2[Q] = 0 \tag{5.170}$$

It will be a very difficult task to carry out all the integrations needed to evaluate the matrices $[P]$ and $[Q]$ by hand, and therefore one can write a computer program to determine them and obtain the values of λ^2.

The convergency of the natural frequencies for the first three modes with 1-, 2-, 3-, 4- and 5-term solutions is given in Fig. 5.16. The variation of the natural frequencies in the first three modes for various combinations of the taper parameters is given in Figs. 5.17 to 5.19, along with experimentally observed values.

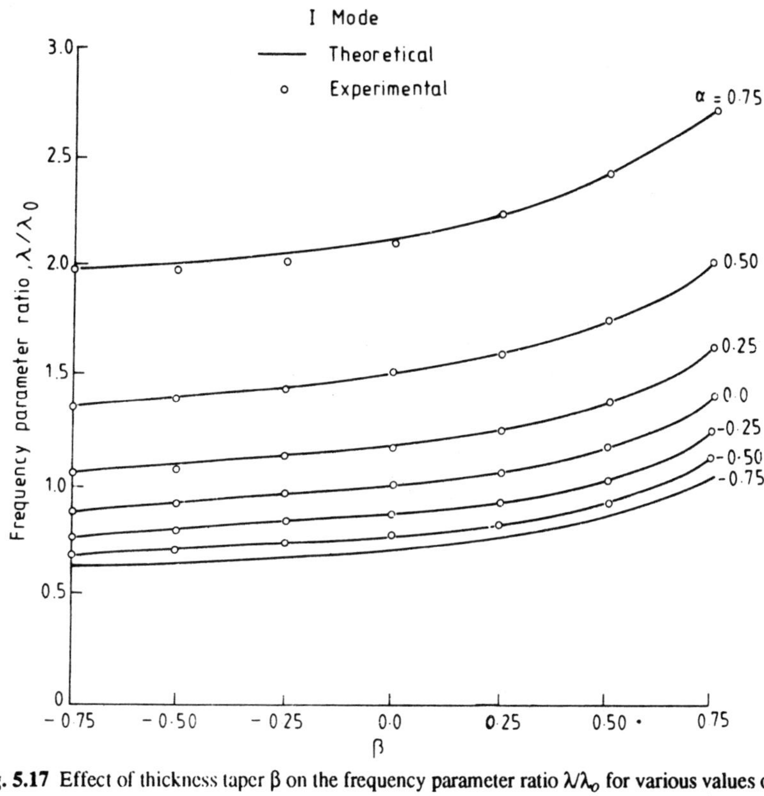

Fig. 5.17 Effect of thickness taper β on the frequency parameter ratio λ/λ_o for various values of breadth taper α, first mode vibration.

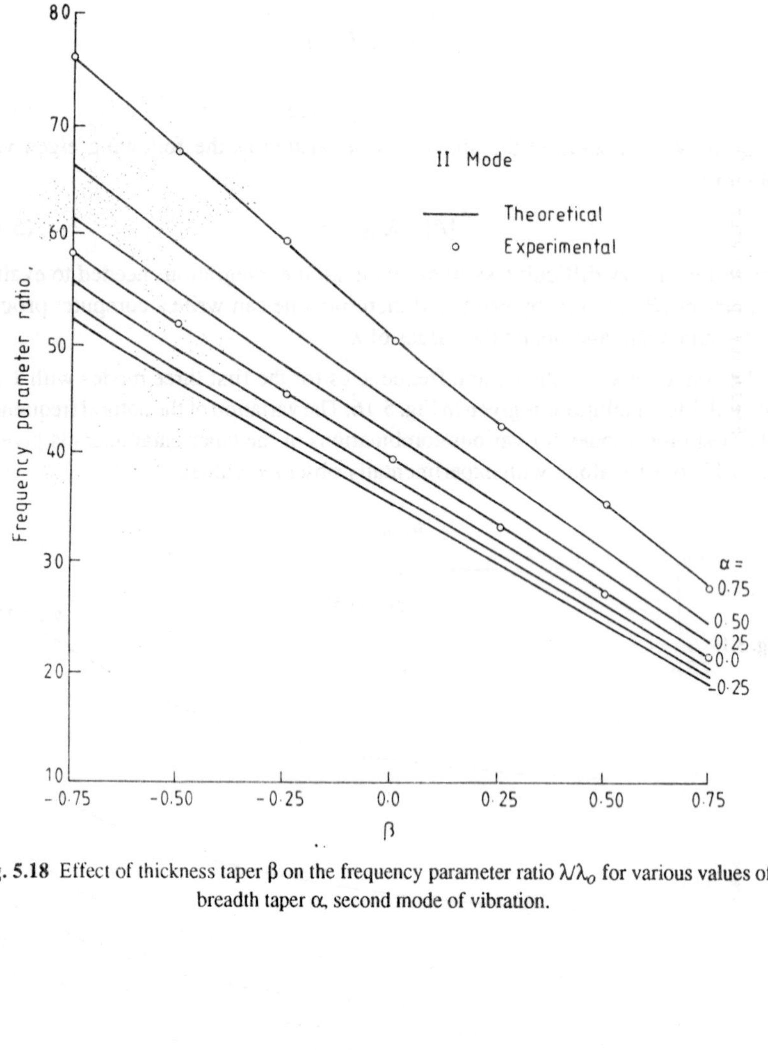

Fig. 5.18 Effect of thickness taper β on the frequency parameter ratio λ/λ_o for various values of breadth taper α, second mode of vibration.

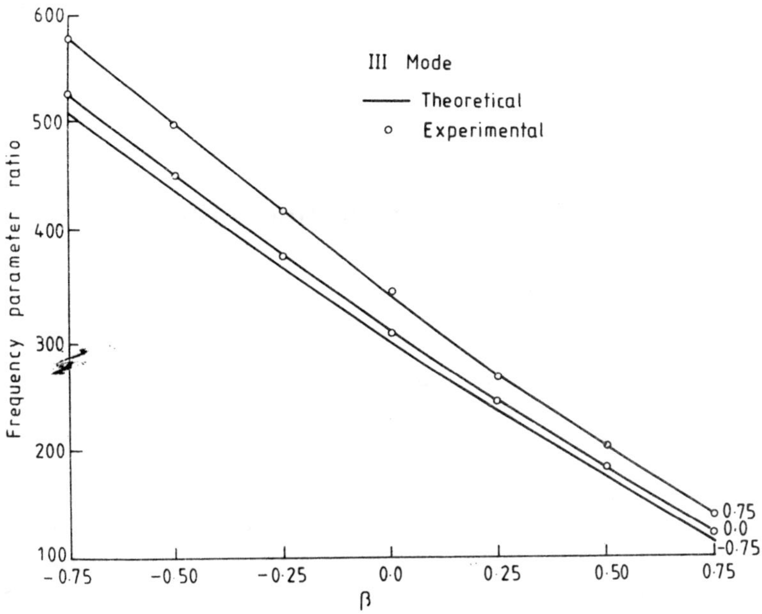

Fig. 5.19 Effect of thickness taper β on the frequency parameter ratio λ/λ_o for various values of breadth taper α, third mode of vibration

6

Longitudinal Vibrations of Bars

The simplest structures are those which can be mathematically modelled by just one coordinate. Such structures are called bars. In Fig. 6.1, the one dimensional structure is represented by the line of centroids of the bar in the undeformed configuration along the x axis, z axis vertically downwards and y axis coming out of the plane of the paper. Thus, y and z axes form the principal directions of the cross-section. Such a bar in Fig. 6.1 can execute longitudinal vibrations along the direction of x axis, angular or torsional vibration about x axis or lateral vibration transverse to the x axis in the direction of z or y axes. These motions could be coupled among themselves if a suitable mechanism for such a coupling exists. For example, an asymmetric cross-section can couple torsional and lateral motions or a pre-twist along the bar can couple two lateral bending motions. In this chapter, we will consider longitudinal motion of the bar.

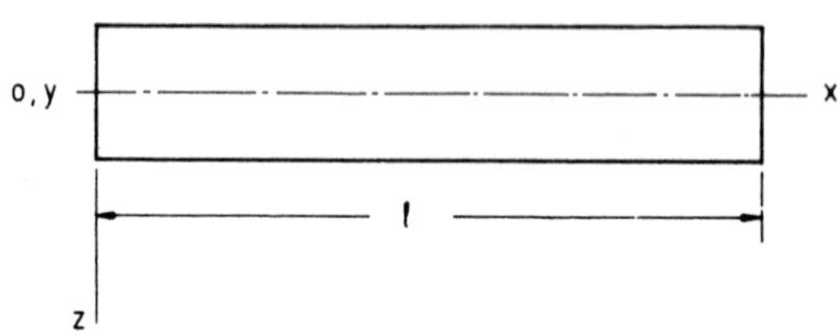

Fig. 6.1 Coordinate system adopted for vibration of bars

6.1 ELEMENTARY THEORY

Consider a bar of length l in Fig 6.2, with its cross-sectional area A (x) at any distance along the x axis. Let ρ be the mass density and E be the Young's modulus of the material of the bar. The bar is excited by a time dependent external force F (t). To start with we will assume the simplest possible displacement field, viz.,

$$u_x = u(x,t)$$

$$u_y = u_z = 0 \tag{6.1}$$

The above assumption indicates that each cross-section of the bar has a longitudinal displacement u which is a function of only one space coordinate and time. The displacement in the y and z directions are ignored.

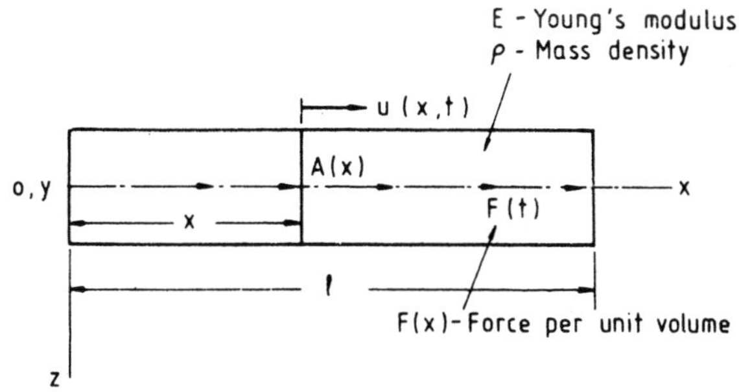

Fig. 6.2 A bar in longitudinal vibration

We can then calculate the strain field which is

$$\varepsilon_{xx} = u_{,x}$$

$$\varepsilon_{xy} = \varepsilon_{xz} = \varepsilon_{yz} = 0$$

$$\varepsilon_{yy} = \varepsilon_{zz} = -\mu u_{,x} \tag{6.2}$$

To retain the simplicity of the theory we can assume the last two normal strain components in the third equation of (6.2) to be zero. Then using Hooke's law we can determine the stress field as

$$\tau_{xx} = (\lambda + 2G)u_{,x}$$

$$\tau_{xy} = \tau_{xz} = \tau_{yy} = \tau_{yz} = \tau_{zz} = 0 \tag{6.3}$$

Consistent with the assumption made for the strain field, i.e., $\mu = 0$ i.e., $\lambda + 2G = E$, the above stress field is

$$\tau_{xx} = Eu_{,x}$$

$$\tau_{xy} = \tau_{xz} = \tau_{yy} = \tau_{yz} = \tau_{zz} = 0 \tag{6.3a}$$

We can now calculate the strain energy, kinetic energy and the work done by F, the external force per unit area of cross-section and unit distance along the axis, as follows.

$$U = \frac{1}{2}\int_0^l EAu_{,x}^2 dx$$

$$T = \frac{1}{2}\int_0^l \rho A\dot{u}^2 dx$$

$$W = \int_0^l FAu\,dx \tag{6.4}$$

Using Hamilton's principle, we obtain the differential equation and boundary conditions as

$$(EAu_{,x})_{,x} - \rho A\ddot{u} + F(t)A = 0$$

$$EAu_{,x}\delta u \,|_0^l = 0 \tag{6.5}$$

The elementary theory of longitudinal vibration of bars is defined in equations (6.5) above. This theory, however, is not correct from the view point of theory of elasticity. For example, any displacement in the direction of x axis is always accompanied by Poisson displacements in y and z direction. In calculating the strain and stress fields also we have deliberately set the Poisson ratio equal to zero, i.e., only axial stress and strain are present. The boundary conditions on stress-free surfaces are satisfied. However, for many practical purposes, for example, in the case of slender rods, the simple elementary theory gives good results from engineering point of view and hence is widely used.

6.2 RAYLEIGH AND BISHOP THEORIES

Rayleigh in 1894, accounted for coupling of cross-sectional deformation with axial deformation in axisymmetric motion. He included both the axial and shear stresses as also strains, but the boundary conditions at the free lateral surfaces are violated; i.e., his theory accounted for axial inertia, but ignored the contribution of shear stiffness in the strain energy. Bishop's theory (1952) considered the effect of shear stiffness while calculating the strain energy. We will derive these theories below.

The displacement field given in (6.1) is now modified to include the cross-sectional deformations in y and z directions. i.e.,

$$u_x = u(x,t)$$

$$u_y = -\mu y u_{,x}$$

$$u_z = -\mu z u_{,x} \tag{6.6}$$

The strain field can be obtained as

$$\varepsilon_{xx} = u_{,x}$$

$$\varepsilon_{xy} = -\frac{1}{2}\mu y u_{,xx}$$

$$\varepsilon_{xz} = -\frac{1}{2}\mu z u_{,xx}$$

$$\varepsilon_{yy} = -\mu u_{,x}$$

$$\varepsilon_{yz} = 0$$

$$\varepsilon_{zz} = -\mu u_{,x} \tag{6.7}$$

By using Hooke's law, we get the stress field as

$$\tau_{xx} = Eu_{,x}$$

$$\tau_{xy} = -\mu G y u_{,xx}$$

$$\tau_{xz} = -\mu G z u_{,xx}$$

$$\tau_{yy} = \tau_{yz} = \tau_{zz} = 0 \tag{6.8}$$

The strain energy of the bar is then given by

$$U = \frac{1}{2}\int_0^l (EAu_{,x}^2 + \mu^2 GI_p u_{,xx}^2)dx \tag{6.9}$$

where

$$I_p = \int\int (y^2 + z^2)dy\,dz$$

is the second polar moment of cross-sectional area.

Similarly, the kinetic energy of the bar is

$$T = \frac{1}{2}\int_0^l (\rho A\dot{u}^2 + u^2\rho I_p \dot{u}_{,x}^2)dx \tag{6.10}$$

The work done by the external force is same as in equation (6.4).

$$W = \int_0^l FAu\,dx$$

Applying the Hamilton's principle we get Bishop's equations of motion and boundary conditions.

$$\{\mu^2 GI_p u_{,xx}\}_{,xx} - \{\mu^2\rho I_p \ddot{u}_{,x}\}_{,x} - \{EAu_{,x}\}_{,x} + \rho A\ddot{u} = FA \tag{6.11}$$

The corresponding boundary conditions are

$$\left\{EAu_{,x} + \mu^2\rho I_p \ddot{u}_{,x} - \mu^2(GI_p u_{,xx})_{,x}\right\}\delta u\,|_0^l + \mu^2 GI_p u_{,xx}\delta u_{,x}\,|_0^l = 0 \tag{6.12}$$

In deriving the above, if we ignored the shear stress terms in the strain energy, we get Rayleigh's equations. Thus,

$$-\{\mu^2\rho I_p \ddot{u}_{,x}\}_{,x} - \{EAu_{,x}\}_{,x} + \rho A\ddot{u} = FA \tag{6.11a}$$

The corresponding boundary conditions are

$$\{EAu_{,x} + \mu^2\rho I_p \ddot{u}_{,x}\}\delta u\,|_0^l = 0 \tag{6.12a}$$

It will be more convenient to adopt operator notation $\partial_x = \partial/\partial x$ for differentiation in equations (6.11) and (6.12). In this notation, we have

$$(\rho A - \partial_x\mu^2\rho I_p\partial_x)\ddot{u} + (-\partial_x EA\partial_x + \partial_x^2\mu^2 GI_p\partial_x^2)u = FA \tag{6.13}$$

$$(\mu^2\rho I_p\partial_x\ddot{u} + EA\partial_x u - \partial_x\mu^2 GI_p\partial_x^2 u)\delta u\,|_0^l$$

$$+ (\mu^2 GI_p\partial_x^2 u)\delta\partial_x u\,|_0^l = 0 \tag{6.14}$$

We nondimensionalize the above equations by choosing reference quantities for each of the parameters and let us set

$$x \leftarrow \frac{x}{l}$$

$$t \leftarrow \frac{t}{t_0}$$

where t_0 is defined by the square of the bar velocity given by

$$\frac{l^2}{t_0^2} = \frac{E_0}{\rho_0}$$

$$u \leftarrow \frac{u}{l}$$

$$\rho \leftarrow \frac{\rho}{\rho_0}$$

$$A \leftarrow \frac{A}{A_0}$$

$$I_p \leftarrow \frac{I_p}{I_{p0}}$$

$$E \leftarrow \frac{E}{E_0}$$

$$G \leftarrow \frac{G}{G_0}$$

$$F \leftarrow \frac{F}{F_0}$$

$$F_0 = \frac{E_0}{l}$$

$$r_1^2 = \mu^2 \frac{r_0^2}{l^2}$$

$$r_0^2 = \frac{I_{p0}}{A_0}$$

$$r_2^2 = \frac{G_0}{E_0} \tag{6.15}$$

Then, the nondimensionalized equation and the boundary conditions for the bar are

$$(\rho A - i_1 r_1^2 \partial_x \rho I_p \partial_x)\ddot{u} + (-\partial_x EA \partial_x + i_2 r_1^2 r_2 \partial_x^2 G I_p \partial_x^2) u = FA \tag{6.13a}$$

$$(i_1 r_1^2 \rho I_p \partial_x \ddot{\mu} + EA \partial_x u - i_2 r_1^2 r_2 \partial_x G I_p \partial_x^2 u) \delta u \,|_0^1$$

$$+ (i_2 r_1^2 r_2 G I_p \partial_x^2 u)\, \delta\, \partial_x u \,|_0^1 = 0 \tag{6.14a}$$

In the above equations (6.13*a*) and (6.14*a*) the indicators i_1, i_2 are introduced with the terms of lateral inertia and shear stiffness corresponding to Rayleigh and Bishop contributions to the elementary theory. If these indicators are set zero, the Rayleigh and Bishop theories in the equations (6.13*a*) and (6.14*a*) reduce to the elementary theory and if they are equal to 1, the corresponding effects are considered in the analysis.

Equations (6.13*a*) can be written in the conventional way

$$M\ddot{u} + Ku = FA \tag{6.13b}$$

where

$$M = \rho A - i_1 r_1^2 \partial_x \rho I_p \partial_x$$

$$K = -\partial_x EA \partial_x + i_2 r_1^2 r_2^2 \partial_x^2 G I_p \partial_x^2$$

We will also write here the nondimensionalized quantities of the energies with respect to

$$U_0 = T_0 = W_0 = E_0 A_0 l \tag{6.16}$$

giving

$$U = \frac{1}{2}\int_0^1 (EAu_{,x}^2 + i_2 r_1^2 r_2 G I_p u_{,xx}^2)dx$$

$$T = \frac{1}{2}\int_0^1 (\rho A \dot{u}^2 + i_1 r_1^2 \rho I_p \dot{u}_{,x}^2)\, dx$$

$$W = \int_0^1 FAu dx \tag{6.17}$$

Using the above theories of Rayleigh and Bishop, natural frequencies of bars are determined in the following sections.

6.3 NATURAL FREQUENCIES OF RAYLEIGH'S BAR

The equation of motion and boundary conditions for free vibrations, from (6.13*a*) and (6.14*a*) with $i_1 = 1$ $i_2 = 0$ are

$$(\rho A - r_1^2 \partial_x \rho I_p \partial_x)\ddot{u} + (-\partial_x EA \partial_x)\, u = 0$$

$$(r_1^2 \rho I_p \partial_x \ddot{u} + EA\partial_x u)\, \delta u \,|_0^1 = 0$$

Let the bar be of uniform cross-section. Then we can choose the reference quantities to be that of the bar itself, i.e.,

$$\rho = A = I_p = E = G = 1$$

and the above equations become

$$(1 - r_1^2 \partial_x^2)\ddot{u} - \partial_x^2 u = 0 \tag{6.18}$$

$$(r_1^2 \partial_x \ddot{u} + \partial_x u)\delta u \,|_0^1 = 0 \tag{6.19}$$

For free vibration of the bar we can write

$$u = X \cos pt$$

Then, equations (6.18) and (6.19) become

$$(1 - r_1^2 d_x^2)\, p^2 X + d_x^2 X = 0 \tag{6.18a}$$

$$\{(1 - r_1^2 p^2)\, d_x X\}\, \delta X \,|_0^1 = 0 \tag{6.19a}$$

The above equations (6.18*a*) and (6.19*a*) lead to the following differential equation and the boundary conditions

$$(1 - r_1^2 p^2) \frac{d^2 X}{dx^2} + p^2 X = 0 \tag{6.18b}$$

(a) *Fixed-Fixed*

$$X(0) = 0; X(1) = 0$$

(b) *Fixed-Free*

$$X(0) = 0; (1 - r_1^2 p^2) X'(1) = 0$$

(c) *Free-Free*

$$(1 - r_1^2 p^2) X'(0) = 0; (1 - r_1^2 p^2) X'(1) = 0 \tag{6.19b}$$

The solution of (6.18*b*) is

$$X = C \cos \sqrt{\frac{p^2}{1 - r_1^2 p^2}}\, x + D \sin \sqrt{\frac{p^2}{1 - r_1^2 p^2}}\, x \tag{6.20}$$

For fixed-fixed boundary condition in (a) above, we have

$$C = 0$$

and

$$\sin \sqrt{\frac{p^2}{1 - r_1^2 p^2}} = 0$$

Hence

$$\frac{p^2}{1 - r_1^2 p^2} = (n\pi)^2$$

i.e.,

$$p^2 = \frac{(n\pi)^2}{1 + r_1^2 (n\pi)^2} \qquad n = 1, 2, 3, \ldots \tag{6.21a}$$

The corresponding mode shape is

$$X = \sin n\pi x \tag{6.21b}$$

With the help of t_0, the natural frequencies in rad/s can be obtained from the following

$$p^2 = \frac{(n\pi)^2}{1 + r_1^2(n\pi)^2} \frac{E}{\rho l^2} \qquad n = 1, 2, 3, .. \tag{6.22}$$

In a similar manner the natural frequencies and mode shapes for other boundary conditions can be obtained as in Table 6.1.

Table 6.1

Frequencies and mode shapes of longitudinal vibrations of rods with the effect of lateral inertia

	Elementary theory		*Rayleigh's correction*	
Boundary conditions	*frequency*	*mode*	*frequency*	*mode*
Fixed-fixed	$n\pi$, $n = 1, 2, 3...$	$\sin n\pi x$	$\frac{n\pi}{\sqrt{1 + r_1^2(n\pi)^2}}$	$\sin n\pi x$
Free-free	$n\pi$, $n = 0, 1, 2...$	$\cos n\pi x$	$\frac{n\pi}{\sqrt{1 + r_1^2(n\pi)^2}}$	$\cos n\pi x$
Fixed-free	$\frac{n\pi}{2}, n = 1, 3, 5...$	$\sin \frac{n\pi x}{2}$	$\frac{\frac{n\pi}{2}}{\sqrt{1 + r_1^2\left(\frac{n\pi}{2}\right)^2}}$	$\sin \frac{n\pi x}{2}$

From Table 6.1 it can be seen that the lateral inertia does not alter the mode shapes, while the frequencies are reduced by a factor of

$$\sqrt{1 + r_1^2 p^2}$$

under its influence. The effect of lateral inertia is not appreciable for very long bars and low frequencies of vibration. However, it becomes increasingly important as the frequency of vibration increases.

6.4 NATURAL FREQUENCIES OF BISHOP'S BAR

The general equation of motion of the bar accounting for both the lateral inertia and shear stiffness is given by

$$(\rho A - r_1^2 \partial_x \rho I_p \partial_x)\ddot{u} + (-\partial_x EA \partial_x + r_1^2 r_2 \partial_x^2 GI_p \partial_x^2)\, u = 0 \tag{6.23}$$

$$(r_1^2 \rho I_p \partial_x \ddot{u} + EA \partial_x u - r_1^2 r_2 \partial_x GI_p \partial_x^2 u)\, \delta u \,|_0^1$$

$$+ (r_1^2 r_2 GI_p \partial_x^2 u)\, \delta \partial_x u \,|_0^1 = 0 \tag{6.24}$$

For bars of uniform cross-section, equations (6.23) and (6.24) become

$$(1 - r_1^2 \partial_x \partial_x)\ddot{u} + (-\partial_x \partial_x + r_1^2 r_2 \partial_x^2 \partial_x^2)u = 0 \tag{6.23a}$$

$$(r_1^2\partial_x\ddot{u} + \partial_x u - r_1^2 r_2 \partial_x \partial_x^2 u)\,\delta u\,|_0^1$$

$$+ (r_1^2 r_2 \partial_x^2 u)\,\delta\,\partial_x u\,|_0^1 = 0 \tag{6.24a}$$

For free vibration of the bar we can write $u = X \cos pt$. Then, equations (6.23*a*) and (6.24*a*) become

$$r_1^2 r_2 d_x^4 X - (1 - r_1^2 p^2) d_x^2 X - p^2 X = 0 \tag{6.23b}$$

$$\{(1 - r_1^2 p^2) d_x X - r_1^2 r_2 d_x^3 X\}\,\delta X\,|_0^1 + (r_1^2 r_2 d_x^2 X)\delta d_x X\,|_0^1 = 0 \tag{6.24b}$$

From equation (6.24b) it can be readily seen that we have to distinguish more possible number of boundary conditions. For example, for fixed-fixed case, there are three possibilities viz.,

(a) both ends rigidly fixed so that the displacements and slopes vanish at the edges.

$$X(0) = X(1) = X'(0) = X'(1) = 0$$

(b) both ends loosely fixed so that the displacements and shear strains vanish at the edges.

$$X(0) = X(1) = X''(0) = X''(1) = 0$$

(c) one edge rigidly fixed and the other loosely fixed.

$$X(0) = X(1) = X'(0) = X''(1) = 0$$

Similarly, the other types of boundary conditions can be further classified, leading to ten possible boundary conditions in all.

The general eigen value problem for Bishop's bar can be solved by Laplace transformation. We can write equations (6.23b) and (6.24b) as

$$d_x^4 X - b d_x^2 X - cX = 0 \tag{6.23c}$$

$$(b d_x X - d_x^3 X)\,\delta X\,|_0^1 + d_x^2 X\,\delta d_x X\,|_0^1 = 0 \tag{6.24c}$$

where

$$b = \frac{(1 - r_1^2 p^2)}{r_1^2 r_2}$$

$$c = \frac{p^2}{r_1^2 r_2} \tag{6.25}$$

The Laplace transform of equation (6.23c) is

$$(s^4 - bs^2 - c)\overline{X} = (s^3 - bs)X(0) + s^2 X'(0)$$

$$+ sX''(0) + \{X'''(0) - bX'(0)\} \tag{6.26}$$

Let us define the following positive quantities

$$s_1^2 = \frac{1}{2}(\sqrt{b^2+4c}+b)$$

$$s_2^2 = \frac{1}{2}(\sqrt{b^2+4c}-b) \tag{6.27}$$

Using (6.25), the second equation in (6.27) above gives

$$p^2 = \frac{s_2^2(1+r_1^2r_2s_2^2)}{1+r_1^2s_2^2} \tag{6.28}$$

The quantities in (6.25) are related to the roots of the characteristic polynomial and satisfy the following relations.

$$(s^2-s_1^2)(s^2+s_2^2) = (s^4-bs^2-c) = 0$$

$$b = s_1^2 - s_2^2$$

$$c = s_1^2 s_2^2$$

$$s_1^4 - b_1^2 - c = 0$$

$$s_2^4 + b_2^2 - c = 0 \tag{6.29}$$

The transfer function $T(x)$ of the system is

$$T(x) = L^{-1}\frac{1}{s^4-bs^2-c}$$

$$= L^{-1}\frac{1}{(s^2-s_1^2)(s^2+s_2^2)}$$

The above can be rearranged as

$$T(x) = L^{-1}\left\{\frac{A}{s^2-s_1^2}+\frac{B}{s^2+s_2^2}\right\}$$

$$= \frac{1}{(s_1^2+s_2^2)(s^2-s_1^2)} - \frac{1}{(s_1^2+s_2^2)(s^2+s_2^2)}$$

i.e.,

$$T(x) = \frac{1}{s_1^2+s_2^2}\left\{\frac{\sinh s_1x}{s_1} - \frac{\sin s_2x}{s_2}\right\}$$

Hence

$$T(x) = \frac{1}{s_1s_2(s_1^2+s_2^2)}(s_2\sinh s_1x - s_1\sin s_2x) \tag{6.30}$$

so that the mode shape and its derivatives can be written as

$$X = [(d_x^3 - bd_x)X(0) + d_x^2X'(0) + d_xX''(0) + \{X'''(0) - bX'(0)\}]T(x)$$

$$X' = [cX(0) + d_x^3X'(0) + d_x^2X''(0) + d_x\{X'''(0) - bX'(0)\}]T(x)$$

$$X'' = [cd_xX(0) + (bd_x^2 + c)X'(0) + d_x^3X''(0) + d_x^2\{(X'''(0) - bX'(0)\}]T(x)$$

$$X''' - bX' = [(cd_x^2 - bc)X(0) + cd_xX'(0) + cX''(0)$$

$$+ (d_x^3 - bd_x)\{X'''(0) - bX'(0)\}]T(x) \tag{6.31}$$

Let us consider the case of a bar with the boundary conditions (b)

$$X(0) = X(1) = X''(0) = X''(1) = 0$$

From (6.30) we note that the conditions at $x = 0$ are satisfied. At $x = 1$, we get

$$[d_x^2X'(0) + \{(X'''(0) - bX'(0)\}]T(1) = 0$$

$$[(bd_x^2 + c)X'(0) + d_x^2\{(X'''(0) - bX'(0)\}]T(1) = 0$$

i.e.,

$$\begin{bmatrix} d_x^2T(1) & T(1) \\ (bd_x^2 + c)T(1) & d_x^2T(1) \end{bmatrix} \begin{Bmatrix} X'(0) \\ X'''(0) - bX'(0) \end{Bmatrix} = 0 \tag{6.32}$$

For nontrivial solution of the above characteristic equation, the determinant of the matrix should vanish, i.e.,

$$\{d_x^2T(1)\}^2 - bd_x^2T(1)T(1) - c\{T(1)\}^2 = 0$$

This gives

$$\sin s_2 = 0 \tag{6.33a}$$

i.e.,

$$s_2 = n\pi \tag{6.33b}$$

From (6.28), we therefore get the frequency expression as

$$p_n^2 = \frac{(n\pi)^2(1 + r_1^2r_2(n\pi)^2)}{1 + r_1^2(n\pi)^2} \tag{6.34}$$

To determine the mode shape, let

$$X'''(0) - bX'(0) = 1$$

Then from equation (6.32) we have

$$X'(0) = -\frac{T(1)}{d_x^2T(1)}$$

Therefore, from the first equation of (6.31), we have

$$X = -\frac{T(1)}{d_x^2 T(1)} d_x^2 T(x) + T(x)$$

In view of (6.33*a*) and (6.33*b*) the above becomes

$$X = \sin s_2 x = \sin n\pi x \qquad n = 1, 2, 3 \ldots \tag{6.35}$$

For the present case, the mode shape is still same as the elementary theory (*see* Table 6.1). However, the natural frequency expression has changed from Rayleigh's theory as well. Let us write down the expressions for the natural frequency from the elementary theory, Rayleigh theory and Bishop theory respectively below.

$$\{p_n^2\}_{\text{elementary theory}} = (n\pi)^2 \tag{6.34a}$$

$$\{p_n^2\}_{\text{Rayleigh theory}} = \frac{(n\pi)^2}{1 + r_1^2 (n\pi)^2} \tag{6.34b}$$

$$\{p_n^2\}_{\text{Bishop theory}} = \frac{(n\pi)^2 (1 + r_1^2 r_2 (n\pi)^2)}{1 + r_1^2 (n\pi)^2} \tag{6.34c}$$

From equation (6.34b) we note that the effect of lateral inertia is to decrease the natural frequency estimated from the elementary theory, whereas from (6.34c) we can see that the shear stiffness increases the natural frequency from that of the elementary theory. This is because, the lateral inertia increases the kinetic energy whereas the shear stiffness increases the strain energy.

Writing equation (6.34c) as

$$\frac{p_n^2}{(n\pi)^2} = \frac{1 + r_1^2 r_2 (n\pi)^2}{1 + r_1^2 (n\pi)^2}$$

i.e.,

$$\frac{p_n^2}{(n\pi)^2} = \frac{\left(1 + \frac{\mu^2 I_p}{A l^2} \sqrt{\frac{G}{E}} (n\pi)^2\right)}{1 + \frac{\mu^2 I_p}{A l^2} (n\pi)^2} \tag{6.36}$$

The frequency ratios obtained for $\mu = 0.3$, $E/G = 0.4$ as a function of the slenderness ratio $\sqrt{I_p / A l^2}$ are given in Fig. 6.3, showing the effect of lateral inertia and shear stiffness on the natural frequencies of longitudinal vibration of bars held at both the ends. From Fig. 6.3 it is clear that the effects of both lateral inertia and shear are negligible for very long bars. However, as the slenderness ratio decreases these effects may be of importance in estimating the natural frequencies. For various possible boundary conditions, the natural frequencies and corresponding mode shapes are given below.

Case 1

Boundary conditions

$$X''(0) = 0; \quad X''(1) = 0$$

$$X'''(0) - bX'(0) = 0; \quad X'''(1) - bX'(1) = 0$$

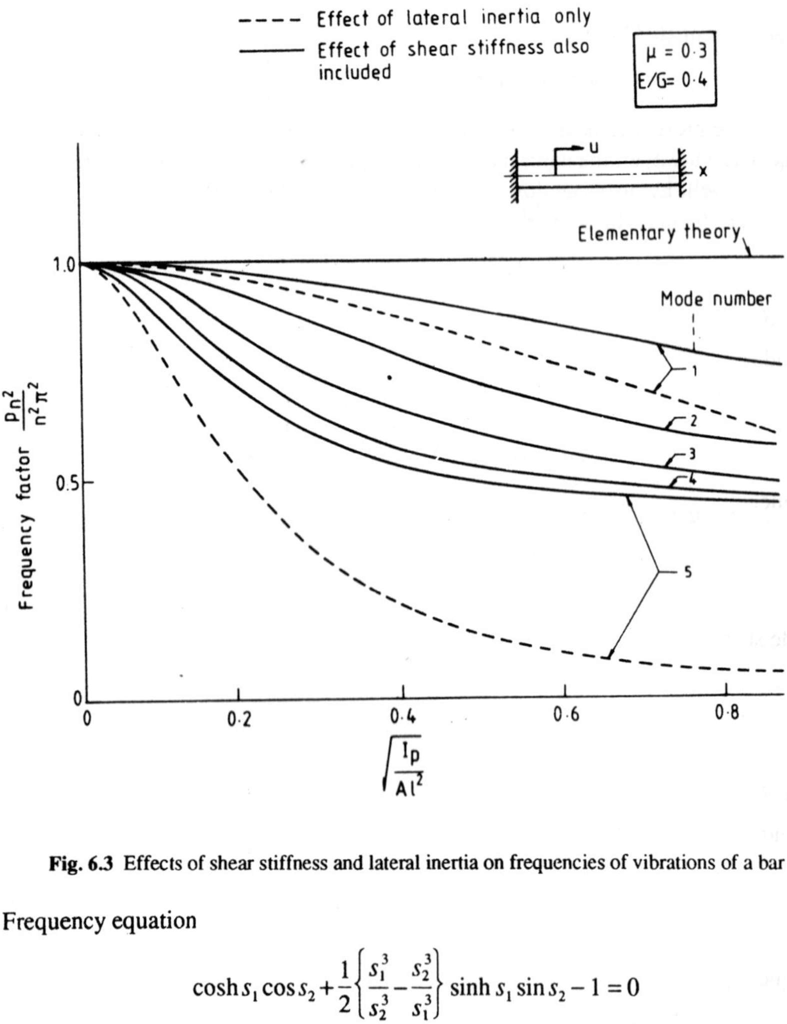

Fig. 6.3 Effects of shear stiffness and lateral inertia on frequencies of vibrations of a bar

Frequency equation

$$\cosh s_1 \cos s_2 + \frac{1}{2}\left\{\frac{s_1^3}{s_2^3} - \frac{s_2^3}{s_1^3}\right\} \sinh s_1 \sin s_2 - 1 = 0$$

Mode shape

$$s_1 s_2(\cosh s_1 - \cos s_2)\left\{\frac{s_1^2 - s_2^2}{s_1 s_2}(s_2^2 \cosh s_1 x + s_1^2 \cos s_2 x)\right.$$

$$\left. - (s_1 \sinh s_1 x + s_2 \sin s_2 x)\right\}$$

$$+ (s_2 \sinh s_1 - s_1 \sin s_2)(s_2^2 \cosh s_1 x + s_1^2 \cos s_2 x)$$

Case 2

Boundary conditions

$$X(0) = 0\,;\, X(1) = 0$$

$$X'(0) = 0\,;\, X'(1) = 0$$

Frequency equation

$$\cosh s_1 \cos s_2 - \frac{1}{2}\left\{\frac{s_1}{s_2} - \frac{s_2}{s_1}\right\} \sinh s_1 \sin s_2 - 1 = 0$$

Mode shape

$$(s_2 \sinh s_1 - s_1 \sin s_2)(\cosh s_1 x - \cos s_2 x)$$

$$-(\cosh s_1 - \cos s_2)(s_2 \sinh s_1 x - s_1 \sin s_2 x)$$

Case 3

Boundary conditions

$$X\,(0) = 0; \quad X''(1) = 0$$

$$X'\,(0) = 0; \quad X'''(1) - bX'(1) = 0$$

Frequency equation

$$\left\{\frac{s_1^2}{s_2^2} + \frac{s_2^2}{s_1^2}\right\} \cosh s_1 \cos s_2 + \left\{\frac{s_1}{s_2} - \frac{s_2}{s_1}\right\} \sinh s_1 \sin s_2 + 2 = 0$$

Mode shape

$$s_1 s_2\,(s_1 \sinh s_1 + s_2 \sin s_2)\,(\cosh s_1 x - \cos s_2 x)$$

$$-(s_1^2 \cosh s_1 + s_2^2 \cos s_2)\,(s_2 \sinh s_1 x - s_1 \sin s_2 x)$$

Case 4

Boundary conditions

$$X'(0) = 0; \quad X'''(1) - bX'(1) = 0$$

$$X'''(0) - bX'(0) = 0; \quad X''(1) = 0$$

Frequency equation

$$s_1^3 \tan s_2 = s_2^3 \tanh s_1$$

Mode shape

$$s_1^2 \cosh s_1 \cos s_2 x + s_2^2 \cos s_2 \cosh s_1 x$$

Case 5

Boundary conditions

$$X(0) = 0; \quad X'''(1) - bX'(1) = 0$$

$$X''(0) = 0; \quad X''(1) = 0$$

Frequency equation

$$s_2^3 \tan s_2 = s_1^3 \tanh s_1$$

Mode shape

$$(s_1^2 + s_2^2) \cosh s_1 \cos s_2 x + 2 s_1 s_2 \cos s_2 \sinh s_1 x$$

Case 6

Boundary conditions

$$X'(0) = 0; \quad X'(1) = 0$$

$$X'''(0) - bX'(0) = 0; \quad X'''(1) - bX'(1) = 0$$

Frequency equation

$$\sin s_2 = 0$$

Mode shape

$$\cos n\pi x$$

Case 7

Boundary conditions

$$X(0) = 0; \quad X(1) = 0$$

$$X''(0); \quad X''(1) = 0$$

Frequency equation

$$\sin s_2 = 0$$

Mode shape

$$\sin n\pi x$$

Case 8

Boundary conditions

$$X(0) = 0; \quad X'(1) = 0$$

$$X''(0) = 0; \quad X'''(1) - bX'(1) = 0$$

Frequency equation

$$\cos s_2 = 0$$

Mode shape

$$\sin \frac{n\pi}{2} x \qquad n = 1, 3, 5, \ldots$$

Case 9

Boundary conditions

$$X(0) = 0; \quad X'''(1) - bX'(1) = 0$$

$$X'(0) = 0; \quad X'(1) = 0$$

Frequency equation

$$s_1 \tanh s_1 + s_2 \tan s_2 = 0$$

Mode shape

$$s_1 s_2 (\cosh s_1 - \cos s_2)(\cosh s_1 x - \cos s_2 x)$$

$$- (s_1 \sinh s_1 + s_2 \sin s_2)(s_2 \sinh s_1 x - s_1 \sin s_2 x)$$

Case 10

Boundary conditions

$$X(0) = 0; \quad X(1) = 0$$

$$X'(0); \quad X''(1) = 0$$

Frequency equation

$$s_2 \tanh s_1 = s_1 \tan s_2$$

Mode shape

$$s_1 s_2 (s_1 \sinh s_1 + s_2 \sin s_2)(\cosh s_1 x - \cos s_2 x)$$

$$- (s_1^2 \cosh s_1 + s_2^2 \cos s_2)(s_2 \sinh s_1 x - s_1 \sin s_2 x)$$

From the above, it is clear that boundary conditions play an important role in the determination of natural frequencies. Quite often it is very difficult to simulate the exact boundary conditions in an experiment and therefore one obtains considerable discrepancy between the theoretical results and experimental values.

6.5 PROPERTIES OF MASS AND STIFFNESS OPERATORS

The system is defined by equations (6.13) and (6.14) which can be written as

$$M\ddot{u} + Ku = 0$$

$$M = 1 - r_1^2 \partial_x^2$$

$$K = -\partial_x^2 + r_1^2 r_2 \partial_x^4$$

$$(r_1^2 \partial_x \ddot{u} + \partial_x u - r_1^2 r_2 \partial_x^3 u)\delta u \,|_0^1 + r_1^2 r_2 \partial_x^2 u \,\delta \partial_x u \,|_0^1 = 0 \qquad (6.37)$$

The associated eigen value problem is

$$(K - p^2 M)\, u = 0$$

$$M = 1 - r_1^2 d_x^2$$

$$K = -d_x^2 + r_1^2 r_2 d_x^4$$

$$\{(1 - r_1^2 p^2)\, d_x u - r_1^2 r_2 d_x^3 u\}\; \delta u \,|_0^1 + r_1^2 r_2 d_x^2 u \,\delta d_x u \,|_0^1 = 0 \qquad (6.37a)$$

Some of the important properties of the mass and stiffness operators are discussed below.

(a) *Self-adjoint system*

Consider any two arbitrary comparison functions u and v. Then, the system is self-adjoint, if

$$\int_V uKv\,dV = \int_V vKu\,dV \tag{6.38a}$$

$$\int_V uMv\,dV = \int_V vMu\,dV \tag{6.38b}$$

Equation (6.38a) implies that operator K is symmetric and similarly M is symmetric, if equation (6.38b) is satisfied.

(b) *Positive-definite system*

For any such comparison function u, if

$$\int_V uKu\,dV \geq 0 \tag{6.39}$$

the operator K is said to be positive. The operator K is positive-definite if the integral in the above is zero only if u is identically zero. If both K and M operators are positive-definite, the system is said to be positive-definite. If K is positive and M is positive-definite, the system is semi-definite. We can establish the properties of a system by means of integration by parts.

Let the two functions u and v satisfy the following

$$[u'v - v'u]_0^1 = 0$$

$$[u'''v - v'''u - u''v' + v''u']_0^1 = 0 \tag{6.40}$$

Consider the mass operator in the following scalar product

$$\int_0^1 Muv\,dx = \int_0^1 \left(u - r_1^2 \frac{d^2u}{dx^2}\right) v\,dx$$

$$= \int_0^1 uv\,dx - \int_0^1 r_1^2 \frac{d^2u}{dx^2} v\,dx$$

Integrating the second term by parts

$$\int_0^1 Muv\,dx = \int_0^1 uv\,dx - r_1^2 \frac{du}{dx} v \,\Big|_0^1 + \int_0^1 r_1^2 \frac{du}{dx}\frac{dv}{dx}\,dx \tag{6.41}$$

The non-integral term in the above equation (6.41) vanishes in view of the conditions (6.40) stipulated above. The first term on the right hand side of (6.41) is symmetric and hence it follows that the mass operator is symmetric over the class of functions stipulated above.

Similarly, consider the stiffness operator and the following scalar product:

$$\int_0^1 Kuv\,dx = \int_0^1 \left(-\frac{d^2u}{dx^2} + r_1^2 r_2 \frac{d^4u}{dx^4}\right) v\,dx$$

$$= -\int_0^1 \frac{d^2u}{dx^2} v\,dx + \int_0^1 r_1^2 r_2 \frac{d^4u}{dx^4} v\,dx$$

Integrating the above by parts, we have

$$\int_0^1 Kuv\,dx = \int_0^1 \frac{du}{dx}\frac{dv}{dx} dx + r_1^2 r_2 \int_0^1 \frac{d^2u}{dx^2}\frac{d^2v}{dx^2} dx$$

$$-\frac{du}{dx}\Big|_0^1 + r_1^2 r_2 \frac{d^3u}{dx^3} v \Big|_0^1 - r_1^2 r_2 \frac{d^2u}{dx^2}\frac{dv}{dx}\Big|_0^1 \tag{6.42}$$

The last three terms in the above vanish in view of the conditions stipulated on u and v in (6.40) and the first two terms are symmetric. Therefore, the stiffness operator is also symmetric, and the system is self-adjoint.

Now consider equations (6.41) to find the positive-definitiveness of the mass operator.

$$\int_0^1 Muu\,dx = \int_0^1 uu\,dx + \int_0^1 r_1^2 \frac{du}{dx}\frac{du}{dx} dx$$

$$= \int_0^1 u^2 dx + \int_0^1 r_1^2 \left\{\frac{du}{dx}\right\}^2 dx > 0 \quad u \neq 0$$

$$= 0 \quad u = 0 \tag{6.43a}$$

Similarly,

$$\int_0^1 Kuu\,dx = \int_0^1 \frac{du}{dx}\frac{du}{dx} dx + \int_0^1 r_1^2 r_2 \frac{d^2u}{dx^2}\frac{d^2u}{dx^2} dx$$

$$= \int_0^1 \left(\frac{du}{dx}\right)^2 dx - \int_0^1 r_1^2 r_2 \left\{\frac{d^2u}{dx^2}\right\}^2 dx > 0\, u \neq 0$$

$$= 0 \quad u = 0 \tag{6.43b}$$

Hence both M and K are positive-definite.

(c) *Orthogonality of modes*

If the stiffness and mass matrices are symmetric and positive-definite, the eigen values are always positive. The eigen functions corresponding to the distinct eigen values simultaneously diagonalize the mass and stiffness operators, thus enabling the solution of free and forced vibration problems by eigen function expansion process.

Let

$$p_i^2,\ u_i(x)$$

$$p_j^2,\ u_j(x)$$

be two distinct eigen values and the corresponding eigen functions for any of the ten cases of boundary conditions in section 6.4. The eigen functions satisfy the eigen value problem and hence

$$p_i^2 M u_i = K u_i$$

$$P_j^2 M u_j = K u_j \qquad \text{(no sum)}$$

Taking scalar products of the above with respect to u_j and u_i, we get

$$\int_0^1 p_i^2 M u_i u_j dx = \int_0^1 K u_i u_j dx$$

$$\int_0^1 p_j^2 M u_j u_i dx = \int_0^1 K u_j u_i dx$$

Since the mass and stiffness operators are symmetric, subtraction leads to

$$\int_0^1 (p_i^2 - p_j^2) M u_i u_j dx = 0$$

i.e.,

$$\int_0^1 M u_i u_j dx = 0 \quad p_i \neq p_j$$

and

$$\int_0^1 K u_i u_j dx = 0 \quad p_i \neq 0$$

Thus

$$\int_0^1 M u_i u_j dx = 0 \quad i \neq j$$

and

$$\int_0^1 K u_i u_j dx = 0 \quad i \neq j \tag{6.44}$$

Therefore, the eigen functions corresponding to distinct eigen values diagonalize the mass and stiffness operators, and the terms on the diagonal can be computed from

$$m_j \delta_{ij} = \int_0^1 M u_i u_j dx \tag{6.45}$$

$$k_j \delta_{ij} = \int_0^1 K u_i u_j dx \tag{6.46}$$

where m_j, k_j are the generalised mass and stiffness in the jth mode. These generalised masses and stiffnesses are related to the jth eigen value by

$$k_j = p_j^2 m_j \quad \text{(no sum)} \tag{6.47}$$

6.6 FREE VIBRATIONS OF BARS IN LONGITUDINAL MOTION

The differential equation of motion of the system is given by

$$M\ddot{u} + Ku = F \tag{6.48}$$

where

$$M = 1 - r_1^2 \partial_x^2$$

$$K = -\partial_x^2 + r_1^2 r_2^2 \partial_x^4$$

Expressing the solution in the form

$$u = u_i(x) q_i(t) \tag{6.49}$$

equation (6.48) becomes

$$Mu_i(x)\ddot{q}_i(t) + Ku_i(x)q_i(t) = F \tag{6.48a}$$

Forming the scalar product with u_j, the above equation gives

$$\int_0^1 Mu_i(x)u_j(x)\ddot{q}_i(t)dx + \int_0^1 Ku_i(x)u_j(x)q_i(t)dx = \int_0^1 Fu_j(x)dx \tag{6.50}$$

In view of equations (6.45) and (6.46), equation (6.50) above becomes

$$m_j\ddot{q}_j + k_j q_j = f_j \quad \text{(no sum)} \tag{6.50a}$$

where f_j is the generalized force in jth mode given by

$$f_j = \int_0^1 Fu_j(x)dx$$

The complete solution of equation (6.50a) can be obtained by using Duhamel's integral

$$q_j(t) = q_j(0)\cos p_j t + \frac{\dot{q}_j(0)}{p_j}\sin p_j t$$

$$+ \frac{1}{m_j p_j}\int_0^t f_j(\tau)\sin p_j(t-\tau)d\tau \quad \text{(no sum)} \tag{6.51}$$

Then, the dynamic displacement is given by modal expansion method

$$u(x,t) = \sum_{1,2,\ldots}^{\infty} \{q_j(0)\cos p_j t + \frac{\dot{q}_j(0)}{p_j}\sin p_j t$$

$$+ \frac{1}{m_j p_j}\int_0^t f_j(\tau)\sin p_j(t-\tau)d\tau\} u_j(x) \quad \text{(no sum)} \tag{6.52}$$

For the case of a free-free bar, the rigid body displacement should be superimposed over the deformation given by the above equation (6.52).

The initial conditions over the displacement are generally prescribed for the system and should be transformed to the generalized coordinates in order to use the equation (6.51). This can be achieved by expanding the prescribed initial conditions in terms of orthogonal modes. Thus, if $f(x)$ and $g(x)$ are, respectively the nondimensional initial displacement and velocity of the bar, then

$$u(x,0) = f(x) = q_i(0)u_i(x)$$

$$\dot{u}(x,0) = g(x) = \dot{q}_i(0)u_i(x) \tag{6.53}$$

Taking scalar product of the above with respect to Mu_j, we have

$$\int_0^1 f(x)Mu_j(x)dx = \int_0^1 q_i(0)Mu_j(x)u_i(x)dx$$

$$\int_0^1 g(x)\,Mu_j(x)\,dx = \int_0^1 \dot{q}_i(0)\,Mu_j(x)u_i(x)\,dx \tag{6.53a}$$

With the help of orthogonal properties of the system, equation (6.53a) gives us

$$q_j(0) = \frac{1}{m_j}\int_0^1 f(x)Mu_j(x)dx$$

$$\dot{q}_j(0) = \frac{1}{m_j}\int_0^1 g(x)Mu_j(x)\,dx \quad j = 1, 2 \ldots \text{(no sum)} \tag{6.54}$$

Using the above equations (6.54) in (6.51) and (6.52), complete solution of the system can be obtained.

As an example, let us consider the case of free vibrations of a bar held at both the ends. The equation of motion and the boundary conditions are

$$M\ddot{u} + Ku = 0$$

$$M = 1 - r_1^2\partial_x^2$$

$$K = -\partial_x^2 + r_1^2 r_2 \partial_x^4$$

$$u(0,t) = 0; \quad u(1,t) = 0$$

$$\partial_x^2 u(0,t) = 0; \quad \partial_x^2 u(1,t) = 0 \tag{6.55}$$

The initial conditions are

$$u(x,0) = f(x)$$

$$\dot{u}(x,0) = g(x) \tag{6.56}$$

The eigen values of the system are obtained earlier, (*see* equation 6.34c)

$$p_i^2 = \frac{(i\pi)^2\{1 + r_1^2 r_2 (i\pi)^2\}}{1 + r_1^2 (i\pi)^2}$$

The corresponding mode shapes are

$$u_i(x) = \sin i\pi x$$

The generalised mass and stiffness in the ith mode are

$$m_i = \int_0^1 (1 - r_1^2\partial_x^2)\sin i\pi x \sin i\pi x dx$$

$$= \int_0^1 \{1 + r_1^2(i\pi)^2\}\sin^2 i\pi x dx$$

$$= \frac{1}{2}\{1 + r_1^2(i\pi)^2\}$$

$$k_i = \int_0^1 (-\partial_x^2 + r_1^2 r_2 \partial_x^4)\sin i\pi x \sin i\pi x dx$$

$$= \frac{1}{2}(i\pi)^2\{1 + r_1^2 r_2 (i\pi)^2\} \quad i = 1, 2, \ldots \quad \text{(no sum)} \tag{6.57}$$

Consider the free vibrations produced in a bar which is moving along with a velocity v_0 before the start of vibration due to subsequent arrest. Thus, the initial conditions are

$$u(x, 0) = f(x) = 0$$

$$\dot{u}(x, 0) = g(x) = v_0 \tag{6.58}$$

With the help of equations (6.54), the initial conditions over the generalised coordinates are

$$q_j(0) = 0 \quad j = 1, 2, \ldots$$

$$\dot{q}_j(0) = \frac{1}{m_i}\int_0^1 (1 - r_1^2\partial_x^2)\sin i\pi x v_0 dx$$

$$= \frac{4v_0}{j\pi} \quad j = 1, 3, 5 \ldots$$

$$= 0 \quad j = 2, 4, 6 \ldots \tag{6.59}$$

The free vibration solution is therefore

$$q_j(t) = \frac{\dot{q}_j(0)}{p_j}\sin p_j t \quad j = 1, 3, 5, \ldots$$

$$= 0 \quad j = 2, 4, 6, \ldots \quad \text{(no sum)} \tag{6.60}$$

In the plane of the bar, equation (6.60) above gives

$$u(x, t) = 4v_0 \sum_{j = 1, 3, \ldots}^{\infty} \frac{1}{p_j(j\pi)}\sin j\pi x \sin p_j t \tag{6.61}$$

Equation (6.61) above is plotted for the displacement at the centrs of the beam for $r_1 = 0.12$, $r_2 = 0.40$ by considering the elementary theory, effect of shear stiffness, effect of the lateral inertia and combined effect of shear and inertia in Fig. 6.4. First 30 terms in the series were included in this plot. From the figure it can be seen that the effect of inertia is to delay the occurrence of the peak, while the influence of shear is to make it occur earlier. Because the effect of inertia is approximately three times that of the effect of shear, the appearance of the peak value is delayed if both the effects are included. Figure 6.4 also shows that the higher order effects reduce the amplitude of vibration. Further, under the influence of shear and lateral inertia, the strain changes smoothly, in contrast to a jump phenomenon that is exhibited by the elementary theory of bars.

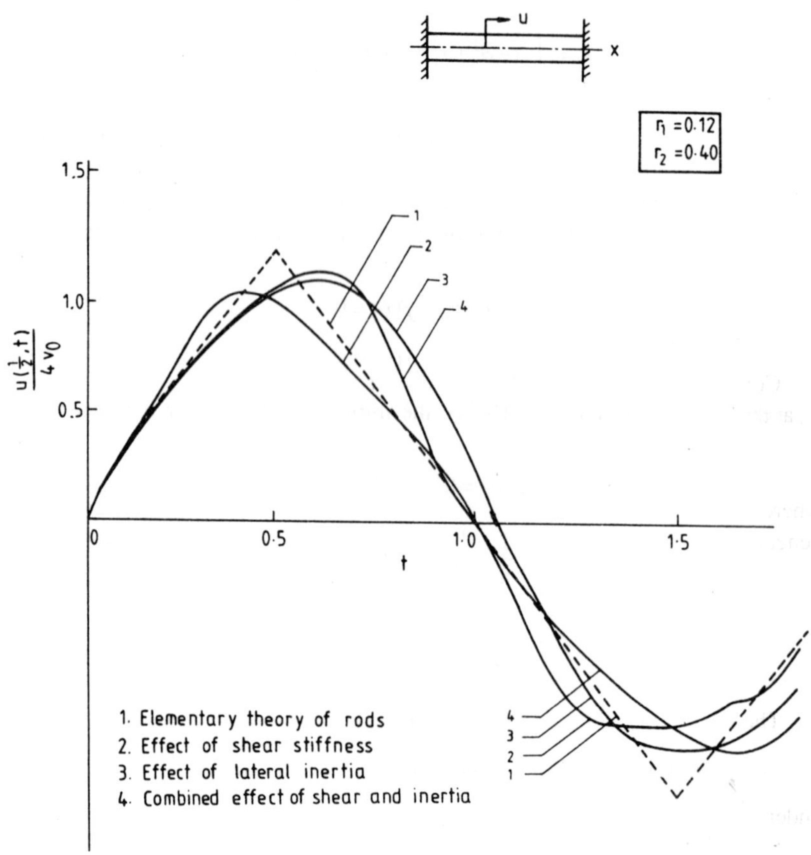

Fig. 6.4 Effects of shear stiffness and lateral inertia on free vibrations of a bar

6.7 FORCED VIBRATION OF BARS

Let us consider the case of a bar held at one end and free at the other end with a given excitation. The system equations are

$$M\ddot{u} + Ku = F$$

$$M = 1 - r_1^2\partial_x^2$$

$$K = -\partial_x^2 + r_1^2 r_2 \partial_x^4$$

$$u(0,t) = 0 \quad \partial_x u(1,t) = 0$$

$$\partial_x^2 u(0,t) = 0 \quad (r_1^2\partial_x\partial_t^2 + \partial_x - r_1^2 r_2 \partial_x^3)u(1,t) = 0 \tag{6.62}$$

This is the case (8) in section 6.4 and the system characteristics can be obtained as before

$$p_i^2 = \left\{\frac{i\pi}{2}\right\}^2 \left[\frac{1 + r_1^2 r_2 \left\{\frac{i\pi}{2}\right\}^2}{1 + r_1^2 \left\{\frac{i\pi}{2}\right\}^2}\right]$$

$$u_i(x) = \sin\left\{\frac{i\pi}{2}\right\}x$$

$$m_i = \frac{1}{2}\left[1 + r_1^2\left\{\frac{i\pi}{2}\right\}^2\right]$$

$$k_i = \frac{1}{2}\left\{\frac{i\pi}{2}\right\}^2\left[1 + r_1^2 r_2\left\{\frac{i\pi}{2}\right\}^2\right] \quad i = 1, 3, 5, \ldots \tag{6.63}$$

Consider the steady dynamic response of a bar due to a suddenly applied force F_0 at the end of the bar $x = 1$. The forcing function can be represented by

$$F = F_0 H(t)\delta(x - 1) \tag{6.64}$$

where $H(t)$ is the Heaviside unit step function and δ is the Dirac delta function. The generalised force is then given by

$$f_j = \int_0^1 F_0 H(t)\delta(x - 1)u_j(x)dx$$

$$= H(t)u_j(1) \tag{6.65}$$

The steady state solution of the equation of motion in the generalised plane

$$m_j\ddot{q}_j + k_j q_j = f_j$$

under the above force in (6.65) is

$$q_j(t) = \frac{1}{m_j p_j}\int_0^t f_j(\tau)\sin p_j(t - \tau)d\tau$$

$$= \frac{u_j(1)}{m_j p_j^2}(1 - \cos p_j t) \tag{6.66}$$

Hence, the steady state dynamic response of the bar under the suddenly applied force is

$$u(x,t) = \sum_{1,3,\ldots}^{\infty} \frac{u_j(1)}{m_j p_j^2}(1-\cos p_j t)u_j(x) \tag{6.66a}$$

The displacement of the free end of the bar is given by

$$u(1,t) = \sum_{1,3,\ldots}^{\infty} \frac{1}{m_j p_j^2}(1-\cos p_j t)$$

$$= \frac{8}{\pi^2}\sum_{1,3,\ldots}^{\infty} \frac{1-\cos p_j t}{j^2\left(1+r_1^2 r_2\left\{\frac{i\pi}{2}\right\}^2\right)} \tag{6.67}$$

For the elementary theory of rods the above reduces to

$$u(1,t) = \frac{8}{\pi^2}\sum_{1,3,\ldots}^{\infty} \frac{1-\cos\frac{j\pi}{2}t}{j^2} \tag{6.67a}$$

The tip displacement of the bar under the influence of a suddenly applied force as given by equation (6.67) is plotted in Fig. 6.5 for $G/E = 0.4$ and $r_1 = 0.12$. This figure shows the influence of higher order effects on the steady vibrations of the tip of the bar. When the higher order effects are ignored, the suddenly applied force produces a maximum deflection which is twice as great as the one obtained by the gradual application of force since

$$\max\left(1-\cos\frac{j\pi}{2}t\right) = 2 \quad \text{at} \quad t = 2$$

This can be observed in Fig. 6.5a. If the higher order effects are included, the maximum deflection under the influence of the suddenly applied force is less than twice the one obtained by the gradual application of force. Also the slope of the displacement-time curve does not suffer a discontinuity at the peak. The effect of inclusion of higher order effects is to increase the stiffness of the system and therefore the maximum deflection decreases with the higher order effects considered in the analysis. Figure 6.5b shows that the maximum deflection uniformly decreases with the increase in slenderness ratio.

6.8 NATURAL FREQUENCIES OF NONUNIFORM BARS

Let us consider the case of a nonuniform bar in which the cross-sectional dimension varies along the length of the bar. The general equation of motion for free vibration of a tapered bar is given by (6.13*b*).

$$M\ddot{u} + Ku = 0 \tag{6.13b}$$

where

$$M = \rho A - r_1^2 \partial_x \rho I_p \partial_x$$

$$K = -\partial_x EA\partial_x + r_1^2 r_2 \partial_x^2 G I_p \partial_x^2$$

where the parameters like A, E etc. are all nondimensional according to equation (6.15).

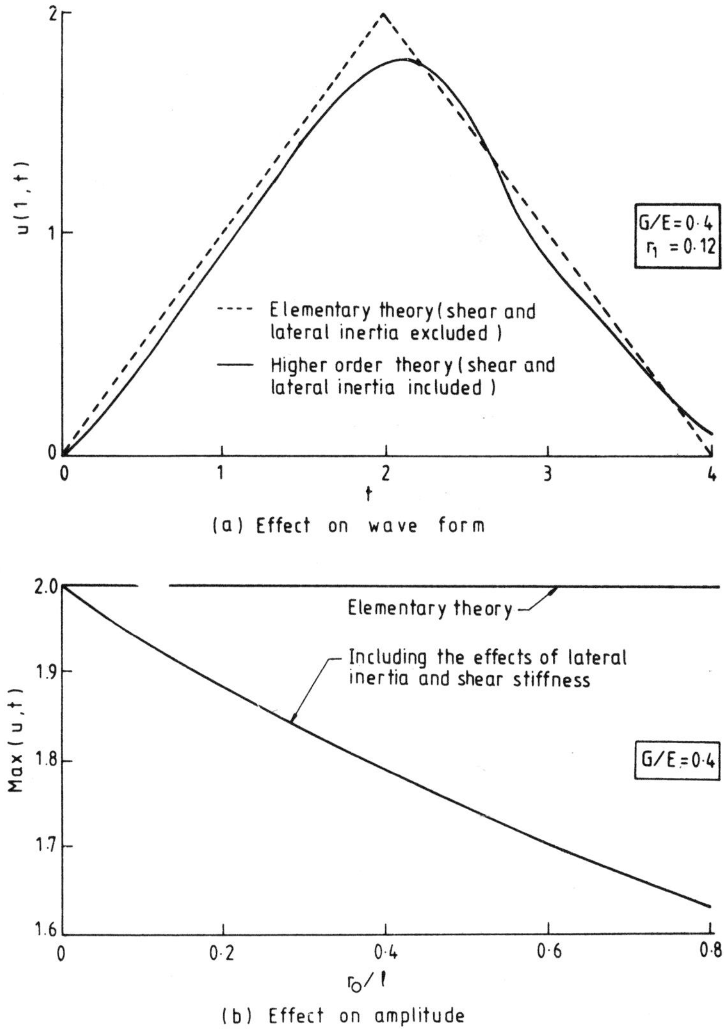

Fig. 6.5 Effects of shear stiffness and lateral inertia on forced vibrations of a bar

In a case like the above where the parameters such as the area of cross-section, and the second polar moment of area vary along the length of the bar, it is not possible in general to obtain an exact solution for the natural frequencies, mode shapes, free and forced vibration response. We will have to adopt an approximate procedure such as Galerkin or Ritz method discussed in chapter 5. Let us assume for free vibration

$$u(x,t) = u(x)\sin pt$$

Then, equation (6.13b) reduces to

$$(K - p^2 M)u = 0 \tag{6.68}$$

Let us use Galerkin method (see section 5.15) to solve the above differential equation of motion. Assume an N-term solution for u (x) as

$$u(x) = q_i u_i(x) \quad i = 1, 2, \ldots N \tag{6.69}$$

where q_i are the arbitrary parameters to be determined by the orthogonalization process of Galerkin. The error in the differential equation is obtained as

$$\in = (-\partial_x EA\partial_x + r_1^2 r_2 \partial_x^2 GI_p \partial_x^2)\, q_i u_i(x) - p^2(\rho A - r_1^2 \partial_x \rho I_p \partial_x)\, q_i u_i(x) \tag{6.70}$$

i.e.,

$$\in = (-EAu_i')' q_i + r_1^2 r_2 (GI_p u_i'')'' q_i - p^2(\rho A u_i q_i - r_1^2 (\rho I_p u_i')' q_i\} \tag{6.70a}$$

According to Galerkin process, we get

$$\int_0^1 [(-EAu_i')' q_i + r_1^2 r_2 (GI_p u_i'')'' q_i$$

$$- p^2\{\rho A u_i q_i - r_1^2 (\rho I_p u_i')' q_i\}]\, q_j u_j dx = 0 \tag{6.71}$$

The above can be written as

$$(K - p^2 M)\, q = 0 \tag{6.72}$$

where K and M are now descretized stiffness and mass matrix operators given by

$$K = [k_{ij}]$$

$$k_{ij} = \int_0^1 (-EAu_i')' u_j dx + \int_0^1 r_1^2 r_2 (GI_p u_i'')'' u_j dx \tag{6.73}$$

$$M = [m_{ij}]$$

$$m_{ij} = \int_0^1 \rho A u_i u_j dx - \int_0^1 r_1^2 (\rho I_p u_i')' u_j dx \tag{6.74}$$

We could also have started from the energy expressions given in (6.17), which are

$$U = \frac{1}{2}\int_0^1 (EAu_{,x}^2 + r_1^2 r_2 GI_p u_{,xx}^2) dx$$

$$T = \frac{1}{2}\int_0^1 (\rho A \dot{u}^2 + r_1^2 \rho I_p \dot{u}_{,x}^2) dx$$

Hence, the Lagrangian expression is

$$L = T - U$$

$$= \frac{1}{2}\int_0^1 (\rho A \dot{u}^2 + r_1^2 \rho I_p \dot{u}_{,x}^2)\, dx - \frac{1}{2}\int_0^1 (EAu_{,x}^2 + r_1^2 r_2 GI_p u_{,xx}^2)\, dx$$

Taking the average energy values over a long period of time and using equation (6.69), Ritz method of section 5.14 can be applied to give

$$\left[\int_0^1 p^2(\rho A u_i^2 + r_1^2 \rho I_p u_{i,x}^2)dx - \int_0^1 (EA u_{i,x}^2 + r_1^2 r_2 G I_p u_{i,xx}^2)dx\right] q_i = 0$$

This leads to the eigen value problem of the type (6.72), where the stiffness and mass operators are given by

$$K = [k_{ij}]$$

$$k_{ij} = \int_0^1 EA u_i' u_j' dx + \int_0^1 r_1^2 r_2 G I_p u_i'' u_j'' dx \tag{6.73a}$$

$$M = [m_{ij}]$$

$$m_{ij} = \int_0^1 \rho A u_i u_j dx + \int_0^1 r_1^2 \rho I_p u_i' u_j' dx \tag{6.74a}$$

With the help of integration by parts, we can see that equations (6.73–74) and (6.73*a*–74*a*) are the same for the cases of fixed-fixed and fixed-free boundary conditions, cases 7 and 8 respectively of Bishop's bar in section 6.4. We can solve the eigen value problem on a computer to determine the natural frequencies. Let us consider a hollow tapered cylindrical bar (*see* Fig. 6.6). Let R be the outer radius and r be the inner radius defined by

$$R(x) = R(0) + \{R(l) - R(0)\}\frac{x}{l}$$

$$r(x) = r(0) + \{r(l) - r(0)\}\frac{x}{l}$$

Also, let the taper parameters be defined by

$$\alpha_1 = \frac{R(l) - R(0)}{R(0)}$$

$$\alpha_2 = \frac{r(l) - r(0)}{r(0)}$$

and

$$\alpha_3 = \frac{r(0)}{R(0)}$$

Let the reference area and the polar second moment of area be

$$A_0 = \pi R^2(0)$$

$$I_{p0} = \frac{\pi}{2} R^4(0)$$

Then, the nondimensional values of the sectional properties are

$$A(x) = (1 + \alpha_1 x)^2 - \alpha_3^2 (1 + \alpha_2 x)^2$$

$$I_p(x) = (1 + \alpha_1 x)^4 - \alpha_3^4 (1 + \alpha_2 x)^4$$

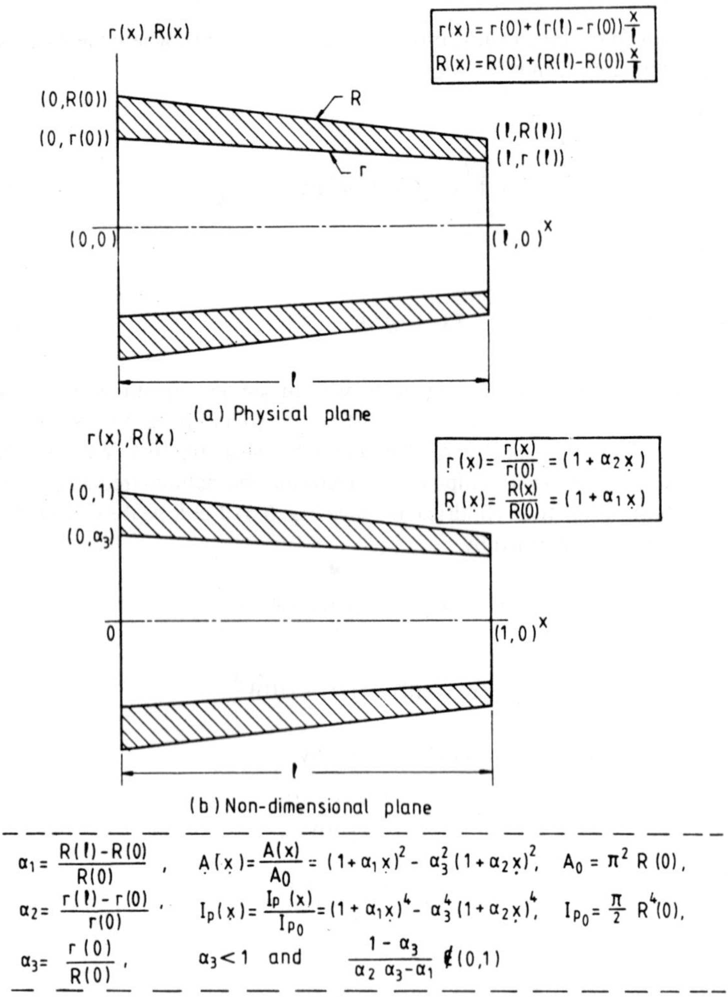

Fig. 6.6 Geometry of hollow tapered cylindrical bar

Using the mode shapes of the corresponding uniform bars as approximating functions, the natural frequencies are determined with a five-term approximation and the results are given in Figs. 6.7 to 6.12 for the cases of fixed-fixed and fixed-free boundary conditions.

6.9 VIBRATION OF UNIFORM VISCOELASTIC BARS

Let the bar be represented by a Kelvin solid so that we can write the nondimensional Young's and shear moduli as

$$E(\partial_t) = (1 + t_1\partial_t)$$

$$G(\partial_t) = (1 + t_1\partial_t) \tag{6.75}$$

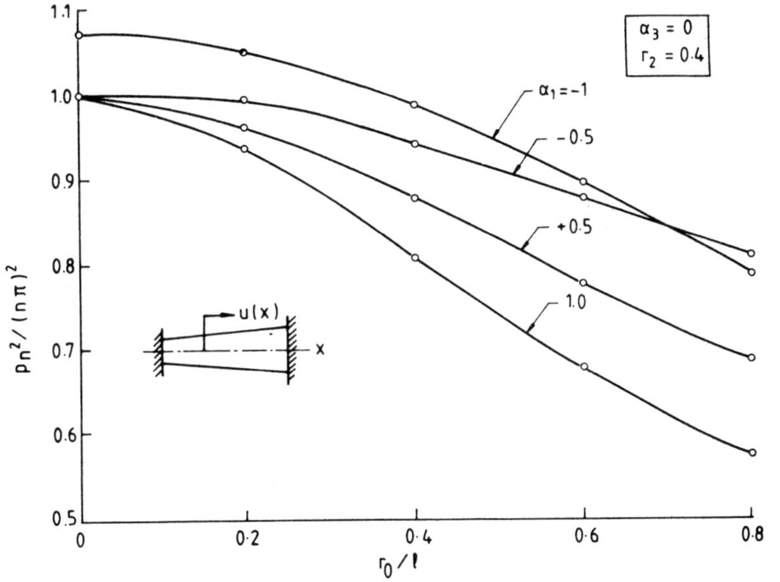

Fig. 6.7 Effects of shear stiffness and lateral inertia on frequencies of vibration of a fixed–fixed tapered bar–1st mode

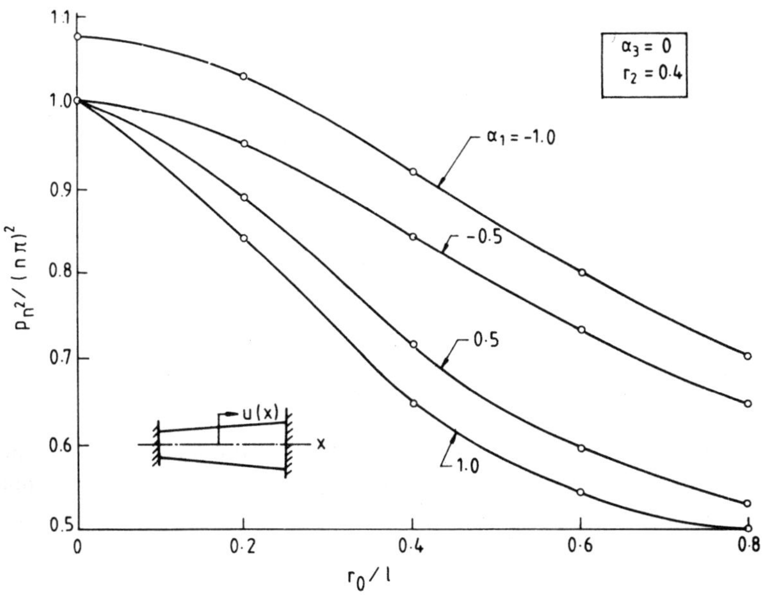

Fig. 6.8 Effects of shear stiffness and lateral inertia on frequencies of vibration of a fixed–fixed tapered bar–2nd mode

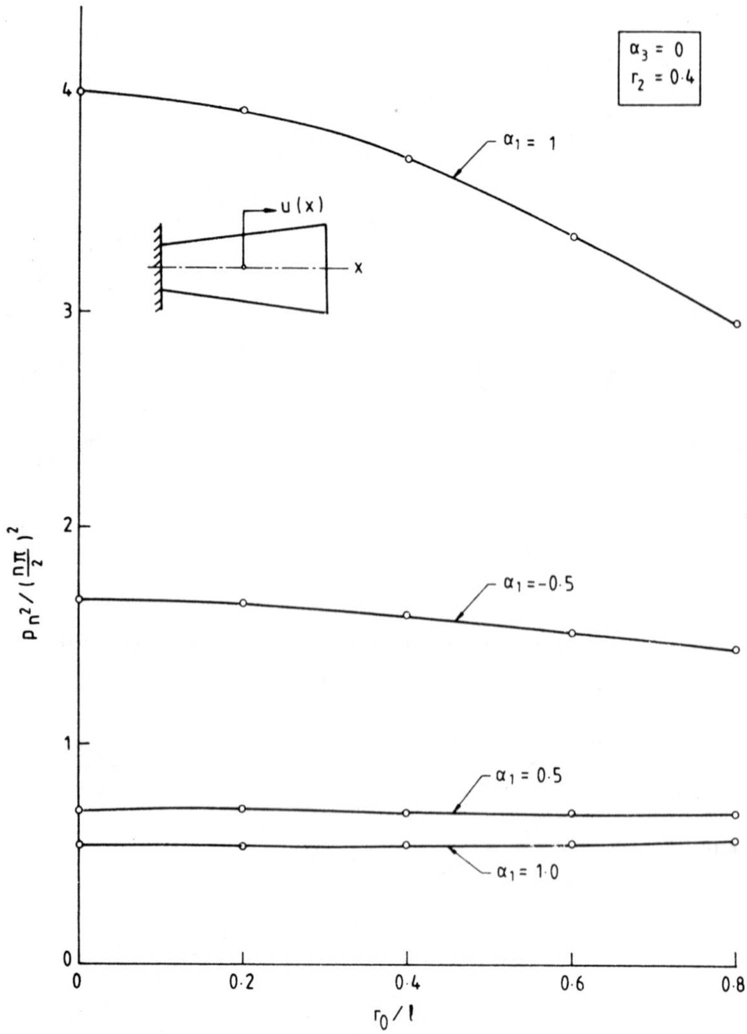

Fig. 6.9 Effects of shear stiffness and lateral inertia on frequencies of vibration of a fixed–free tapered bar–1st mode

where t_1 is the relaxation time constant nondimensionalized according to equation (6.15). Then the nondimensional differential equations governing the longitudinal motion and energies of a uniform viscoelastic bar are

$$M\ddot{u} + E(\partial_t)Ku = F \tag{6.76}$$

where

$$M = (1 - r_1^2\partial_x^2)$$

$$K = (-\partial_x^2 + r_1^2 r_2 \partial_x^4)$$

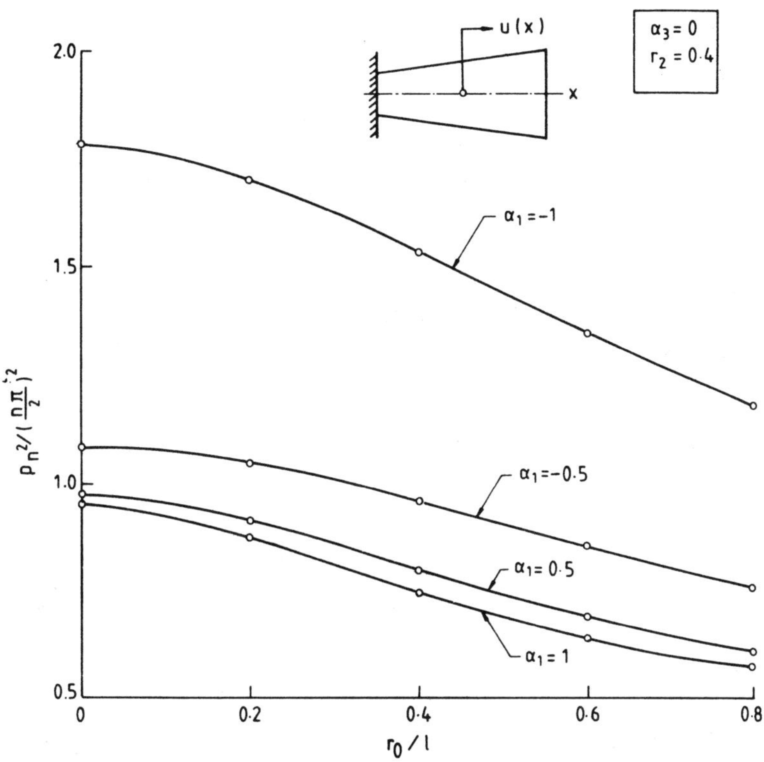

Fig. 6.10 Effects of shear stiffness and lateral inertia on frequencies of vibration of a fixed–free tapered bar–2nd mode

and

$$\{r_1^2\partial_x\ddot{u} + E\,(\partial_t)\,\partial_x u - r_1^2 r_2 E\,(\partial_t)\partial_x^3 u\}\,\delta u\,|_0^1$$

$$+\,(r_1^2 r_2 E\,(\partial_t)\,\partial_x^2 u\}\,\delta\partial_x u\,|_0^1 = 0 \tag{6.77}$$

$$U = \frac{1}{2}E(\partial_t)\int_0^1 (u_{,x}^2 + r_1^2 r_2 u_{,xx}^2)\,dx$$

$$T = \frac{1}{2}\int_0^1 (\dot{u}^2 + r_1^2\dot{u}_{,x}^2)\,dx$$

$$W = \int_0^1 F\,u\,dx \tag{6.78}$$

To find the solution of equation (6.76), let us use the eigen function expansion process by using equation (6.69), which is

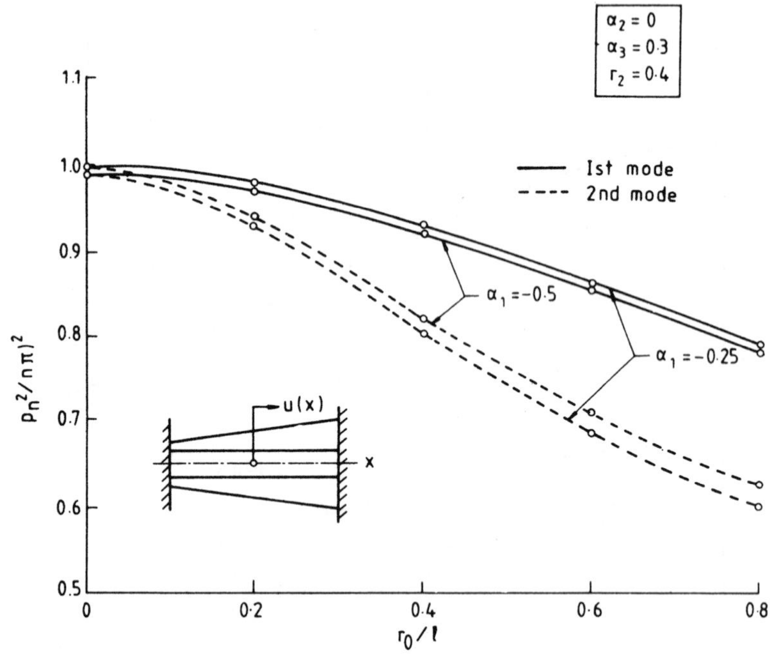

Fig. 6.11 Effects of shear stiffness and lateral inertia on frequencies of vibration of a fixed–fixed hollow tapered bar

$$u = q_i u_i \qquad (u_i - \text{eigen function})$$

$$(q_i - \text{arbitrary time function})$$

$$i = 1, 2, \ldots \tag{6.69}$$

Substituting equation (6.69) above in the differential equation (6.76), multiplying with u_j, integrating over the length of the bar and using the symmetry relations of the operators, we get

$$\int_0^1 M u_i(x) u_j(x) \ddot{q}_i(t) dx + E(\partial_t) \int_0^1 K u_i(x) u_j(x) q_i(t)\, dx = \int_0^1 F u_j(x) dx$$

i.e.,

$$m_i \ddot{q}_i + E(\partial_t)\, k_i q_i = f_i \tag{6.79}$$

where the generalised mass, stiffness and force in ith mode are given by

$$m_i = \int_0^1 M u_i(x)\, u_i(x) dx \quad \text{(no sum)}$$

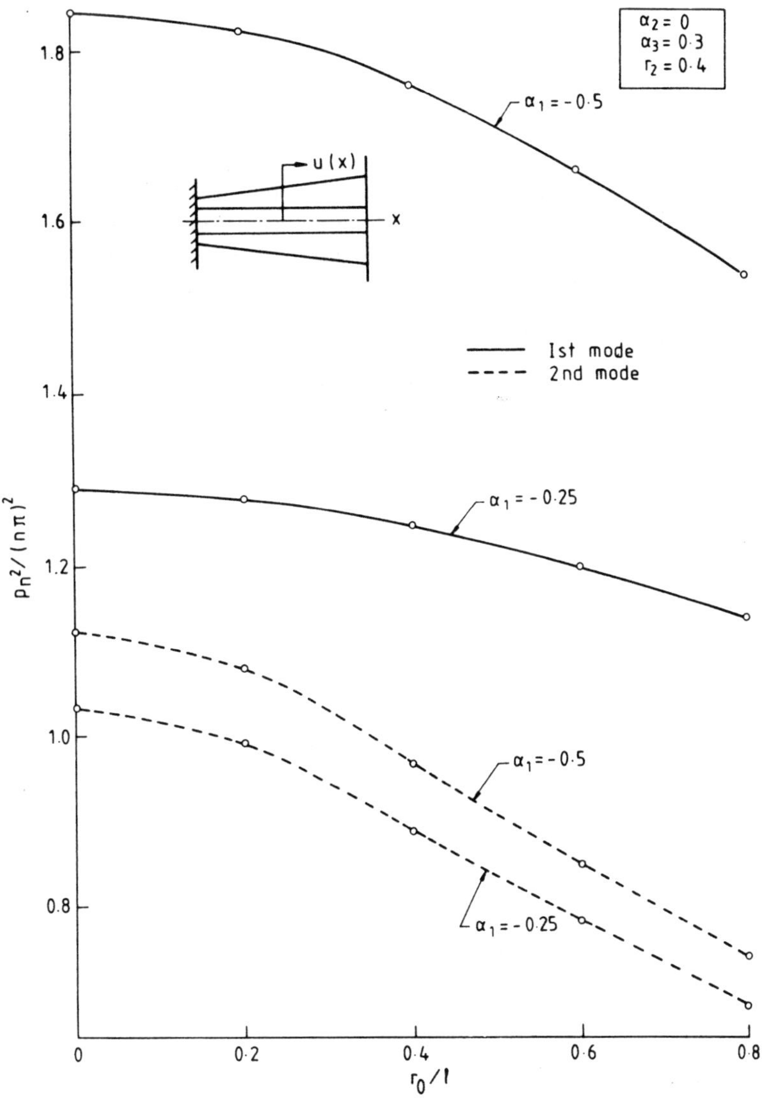

Fig. 6.12 Effects of shear stiffness and lateral inertia on frequencies of vibration of a fixed–free hollow tapered bar

$$k_i = \int_0^1 K u_i(x) u_i(x)\, dx \quad \text{(no sum)}$$

$$f_i = \int_0^1 F u_i(x)\, dx \tag{6.80}$$

As done in section 6.6, we can transform the initial conditions over the bar coordinates into initial conditions over q_i by equation (6.54)

$$q_i(0) = \frac{1}{m_i}\int_0^1 u(x,0)Mu_i(x)dx$$

$$\dot{q}_i(0) = \frac{1}{m_i}\int_0^1 \dot{u}(x,0)Mu_i(x)dx \quad \text{(no sum)} \tag{6.81}$$

Solving equations (6.79) and substituting the resulting expression for the generalized coordinate into equation (6.69) leads to the complete solution of the problem under appropriate boundary and initial conditions. We will first list the eigen values, eigen functions, generalised masses and stiffnesses for three most common boundary conditions of the undamped bar.

Case 1 Fixed-fixed conditions

Eigen values

$$p_i^2 = (i\pi)^2 \frac{1 + r_1^2 r_2 (i\pi)^2}{1 + r_1^2 (i\pi)^2}$$

Mode shapes

$$\sin i\pi x$$

Generalised mass

$$m_i = \frac{1}{2}\{1 + r_1^2\,(i\pi)^2\}$$

Generalised stiffness

$$k_i = \frac{1}{2}(i\pi)^2\,\{1 + r_1^2 r_2 (i\pi)^2\} \tag{6.82}$$

Case 2 Fixed-free conditions

Eigen values

$$p_i^2 = \left(\frac{i\pi}{2}\right)^2 \frac{1 + r_1^2 r_2 \left(\frac{i\pi}{2}\right)^2}{1 + r_1^2 \left(\frac{i\pi}{2}\right)^2}$$

Mode shapes

$$\sin\frac{i\pi}{2}x$$

Generalised mass

$$m_i = \frac{1}{2}\left\{1 + r_1^2\left(\frac{i\pi}{2}\right)^2\right\}$$

Generalised stiffness

$$k_i = \frac{1}{2}\left(\frac{i\pi}{2}\right)^2\left\{1 + r_1^2 r_2\left(\frac{i\pi}{2}\right)^2\right\} \quad i = 1,3,5\ldots \tag{6.83}$$

Case 3 Free-free conditions

Eigen values

$$p_i^2 = (i\pi)^2 \frac{1 + r_1^2 r_2 (i\pi)^2}{1 + r_1^2 (i\pi)^2}$$

Mode shapes

$$\cos i\pi x$$

Generalised mass

$$m_i = \frac{1}{2}\{1 + r_1^2 (i\pi)^2\}$$

Generalised stiffness

$$k_i = \frac{1}{2}(i\pi)^2\} \{1 + r_1^2 r_2 (i\pi)^2\} \tag{6.84}$$

Using (6.75), equation (6.79) becomes

$$m_i \ddot{q}_i + (1 + t_1 \partial_t) k_i q_i = f_i$$

i.e.,

$$\ddot{q}_i + 2\beta_i \dot{q}_i + p_i^2 q_i = \frac{f_i}{m_i} \tag{6.85}$$

Damping in ith mode is given by

$$2\beta_i = t_1 p_i^2 \tag{6.85a}$$

We see from equation (6.85) that the viscoelastic bar in ith mode is governed by the equation of motion which corresponds to the damped spring mass system. Solution of equation (6.85) is

$$q_i(t) = e^{-\beta_i t}\left[q_i(0) \cos p_\epsilon t + \{\beta_i q_i(0) + \dot{q}_i(0)\} \frac{\sin p_\epsilon t}{p_\epsilon} \right]$$

$$+ \frac{1}{m_i p_\epsilon} \int_0^t f_i(\tau) e^{-\beta_i (t-\tau)} \sin p_\epsilon (t - \tau)\, d\tau \tag{6.86}$$

where the damped natural frequency of the bar is

$$p_\epsilon^2 = p_i^2 - \beta_i^2$$

The first two terms in equation (6.86) represent the damped free vibrations while the third term gives the forced vibration. The term β_i, represents the decay factor of the amplitude of vibration in ith mode and $p_i t_1$ is the loss factor in the ith mode.

If the damping measured by the relaxation time t_1 in (6.85) is small, then (6.86) signifies that the amplitude decreases slowly. The behaviour of the viscoelastic bar can however be quite complicated, because the higher modes can be overdamped. Let the equivalent viscous damping ratio be ξ, then,

$$2\beta_i = 2\xi p_i = t_1 p_i^2$$

Therefore

$$\xi = \frac{1}{2} t_1 p_i \tag{6.87}$$

Hence, ith mode of the system is overdamped if

$$t_1 > \frac{2}{p_i} \tag{6.88a}$$

It is critically damped if

$$t_1 = \frac{2}{p_i} \tag{6.88b}$$

and it is underdamped if

$$t_1 < \frac{2}{p_i} \tag{6.88c}$$

We see from equation (6.88a) that however small t_1 may be, there will always be higher modes which will be overdamped. Therefore, a viscoelastic bar will exhibit a complicated dynamic response when compared with an ordinary viscous damped single degree freedom system.

6.10 FREE VIBRATIONS OF VISCOELASTIC BARS

Example 1

Consider the case of a viscoelastic bar compressed by axial forces applied at the ends and suddenly released at time $t = 0$. Assuming that the middle of the free-free bar is stationary, the initial conditions are

$$u(x,0) = \varepsilon\left(\frac{1}{2} - x\right)$$

$$\dot{u}(x,0) = 0$$

where ε is the axial strain. With the help of case (3) in the previous section 6.9, for free-free bar conditions and equation (6.81), we can transform the above initial conditions to generalised coordinates as

$$q_i(0) = \frac{4\varepsilon}{(i\pi)^2} \quad i \text{ odd}$$

$$= 0 \quad i \text{ even}$$

$$\dot{q}_i(0) = 0$$

From (6.86), we have the generalised coordinate in ith mode as

$$q_i(t) = e^{-\beta_i t} \frac{4\varepsilon}{(i\pi)^2}\left\{\cos p_\epsilon t + \beta_i \frac{\sin p_\epsilon t}{p_\epsilon}\right\} \quad i = 1, 3, \ldots$$

From (6.69), the dynamic displacement of the Kelvin viscoelastic bar is

$$u(x,t)=\frac{4\varepsilon}{\pi^2}\sum_{i=1,3,\ldots}^{\infty}\frac{e^{-\beta_i t}}{i^2}\left\{\cos p_\epsilon t+\beta_i\frac{\sin p_\epsilon t}{p_\epsilon}\right\}\cos i\pi x$$

For an ordinary bar, we have

$$t_1=0 \quad p_i=p_\epsilon=i\pi$$

and therefore the response simplifies to

$$u(x,t)=\frac{4\varepsilon}{\pi^2}\sum_{i=1,3,5\ldots}^{\infty}\frac{\cos i\pi x\cos i\pi t}{i^2}$$

If there is any rigid body motion, it should be separately accounted.

Example 2

Let us now consider a Kelvin viscoelastic fixed-free bar subjected to an axial force at $x = 1$ which is suddenly released at time $t = 0$. The given initial conditions can be written as

$$u(x,0)=\varepsilon x$$

$$\dot{u}(x,0)=0$$

With the help of case (2) in the previous section 6.9 for fixed-free bar conditions and equation (6.81), we can transform the above initial conditions to generalized coordinates as

$$q_i(0)=\frac{8\varepsilon}{\pi^2}\frac{(-1)^{\frac{i-1}{2}}}{i^2}$$

$$\dot{q}_i(0)=0 \qquad i=1,3,\ldots$$

From (6.86), we have the generalised coordinate in the ith mode as

$$q_i(t)=e^{-\beta_i t}\frac{8\varepsilon}{\pi^2}\frac{(-1)^{\frac{i-1}{2}}}{i^2}\left\{\cos p_\epsilon t+\beta_i\frac{\sin p_\epsilon t}{p_\epsilon}\right\} \quad i=1,3,\ldots$$

The dynamic displacement of the viscoelastic bar is therefore

$$u(x,t)=\frac{8\varepsilon}{\pi^2}\sum_{i=1,3,\ldots}^{\infty}e^{-\beta_i t}\frac{(-1)^{\frac{i-1}{2}}}{i^2}\left\{\cos p_\epsilon t+\beta_i\frac{\sin p_\epsilon t}{p_\epsilon}\right\}\sin\frac{i\pi}{2}x$$

For an ordinary bar, we have

$$t_1=0 \qquad p_i=p_\epsilon=\frac{i\pi}{2}$$

and therefore the response simplifies to

$$u(x,t) = \frac{8\varepsilon}{\pi^2} \sum_{i=1,3,\ldots}^{\infty} \frac{(-1)^{\frac{i-1}{2}}}{i^2} \sin\frac{i\pi}{2}x \cos\frac{i\pi}{2}t$$

Example 3

Now consider a fixed-fixed bar subjected to a central axial force which is suddenly removed. When this central axial force is removed, the strain produced in the left half of the bar is equal to that in the right half of the bar. Therefore, the initial conditions are

$$u(x,0) = \varepsilon x \qquad 0 \le x \le \frac{1}{2}$$

$$= \varepsilon\,(1-x) \frac{1}{2} \le x \le 1$$

$$\dot{u}(x,0) = 0$$

With the help of case (1) in the previous section 6.9, for fixed-fixed bar conditions and equation (6.81), we can transform the above initial conditions to generalised coordinates as

$$q_i(0) = \frac{4\varepsilon}{\pi^2} \frac{(-1)^{\frac{i-1}{2}}}{i^2} \qquad i \quad \text{odd}$$

$$= 0 \qquad i \quad \text{even}$$

$$\dot{q}_i(0) = 0 \qquad i = 1, 2, \ldots$$

The dynamic displacement of the viscoelastic bar is therefore

$$u(x,t) = \frac{4\varepsilon}{\pi^2} \sum_{i=1,3,\ldots}^{\infty} e^{-\beta_i t} \frac{(-1)^{\frac{i-1}{2}}}{i^2} \left\{ \cos p_\epsilon t + \beta_i \frac{\sin p_\epsilon t}{p_\epsilon} \right\} \sin i\pi x$$

For an ordinary bar, we have

$$t_1 = 0 \quad p_i = p_\epsilon = i\pi$$

and therefore the response simplifies to

$$u(x,t) = \frac{4\varepsilon}{\pi^2} \sum_{i=1,3,\ldots}^{\infty} \frac{(-1)^{\frac{i-1}{2}}}{i^2} \sin i\pi x \cos i\pi t$$

Example 4

Consider a bar moving with a velocity v suddenly stopped at the end $x = 0$. After the bar is stopped, we can assume that the end $x = 0$ can be taken as fixed and the other end free. The initial conditions are therefore

$$u(x,0) = 0$$

$$\dot{u}(x,0) = v$$

This leads to

$$q_i(0) = 0$$

$$\dot{q}_i(0) = \frac{4v}{i\pi} \quad i = 1, 3, \ldots$$

The dynamic displacement of the viscoelastic bar is therefore

$$u(x,t) = \frac{4v}{\pi} \sum_{i=1,3,\ldots}^{\infty} e^{-\beta_i t} \frac{1}{i} \frac{\sin p_\epsilon t}{\pi} \sin\frac{i\pi}{2}x$$

For an ordinary bar, we have

$$t_1 = 0 \quad p_i = p_\epsilon = \frac{i\pi}{2}$$

and therefore the response simplifies to

$$u(x,t) = \frac{8v}{\pi^2} \sum_{i=1,3,\ldots}^{\infty} \frac{1}{i^2} \sin\frac{i\pi}{2}x \sin\frac{i\pi}{2}t$$

6.11 FORCED VIBRATIONS OF A VISCOELASTIC BAR

Example 1

Consider a fixed-free bar subjected to a suddenly applied axial force at $x = 1$. The initial conditions of the bar are

$$u(x,0) = 0$$

$$\dot{u}(x,0) = 0$$

$$F = P\delta(x-1)H(t)$$

where P is the intensity of the external force, δ is the Dirac delta function and $H(t)$ is Heaviside unit step function. In view of the first two initial conditions above, we find that

$$q_i = 0$$

$$\dot{q}_i = 0$$

The generalised force is given by

$$f_i = \int_0^1 P\delta(x-1)H(t)u_i(x)dx$$

$$= PH(t)u_i(1)$$

$$= PH(t)\sin\frac{i\pi}{2} \quad i = 1, 3, \ldots$$

From equation (6.86) we have

$$q_i(t)=\frac{1}{m_i p_\varepsilon}\int_0^t P\sin\frac{i\pi}{2}H(\tau)e^{-\beta_i(t-\tau)}\sin p_\varepsilon(t-\tau)d\tau$$

Using the eigen function expansion process, we get the dynamic displacement as

$$u(x,t)=P\sum_{i=1,3,\ldots}^{\infty}\frac{\sin\frac{i\pi}{2}\sin\frac{i\pi}{2}x}{m_i(\beta_i^2+p_\epsilon^2)}\left\{1-e^{-\beta_i t}\left(\cos p_\epsilon t+\beta_i\frac{\sin p_\epsilon t}{p_\epsilon}\right)\right\} \qquad (a)$$

For an elastic bar, according to the elementary theory

$$\beta_i=0$$

$$p_i=p_\varepsilon=\frac{i\pi}{2}$$

and the response of the ordinary bar at the free end simplifies to

$$u(1,t)=\frac{8}{\pi^2}P\sum_{i=1,3,}^{\infty}\frac{1-\cos\frac{i\pi}{2}t}{i^2} \qquad (b)$$

Figure 6.13 gives the response of the bar given in equations (*a*) and (*b*) above. The effect of viscoelasticity is to make the response oscillate about the static equilibrium position with lesser amplitude.

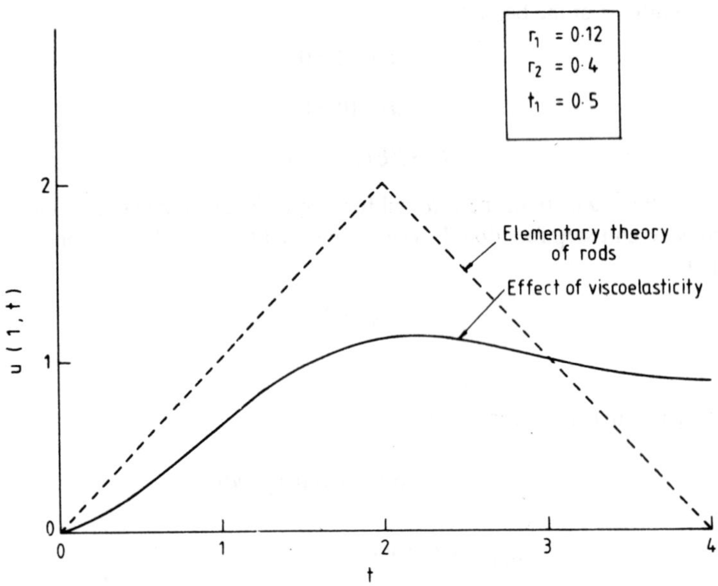

Fig. 6.13 Effects of viscoelasticity on forced vibration of a cantilever bar

Example 2

Now consider a free-free bar subjected to a force P suddenly applied at $x = 1$. The initial conditions are the same as in the example 1, viz.,

$$u(x,0) = 0$$

$$\dot{u}(x,0) = 0$$

$$F = P\delta(x-1)H(t)$$

Once again, we have the initial conditions in generalised coordinates

$$q_i = 0$$

$$\dot{q}_i = 0$$

The generalised force is given by

$$f_i = \int_0^1 P\delta(x-1)H(t)u_i(x)dx$$

$$= PH(t)u_i(1)$$

$$= PH(t)\cos i\pi \qquad i = 1,2,\ldots$$

From equation (6.86) we have

$$q_i(t) = \frac{1}{m_i p_\epsilon}\int_0^t P\cos i\pi\, H(\tau)\, e^{-\beta_i(t-\tau)}\sin p_\epsilon\,(t-\tau)\,d\tau$$

Using the eigen function expansion process, and including the rigid body displacement, we get the dynamic displacement as

$$u(x,t) = \frac{Pt^2}{2} + P\sum_{i=1,2,\ldots}^{\infty}\frac{\cos i\pi\cos i\pi x}{m_i(\beta_i^2 + p_\epsilon^2)}\left\{1 - e^{-\beta_i t}\left(\cos p_\epsilon t + \beta_i\frac{\sin p_\epsilon t}{p_\epsilon}\right)\right\}$$

For an elastic bar, according to the elementary theory

$$\beta_i = 0$$

$$p_i = p_\epsilon = i\pi$$

and the response of the ordinary bar simplifies to

$$u(x,t) = \frac{Pt^2}{2} + \frac{2}{\pi^2}P\sum_{i=1,2,\ldots}^{\infty}\frac{1}{i^2}\cos i\pi x\cos i\pi t(1-\cos i\pi t)$$

Example 3

As another example, consider the case of a fixed-free bar subjected to a sinusoidal axial excitation force at $x = 1$. Then the initial conditions are

$$u(x,0) = 0$$

$$\dot{u}(x,0) = 0$$

$$F = P\delta(x-1)\sin\omega t$$

Once again, we have the initial conditions and the excitation force in generalized coordinates as

$$q_i = 0$$

$$\dot{q}_i = 0$$

$$f_i = \int_0^1 P\delta(x-1)\sin\omega t\, u_i(x)dx$$

$$= P\sin\omega t\, u_i(1)$$

$$= P\sin\omega t \sin\frac{i\pi}{2} \quad i = 1,3,\ldots$$

As before we can obtain the dynamic displacement of the bar by using the eigen function expansion process as

$$u\,(x,t) = P \sum_{i=1,3,\ldots}^{\infty} \frac{\sin\frac{i\pi}{2}\sin\frac{i\pi}{2}x(p_i^2-\omega^2)}{m_i\{(p_i^2-\omega^2)^2+(2\beta_i-\omega)^2\}}\sin\omega t$$

For an elastic bar, according to the elementary theory

$$\beta_i = 0$$

$$p_i = p_\epsilon = \frac{i\pi}{2}$$

and the response for the special case of a bar according to the elementary theory simplifies to

$$u(x,t) = 2P \sum_{i=1,3,\ldots}^{\infty} \frac{\sin\frac{i\pi}{2}\sin\frac{i\pi}{2}x}{p_i^2-\omega^2}\sin\omega t$$

Example 4

Let us consider the solution of a general viscoelastic bar with its ith mode response defined by

$$m_i\ddot{q}_i + E\,(\partial_t)k_i q_i = f_i \tag{6.79a}$$

with the initial conditions

$$u(x,0) = 0$$

$$\dot{u}(x,0) = 0$$

and the excitation force to be sinusoidal with frequency ω such that the generalised force is

$$f_i = P\sin\omega t\sin\frac{i\pi}{2} \quad i = 1,3,\cdots$$

The solution of (6.79a) is

$$q_i(t) = \frac{f_i}{E(j\omega)k_i - m_i\omega^2}$$

where

$$j = \sqrt{-1}$$

The above solution can be written as

$$q_i(t) = \frac{P \sin \omega t \sin \frac{i\pi}{2}}{m_i p_i^2 \left\{ E(j\omega) - \left(\frac{\omega}{p_i}\right)^2 \right\}} \qquad i = 1, 3, \ldots$$

The dynamic response can then be written as

$$u(x,t) = P \sum_{i=1,3,\ldots}^{\infty} \frac{\sin \omega t \sin \frac{i\pi}{2}}{m_i p_i^2 \left\{ E(j\omega) - \left(\frac{\omega}{p_i}\right)^2 \right\}} \sin \frac{i\pi}{2} x$$

Consider now the viscoelastic bar to be a standard linear solid with its viscoelasticity modulus

$$E(\partial_t) = \frac{1 + t_1 \partial_t}{1 + t_2 \partial_t}$$

i.e.,

$$E(j\omega) = \frac{1 + jt_1\omega}{1 + jt_2\omega}$$

Then, the dynamic magnification factor for the ith mode is given by

$$H_i(\omega) = \left| \frac{P \sin \frac{i\pi}{2}}{m_i p_i^2} \frac{1 + j\omega t_2}{1 - r_i + j\omega (t_1 - r_1 t_2)} \right|$$

where the frequency ratio r_i is

$$r_i = \left(\frac{\omega}{p_i}\right)^2$$

For resonance in ith mode, i.e., $r_i = 1$, the dynamic magnification factor reduces to

$$H_i(\omega) = \left| \frac{P \sin \frac{i\pi}{2}}{m_i p_i^2} \frac{1 + j\omega t_2}{j\omega (t_1 - t_2)} \right|$$

At this frequency, the behaviour of the bar depends on the ratio of relaxation times. If $t_1 = t_2$, then resonance occurs, as it should, since $E(\partial_t)$ becomes a constant. The closer the two relaxation times, the larger is the dynamic magnification factor. Thus, in order to reduce the maximum value of dynamic magnification factor, the two relaxation time constants should be separated as far away as possible.

Figure 6.14 shows the dynamic magnification factor for the first mode of vibration of a fixed-free viscoelastic Bishop bar of the standard linear type. From the figure it is clear that the effect of the damping parameter t_1 is to increase the frequency of resonance. At high value of t_1, i.e., $t_1 = 2$, the response curve 3, is well damped.

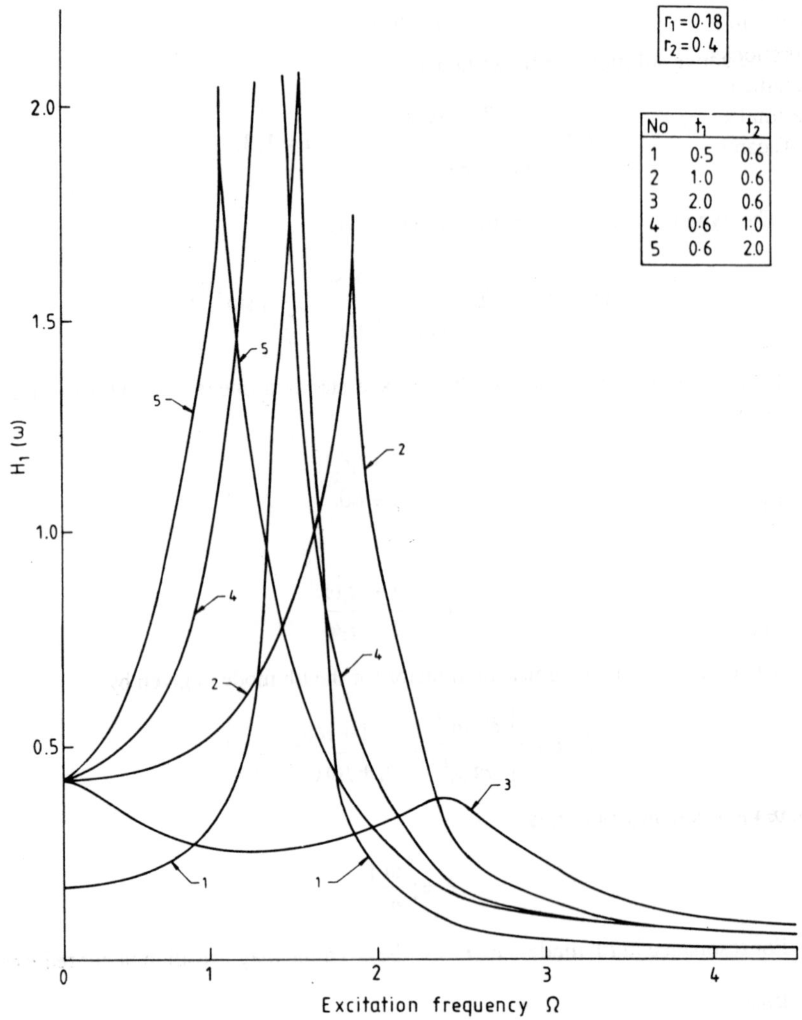

Fig. 6.14 Effects of shear stiffness and lateral inertia on the dynamic magnification factor of a uniform viscoelastic bar

6.12 VISCOELASTICALLY COATED BARS

Common structural materials do not possess very good viscoelastic properties and therefore it is not easy to suppress their large amplitude vibrations under resonant or near resonant conditions. Viscoelastic materials, which have good damping properties, unfortunately are not good for carrying loads and by themselves cannot be used as structural members. One way to incorporate both the requirements is to use a metallic core with a viscoelastic coating. Let a metallic bar with properties A_1, E_1, ... be encased by a thin viscoelastic coating with properties A_2, E_2, $E(\partial_t)$, ... as shown in Fig. 6.15. Such a coated bar can be used in reducing the transmission of vibrations to the nearby structures. Assuming that the coating is rigidly bonded to the metallic core, so that it undergoes the same axial displacement as the core material, the total energies of the system following Bishop's theory, from (6.9), (6.10) and (6.4) can be written as

$$U = \frac{1}{2}\int_0^l [\{E_1A_1 + E_2A_2E(\partial_t)\}u_{,x}^2 +$$

$$+\{\mu_1^2G_1I_{p1} + \mu_2^2G_2I_{p2}G(\partial_t)\}u_{,xx}^2 dx$$

$$T = \frac{1}{2}\int_0^l \{(\rho_1A_1 + \rho_2A_2)\dot{u}^2 + (\mu_1^2\rho_1I_{p1} + \mu_2^2\rho_2I_{p2})\dot{u}_{,x}^2\}\, dx$$

$$W = \int_0^l FAu\,dx \tag{6.89}$$

Let the viscoelastic material behaviour correspond to that of an incompressible Kelvin model with its modulus given by

$$E(\partial_t) = (1 + t_1\partial_t) = G(\partial_t)$$

Then, the strain energy is

$$U = \frac{1}{2}\int_0^l [\{E_1A_1 + E_2A_2(1 + t_1\partial_t)\}u_{,x}^2 +$$

$$+\{\mu_1^2G_1I_{pi} + \mu_2^2G_2I_{p_2}(1 + t_1\partial_t)\}u_{,xx}^2]dx \tag{6.90a}$$

We can rewrite equation (6.90*a*) as

$$U = \frac{1}{2}\int_0^l [\{(E_1A_1 + E_2A_2) + E_2A_2t_1\partial_t\}u_{,x}^2$$

$$+\{(\mu_1^2G_1I_{p1} + \mu_2^2G_2I_{p2}) + \mu_2^2G_2I_{p2}t_1\partial_t\}u_{,xx}^2]dx \tag{6.90b}$$

Rearranging equation (6.90*b*) as

$$U = \frac{1}{2}\int_0^l [EA(1 + t_1^*\partial_t)u_{,x}^2 + \mu^2GI_p(1 + t_1^{**}\partial_t)u_{,xx}^2]dx \tag{6.90}$$

where

$$EA = E_1A_1 + E_2A_2 \tag{6.91}$$

$$t_1^* = \frac{E_2A_2t_1}{E_1A_1 + E_2A_2} \tag{6.92}$$

$$\mu^2 G I_p = \mu_1^2 G_1 I_{p1} + \mu_2^2 G_2 I_{p2} \tag{6.93}$$

$$t_1^{**} = \frac{\mu_2^2 G_2 I_{p2}}{\mu_1^2 G_1 I_{p1} + \mu_2^2 G_2 I_{p2}} t_1 \tag{6.94}$$

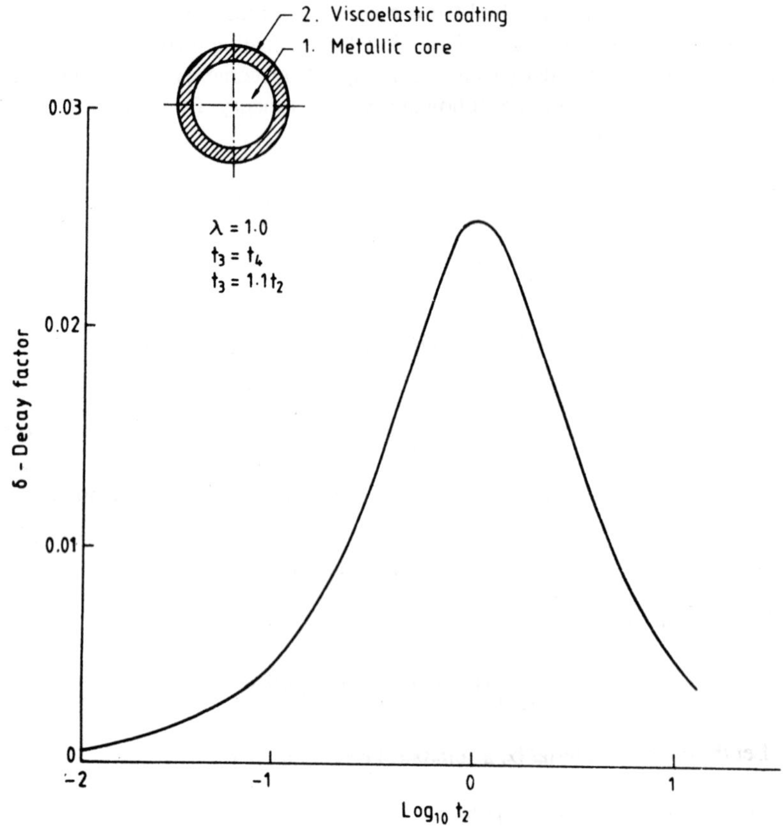

Fig. 6.15 Effect of viscoelastic coating on decay factor–δ

From (6.90) we can see that behaviour of the viscoelastically coated bars is very much similar to that of uncoated viscoelastic bars. Neglecting the shear stiffness, the behaviour of the coated bar is equivalent to that of an uncoated bar with an axial stiffness defined in (6.91) and relaxation time constant of (6.92). The effect of coating is therefore to increase the axial stiffness and simultaneously reduce the relaxation time constant.

Assuming the coating to be incompressible, so that

$$\mu_2 \leftarrow \frac{1}{2}$$

$$G_2 = \frac{E_2}{3}$$

and nondimensionalizing with respect to the base variables, as in (6.15), we get

$$x \leftarrow \frac{x}{l}$$

$$t \leftarrow \frac{t}{t_0}$$

$$t_0^2 = \frac{(\rho_1 A_1 + \rho_2 A_2)l^2}{E_1 A_1 + E_2 A_2}$$

$$u \leftarrow \frac{u}{l}$$

$$F \leftarrow \frac{F}{F_0}$$

$$F_0 = \frac{E_1 A_1 + E_2 A_2}{(A_1 + A_2)l}$$

$$U \leftarrow \frac{U}{U_0}$$

$$T \leftarrow \frac{T}{T_0}$$

$$W \leftarrow \frac{W}{W_0}$$

$$U_0 = T_0 = W_0 = (E_1 A_1 + E_2 A_2)l \tag{6.95}$$

Let the coating material be a standard linear solid so that

$$E(\partial_t) = \frac{1 + t_1 \partial_t}{1 + t_2 \partial_t} = G(\partial_t)$$

The nondimensionalised energies can then be obtained as

$$2U = \frac{1 + t_3 \partial_t}{1 + t_2 \partial_t} \int_0^1 u_{,x}^2 dx + r_1^2 \frac{1 + t_4 \partial_t}{1 + t_2 \partial_t} \int_0^1 u_{,xx}^2 dx$$

$$2T = \int_0^1 \dot{u}^2 dx + r_2^2 \frac{1 + t_4 \partial_t}{1 + t_2 \partial_t} \int_0^1 \dot{u}_{,x}^2 dx$$

$$W = \int_0^1 F u dx \tag{6.96}$$

where

$$r_1^2 = \frac{\mu_1^2 G_1 I_{p1} + \mu_2^2 G_2 I_{p2}}{(E_1 A_1 + E_2 A_2) l^2}$$

$$r_2^2 = \frac{\mu_1^2 \rho_1 I_{p1} + \mu_2^2 \rho_2 I_{p2}}{(\rho_1 A_1 + \rho_2 A_2) l^2}$$

$$t_3 = \frac{E_1 A_1 t_2 + E_2 A_2 t_1}{E_1 A_1 + E_2 A_2}$$

$$t_4 = \frac{\mu_1^2 G_1 I_{p1} t_2 + \mu_2^2 G_2 I_{p2} t_1}{\mu_1^2 G_1 I_{p1} + \mu_2^2 G_2 I_{p2}} \tag{6.97}$$

Application of Hamilton's principle to the energies in equation (6.96) for free vibrations leads to

$$(1 + t_2 \partial_t)\ddot{u} - \{1 + t_3 \partial_t + r_2^2 (1 + t_2 \partial_t) \partial_t^2\} \partial_x^2 u + r_1^2 (1 + t_4 \partial_t) \partial_x^4 u = 0 \tag{6.98}$$

$$[\{1 + t_3 \partial_t + r_2^2 (1 + t_2 \partial_t) \partial_t^2\} \partial_x u - r_1^2 (1 + t_4 \partial_t) \partial_x^3 u] \delta u \,|_0^1 +$$

$$+ r_1^2 (1 + t_4 \partial_t) \, \partial_x^2 u \, \delta \partial_x u \,|_0^1 = 0 \tag{6.99}$$

When $r_1 = r_2 = 0$, we have the special case of an elementary bar coated with a viscoelastic material.

$$(1 + t_2 \partial_t)\ddot{u} - (1 + t_3 \partial_t) \, \partial_x^2 u = 0 \tag{6.100}$$

$$(1 + t_3 \partial_t) \, \partial_x u \, \delta u \,|_0^1 = 0 \tag{6.101}$$

We can assume a solution for (6.100) as

$$u(x,t) = u_0 e^{px} e^{qt} \tag{6.102}$$

where p and q are complex numbers and substitute the same to obtain the following characteristic equation.

$$-p^2 q^0 - t_3 p^2 q^1 + q^2 + t_2 q^3 = 0 \tag{6.103}$$

Since in the cases of practical interest, the amplitude will be exponentially damped, it will be convenient to consider the solutions for which, $p = i\,\lambda$ where λ denotes the wavelength of the bar. Then

$$\lambda^2 q^0 + t_3 \lambda^2 q^1 + q^2 + t_2 q^3 = 0 \tag{6.104}$$

There will be in general three complex roots of the above polynomial, which are the complex frequencies, satisfying the above polynomial. Let each root be of the form

$$q = \delta + j\omega \tag{6.105}$$

Then equation (6.102) becomes

$$u(x,t) = u_0 e^{i\lambda x} e^{(\delta + j\omega)t} \tag{6.102a}$$

The amount of damping of the wave is measured by the real part δ in the above. Let us consider the following data

$$\lambda = 1$$

$$t_3 = 1.1 t_2$$

$$t_3 = t_4$$

The values of the decay factor δ are plotted in Fig. 6.15 as a function of the relaxation time constant t_2. From the graph it can be observed that the decay factor reaches a maximum when $t_2 = 1$.

7

Torsional Vibrations of Rods

We will continue to discuss the case of bars, which are described by only one axial coordinate, i.e., one dimensional structures. We will consider the case of torsional vibration of such bars about the x axis in this chapter. For distinguishing the nomenclature from the bars executing longitudinal vibration, one dimensional structural members vibrating in torsion are called rods.

7.1 ELEMENTARY THEORY OF CIRCULAR RODS

Consider the rod of length l shown in Fig. 7.1. The cross-sectional area is A and the second polar moment of area is denoted I_p. The mass density of the rod is ρ. The rod is acted on by an external moment $M(x, t)$ and executes torsional vibrations about the axis x.

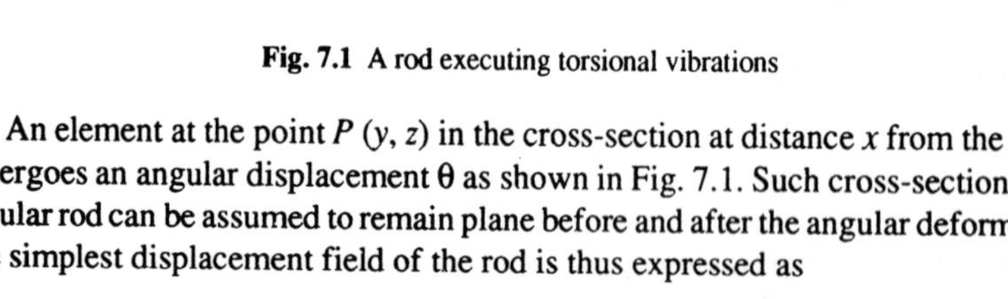

Fig. 7.1 A rod executing torsional vibrations

An element at the point $P(y, z)$ in the cross-section at distance x from the origin undergoes an angular displacement θ as shown in Fig. 7.1. Such cross-sections for a circular rod can be assumed to remain plane before and after the angular deformation. The simplest displacement field of the rod is thus expressed as

$$u_x = 0$$

$$u_y = -z\theta(x,t)$$

$$u_z = y\theta(x,t) \tag{7.1}$$

The strain field is

$$\varepsilon_{xx} = \varepsilon_{yy} = \varepsilon_{zz} = \varepsilon_{yz} = 0$$

$$\varepsilon_{xy} = -\frac{1}{2} z \theta_{,x}$$

$$\varepsilon_{xz} = \frac{1}{2} y \theta_{,x} \tag{7.2}$$

The stress field is therefore given by

$$\tau_{xx} = \tau_{yy} = \tau_{zz} = \tau_{yz} = 0$$

$$\tau_{xy} = -Gz\,\theta_{,x}$$

$$\tau_{xz} = G y\,\theta_{,x} \tag{7.3}$$

Thus, the strain energy is

$$U = \frac{1}{2}\int_0^l \int_A G\,\theta_{,x}^2\,(z^2 + y^2)\,dA\,dx$$

$$= \frac{1}{2}\int_0^l G I_p \theta_{,x}^2\,dx \tag{7.4}$$

The kinetic energy is given by

$$T = \frac{1}{2}\int_0^l \int_A \rho\,(z^2\dot{\theta}^2 + y^2\dot{\theta}^2) dA\,dx$$

$$= \frac{1}{2}\int_0^l \rho I_p\,\dot{\theta}^2\,dx \tag{7.5}$$

The work done by M, the external moment per unit length, is

$$W = \int_0^l M\,\theta\,dx \tag{7.6}$$

Application of Hamilton's principle leads to

$$\delta \int_{t1}^{t2} \int_0^l \left[\frac{1}{2} G\,I_p \theta_{,x}^2 - \frac{1}{2}\rho I_p \dot{\theta}^2 - M\theta\right] dx\,dt = 0 \tag{7.7}$$

Taking the variation in (7.7) above, we get

$$\rho\,I_p \ddot{\theta} - (G\,I_p \theta_{,x})_{,x} = M\,(x,t) \tag{7.8}$$

$$G\,I_p \theta_{,x} \delta\theta\,|_0^l = 0 \tag{7.9}$$

We can rewrite (7.8) and (7.9) as

$$\rho\,I_p\,\ddot{\theta} - (C\theta_{,x})_{,x} = M\,(x,t) \tag{7.8a}$$

$$C\theta_{,x}\,\delta\theta\,|_0^l = 0 \tag{7.9a}$$

where C is the torsional rigidity of the rod given by

$$C = G\,I_p \tag{7.10}$$

A comparison of the above with equation (6.5) shows that the governing differential equation and the boundary conditions of the rod in torsional vibration are very similar to those of the bar in longitudinal vibration for the elementary theory.

The solution for free and forced torsional vibration of rods is same as that of the longitudinal vibration of bars with the following parameter equivalence for both the motions.

Table 7.1
Equivalence of bars and rods in elementary theory

Parameter	*Longitudinal vibration*	*Torsional vibration*
Mass	ρA	ρI_p
Stiffness	$E A$	$C = G I_p$
Force	$F A$	M

7.2 ST VENANT'S THEORY OF NONCIRCULAR RODS

For noncircular rods, the assumption that the cross-sections remain plane after deformation is not valid. The cross-section undergoes warping, thus generating an axial displacement of the rod. As an example, a rod with rectangular cross-section shown in Fig. 7.2 shows the distortions after twist which are greatest at the middle of the sides, i.e., at the points which are nearest to axis of the rod. The warping function is denoted by ϕ (y, z) and the displacement in the axial direction, according to St, Venant, is $\phi\,(y,z)\,\theta_{,x}$. St. Venant has shown this to be correct by using a semi inverse method, the details of which are not given here. Thus the displacement field for a rod of any arbitrary cross-section is expressed as

$$u_x = \phi\,(y,z)\,\theta_{,x}$$

$$u_y = -z\theta\,(x,t)$$

$$u_z = y\theta\,(x,t) \tag{7.11}$$

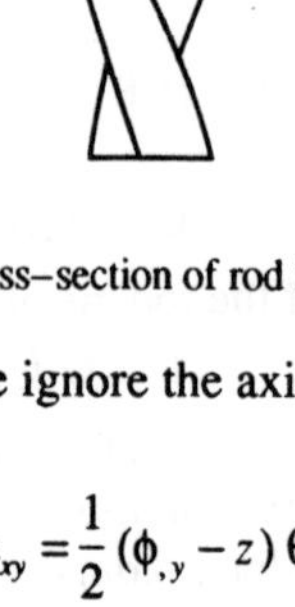

Fig. 7.2 Cross–section of rod in torsion

As in the elementary theory, we ignore the axial strain ε_{xx} and modify the shear strains in equation (7.2) as

$$\varepsilon_{xy} = \frac{1}{2}(\phi_{,y} - z)\,\theta_{,x}$$

$$\varepsilon_{xz} = \frac{1}{2}(\phi_{,z} + y)\,\theta_{,x} \tag{7.12}$$

The corresponding stress field is

$$\tau_{xy} = G\ (\phi_{,y} - z)\ \theta_{,x}$$

$$\tau_{xz} = G\ (\phi_{,z} + y)\ \theta_{,x} \qquad (7.13)$$

The potential energy now is

$$U = \frac{1}{2}\int_0^l \int_A G\ \{(\phi_{,y} - z)^2 + (\phi_{,z} + y)^2\}\ \theta_{,x}^2\ dA\ dx$$

$$= \frac{1}{2}\int_0^l C\theta_{,x}^2\ dx \qquad (7.14)$$

where the torsional rigidity for noncircular section rod is given by

$$C = \int_A G\ \{(\phi_{,y} - z)^2 + (\phi_{,z} + y)^2\}\ dA \qquad (7.15)$$

If we ignore the axial inertia, the kinetic energy expression remains same as equation (7.5) and the work done by external moment is given in equation (7.6). Application of Hamilton's principle now leads to

$$\delta \int_{t1}^{t2}\left[\int_v \frac{1}{2} G\ \{(\phi_{,y} - z)^2 + (\phi_{,z} + y)^2\}\ \theta_{,x}^2\ dA\ dx\right]$$

$$\left[-\int_0^l \frac{1}{2}\rho I_p\ \dot{\theta}^2\ dx - \int_0^l M\theta\ dx\right] dt = 0 \qquad (7.16)$$

The second and third integrals in (7.16) above are same as those in equation (7.7) and lead to the same inertia and forcing terms in the differential equation (7.8). The first integral I_1 in (7.16) above is separately considered below.

$$I_1 = \delta \int_{t1}^{t2} \int_v \frac{1}{2} G\ \{(\phi_{,y} - z)^2 + (\phi_{,x} + y)^2\}\theta_{,x}^2\ dA\ dx\ dt \qquad (7.17a)$$

Taking the variation in (7.17a) above, we get

$$I_1 = \int_{t1}^{t2} \int_0^l \int_A G\ \{(\phi_{,y} - z)^2 + (\phi_{,z} + y)^2\}\ \theta_{,x}\ \delta\,\theta_{,x}\ dA\ dx\ dt$$

$$+ \int_{t1}^{t2} \int_v G\theta_{,x}^2\ \{(\phi_{,y} - z)\ \delta\,\phi_{,y} + (\phi_{,z} + y)\ \delta\phi_{,z}\}\ dA\ dx\ dt \qquad (7.17b)$$

Inview of (7.15) the first integral in the above equation (7.17b) is

$$I_1' = \int_{t1}^{t2} \int_0^l \int_A C\theta_{,x}\delta\theta_{,x} dx\ dt$$

which leads to the stiffness term in the differential equation (7.8a). We also note from (7.17b) that $\delta\phi$ and $\delta\theta$ are independent; the first integral joins with the rest of the terms in the variational process with $\delta\theta$ and leads to the differential equation

$$\rho I_p \ddot{\theta} - (C\theta_{,x})_{,x} = M\ (x, t) \qquad (7.18)$$

where C is given by equation (7.15) for noncircular rod. The second term in (7.17b) is set independently to zero to give

$$\int_v G\theta_{,x}^2 \{(\phi_{,y} - z)\,\delta\phi_{,y} + (\phi_{,z} + y)\,\delta\phi_{,z}\}\; dy\, dz\, dx = 0 \qquad (7.17c)$$

Integrating (7.17c) above by parts

$$g\left[\int_s (\phi_{,y} - z)\, a_v \delta\phi ds - \int\int \phi_{,yy}\delta\,\phi\, dy\, dz\right.$$

$$\left. + \int_s (\phi_{,z} + y) a_{nz} \delta\,\phi\, ds - \int\int \phi_{,zz}\delta\,\phi\, dy\, dz\right] = 0 \qquad (7.19)$$

where

$$g = \int_0^l G\theta_{,x}^2\, dx \qquad (7.19a)$$

i.e.,

$$\nabla^2 \phi = 0 \qquad (7.20)$$

$$(\phi_{,y} - z)\, a_v + (\phi_{,z} + y)\, a_{nz} = 0 \quad \text{(on boundary)} \qquad (7.21)$$

Equation (7.20) is Laplace equation and its solution subject to the boundary conditions in (7.21) defines the warping function ϕ. For a noncircular rod, therefore, we first solve equation (7.20) and then determine C by using equation (7.15). Otherwise, the governing differential equations remain same as those of a circular rod.

7.3 LOVE'S THEORY OF RODS

Love's theory includes the axial inertia term arising out of the warping deformation in St Venant's theory. Thus, the kinetic energy expression is

$$T = \frac{1}{2}\int_0^l \int_A \rho(z^2\dot{\theta}^2 + y^2\dot{\theta}^2 + \phi^2\theta_{,xt}^2)\, dA\, dx$$

$$= \frac{1}{2}\int_0^l \rho I_p\, \dot{\theta}^2\, dx + \frac{1}{2}\int_0^l \int_A \phi^2\, dA\, \rho\theta_{,xt}^2 dx$$

The second term in the above integral is the effect of axial inertia, which is

$$I_L = \frac{1}{2}\int_0^l \rho\left(\int_A \phi^2\, d\,A\right)\theta_{,xt}^2\, dx$$

The corresponding term in Hamilton's principle application is

$$\delta\int_{t1}^{t2} I_L\, dt = \delta\int_{t1}^{t2}\frac{1}{2}\int_0^l \rho\left(\int_A \phi^2\, dA\right)\theta_{,xt}^2\, dx\, dt \qquad (a)$$

Taking the variation in (*a*) above, we get

$$\delta I_L = \int_{t1}^{t2}\int_0^l \rho\left(\int_A \phi^2\, dA\right)\theta_{,xt}\,\delta\theta_{,xt} dx\, dt + \int_{t1}^{t2}\int_0^l \rho\theta_{,xt}^2\left(\int_A \phi\,\delta\,\phi\, dA\right) dx\, dt \qquad (b)$$

The first integral in (*b*) above is independent of the second and is to be combined with the terms of the St. Venant's theory for torsional vibration. The second one belongs to the set of terms of the Laplace differential equation defining the warping function. The first integral is

$$\delta I_{L1} = \int_{t1}^{t2} \int_0^l \rho I_{\phi\phi} \theta_{,xt}\, \delta\theta_{,xt}\, dx\, dt \tag{c}$$

where

$$I_{\phi\phi} = \int_A \phi^2 dA \tag{7.22}$$

Integrating first with respect to time equation (*c*) becomes

$$\delta I_{L1} = -\int_{t1}^{t2} \int_0^l (\rho I_{\phi\phi} \theta_{,xt})_{,t}\, \delta\theta_{,x}\, dx\, dt$$

Now integrating by parts with respect to *x*, we get

$$\delta I_{L1} = -\int_{t1}^{t2} (\rho I_{\phi\phi} \theta_{,xt})_{,t}\, \delta\theta \,|_0^l\, dt + \int_{t1}^{t2} \int_0^l (\rho I_{\phi\phi} \theta_{,xt})_{,xt} \delta\theta\, dx\, dt \tag{d}$$

In view of (*d*) above, the differential equation of motion and the boundary conditions now get modified to

$$\rho I_p \ddot{\theta} - (\rho I_{\phi\phi} \theta_{,xt})_{,xt} - (C\theta_{,x})_{,x} = M\,(x,t) \tag{7.23}$$

$$(C\theta_{,x} + \rho I_{\phi\phi} \ddot{\theta}_{,x})\, \delta\theta \,|_0^l = 0 \tag{7.24}$$

Now, we consider the second integral in equation (b), which is

$$\delta I_{L2} = \int_{t1}^{t2} \int_0^l \rho\, \theta_{,xt}^2 \left(\int_A \phi\, \delta\phi\, dA \right) dx\, dt = \int_{t1}^{t2} r \int_A \phi\, \delta\phi\, dA\, dt \tag{e}$$

where

$$r = \int_0^l \rho\, \theta_{,xt}^2 \tag{7.25}$$

We can now combine (*e*) with (7.19) to get

$$g \left[\int_s (\phi_{,y} - z)\, a_v \delta\phi\, ds - \int\int \phi_{,yy} \delta\phi\, dy\, dz + \int_s (\phi_{,z} + y)\, a_{nz}\, \delta\phi\, ds \right.$$

$$\left. - \int\int \phi_{,zz}\, \delta\phi\, dy\, dz \right] - r \int_A \phi\, \delta\phi\, dA = 0 \tag{7.26a}$$

i.e.,

$$\nabla^2 \phi + \frac{r}{g} \phi = 0 \tag{7.26}$$

The boundary conditions remain same as (7.21).

$$(\phi_{,y} - z)\, a_v + (\phi_{,z} + y)\, a_{nz} = 0 \quad \text{(on boundary)} \tag{7.21a}$$

In practice, however, it is easy to solve equation (7.20) rather than equation (7.26) and the axial inertia effect on the solution for the warping function may be neglected.

7.4 TIMOSHENKO-GERE THEORY

The displacement field for the rod used in St Venant's theory or Love's theory is

$$u_x = \phi\,(y,z)\,\theta_{,x}$$

$$u_y = -z\theta\,(x,t)$$

$$u_z = y\theta\,(x,t) \tag{7.11}$$

In the theories discussed so far we ignored the axial strain ε_{xx} and also the normal strains in y and z directions, ε_{yy} ε_{zz}. Timoshenko's theory accounts for the normal strain ε_{xx} in the torsion of rods. The strain field thus is given by

$$\varepsilon_{xx} = \phi\theta_{,xx}$$

$$\varepsilon_{yy} = 0$$

$$\varepsilon_{zz} = 0$$

$$\varepsilon_{xy} = \frac{1}{2}\,(\phi_{,y} - z)\,\theta_{,x}$$

$$\varepsilon_{xz} = \frac{1}{2}\,(\phi_{,z} + y)\,\theta_{,x}$$

$$\varepsilon_{yz} = 0 \tag{7.27}$$

In the above equations (7.27) the influence of Poisson's ratio is neglected. The stress field is then given by

$$\tau_{xx} = E\,\phi\,\theta_{,xx}$$

$$\tau_{yy} = \tau_{zz} = 0$$

$$\tau_{xy} = G\,(\phi_{,y} - z)\,\theta_{,x}$$

$$\tau_{xz} = G\,(\phi_{,z} + y)\,\theta_{,x}$$

$$\tau_{yz} = 0 \tag{7.28}$$

The strain energy is given by

$$U = \frac{1}{2}\int_0^l \int_A G\,\{(\phi_{,y} - z)^2 + (\phi_{,z} + y)^2\}\,\theta_{,x}^2\,dA\,dx + \frac{1}{2}\int_0^l \int_A E\phi^2\theta_{,xx}^2\,dx\,dA \tag{7.29}$$

The kinetic energy and work done by the external moment remain same as in the case of Love's theory. In equation (7.29) the first integral has been accounted in Love's theory. The additional term due to the axial strain ε_{xx} is considered below.

The appropriate integral in the variational equation is

$$\delta I = \frac{1}{2}\delta \int_{t1}^{t2}\int_0^l \int_A E\phi^2\theta_{,xx}^2 dx\, dA\, dt \tag{a}$$

Taking the variation in (*a*) above, we get

$$\delta I = \int_{t1}^{t2}\int_0^l \int_A E\,\phi\,\delta\phi\,\theta_{,xx}^2\, dx\, dA\, dt + \int_{t1}^{t2}\int_0^l \int_A E\,\phi^2\theta_{,xx}\,\delta\theta_{,xx} dx\, dA\, dt \tag{7.30}$$

Once again we have two integrals above, the first one corresponding to the warping function equation and the second one to the torsional vibration of the rod. Considering the second integral in (7.30), we have

$$\delta I_2 = \int_{t1}^{t2}\int_0^l EI_{\phi\phi}\theta_{,xx}\delta\theta_{,xx} dx\, dt \tag{b}$$

Integrating by parts, (*b*) gives

$$\delta I_2 = \int_{t1}^{t2} [EI_{\phi\phi}\theta_{,xx}\delta\,\theta_{,x}\,|_0^l - (EI_{\phi\phi}\theta_{,xx})_{,x}\delta\theta\,|_0^l$$

$$+ \int_0^l (EI_{\phi\phi}\theta_{,xx})_{,xx}\delta\,\theta\, dx]\, dt \tag{7.31}$$

In view of (7.31) above, the differential equation of motion (7.23) and the boundary conditions (7.24) now get modified to

$$\rho I_p\ddot{\theta} - (\rho I_{\phi\phi}\theta_{,xt})_{,xt} - (C\theta_{,x})_{,x} + (EI_{\phi\phi}\theta_{,xx})_{,xx} = M\,(x,t) \tag{7.32}$$

$$\{C\theta_{,x} + \rho I_{\phi\phi}\ddot{\theta}_{,x} - (EI_{\phi\phi}\theta_{,xx})_{,x}\}\ \delta\theta\,|_0^l = 0$$

$$EI_{\phi\phi}\theta_{,xx}\delta\theta_{,x}\,|_0^l = 0 \tag{7.33}$$

Consider now the first integral in (7.30)

$$\delta I_1 = \int_{t1}^{t2}\int_0^l \int_A E\,\phi\,\theta_{,xx}^2\,\delta\phi\, dx\, dA\, dt$$

We can combine this with equation (7.26*a*) to give the following

$$g\left[\int_s (\phi_{,y} - z)\, a_v\delta\phi\, ds - \int\int \phi_{,yy}\delta\phi\, dy\, dz + \int_s (\phi_{,z} + y)\, a_{nz}\delta\phi\, ds\right.$$

$$\left. - \int\int \phi_{,zz}\delta\phi\, dy\, dz\right] - r\int_A \phi\,\delta\phi\, dA + d\int_A \phi\,\delta\phi\, dA = 0$$

where

$$d = \int_0^l E\theta_{,xx}^2 dx \tag{7.34}$$

Therefore, equation (7.26) gets modified to

$$g\nabla^2\phi + (r - d)\,\phi = 0 \tag{7.35}$$

The boundary conditions remain same as (7.21)

$$(\phi_{,y} - z)\, a_v + (\phi_{,z} + y)\, a_{nz} = 0 \quad \text{(on boundary)} \tag{7.21}$$

In practice, however, it is easy to solve equation (7.20) rather than equation (7.26) or (7.35) and the axial inertia and strain effects on the solution for the warping function may be neglected. For a detailed analysis of this equation (7.20), any book on Theory of Elasticity may be referred and we will assume here that ϕ is defined by such analysis. We now proceed to nondimensionalize the governing differential equation of motion (7.32) and the boundary conditions (7.33), by choosing reference quantities for each of the parameters and let us set

$$x \leftarrow \frac{x}{l}$$

$$t \leftarrow \frac{t}{t_0}$$

where t_0 is defined by the square of the rod velocity given by

$$\frac{l^2}{t_0^2} = \frac{C_0}{\rho_0 I_{p0}}$$

$$\rho \leftarrow \frac{\rho}{\rho_0}$$

$$I_p \leftarrow \frac{I_p}{I_{p0}}$$

$$I_{\phi\phi} \leftarrow \frac{I_{\phi\phi}}{I_{\phi\phi 0}}$$

$$I_{\phi\phi 0} = I_{p0} l^2$$

$$E \leftarrow \frac{E}{E_0}$$

$$M \leftarrow \frac{M}{M_0}$$

$$M_0 = \frac{C_0}{l^2}$$

$$r_2 = \frac{E_0 I_{\phi\phi 0}}{C_0 l^2} \tag{7.36}$$

With the help of the above, equations (7.32) and (7.33) become

$$\rho I_p \ddot{\theta} - (\rho I_{\phi\phi} \theta_{,xt})_{,xt} - (C\theta_{,x})_{,x} + r_2 (EI_{\phi\phi} \theta_{,xx})_{,xx} = M(x,t) \tag{7.37}$$

$$\{C\theta_{,x} + \rho I_{\phi\phi} \ddot{\theta}_{,x} - r_2 (EI_{\phi\phi} \theta_{,xx})_{,x}\} \, \delta\theta \, |_0^1 = 0$$

$$(EI_{\phi\phi} \theta_{,x})_{,x} \delta\theta_{,x} \, |_0^1 = 0 \tag{7.38}$$

Referring to the nondimensional equations governing the longitudinal motion of a Bishop bar in chapter 6, section 1, equations (6.13*a*) and (6.14*a*) are written down here once again for the purpose of comparison with the torsional motion of a rod according to Timoshenko-Gere theory.

$$(\rho A - r_1^2 \partial_x \rho I_p \partial_x)\ddot{u} + (-\partial_x EA \partial_x + r_1^2 r_2^2 \partial_x^2 G I_p \partial_x^2) u = FA \tag{6.13a}$$

$$(r_1^2 \rho I_p \partial_x \ddot{u} + EA \partial_x u - r_1^2 r_2 \partial_x G I_p \partial_x^2 u)\, \delta u \,|_0^1 +$$

$$+(r_1^2 r_2 G I_p \partial_x^2 u) \delta \partial_x u \,|_0^1 = 0 \tag{6.14a}$$

A comparison of the differential equations (7.37) against (6.13*a*) and boundary conditions (7.38) against (6.14*a*), for the two cases show that they are similar to each other in all respects as per the commonalities of the parameters listed in Table 7.2.

Table 7.2

Comparison of parameters for longitudinal vibrations of bars and torsional vibrations of rods

Longitudinal Vibration	*Torsional Vibration*
A	I_p
I_p	$I_{\phi\phi}$
EA	C
GI_p	$EI_{\phi\phi}$
r_1 *See equation (6.15)*	$r_1 = 1$
r_2 *See equation (6.15)*	r_2 *See equation (7.36)*

Hence, the solution of torsional vibrations of rods according to Timoshenko-Gere theory is same as that given in chapter 6 for the longitudinal vibrations of bars according to Bishop theory. The natural frequencies and mode shapes for different boundary conditions discussed in section 6.4 for 10 different cases are applicable to the case of torsional vibrations of rods along with the orthogonal properties in section 6.5. The analysis of free and forced vibrations of rods in torsion can also be accomplished just as in the previous chapter for longitudinal vibrations. The analysis for tapered rods is also similar to that given in chapter 6. We should keep in our mind that the warping function for torsion of rods is to be separately determined before making use of this analysis.

7.5 TORSIONAL RIGIDITY OF NONCIRCULAR RODS

Before obtaining the solution for the torsional vibration, it is necessary to determine the torsional rigidity C of the rod. This is obtained from the solution of Laplace equation derived earlier according to St Venant's theory.

$$\nabla^2 \phi = 0 \tag{7.20a}$$

$$\tau_{xy} a_v + \tau_{xx} a_{nz} = 0 \quad \text{(on boundary)} \tag{7.21a}$$

The boundary conditions in (7.21) are complicated and therefore it is difficult to obtain the solution of (7.20). We can write the above equations with another independent variable, namely, the stress function instead of the warping function, which will simplify the boundary conditions. This is done as follows.

Consider a rod shown in Fig. 7.3 where, the angle of rotation at distance x is given by

$$\theta(x) = \Theta x \tag{7.39}$$

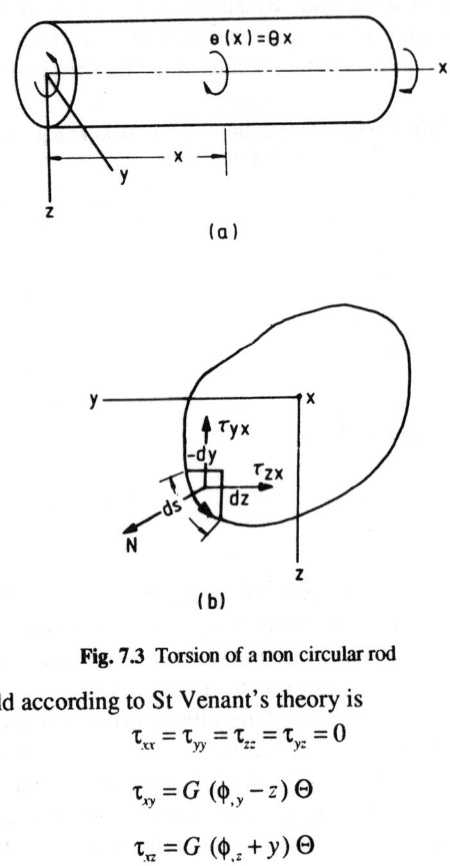

Fig. 7.3 Torsion of a non circular rod

The stress field according to St Venant's theory is

$$\tau_{xx} = \tau_{yy} = \tau_{zz} = \tau_{yz} = 0$$

$$\tau_{xy} = G\,(\phi_{,y} - z)\,\Theta$$

$$\tau_{xz} = G\,(\phi_{,z} + y)\,\Theta \tag{7.40}$$

In view of the first equation in the above, the equations of equilibrium, (A3.6, Appendix 3), give

$$\frac{\partial \tau_{xy}}{\partial x} = 0$$

$$\frac{\partial \tau_{xz}}{\partial x} = 0$$

$$\frac{\partial \tau_{xy}}{\partial y}+\frac{\partial \tau_{xz}}{\partial z}=0 \tag{7.41}$$

The first two in the above equation (7.41) are already satisfied in (7.40). To satisfy the third equation, we can express

$$\tau_{xy}=\frac{\partial \Phi}{\partial z}$$

$$\tau_{xz}=-\frac{\partial \Phi}{\partial y} \tag{7.42}$$

With the help of (7.40), equations (7.42) can be written as

$$\frac{\partial \Phi}{\partial z}=G\Theta\,(\phi_{,y}-z) \tag{a}$$

$$-\frac{\partial \Phi}{\partial y}=G\Theta\,(\phi_{,z}+y) \tag{b}$$

Differentiating (*a*) with respect to z and (*b*) with respect to y, we get

$$\frac{\partial^2 \Phi}{\partial z^2}=G\Theta\,(\phi_{,zy}-1) \tag{c}$$

$$-\frac{\partial^2 \Phi}{\partial y^2}=G\Theta\,(\phi_{,yz}+1) \tag{d}$$

Subtracting (*d*) from (*c*), we get

$$\nabla^2\Phi=-2G\Theta \tag{7.43}$$

With the help of (7.42), the boundary conditions (7.21) become

$$\frac{\partial \Phi}{\partial z}a_{v}+\frac{\partial \Phi}{\partial y}a_{nz}=0$$

or

$$\frac{\partial \Phi}{\partial s}=0$$

i.e., the stress function Φ is constant on the boundary of the cross-section, and we can conveniently choose this to be zero. Thus, the boundary condition for the problem can be written as

$$\Phi=0 \quad \text{on the boundary} \tag{7.44}$$

Now, let us consider the ends of the twisted rod, whose normals are parallel to the x axis. The resultant forces on these end faces can be determined as

$$\iint \tau_{xy}\,dy\,dz=\iint \frac{\partial \Phi}{\partial z}\,dy\,dz$$

$$= \int dy \int \frac{\partial \Phi}{\partial z} dz = 0$$

$$\iint \tau_{xz}\, dy\, dz = \iint \frac{\partial \Phi}{\partial y} dy\, dz$$

$$= \int dz \int \frac{\partial \Phi}{\partial y} dy = 0$$

Thus, the resultant forces distributed over the ends of the rod are zero. However, there is resultant couple, M_t, which is

$$M_t = -\iint \frac{\partial \Phi}{\partial z} z\, dy\, dz - \iint \frac{\partial \Phi}{\partial y} y\, dy\, dz$$

Integrating by parts

$$M_t = -\int \Phi z\, dy + \iint \Phi\, dy\, dz - \int \Phi y\, dz + \iint \Phi\, dy\, dz$$

$$= 2\iint \Phi\, dy\, dz \tag{7.45}$$

There are very few cross-sections for which we can get an exact solution of the torsion problem. One such section is an elliptical cross-section, shown in Fig. 7.4. Let the boundary of the cross-section be given by

$$\frac{y^2}{a^2} + \frac{z^2}{b^2} - 1 = 0 \tag{e}$$

Consider the stress function

$$\Phi = m\left(\frac{y^2}{a^2} + \frac{z^2}{b^2} - 1\right) \tag{f}$$

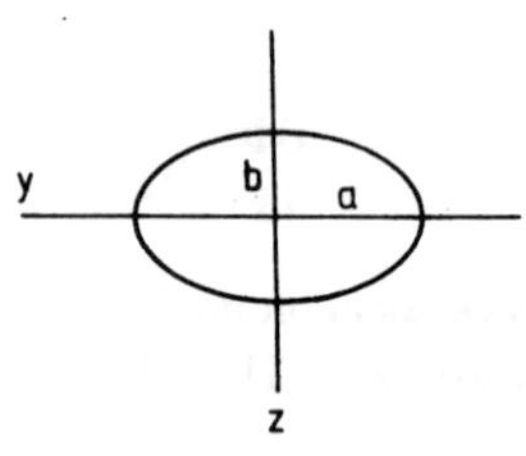

Fig. 7.4 Elliptical cross–section rod

where m is a constant. Substituting the above in equation (7. 43), we find the constant m.

$$m = -\frac{a^2 b^2}{(a^2 + b^2)} G\Theta$$

Hence

$$\Phi = -\frac{a^2b^2}{(a^2+b^2)} G\Theta\left(\frac{y^2}{a^2}+\frac{z^2}{b^2}-1\right) \qquad (g)$$

From (7.45) we get

$$M_t = -2\frac{a^2b^2}{(a^2+b^2)} G\Theta\left(\frac{1}{a^2}\int\int y^2 dy\, dz + \frac{1}{b^2}\int\int z^2 dy dz - \int\int dy\, dz\right) \qquad (h)$$

We note

$$\int\int y^2 dy\, dz = I_x = \frac{1}{4}\pi b a^3$$

$$\int\int z^2 dy\, dz = I_y = \frac{1}{4}\pi a b^3$$

$$\int\int dy\, dz = \pi ab \qquad (i)$$

Using (i) above, equation (h) becomes

$$M_t = -2\frac{a^2b^2}{(a^2+b^2)} G\Theta\left(\frac{1}{4}\pi b a + \frac{1}{4}\pi ab - \pi ab\right)$$

$$= \frac{\pi a^3 b^3}{(a^2+b^2)} G\Theta \qquad (j)$$

Hence, the torsional rigidity of an elliptical cross-section is given by

$$C = \frac{M_t}{\Theta} = \frac{\pi a^3 b^3 G}{(a^2+b^2)} \qquad (k)$$

7.6 PRANDTL'S MEMBRANE ANALOGY

We are not so much accustomed to visualizing the warping of a noncircular cross-section or the stress function to obtain the solution of Laplace or Poisson equation considered earlier. Prandtl introduced a membrane analogy for the torsion problem, which enables us to solve the Poisson equation. Consider a homogeneous membrane supported at the edges corresponding to the outline of the cross-section of the rod and subjected to a uniform tension *S* per unit length as shown in Fig. 7.5. Let this membrane be subjected to a uniform lateral load *q* per unit area. Then, from Fig. 7.5 for the equilibrium of the element *abcd*, We have

$$q\, dy\, dz + S\frac{\partial^2 x}{\partial y^2} dy\, dz + S\frac{\partial^2 x}{\partial z^2} dy\, dz = 0$$

i.e.,

$$\frac{\partial^2 x}{\partial y^2} + \frac{\partial^2 x}{\partial z^2} = -\frac{q}{S} \qquad (7.46)$$

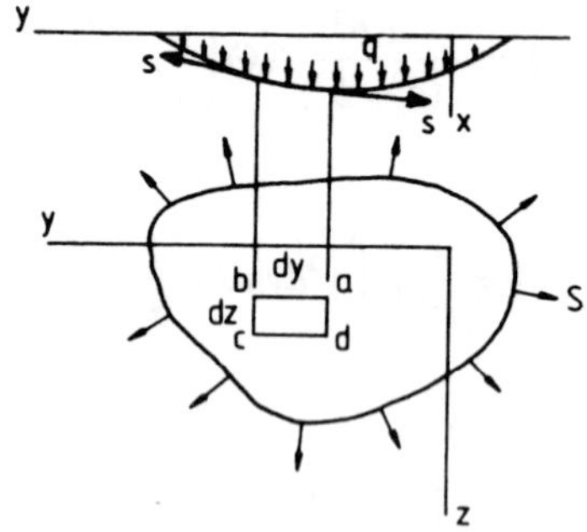

Fig. 7.5 Membrane analogy

The boundary condition is

$$x = 0 \quad \text{on the boundary} \tag{7.47}$$

Therefore, equation (7.46) and (7.47) are analogical to equations (7.43) and (7.44). The deflection x of the membrane is identical to the stress function Φ and q/S is identical to $2G\,\Theta$. We can easily visualize the deflections of the membrane and several important conclusions can be made regarding the stress distribution in torsion of a rod. Consider a contour line shown in Fig. 7.6 through the point P, i.e., the deflection on this line is constant. Therefore

$$\frac{\partial x}{\partial s} = 0$$

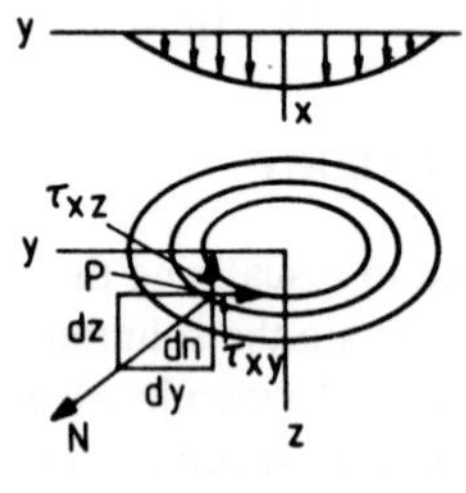

Fig. 7.6 Stress contour lines

The corresponding equation for the stress function Φ is

$$\frac{\partial \Phi}{\partial s} = \left(\frac{\partial \Phi}{\partial z}\frac{\partial z}{\partial s} + \frac{\partial \Phi}{\partial y}\frac{\partial y}{\partial s}\right)$$

$$= \tau_{xy}\frac{\partial z}{\partial s} - \tau_{xz}\frac{\partial y}{\partial s} = 0 \tag{a}$$

The above equation (a) shows that the stress normal to the contour line at a point P is zero and therefore the shearing stress in the twisted rod is in the direction of the tangent to contour line at the point P. In other words, the stress contour lines are lines of shear stress. The magnitude of the shearing stress τ at the point P is

$$\tau = \tau_{xz} a_{y} - \tau_{xy} a_{nz}$$

$$= -\left(\frac{\partial \Phi}{\partial y}\frac{\partial y}{\partial n} + \frac{\partial \Phi}{\partial z}\frac{\partial z}{\partial n}\right)$$

$$= -\frac{d\Phi}{dn} \qquad (b)$$

Hence, the magnitude of the shearing stress at P is given by the maximum slope of the membrane at this point. It can also be concluded that the maximum shear stress acts at the points where the contour lines are closest to each other. Also, from equation (7.45) we conclude that double the volume bounded by the deflected membrane and the yz plane, gives the torque. In all the conversions, we should replace q/S by $2G\,\Theta$.

7.7 TORSION OF A ROD WITH RECTANGULAR CROSS-SECTION

Case 1: Narrow rectangular cross-section

Consider first a narrow rectangular section of width a and depth b (*see* Fig. 7.7). The membrane analogy is used and we consider the deflection curve to be cylindrical by neglecting the effect of the short sides of the rectangle. The deflection curve is a parabola under the tension S of a uniformly loaded string.

$$\delta = \frac{qa^2}{8S} \qquad (a)$$

The volume bounded by the deflected membrane is

$$V = \frac{2}{3} a\,\delta\, b$$

$$= \frac{qa^3 b}{12S} \qquad (b)$$

Hence

$$M_t = \frac{1}{3} a^3 b\, G\,\Theta$$

$$C = \frac{1}{3} a^3 b\, G \qquad (c)$$

Case 2: General rectangular cross-section

If the cross-section is not narrow as in the above case 1, we cannot ignore the effect of short sides as in Fig. 7.7. The deflections are to be obtained from

$$\frac{\partial^2 x}{\partial y^2} + \frac{\partial^2 x}{\partial z^2} = -\frac{q}{S} \qquad (7.46)$$

$$x = 0 \quad \text{on the boundary} \qquad (7.47)$$

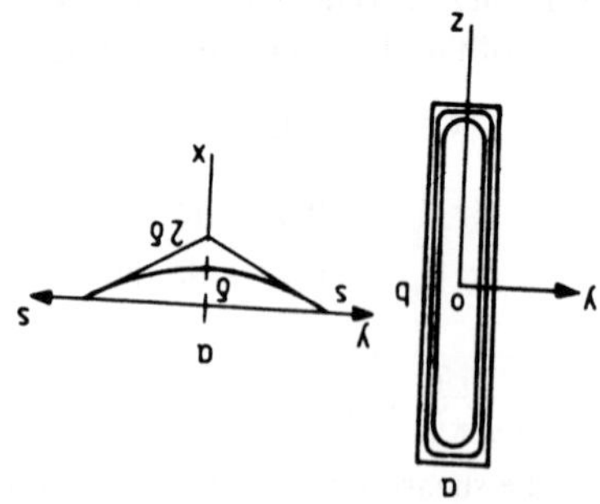

Fig. 7.7 Narrow rectangular section

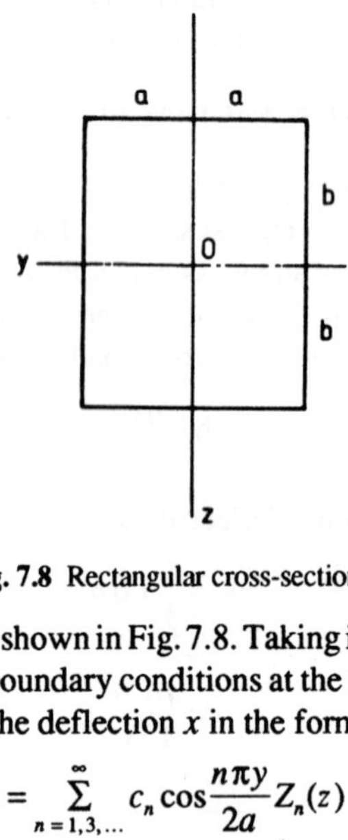

Fig. 7.8 Rectangular cross-section

Let the cross-section be as shown in Fig. 7.8. Taking into account of the symmetry with respect to z axis and the boundary conditions at the sides $y = \pm a$ of the rectangle are satisfied, we can express the deflection x in the form

$$x = \sum_{n=1,3,\ldots}^{\infty} c_n \cos\frac{n\pi y}{2a} Z_n(z) \tag{d}$$

in which c are constant coefficients. Substituting (d) in (7.46), we get

$$\sum_{n=1,3,\ldots}^{\infty} c_n \cos\frac{n\pi y}{2a}\left\{-\frac{n^2\pi^2}{4a^2}Z_n + Z_n''\right\} = -\frac{q}{S} \tag{e}$$

Making use of the relation

$$\sum_{n=1,3,\ldots}^{\infty} \frac{4}{n\pi}(-1)^{\frac{n-1}{2}} \cos\frac{n\pi y}{2a} \doteq 1$$

equation (e) becomes

$$Z_n'' - \frac{n^2\pi^2}{4a^2}Z_n = -\frac{q}{S}\frac{4}{n\pi c_n}(-1)^{\frac{n-1}{2}} \tag{f}$$

The solution of the above equation (*f*) is

$$Z_n = A \sinh\frac{n\pi z}{2a} + B\cosh\frac{n\pi z}{2a} + \frac{16qa^2}{Sn^2\pi^2 c_n}(-1)^{\frac{n-1}{2}} \tag{g}$$

We note that

$$[Z_n]_{z=\pm b} = 0$$

and equation (*g*) becomes

$$Z_n = \frac{16qa^2}{Sn^2\pi^2 c_n}(-1)^{\frac{n-1}{2}}\left\{1 - \frac{\cosh\frac{n\pi z}{2a}}{\cosh\frac{n\pi b}{2a}}\right\} \tag{h}$$

From equation (*d*), we now obtain

$$x = \frac{16qa^2}{S\pi^2}\sum_{n=1,3,\ldots}^{\infty}\frac{1}{n^2}(-1)^{\frac{n-1}{2}}\left\{1 - \frac{\cosh\frac{n\pi z}{2a}}{\cosh\frac{n\pi b}{2a}}\right\}\cos\frac{n\pi y}{2a} \tag{i}$$

Therefore, the stress function is given by

$$\Phi = \frac{32G\Theta a^2}{\pi^2}\sum_{n=1,3,\ldots}^{\infty}\frac{1}{n^2}(-1)^{\frac{n-1}{2}}\left\{1 - \frac{\cosh\frac{n\pi z}{2a}}{\cosh\frac{n\pi b}{2a}}\right\}\cos\frac{n\pi y}{2a} \tag{j}$$

The torque M_t is therefore

$$M_t = \frac{64G\Theta a^2}{\pi^2}\int_{-a}^{a}\int_{-b}^{b}\sum_{n=1,3,\ldots}^{\infty}\frac{1}{n^2}(-1)^{\frac{n-1}{2}}\left\{1 - \frac{\cosh\frac{n\pi z}{2a}}{\cosh\frac{n\pi b}{2a}}\right\}\cos\frac{n\pi y}{2a}\,dy\,dz$$

On evaluating the integrals, we have

$$M_t = \frac{32G\Theta(2a)^3(2b)}{\pi^4}\sum_{n=1,3,\ldots}^{\infty}\frac{1}{n^4} - \frac{64G\Theta(2a)^4}{\pi^5}\sum_{n=1,3,\ldots}^{\infty}\frac{1}{n^5}\tanh\frac{n\pi b}{2a} \tag{k}$$

Noting that

$$\sum_{n=1,3,\ldots}^{\infty}\frac{1}{n^4} = \frac{\pi^4}{96}$$

equation (*k*) becomes

$$M_t = \frac{1}{3}G\Theta(2a)^3(2b)\left\{1 - \frac{192a}{\pi^5 b}\sum_{n=1,3,\ldots}^{\infty}\frac{1}{n^5}\tanh\frac{n\pi b}{2a}\right\} \tag{l}$$

The torsional rigidity can therefore be written as

$$C = k_1 G\,(2a)^3(2b) \tag{7.48}$$

where

$$k_1 = \frac{1}{3}\left\{1 - \frac{192a}{\pi^5 b}\sum_{n=1,3,\ldots}^{\infty}\frac{1}{n^5}\tanh\frac{n\pi b}{2a}\right\} \tag{7.49}$$

The series on the right hand side converges rapidly and C can be evaluated for any ratio a/b. Table 7.3 gives k_1 for different values of b/a.

Table 7.3
Values of k_1 for a rectangular rod

$\frac{b}{a}$	k_1
1.0	0.1406
1.2	0.166
1.5	0.196
2.0	0.229
2.5	0.249
3.0	0.263
4.0	0.281
5.0	0.291
10.0	0.312
∞	0.333

Case 3: Ritz method for determining torsional rigidity

In the cases, where it is not possible to obtain a solution as in the case 2 above, we can use variational principles, e.g., Ritz method. The strain energy of the rod from (7.14) is

$$U = \frac{1}{2G}\int\int(\tau_{xy}^2 + \tau_{xz}^2)\,dy\,dz$$

$$= \frac{1}{2G}\int\int\left\{\left(\frac{\partial\Phi}{\partial y}\right)^2 + \left(\frac{\partial\Phi}{\partial z}\right)^2\right\}dy\,dz$$

The work done by the external torque (*see* equation 7.45), is

$$2\Theta\int\int\Phi\,dy\,dz$$

Hence

$$\pi = \frac{1}{2G}\int\int\left\{\left(\frac{\partial\Phi}{\partial y}\right)^2 + \left(\frac{\partial\Phi}{\partial z}\right)^2\right\}dy\,dz - 2\Theta\int\int\Phi dy\,dz$$

or

$$\pi = \int\int\left[\frac{1}{2}\left\{\left(\frac{\partial\Phi}{\partial y}\right)^2 + \left(\frac{\partial\Phi}{\partial z}\right)^2\right\} - 2\,G\,\Theta\Phi\right]dy\,dz \tag{7.50}$$

Let us choose for the rectangular cross-section, the following approximate solution for Φ

$$\Phi = (y^2 - a^2)(z^2 - b^2)\Sigma\Sigma a_{mn}y^m z^n \tag{m}$$

The above equation (m) satisfies the boundary conditions

$$\Phi(y = \pm a) = 0 \quad \Phi(z = \pm b) = 0$$

A one-term approximation of equation (*m*) is

$$\Phi = a_0 (y^2 - a^2)(z^2 - b^2) \tag{n}$$

Introducing (*n*) in (7.50), we get

$$\pi = \iint \left[\frac{1}{2} a_0^2 \{(4y^2 (z^2 - b^2)^2 + 4z^2 (y^2 - a^2)^2\} \right.$$

$$\left. - 2\, G\, \Theta\, a_0 (y^2 - a^2)(z^2 - b^2) \right] dy\, dz \tag{o}$$

Minimizing π with respect to a_0, we have

$$a_0 = \frac{2\, G\, \Theta \iint (y^2 - a^2)(z^2 - b^2)\, dy\, dz}{\iint \{4y^2 (z^2 - b^2)^2 + 4z^2 (y^2 - a^2)^2\}\, dy\, dz} \tag{p}$$

For simplicity, let us choose a square cross-section, $a = b$, then

$$a_0 = \frac{5G\, \Theta}{8a^2} \tag{q}$$

and

$$\Phi = \frac{5G\, \Theta}{8a^2} (y^2 - a^2)(z^2 - a^2) \tag{r}$$

The torque is then given by

$$M_t = \iint \frac{5G\theta}{4a^2} (y^2 - a^2)(z^2 - a^2)\, dy\, dz$$

$$= 0.1388\, (2a)^4\, G\, \Theta \tag{s}$$

Hence, the torsional rigidity of a square cross-section rod, with one-term approximation is

$$C = 0.1388G\, (2a)^4 \tag{t}$$

From table 7.3, we find that the above is in error by 1.33%. The accuracy can be improved by taking more number of terms in the assumed solution (m).

7.8 NATURAL FREQUENCIES OF TAPERED RODS WITH RECTANGULAR CROSS SECTION

Let us consider the case of a fixed-free rod of rectangular cross-section with a taper along its length l. The breadth of the cross-section is taken to be constant equal to b and the thickness at the root, $x = 0$, is t_0. The thickness at any length x is taken as

$$t = t_0 (1 - \beta X) \tag{7.51}$$

where

$$X = \frac{x}{l}$$

When the taper parameter β is positive, the thickness decreases along the length, the extreme case being zero thickness at the free end, for $\beta = 1$. For simplicity, let us consider the St Venant's theory of rods without any axial displacement and inertia. For free vibrations with harmonic frequency p, the differential equation of motion for the system is

$$(C\theta_{,x})_{,x} + I_p \rho p^2 \theta = 0 \tag{7.52}$$

The polar second moment of area of the cross-section at any distance x is

$$I_p = \frac{b t_0}{12} (1 - \beta X) \{b^2 + t_0^2 (1 - \beta X)^2\} \tag{a}$$

With the help of equation (7.48), we have the torsional rigidity of the cross-section at any distance x as

$$C = \frac{1}{3} G b t_0^3 (1 - \beta X)^3 \left[1 - \frac{192 t_0}{\pi^5 b} (1 - \beta X) \sum_{n=1,3,\ldots} \frac{1}{n^5} \tanh \left\{ \frac{n\pi}{2} \frac{b}{t_0 (1 - \beta X)} \right\} \right] \tag{b}$$

The shape functions can be assumed to be polynomial functions given by

$$\theta = \sum_r C_r F_r (X)$$

$$= \sum_r C_r (D_r X^r - E_r X^{r+2} + X^{r+3} \tag{7.53}$$

where

$$D_r = \frac{2 (r+1)(r+3)}{r^2 (r+1) - r (r-1)(r-2)}$$

$$E_r = \frac{(r+1)(r+2)(r+3) - (r-1)(r-2)(r+3)}{r (r+1)(r+2) - (r-1)(r-2)(r+2)}$$

The shape functions assumed in equation (7.53) satisfy the boundary conditions of a fixed-free rod, viz.,

$$\theta = 0 \quad x = 0$$

$$\theta_{,x} = 0 \quad x = l$$

We will follow the Galerkin method to obtain the natural frequencies of the rod. Substituting the shape functions θ, from equation (7.53) into equation (7.52), the error in the differential equation $\in$ is

$$\in = \left[\frac{1}{3} G b t_0^3 (1 - \beta X)^3 \left\{ 1 - \frac{192 t_0}{\pi^5 b} (1 - \beta X) \right. \right.$$

$$\left. \left. \times \sum_{n=1,3,\ldots} \frac{1}{n^5} \tanh \left(\frac{n\pi}{2} \frac{b}{t_0 (1 - \beta X)} \right) \right\} \left\{ \sum_r C_r F_r (X) \right\}_{,x} \right]_{,x} + I_p \rho p^2 \sum_r C_r F_r (X)$$

i.e.,

$$\in = \frac{1}{3} G\, b\, t_0^3 \Bigg[\Bigg\{ (1-\beta X)^3 - \frac{192\, t_0}{\pi^5\, b} (1-\beta X)^4$$

$$\times \sum_{n=1,3,\ldots} \frac{1}{n^5} \tanh\left[\frac{n\pi}{2} \frac{b}{t_0 (1-\beta X)} \right] \Bigg\} \sum_r C_r F_{r,xx}$$

$$+ \Bigg\{ \frac{768\, t_0}{\pi^5\, b} \beta\, (1-\beta X)^3 \sum_{n=1,3,\ldots} \frac{1}{n^5} \tanh\left\{ \frac{n\pi}{2} \frac{b}{t_0 (1-\beta X)} \right\}$$

$$- \frac{96}{\pi^4} \beta\, (1-\beta X)^2 \sum_{n=1,3,\ldots} \frac{1}{n^4 \cosh^2 \frac{n\pi b}{2 t_0 (1-\beta X)}} - 3\beta\, (1-\beta X)^2 \Bigg\} \frac{1}{l} \sum_r C_r F_{r,x} \Bigg]$$

$$+ \frac{b\, t_0 \rho}{12} p^2 (1-\beta X) \{ b^2 + t_0^2 (1-\beta X)^2 \} \sum_r C_r F_r \tag{c}$$

The application of Galerkin's procedure now gives

$$\int_0^1 \in F_j(X) dX = 0 \quad j = 1, 2, \ldots \tag{d}$$

For a finite solution of N-terms, say 5, the above equation (*d*) leads to the following eigen value problem.

$$[P] + p^2 [Q] = 0 \tag{e}$$

where

$$P_{rj} = \int_0^1 \frac{1}{3} G\, b\, t_0^3 \Bigg[(1-\beta X)^3 - \frac{192 t_0}{\pi^5 b} (1-\beta X)^4$$

$$\times \sum_{n=1,3,\ldots} \frac{1}{n^5} \tanh\left\{ \frac{n\pi}{2} \frac{b}{t_0 (1-\beta X)} \right\} \Bigg] F_{r,xx} F_j dX$$

$$+ \int_0^1 \frac{1}{3} G b t_0^3 \Bigg\{ \frac{768\, t_0}{\pi^5\, b} \beta (1-\beta X)^3 \sum_{n=1,3,\ldots} \frac{1}{n^5} \tanh\left\{ \frac{n\pi}{2} \frac{b}{t_0 (1-\beta X)} \right\}$$

$$- \frac{96}{\pi^4} \beta (1-\beta X)^2 \sum_{n=1,3,\ldots} \frac{1}{n^4 \cosh^2 \frac{n\pi b}{2 t_0 (1-\beta X)}} - 3\beta (1-\beta X)^2 \Bigg\} \frac{1}{l} F_{r,x} F_j dX \tag{f}$$

$$Q_{rl} = \int_0^1 \frac{b\, t_0 \rho}{12} p^2 (1-\beta X) \{ b^2 + t_0^2 (1-\beta X)^2 \} F_r F_j dX \tag{g}$$

The elements in $[P]$ and $[Q]$ can be evaluated by Gaussian Quadrature formulae, with 100 intervals in the range 0 to 1 of X. The eigen values can be obtained by QR transformation process. Most of the computer centres have built-in subroutines in their libraries, which can be directly used for this purpose. With a five-term solution, we can expect good results for the first three modes.

Let the rod have the following properties.

$b = 6.35$ cm
$t_0 = 1.27$ cm
$l = 15.24$ cm
$\rho = 0.00786$ kg/cm^3
$G = 0.75 \times 10^{11}$ Pa

The results obtained with the help of a computer program for the first two modes are given as a function of thickness taper β in Fig. 7.9. It can be observed that the first mode of vibration is very little affected by the taper, whereas, the second mode frequency decreases rapidly as taper is increased.

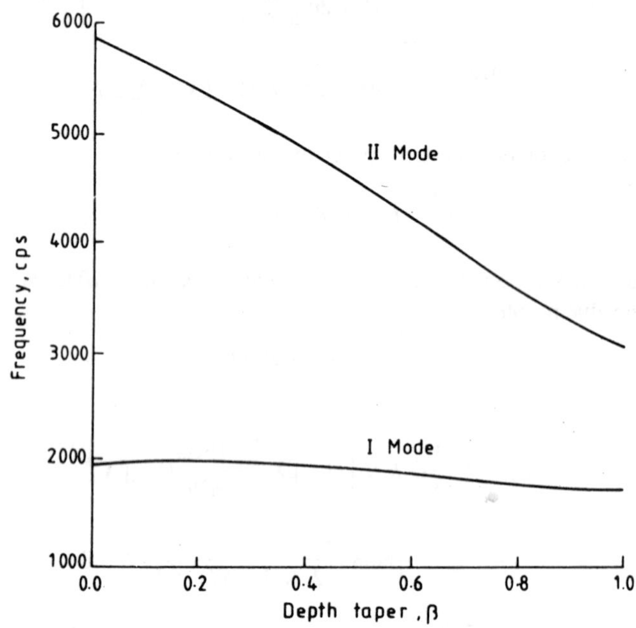

Fig. 7.9 Natural frequencies against thickness taper β for the first and second modes of torsional vibration

7.9 TORSION OF PRE-TWISTED RODS – CARNEGIE'S THEORY

When a rod of thin-walled section is pre-twisted, additional torsional stiffness is present due to the inclination of longitudinal fibres of this rod. Consider a thin blade-like rod of length l, as shown in Fig. 7.10. The line of centroid for the rod is O–O and we consider a longitudinal fibre AB of length l, whose area of cross-section is dA. The right end of the blade is pre-twisted relative to the left end, about the line of centroid O–O by an angle γ_l, as a result of which, the fibre AB is displaced to AB'. The distance of the fibre in the unstressed condition from line of centroid is r_1. The helix angle due to pre-twist β_p of this fibre is

$$\beta_p = \tan^{-1}\frac{\gamma_l r_1}{l} \tag{7.54}$$

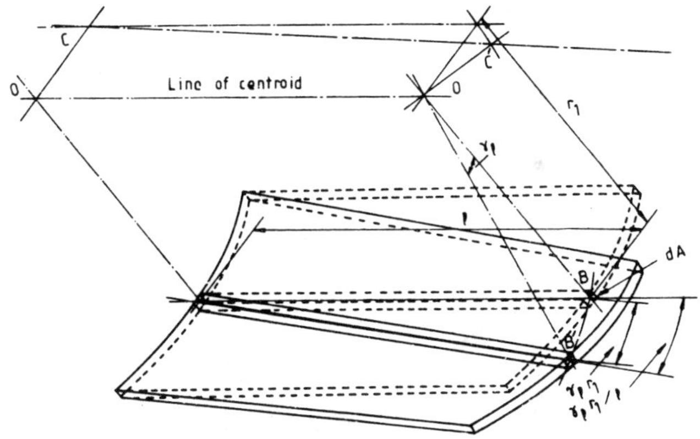

Fig. 7.10 Pre-twisted blade of thin-walled section before elastic twist

Therefore, the fibre length AB' is

$$AB' = l \sec\left(\frac{\gamma_l r_1}{l}\right) \tag{7.55}$$

The line of centre of torsion of the cross-section of the blade is given by C–C, which moves to C–C' after giving the pre-twist. In Fig. 7.11, the pre-twisted blade is given an elastic twist θ, as a result of which AB' moves to AB'', and the blade length increases by Δl. BB'' can be assumed to be

$$BB'' = \gamma_l r_1 + \theta r_2 \tag{7.56}$$

where r_2 is the distance of line of centre of torsion from the fibre location B'. The helix angle β_e after the elastic twist is therefore given by

$$\beta_e = \tan^{-1}\frac{\gamma_l r_1 + \theta r_2}{l} \tag{7.57}$$

and

$$AB'' = l + \Delta l \sec\left(\frac{\gamma_l r_1 + \theta r_2}{l}\right) \tag{7.58}$$

If the helix angle does not exceed 10 degrees and the elastic twist is small, the change in length of the fibre λ is

$$\lambda = \frac{\gamma_l r_1 \theta r_2}{l} + \Delta l + \frac{1}{2}\Delta l\left(\frac{\gamma_l r_1}{l}\right)^2 \tag{7.59}$$

We can approximate the direct stress in the fibre to be

$$f = E\left\{\frac{\gamma_l r_1 \theta r_2}{l^2} + \frac{\Delta l}{l}\right\} \tag{7.60}$$

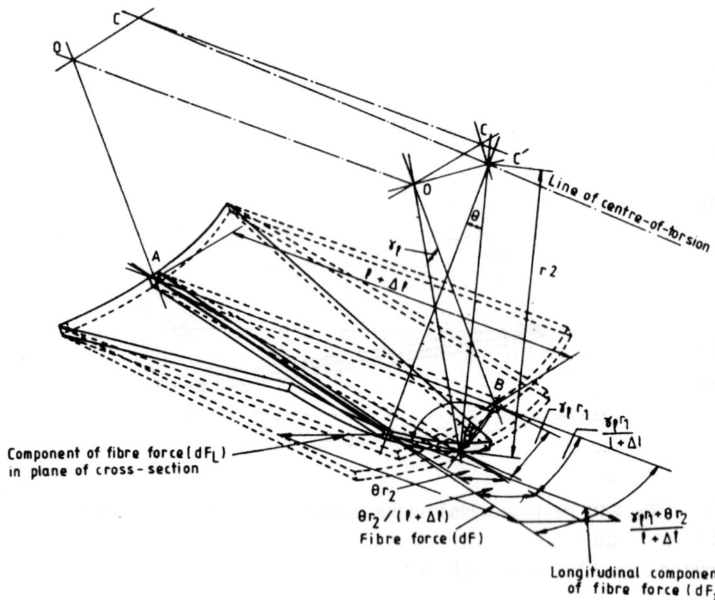

Fig. 7.11 Pre-twisted blade after elastic twist

The force dF acting along the fibre and normal to the blade axis is

$$dF = E\left\{\frac{\gamma_l r_1 \theta r_2}{l^2} + \frac{\Delta l}{l}\right\} dA \cos\frac{\gamma_l r_1 + \theta r_2}{l} \tag{7.61}$$

The longitudinal component of this force is

$$dF_x = E\left\{\frac{\gamma_l r_1 \theta r_2}{l^2} + \frac{\Delta l}{l}\right\} dA \cos^2\frac{\gamma_l r_1 + \theta r_2}{l}$$

$$\approx E\left\{\frac{\gamma_l r_1 \theta r_2}{l^2} + \frac{\Delta l}{l}\right\} dA \tag{7.62}$$

Integrating over the area of the cross-section,

$$F_x = E\int_A \left\{\frac{\gamma_l r_1 \theta r_2}{l^2} + \frac{\Delta l}{l}\right\} dA \tag{7.63}$$

Equating the above to zero since the blade is in equilibrium, we get the change in blade length as

$$\frac{\Delta l}{l} = \frac{\gamma_l \theta}{A l^2}\int_A r_1 r_2 dA \tag{7.64}$$

The lateral component of the fibre force is

$$dF_L = E\frac{\gamma_l \theta}{l^2}\left\{r_1 r_2 - \frac{1}{A}\int_A r_1 r_2 dA\right\} dA \tan\frac{\gamma_l r_1 + \theta r_2}{l} \tag{7.65}$$

This can be approximated to

$$dF_L = E\frac{\gamma_l^2\theta}{l^3}\left\{r_1^2 r_2 - \frac{r_1}{A}\int_A r_1 r_2 dA\right\}dA \tag{7.66}$$

Taking moment of these forces about the centre of torsion, we get

$$M_c = E\frac{\gamma_l^2\theta}{l^3}\int_A\left\{r_1^2 r_2^2 \frac{r_1 r_2}{A}\int_A r_1 r_2 dA\right\}dA \tag{7.67}$$

Therefore, the additional torsional rigidity due to pre-twist, C_γ is given by

$$C_\gamma = \frac{M_c}{\frac{\theta}{l}} = E\frac{\gamma_l^2}{l^2}\int_A\left\{r_1^2 r_2^2 - \frac{r_1 r_2}{A}\int_A r_1 r_2 dA\right\}dA \tag{7.68}$$

If the rod is symmetrical about two principal axes, $r_1 = r_2 = r$ and (7.68) above becomes

$$C_\gamma = E\frac{\gamma_l^2}{l^2}\int_A\left\{r^4 - \frac{r^2 I_p}{A}\right\}dA \tag{7.69}$$

The additional torsional rigidity due to pre-twist given in (7.69) above should be added to St Venant torsional rigidity to determine the vibration characteristics of pre-twisted rods. Consider a thin rectangular cross-section rod as in Fig. 7.7. Then, equation (7.69) gives

$$C_\gamma = E\frac{\gamma_l^2}{l^2}\int_{-\frac{b}{2}}^{\frac{b}{2}}\left\{z^4 - \frac{z^2(b^2+a^2)}{12}\right\}a\,dz \tag{7.69a}$$

After carrying out the integration in (7.69*a*), we get

$$C_\gamma = E\frac{\gamma_l^2 b^5 a}{l^2 180} \tag{7.70}$$

From case 1 of section 7.7, we therefore obtain the total torsional rigidity of a thin rectangular section rod with a total pre-twist of γ_l over length l, as

$$C = \frac{1}{3}ba^3G + E\frac{\gamma_l^2 b^5 a}{l^2 180} \tag{7.70a}$$

For a uniform pre-twisted rod, the procedure outlined before can now be used by using the expression for torsional rigidity in equation (7.70). For tapered rods, however, we cannot obtain any closed form solutions. In the next section, Galerkin procedure is illustrated to determine the natural frequencies of tapered pre-twisted rods.

7.10 TORSIONAL VIBRATION OF PRE-TWISTED TAPERED RODS

Let us consider a doubly tapered cantilever rod fixed at $x = 0$, with a rectangular cross-section, whose breadth and thickness are defined by

$$b = b_0(1 - \alpha X)$$

$$t = t_0(1 - \beta X) \tag{7.71}$$

where

$$X = \frac{x}{l}$$

When the taper parameters α, β are positive, the breadth and thickness decrease along the length, the extreme case being zero breadth and thickness at the free end for $\alpha = \beta = 1$. Let the rod be pre-twisted with γ over the total length l of the rod. For free vibrations with harmonic frequency p, the differential equation of motion for the system is

$$(C\theta_{,x})_{,x} + I_p \rho p^2 \theta = 0 \tag{7.72}$$

Let the rod be sufficiently thin in cross-section so that the additional torsional rigidity due to pre-twist can be estimated by equation (7.70). Then

$$C = \frac{1}{3} G b_0 t_0^3 (1 - \alpha X)(1 - \beta X)^3 \left[1 - \frac{192}{\pi^5} \frac{t_0 (1 - \beta X)}{b_0 (1 - \alpha X)} \right.$$

$$\left. \times \sum_{n=1,3,\ldots} \frac{1}{n^5} \tanh \left\{ \frac{n\pi}{2} \frac{b_0 (1 - \alpha X)}{t_0 (1 - \beta X)} \right\} \right] + \frac{E t_0 b_0^5 \gamma^2}{180 l^2} (1 - \beta X)(1 - \alpha X)^5 \tag{a}$$

The polar second moment of area of the cross-section is

$$I_p = \frac{b_0 t_0}{12} (1 - \alpha X)(1 - \beta X) \{ b_0^2 (1 - \alpha X)^2 + t_0^2 (1 - \beta X)^2 \} \tag{b}$$

The shape functions can be assumed to be polynomial functions in equation (7.53).

$$\theta = \sum_r C_r F_r(X)$$

$$= \sum_r C_r (D_r X^r - E_r X^{r+2} + X^{r+3}) \tag{7.53}$$

where

$$D_r = \frac{2(r+1)(r+3)}{r^2(r+1) - r(r-1)(r-2)}$$

$$E_r = \frac{(r+1)(r+2)(r+3) - (r-1)(r-2)(r+3)}{r(r+1)(r+2) - (r-1)(r-2)(r+2)}$$

which satisfy the boundary conditions of a fixed-free rod

$$\theta = 0 \quad x = 0$$

$$\theta_{,x} = 0 \quad x = l$$

The error function $\in$ can be obtained by substituting the above assumed shape functions and equations (a) and (b) for torsional rigidity and polar second moment area respectively in the differential equation (7.72).

$$\in = \{ [\frac{1}{3} G b_0 t_0^3 (1 - \alpha X)(1 - \beta X)^3 \{ 1 - \frac{192}{\pi^5} \frac{t_0 (1 - \beta X)}{b_0 (1 - \alpha X)}$$

$$\times \sum_{n=1,3,\ldots} \frac{1}{n^5} \tanh\left(\frac{n\pi}{2}\frac{b_0(1-\alpha X)}{t_0(1-\beta X)}\right)\} + \frac{E t_0 b_0^5 \gamma^2}{180 l^2}(1-\beta X)(1-\alpha X)^5]\Sigma C_r F_{r,x}\}_{,x}$$

$$+\left[\frac{b_0 t_0}{12}(1-\alpha X)(1-\beta X)\{b_0^2(1-\alpha X)^2 + t_0^2(1-\beta X)^2\}\right]\rho p^2 \Sigma C_r F_r \qquad (c)$$

Equation (c) above can be expanded to

$$\in = \frac{1}{3} G b_0 t_0^3 (1-\alpha X)(1-\beta X)^3 - \frac{192}{\pi^5}\frac{t_0}{b_0}(1-\beta X)^4$$

$$\times \sum_{n=1,3,\ldots} \frac{1}{n^5}\tanh\left\{\frac{n\pi}{2}\frac{b_0(1-\alpha X)}{t_0(1-\beta X)}\right\} + \frac{E t_0 b_0^5 \gamma^2}{180 l^2}(1-\beta X)(1-\alpha X)^5\}\Sigma C_r F_{r,xx}$$

$$+[\frac{1}{3l} G b_0 t_0^3 \{\frac{768}{\pi^5}\beta(1-\beta X)^3 \frac{t_0}{b_0}\sum_{n=1,3,\ldots}\frac{1}{n^5}\tanh\left\{\frac{n\pi}{2}\frac{b_0(1-\alpha X)}{t_0(1-\beta X)}\right\}$$

$$-\frac{96}{\pi^4}(\beta-\alpha)(1-\beta X)^2 \sum_{n=1,3,\ldots}\frac{1}{n^4}\frac{1}{\cosh^2 \frac{n\pi}{2}\frac{b_0(1-\alpha X)}{t_0(1-\beta X)}}$$

$$-\alpha(1-\beta X)^3 - 3\beta(1-\alpha X)(1-\beta X)^2\} - \frac{E t_0 b_0^5 \gamma^2}{180 l^3}\times$$

$$\{5\alpha(1-\alpha X)^4(1-\beta X) + \beta(1-\alpha X)^5\}]\Sigma C_r F_{r,x}$$

$$+\left[\frac{b_0 t_0}{12}(1-\alpha X)(1-\beta X)\{b_0^2(1-\alpha X)^2 + t_0^2(1-\beta X)^2\}\right]\rho p^2 \Sigma C_r F_r \qquad (d)$$

According to Galerkin's process, we have

$$\int_0^1 \in F_j dX = 0 \quad j = 1, 2, \ldots$$

which leads to

$$[P] - p^2[Q] = 0 \qquad (e)$$

where the elements in the above matrices are

$$P_{rj} = \int_0^1 \frac{1}{3} G b_0 t_0^3 \{(1-\alpha X)(1-\beta X)^3 - \frac{192}{\pi^5}\frac{t_0}{b_0}(1-\beta X)^4$$

$$\times \sum_{n=1,3,\ldots}\frac{1}{n^5}\tanh\left\{\frac{n\pi}{2}\frac{b_0(1-\alpha X)}{t_0(1-\beta X)}\right\} + \frac{E t_0 b_0^5 \gamma^2}{180 l^2}(1-\beta X)(1-\alpha X)^5\} F_{r,xx} F_j dX$$

$$+\int_0^1 [\frac{1}{3l} G b_0 t_0^3 \{\frac{768}{\pi^5}\beta(1-\beta X)^3 \frac{t_0}{b_0}\sum_{n=1,3,\ldots}\frac{1}{n^5}\tanh\left\{\frac{n\pi}{2}\frac{b_0(1-\alpha X)}{t_0(1-\beta X)}\right\}$$

$$-\frac{96}{\pi^4}(\beta-\alpha)(1-\beta X)^2 \sum_{n=1,3,\ldots}\frac{1}{n^4}\frac{1}{\cosh^2 \frac{n\pi}{2}\frac{b_0(1-\alpha X)}{t_0(1-\beta X)}}$$

$$-\alpha(1-\beta X)^3 - 3\beta(1-\alpha X)(1-\beta X)^2\} - \frac{E t_0 b_0^5 \gamma^2}{180 l^3}$$

$$\times\{5\alpha(1-\alpha X)^4(1-\beta X) + \beta(1-\alpha X)^5\}]F_{r,x} F_j dX$$

and

$$Q_{rj} = \int_0^1 \left[\frac{b_0 t_0}{12}(1-\alpha X)(1-\beta X)\{b_0^2(1-\alpha X)^2 + t_0^2(1-\beta X)^2\} \right] \rho p^2 F_r F_j dX$$

Obviously we need to use a computer program to evaluate the above elements and solve the eigen value problem in equation (*e*). A five-term solution may be good enough for obtaining three natural frequencies accurately. Let us consider a cantilever rod with the following data.

$b_0 = 2.54$ cm

$t_0 = 0.46$ cm

$l = 15.24$ cm

$\rho = 0.00786$ kg/cm^3

$E = 1.875 \times 10^{11}$ Pa

$G = 0.750 \times 10^{11}$ Pa

$\beta = 0.436464$

$\gamma = 45°$

The results obtained are given in Fig. 7.12 as a function of breadth taper α, for the first three modes. This figure shows that the natural frequencies increase with breadth taper for a given depth taper of the rod. Experimental results for rods with $\beta = 0.436464$ and $\alpha = 0.0, 0.19, 0.39$ and 0.59 are also given in Fig. 7.12. It is observed that they are higher than the theoretical values, particularly in higher modes. The influence of axial stress component given by the term r_2 is relatively significant as a higher order effect and this is considered below.

For the narrow rectangular cross-section, case 1 of section 7.7, the deflection x, can be written as

$$x = \frac{4\delta}{a^2}\left(\frac{a^2}{4} - y^2\right) \tag{7.73}$$

The slope of the deflection curve then is

$$\frac{dx}{dy} = -\frac{8\delta y}{a^2}$$

$$= -\frac{q}{S} y \tag{7.74}$$

Therefore

$$\tau_{xz} = 2G\Theta y$$

$$= G\Theta\left(\frac{\partial \phi}{\partial z} + y\right) \tag{7.75}$$

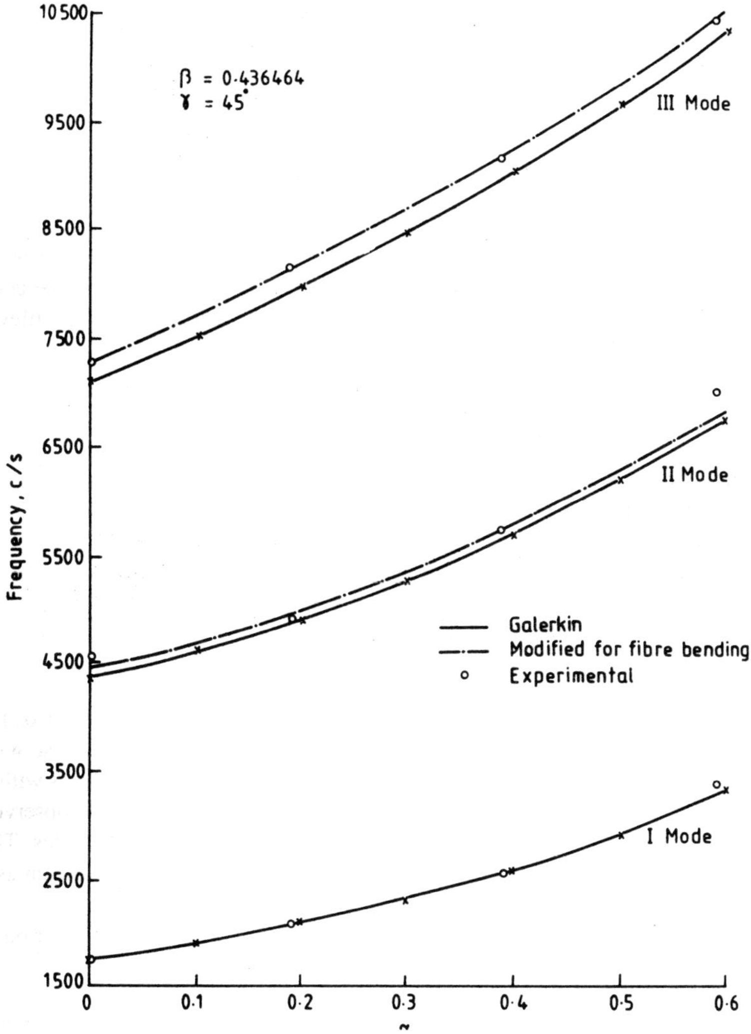

Fig. 7.12 Natural frequencies against width taper for pre-twisted blades

Hence, the warping function for a thin section can be written as

$$\phi = yz \tag{7.76}$$

We can now determine

$$I_{\phi\phi} = \int_A \phi^2 dA$$

$$= \int_{-\frac{b}{2}}^{\frac{b}{2}} \int_{-\frac{t}{2}}^{\frac{t}{2}} y^2 z^2 \, dy dz$$

$$= \frac{b^3 t^3}{144} \tag{f}$$

From Table 7.3, we can interpolate the value of the coefficient k for the present cross-section of the rod at the root ($b/t = 5.525$) to be 0.3189. Hence the St Venant torsional rigidity for cross-section of the rod at the root is

$$C = 0.3189 \times 0.46^3 \times 0.75 \times 10^{11}$$

From equation (7.36), we can now determine r_2

$$r_2 = \frac{1.875 \times 10^{11} \times 2.54^3 \times 0.46^3}{0.3189 \times 2.54 \times 0.46^3 \times 0.75 \times 10^{11} \times 15.24^2 \times 144}$$

$$= 0.0015$$

From equation (6.83), we can write the correction factor for the frequency by considering only the axial stress term as

$$\sqrt{1 + r_2\left(\frac{n\pi}{2}\right)^2} \qquad n = 1, 2, \ldots$$

The correction values for the first three modes from the above works out to 1.0018, 1.0074 and 1.0165. We have, of course, ignored the taper in the rod while arriving at these corrections. Since they are of small order, we may be justified in doing so. Figure 7.12 shows the corrected values accounting for the axial stress in the longitudinal fibres of the rod. We note that these corrections improve the theoretical values and bring them closer to experimental values. Other higher order terms are generally found to be of much less significance and can be ignored in engineering analysis.

7.11 REISSNER AND LO-GOULARD'S THEORY

Reissner proposed nonuniform twist in the rod by modifying the field of Timoshenko, i.e.,

$$u_x = \phi(y, z)\alpha$$

$$u_y = -z\theta(x, t)$$

$$u_z = y\theta(x, t) \tag{7.77}$$

$$\varepsilon_{xx} = \phi\alpha_{,x}$$

$$\varepsilon_{yy} = 0$$

$$\varepsilon_{zz} = 0$$

$$\varepsilon_{xy} = \frac{1}{2}(\phi_{,y}\alpha - z\theta_{,x})$$

$$\varepsilon_{xz} = \frac{1}{2}(\phi_{,z}\alpha + y\theta_{,x})$$

$$\varepsilon_{yz} = 0 \tag{7.78}$$

In equations (7.78) above the influence of Poisson's ratio is neglected. The stress field is then given by

$$\tau_{xx} = E\phi\alpha_{,x}$$

$$\tau_{yy} = \tau_{zz} = 0$$

$$\tau_{xy} = G(\phi_{,y}\alpha - z\theta_{,x})$$

$$\tau_{xz} = G(\phi_{,z}\alpha + y\theta_{,x})$$

$$\tau_{yz} = 0 \tag{7.79}$$

The strain energy is given by

$$U = \frac{1}{2}\int_0^l \int_A G\{(\phi_{,y}\alpha - z\theta_{,x})^2 + (\phi_{,z}\alpha + y\theta_{,x})^2\} dA\,dx$$

$$+\frac{1}{2}\int_0^l \int_A E\phi^2\alpha_{,x}^2 dx\,dA \tag{7.80}$$

Neglecting the axial inertia, the kinetic, energy is given by equation (7.5).

$$T = \frac{1}{2}\int_0^l \int_A \rho(z^2\dot{\theta}^2 + y^2\dot{\theta}^2) dA\,dx$$

$$= \frac{1}{2}\int_0^l \rho I_p \dot{\theta}^2 dx \tag{7.5}$$

The work done by the external moment is given by equation (7.6).

$$W = \int_0^l M\theta dx \tag{7.6}$$

We can now write the variational equation for the rod as

$$\int_{t_1}^{t_2} [\frac{1}{2}\int_0^l \int_A G\{(\phi_{,y}\alpha - z\theta_{,x})^2 + (\phi_{,z}\alpha + y\theta_{,x})^2\} dA\,dx$$

$$+\frac{1}{2}\int_0^l \int_A E\phi^2\alpha_{,x}^2 dx\,dA]\,dt$$

$$-\int_{t_1}^{t_2}\left[\frac{1}{2}\int_0^l \rho I_p \dot{\theta}^2 dx + \int_0^l M\theta dx\right] \quad \text{is stationary} \tag{7.81}$$

The second integral has been dealt with before. Taking the variation of the first integral in (7.81) above we have

$$I = \delta\int_{t_1}^{t_2} [\frac{1}{2}\int_0^l \int_A G\{(\phi_{,y}\alpha - z\theta_{,x})^2 + (\phi_{,z}\alpha + y\theta_{,x})^2\} dA\,dx$$

$$+\frac{1}{2}\int_0^l \int_A E\phi^2\alpha_{,x}^2 dx\,dA]\,dt \tag{a}$$

We note in the above that ϕ, θ and α are independent. Taking variation with respect to all ϕ, θ and α, we have

$$I_1 = \int_{t1}^{t2}\int_0^l\int_A G\{\phi_{,y}^2 + \phi_{,x}^2\}\alpha\delta\alpha dA\,dx\,dt$$

$$+\int_{t1}^{t2}\int_0^l\int_A G\{z^2 + y^2)\theta_{,x}\delta\theta_{,x} dA\,dx\,dt$$

$$+\int_{t1}^{t2}\int_0^l\int_A G(\phi_{,z}y - \phi_{,y}z)\alpha\delta\theta_{,x} dA\,dx\,dt$$

$$+\int_{t1}^{t2}\int_0^l\int_A G(\phi_{,z}y - \phi_{,y}z)\theta_{,x}\delta\alpha dA\,dx\,dt$$

$$+\int_{t1}^{t2}\int_v G\{(\phi_{,y}\alpha - z\theta_{,x})\alpha\delta\phi_{,y} + (\phi_{,z}\alpha + y\theta_{,x})\alpha\delta\phi_{,z}\} dA\,dx\,dt$$

$$+\int_{t1}^{t2}\int_0^l\int_A E\phi\delta\phi\alpha_{,x}^2 dx\,dA\,dt$$

$$+\int_{t1}^{t2}\int_0^l\int_A E\phi^2\alpha_{,x}\delta\alpha_{,x} dx\,dA\,dt \qquad (b)$$

We have seven integrals in (b) above, which are discussed below. The first integral is

$$I_{11} = \int_{t1}^{t2}\int_0^l\int_A G\{\phi_{,y}^2 + \phi_{,z}^2\}\alpha\delta\alpha dA\,dx\,dt$$

i.e.,

$$I_{11} = \int_{t1}^{t2}\int_0^l KG\alpha\delta\alpha dx\,dt$$

where

$$K = \int_A \{\phi_{,y}^2 + \phi_{,z}^2\} dA \qquad (7.82)$$

The second integral in (b) above is

$$I_{12} = \int_{t1}^{t2}\int_0^l\int_A G(z^2 + y^2)\theta_{,x}\delta\theta_{,x} dA\,dx\,dt$$

i.e.,

$$I_{12} = \int_{t1}^{t2}\int_0^l GI_p\theta_{,x}\delta\theta_{,x} dx\,dt$$

Integrating by parts, we get

$$I_{12} = \int_{t_1}^{t2}\left[GI_p\theta_{,x}\delta\theta\,|_0^l - \int_0^l (GI_p\theta_{,x})_{,x}\delta\theta dx\right]dt$$

The third integral is

$$I_{13} = \int_{t1}^{t2}\int_0^l\int_A G(\phi_{,z}y - \phi_{,y}z)\alpha\delta\theta_{,x} dA\,dx\,dt$$

i.e.,

$$I_{13} = \int_{t1}^{t2} \int_0^l GL\alpha\delta\theta_{,x} dx\, dt$$

where

$$L = \int_A (\phi_{,z} y - \phi_{,y} z) dA \tag{7.83}$$

Integrating by parts

$$I_{13} = \int_{t1}^{t2} \left[GL\alpha\delta\theta \,|_0^l - \int_0^l GL\alpha_{,x} \delta\theta dx \right] dt$$

The fourth integral is

$$I_{14} = \int_{t1}^{t2} \int_0^l \int_A G(\phi_{,z} y - \phi_{,y} z)\theta_{,x} \delta\alpha dA\, dx\, dt$$

i.e.,

$$I_{14} = \int_{t1}^{t2} \int_0^l LG\theta_{,x} \delta\alpha dx\, dt$$

The fifth integral is

$$I_{15} = \int_{t1}^{t2} \int_V G\{(\phi_{,y}\alpha - z\theta_{,x})\alpha\delta\phi_{,y} + (\phi_{,z}\alpha + y\theta_{,x})\alpha\delta\phi_{,z}\} dA\, dx\, dt$$

Integrating by parts, we get

$$I_{15} = \int_{t1}^{t2} [G \int_s (\phi_{,y}\alpha - z\theta_{,x}) a_v \delta\phi ds = g_R \int\int \phi_{,yy} \delta\phi dy\, dz$$

$$+ G \int_s (\phi_{,z}\alpha + y\theta_{,x}) a_{nz} \delta\phi ds - g_R \int\int \phi_{,zz} \delta\phi dy\, dz] dt$$

where

$$g_R = \int_0^l G\alpha^2 dx \tag{7.84}$$

i.e.,

$$I_{15} \int_{t1}^{t2} \left[-g_R \nabla^2\phi + \int_s G\{(\phi_{,y}\alpha - z\theta_{,x}) a_v + (\phi_{,z}\alpha + y\theta_{,x}) a_{nz}\} ds \right] \delta\phi dt$$

The sixth integral in (b) is

$$I_{16} = \int_{t1}^{t2} \int_0^l \int_A E\phi\delta\phi\alpha_{,x}^2 dx\, dA\, dt$$

which is

$$I_{16} = \int_{t1}^{t2} \int_A d_R \phi\delta\phi dx\, dA\, dt$$

where

$$d_R = \int_0^l E\alpha_{,x}^2 dx \tag{7.85}$$

The seventh integral is

$$I_{17} = \int_{t1}^{t2} \int_0^l EI_{\phi\phi}\alpha_{,x}\delta\alpha_{,x} dx\, dt$$

Integrating by parts, the above gives

$$I_{17} = \int_{t_1}^{t2} [EI_{\phi\phi}\alpha_{,x}\delta\alpha \,|_0^l +$$

$$-\int_0^l (EI_{\phi\phi}\alpha_{,x})_{,x}\delta\alpha dx] dt$$

Combining fifth and sixth integrals, we get

$$-g_R \nabla^2\phi + d_R\phi = 0 \tag{7.86}$$

$$\{(\phi_{,y}\alpha - z\theta_{,x})\alpha_y + (\phi_{,z}\alpha + y\theta_{,x})a_{nz}\} = 0 \quad \text{(on boundary)} \tag{7.87}$$

Equations (7.86) and (7.87) above define the modified Poisson equation to determine the warping function.

Combining the first four and the last integrals, and adding the inertia and external moment terms, we get

$$\begin{bmatrix} \rho I_p & 0 \\ 0 & 0 \end{bmatrix} \begin{Bmatrix} \ddot{\theta} \\ \ddot{\alpha} \end{Bmatrix} + \begin{bmatrix} -GI_p\partial_x^2 & -GL\partial_x \\ GL\partial_x & GK - EI_{\phi\phi}\partial_x^2 \end{bmatrix} \begin{Bmatrix} \theta \\ \alpha \end{Bmatrix} = \begin{Bmatrix} M \\ 0 \end{Bmatrix} \tag{7.88}$$

$$(GI_p\theta_{,x} + GL\alpha)\delta\theta \,|_0^l = 0$$

$$-EI_{\phi\phi}\alpha_{,x}\delta\alpha \,|_0^l = 0 \tag{7.89}$$

7.12 BARR'S THEORY

Barr included in the above theory of section 7.11, the effect of axial inertia. The kinetic energy in equation (7.5) now gets modified to

$$T = \frac{1}{2}\int_0^l \int_A \rho(z^2\dot{\theta}^2 + y^2\dot{\theta}^2 + \phi^2\dot{\alpha}^2) dA\, dx$$

$$= \frac{1}{2}\int_0^l \rho(I_p\dot{\theta}^2 + I_{\phi\phi}\dot{\alpha}^2) dx \tag{7.90}$$

Consequently, equation (7.88) gets modified to

$$\begin{bmatrix} \rho I_p & 0 \\ 0 & \rho I_{\phi\phi} \end{bmatrix} \begin{Bmatrix} \ddot{\theta} \\ \ddot{\alpha} \end{Bmatrix} + \begin{bmatrix} -GI_p\partial_x^2 & -GL\partial_x \\ GL\partial_x & GK - EI_{\phi\phi}\partial_x^2 \end{bmatrix} \begin{Bmatrix} \theta \\ \alpha \end{Bmatrix} = \begin{Bmatrix} M \\ 0 \end{Bmatrix} \tag{7.91}$$

The corresponding boundary conditions are

$$(GI_p\theta_{,x} + GL\alpha)\delta\theta \,|_0^l = 0$$

$$-(EI_{\phi\phi}\alpha_{,x} + \rho I_{\phi\phi}\dot{\alpha})\delta\alpha \,|_0^l = 0 \tag{7.92}$$

A theory that satisfies theory of elasticity relations is given in the next section.

7.13 A REFINED THEORY OF TORSIONAL VIBRATION OF RODS

We considered various theories of torsional vibration of rods in the previous sections. None of these theories, however, is accurate from a theory of elasticity point of view. For example, Timoshenko-Gere's theory does not satisfy the strain-displacement relations according to Poisson equations. The author's theory removes this discrepancy by assuming the following displacement field.

$$u_x = \phi(y,z)\theta_{,x}$$

$$u_y = -z\theta(x,t) - \mu\theta_{,xx}\int \phi\, dy$$

$$u_z = y\theta(x,t) - \mu\theta_{,xx}\int \phi\, dz \tag{7.93}$$

The corresponding velocities are

$$u_{x,t} = \phi(y,z)\theta_{,x}$$

$$u_{y,t} = -z\theta_{,t} - \mu\theta_{,xxt}\int \phi\, dy$$

$$u_{z,t} = y\theta_{,t} - \mu\theta_{,xxt}\int \phi\, dz \tag{7.94}$$

The strain field from equation (7.93) can be obtained as

$$\varepsilon_{xx} = \phi\theta_{,xx}$$

$$\varepsilon_{yy} = -\mu\phi\theta_{,xx}$$

$$\varepsilon_{zz} = -\mu\phi\theta_{,xx}$$

$$\varepsilon_{xy} = \frac{1}{2}\left[(\phi_{,y} - z)\theta_{,x} - \mu\theta_{,xxx}\int \phi\, dy\right]$$

$$\varepsilon_{yz} = -\frac{1}{2}\mu\theta_{,xx}\left[\left(\int \phi\, dy\right)_{,x} + \left(\int \phi\, dz\right)_{,y}\right]$$

$$\varepsilon_{zx} = \frac{1}{2}\left[(\phi_{,z} + y)\theta_{,x} - \mu\theta_{,xxx}\int \phi\, dz\right] \tag{7.95}$$

Hence, the stress field is

$$\tau_{xx} = E\phi\theta_{,xx}$$

$$\tau_{yy} = \tau_{zz} = 0$$

$$\tau_{xy} = G\left[(\phi_{,y} - z)\theta_{,x} - \mu\theta_{,xxx}\int \phi\, dy\right]$$

$$\tau_{yz} = -\mu\theta_{,xx}G\left[\left(\int \phi\, dy\right)_{,z} + \left(\int \phi\, dz\right)_{,y}\right]$$

$$\tau_{zx} = G\left[(\phi_{,z} + y)\theta_{,x} - \mu\theta_{,xxx}\int \phi\, dz\right] \tag{7.96}$$

The strain energy is

$$U = \frac{1}{2}[\int_0^l (EI_{\phi\phi} + \mu^2 GC_4)\theta_{,xx}^2 dx + \int_0^l C_1 G\theta_{,x}^2 dx$$

$$+ \int_0^l GC_2\mu^2\theta_{,xxt}^2 dx - \int_0^l 2G\mu C_3\theta_{,x}\theta_{,xxt} dx] \tag{7.97}$$

$$T = \frac{1}{2}[\int_0^l \rho I_p \theta_{,t}^2 dx + \int_0^l \rho C_2\mu^2\theta_{,xxt}^2 dx$$

$$+ \int_0^l 2\rho\mu C_5\theta_{,xxt}\theta_{,t} dx + \int_0^l \rho I_{\phi\phi}\theta_{,xt}^2 dx] \tag{7.98}$$

where

$C_1 = C$ St Venant's torsional rigidity

$$C_2 = \int_A \left[\left(\int \phi dy\right)^2 + \left(\int \phi dz\right)^2\right] dA$$

$$C_3 = \int_A \left[(\phi_{,y} - z)\int \phi dy + (\phi_{,z} + y)\int \phi dz\right] dA$$

$$C_4 = \int_A \left[\left(\int \phi dy\right)_{,z} + \left(\int \phi dz\right)_{,y}\right]^2 dA$$

$$C_5 = \int_A \left[z\int \phi dy - y\int \phi dz\right] dA \tag{7.99}$$

It is assumed that the above values of C in (7.99) are constants except that of C_1 for the purpose of deriving the Laplace equation for the warping function. This simplifies the analysis further. In other words, the warping function is assumed to be independent of Poisson's ratio μ and can be taken from Timoshenko-Gere theory.

Variational operations corresponding to each of the terms in (7.97) and (7.98) are given below.

The first integral in U is

$$I_1 = \delta \int_{t1}^{t2} \frac{1}{2}\int_0^l (EI_{\phi\phi} + \mu^2 GC_4)\theta_{,xx}^2 dx\, dt$$

i.e.,

$$I_1 = \int_{t1}^{t2} \int_0^l (EI_{\phi\phi} + \mu^2 GC_4)\theta_{,xx}\delta\theta_{,xx} dx\, dt$$

Integrating by parts

$$I_1 = \int_{t1}^{t2} \left[(EI_{\phi\phi} + \mu^2 GC_4)\theta_{,xx}\delta\theta_{,x} |_0^l - \int_0^l \{(EI_{\phi\phi} + \mu^2 GC_4)\theta_{,xx}\}_{,x} dx\right] dt$$

i.e.,

$$I_1 = \int_{t1}^{t2} [(EI_{\phi\phi} + \mu^2 GC_4)\theta_{,xx}\delta\theta_{,x} |_0^l - \{(EI_{\phi\phi} + \mu^2 GC_4)\theta_{,xx}\}_{,x}\delta\theta |_0^l$$

$$+ \int_0^l \{(EI_{\phi\phi} + \mu^2 GC_4)\theta_{,xx}\}_{,xx}\delta\theta dx] dt$$

The second integral in U is

$$I_2 = \delta \int_{t1}^{t2} \frac{1}{2} \int_0^l C_1 G \theta_{,x}^2 dx\, dt$$

i.e.,

$$I_2 = \int_{t1}^{t2} \int_0^l C_1 G \theta_{,x} \delta\theta_{,x} dx\, dt$$

Integrating by parts

$$I_2 = \int_{t1}^{t2} \left[C_1 G \theta_{,x} \delta\theta \,|_0^l - \int_0^l \{C_1 G \theta_{,x}\}_{,x} \delta\theta dx \right] dt$$

The third integral in U is

$$I_3 = \delta \int_{t1}^{t2} \frac{1}{2} \int_0^l G C_2 \mu^2 \theta_{,xx}^2 dx\, dt$$

i.e.,

$$I_3 = \int_{t1}^{t2} \int_0^l G C_2 \mu^2 \theta_{,xx} \delta\theta_{,xx} dx\, dt$$

Integrating by parts

$$I_3 = \int_{t1}^{t2} \left[G C_2 \mu^2 \theta_{,xx} \delta\theta_{,x} \,|_0^l - \{G C_2 \mu^2 \theta_{,xx}\}_{,x} \delta\theta_{,x} dx \right] dt$$

$$I_3 = \int_{t1}^{t2} [G C_2 \mu^2 \theta_{,xx} \delta\theta_{,x} \,|_0^l - \{G C_2 \mu^2 \theta_{,xx}\}_{,x} \delta\theta_{,x} \,|_0^l$$

$$+ \int_0^l \{G C_2 \mu^2 \theta_{,xx}\}_{,xx} \delta\theta_{,x} dx] dt$$

$$I_3 = \int_{t1}^{t2} [G C_2 \mu^2 \theta_{,xx} \delta\theta_{,x} \,|_0^l - \{G C_2 \mu^2 \theta_{,xx}\}_{,x} \delta\theta_{,x} \,|_0^l$$

$$+ \{G C_2 \mu^2 \theta_{,xx}\}_{,xx} \delta\theta \,|_0^l - \int_0^l \{G C_2 \mu^2 \theta_{,xx}\}_{,xxx} \delta\theta dx] dt$$

The fourth integral in U is

$$I_4 = \delta \int_{t1}^{t2} \int_0^l G \mu C_3 \theta_{,x} \theta_{,xxx} dx\, dt$$

i.e.,

$$I_4 = \int_{t1}^{t2} \int_0^l G \mu C_3 (\delta\theta_{,x} \theta_{,xxx} + \theta_{,x} \delta\theta_{,xxx})\, dx\, dt$$

We split the above into two integrals.

$$I_{41} = \int_{t1}^{t2} \int_0^l G \mu C_3 \delta\theta_{,x}\, \theta_{,xxx} dx\, dt$$

Integrating by parts

$$I_{41} = \int_{t1}^{t2} [G\mu C_3\theta_{,xxx}\delta\theta \,|_0^l - \int_0^l G\mu C_3\theta_{,xxx}\}_{,x}\delta\theta dx]\, dt$$

The second integral of I_4 is

$$I_{42} = \int_{t1}^{t2} \int_0^l G\mu C_3\theta_{,x}\delta\theta_{,xxx} dx\, dt$$

Integrating by parts

$$I_{42} = \int_{t_1}^{t2} \left[G\mu C_3\theta_{,x}\delta\theta_{,xx} \,|_0^l - \int_0^l \{G\mu C_3\theta_{,x}\}_{,x}\delta\theta_{,xx} dx \right] dt$$

i.e.,

$$I_{42} = \int_{t1}^{t2} [G\mu C_3\theta_{,x}\delta\theta_{,xx} \,|_0^l - \{G\mu C_3\theta_{,x}\}_{,x}\delta\theta_{,x} \,|_0^l$$

$$+ \int_0^l \{G\mu C_3\theta_{,x}\}_{,xx}\delta\theta_{,x} dx]\, dt$$

One more integration of the above gives

$$I_{42} = \int_{t1}^{t2} [G\mu C_3\theta_{,x}\delta\theta_{,xx} \,|_0^l - \{G\mu C_3\theta_{,x}\}_{,x}\delta\theta_{,x} \,|_0^l$$

$$+ \{G\mu C_3\theta_{,x}\}_{,xx}\delta\theta \,|_0^l - \int_0^l \{G\mu C_3\theta_{,x}\}_{,xxx}\delta\theta dx]\, dt$$

The first integral in T is

$$I_5 = \frac{1}{2}\delta \int_{t1}^{t2} \int_0^l \rho I_p \theta_{,t}^2 dx\, dt$$

i.e.,

$$I_5 = \int_{t1}^{t2} \int_0^l \rho I_p \theta_{,t}\delta\theta_{,t} dx\, dt$$

Integrating by parts with respect to time

$$I_5 = -\int_{t1}^{t2} \int_0^l \rho I_p \theta_{,tt}\delta\theta dx\, dt$$

The second integral in T is

$$I_6 = \frac{1}{2}\delta \int_{t1}^{t2} \int_0^l \rho C_2\mu^2\theta_{,xxt}^2 dx\, dt$$

i.e.,

$$I_6 = \int_{t1}^{t2} \int_0^l \rho C_2\mu^2\theta_{,xxt}\delta\theta_{,xxt} dx\, dt$$

Integrating by parts with respect to time

$$I_6 = -\int_{t1}^{t2} \int_0^l \rho C_2\mu^2\theta_{,xxtt}\delta\theta_{,xx} dx\, dt$$

Now integrating with respect to x by parts,

$$I_6 = \int_{t1}^{t2}\left[-\rho C_2\mu^2\theta_{,xxtt}\delta\theta_{,x}\,|_0^l + \int_0^l \{\rho C_2\mu^2\theta_{,xxtt}\}_{,x}\delta\theta_{,x}dx\right]dt$$

i.e.,

$$I_6 = \int_{t1}^{t2}[-\rho C_2\mu^2\theta_{,xxtt}\delta\theta_{,x}\,|_0^l + \{\rho C_2\mu^2\theta_{,xxtt}\}_{,x}\delta\theta\,|_0^l$$

$$-\int_0^l \{\rho C_2\mu^2\theta_{,xxtt}\}_{,xx}\delta\theta dx]\,dt$$

The third integral in T is

$$I_7 = \delta\int_{t1}^{t2}\int_0^l \rho\mu C_5\theta_{,xxt}\theta_{,t}dx\,dt$$

i.e.,

$$I_7 = \int_{t1}^{t2}\int_0^l \rho\mu C_5(\delta\theta_{,xxt}\theta_{,t} + \theta_{,xxt}\delta\theta_{,t})dx\,dt$$

We split the above into two integrals, the first of which is

$$I_{71} = \int_{t1}^{t2}\int_0^l \rho\mu C_5\delta\theta_{,xxt}\theta_{,t}dx\,dt$$

Integrating first with respect to time

$$I_{71} = \int_{t1}^{t2}\int_0^l -\rho\mu C_5\theta_{,tt}\delta\theta_{,xx}dx\,dt$$

Integrating now with respect to x

$$I_{71} = \int_{t1}^{t2}\left[-\rho\mu C_5\theta_{,tt}\delta\theta_{,x}\,|_0^l + \int_0^l (\rho\mu C_5\theta_{,tt})_{,x}\delta\theta_{,x}dx\right]dt$$

$$I_{71} = \int_{t1}^{t2}[-\rho\mu C_5\theta_{,tt}\delta\theta_{,x}\,|_0^l + (\rho\mu C_5\theta_{,tt})_{,x}\delta\theta\,|_0^l$$

$$-\int_0^l (\rho\mu C_5\theta_{,tt})_{,xx}\delta\theta dx]\,dt$$

The second integral of I_7 is

$$I_7 = \int_{t1}^{t2}\int_0^l \rho\mu C_5\theta_{,xxt}\delta\theta_{,t}dx\,dt$$

Integrating by parts with respect to time

$$I_7 = \int_{t_1}^{t2}\int_0^l -\rho\mu C_5\theta_{,xxtt}\delta\theta dx\,dt$$

The last integral in T is

$$I_8 = \frac{1}{2}\delta\int_{t1}^{t2}\int_0^l \rho I_{\phi\phi}\theta_{,xt}^2 dx\,dt$$

i.e.,

$$I_8 = \int_{t1}^{t2} \int_0^l \rho I_{\phi\phi} \theta_{,xt} \delta\theta_{,xt} dx\, dt$$

Integrating first with respect to time

$$I_8 = \int_{t1}^{t2} \int_0^l -\rho I_{\phi\phi} \theta_{,xtt} \delta\theta_{,x} dx\, dt$$

Next integrating with respect to x,

$$I_8 = \int_{t1}^{t2} \left[-\rho I_{\phi\phi} \theta_{,xtt} \delta\theta \,|_0^l + \int_0^l (\rho I_{\phi\phi} \theta_{,xtt})_{,x} \delta\theta dx \right] dt$$

Rearranging all the terms in the above integrals, we get the following differential equation of motion and boundary conditions for free torsional vibration of a non-circular rod.

$$(\mu^2 G C_2 \theta_{,xxx})_{,xxx} - \{(EI_{\phi\phi} + \mu^2 G C_4)\theta_{,xx}\}_{,xx} + (\mu G C_3 \theta_{,xxx})_{,x}$$

$$+ (\mu G C_3 \theta_{,x})_{,xxx} - (C_2 \rho \mu^2 \theta_{,xxtt})_{,xx}$$

$$+ (C_1 G \theta_{,x})_{,x} + (\rho I_{\phi\phi} \theta_{,xtt})_{,x} - \rho\mu C_5 \theta_{,xxtt} - (\rho\mu C_5 \theta_{,tt})_{,xx} - \rho I_p \theta_{,tt} = 0 \qquad (7.100a)$$

$$[-\{(EI_{\phi\phi} + \mu^2 G C_4)\theta_{,xx}\}_{,x} - \mu G C_3 \theta_{,xxx} - (\mu G C_3 \theta_{,x})_{,xx} + C_1 G \theta_{,x} + (\mu^2 G C_2 \theta_{,xxx})_{,xx}$$

$$- (C_2 \rho \mu^2 \theta_{,xxtt})_{,x} - (\rho\mu C_5 + \rho I_{\phi\phi})\theta_{,xtt}] \delta\theta \,|_0^1 = 0 \qquad (7.100b)$$

$$[(EI_{\phi\phi} + \mu^2 G C_4)\theta_{,xx} + (\mu G C_3 \theta_{,x})_{,x} - (\mu^2 G C_2 \theta_{,xxx})_{,x}$$

$$+ (C_2 \rho \mu^2 \theta_{,xxtt} + \rho\mu C_5)\theta_{,tt}] \delta\theta_{,x} \,|_0^1 = 0 \qquad (7.100c)$$

$$(\mu^2 G C_2 \theta_{,xxx} - \mu G C_3 \theta_{,x}) \delta\theta_{,xx} \,|_0^l = 0 \qquad (7.100d)$$

The differential equation (7.100*a*) with the boundary conditions in (7.100*b*) to (7.100*d*) above gives an accurate theory for small amplitudes of vibrations of a rod with uniform twist with the deformations following linear principles of elasticity. This differential equation of motion and the corresponding boundary conditions are more complicated and a solution to the problem cannot be easily obtained. However, such a complicated theory may not produce any significantly different results from those of St Venant and Timosehenko-Gere theories, particularly from engineering point of view. Therefore, the above theory can be considered to be more of an academic exercise to understand the assumptions made in simpler theories and also to know the possible small order errors that may occur in such solutions.

7.14 STRESS ANALYSIS

Consider the case of a circular rod, fixed at $x = 0$ and free at $x = l$, acted upon by an external sinusoidal torque $M_0 \sin \omega t$, shown in Fig. 7.13. We will determine the stress response of this problem by an exact solution and by approximate methods of Lagrange and Reissner's principle.

Case 1: Exact solution

The differential equation for this problem is given by equation (7.8a)

$$\rho I_p \ddot{\theta} - (C\theta_{,x})_{,x} = M_0\}_{x=l} \sin \omega t \qquad (7.8a)$$

i.e.,

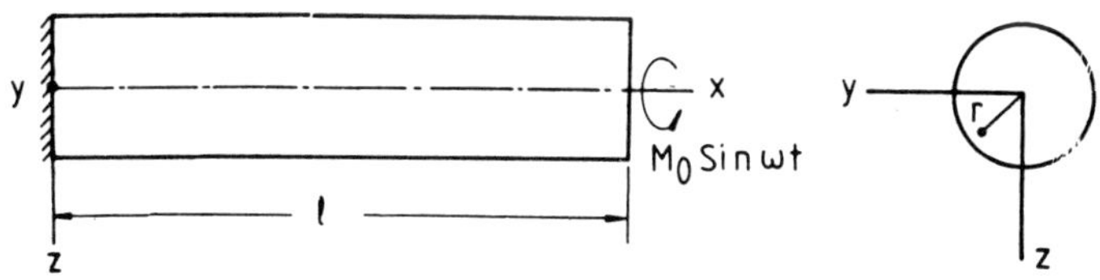

Fig. 7.13 Circular cantilever rod with excitation at free end

$$\ddot{\theta} - a^2\theta_{,xx} = \frac{M_0}{\rho I_p}\sin\omega t \tag{7.101}$$

where

$$a^2 = \frac{G}{\rho} \tag{7.102}$$

For fixed-free rod, the natural frequencies and corresponding mode shapes are given in Table 6.1, which in dimensional form are

$$p_i = \frac{i\pi a}{2l} \quad i = 1, 3, 5, 7 \tag{7.103}$$

$$X_i(x) = \sin\frac{i\pi x}{2l} \tag{7.104}$$

For the forced vibration problem, let

$$\theta = \sum_i X_i q_i(t) \tag{7.105}$$

Substituting equation (7.105) above in (7.101), we have

$$\sum_i X_i \ddot{q}_i - a^2 \sum_i X_{i,xx} q_i = \frac{M_0}{\rho I_p}\sin\omega t \tag{a}$$

Multiplying the above with X_j and integrating over the length of the rod, we get

$$\int_0^l \sin\frac{j\pi x}{2l}\sum\sin\frac{i\pi x}{2l}\ddot{q}_i + a^2\int_0^l \sin\frac{j\pi x}{2l}\sum\left\{\frac{i\pi}{2l}\right\}^2 \sin\frac{i\pi x}{2l} q_i$$

$$= \int_0^l \sin\frac{j\pi x}{2l}\frac{M_0}{\rho I_p}\sin\omega t \tag{b}$$

i.e.,

$$\ddot{q}_j + \frac{i^2\pi^2 a^2}{4l^2} q_j = \frac{2}{I_p \rho l}(-1)^{\frac{j-1}{2}} M_0 \sin\omega t \tag{7.106}$$

The steady-state solution of the above is

$$q_j = \frac{M_0}{\rho I_p l}\frac{2(-1)^{\frac{j-1}{2}}}{p_j^2 - \omega^2}\sin\omega t \tag{7.107}$$

Hence

$$\theta = \frac{M_0}{\rho I_p l}\sum_{j=1,3,\ldots}\frac{2(-1)^{\frac{j-1}{2}}}{p_j^2 - \omega^2}\sin\frac{j\pi x}{2l}\sin\omega t \tag{7.108}$$

Let ξ_j be the modal damping in jth mode. Then, equation (7.108) above can be modified to

$$\theta = \frac{M_0}{\rho I_p l} \sum_{j=1,3,\ldots} \frac{2(-1)^{\frac{j-1}{2}}}{\sqrt{(p_j^2 - \omega^2)^2 + 4\xi_j^2 p_j^2 \omega^2}} \sin\frac{j\pi x}{2l} \sin(\omega t - \psi_j) \tag{7.109}$$

Therefore, the torsional stress at any radius r in the rod is

$$\tau = Gr\frac{d\theta}{dx}$$

$$= \frac{M_0 G r}{\rho I_p l^2} \sum_{j=1,3,\ldots} \frac{j\pi(-1)^{\frac{j-1}{2}}}{\sqrt{(p_j^2 - \omega^2)^2 + 4\xi_j^2 p_j^2 \omega^2}} \cos\frac{j\pi x}{2l} \sin(\omega t - \psi_j) \tag{7.110}$$

Case 2: Lagrange solution

The kinetic energy of the rod is

$$T = \frac{1}{2}\int_0^l \rho I_p \dot{\theta}^2 dx \tag{7.111}$$

The strain energy is

$$U = \frac{1}{2}\int_0^l G I_p \theta'^2 dx \tag{7.112}$$

The work done by the external moment is

$$W = \int_0^l \theta M_0 (x = l) \sin\omega t \tag{7.113}$$

Let us assume the solution for θ in the form of a polynomial function

$$\theta = \sum_{i=1}^{N} g_i(X) q_i(t) \tag{7.114}$$

where

$$g_i(X) = X^i - \frac{i}{i+1} X^{i+1} \tag{7.115}$$

$$X = \frac{x}{l}$$

Substituting (7.114) in (7.111) to (7.113), we get

$$T = \frac{1}{2}\int_0^l \rho I_p \left\{\sum_{i=1}^{N} g_i \dot{q}_i\right\}^2 dx \tag{7.111a}$$

$$U = \frac{1}{2}\int_0^l G I_p \left\{\sum_{i=1}^{N} g'_i q_i\right\}^2 dx \tag{7.112a}$$

$$W = \int_0^l \sum_{i=1}^{N} g_i q_i M_0 (x = l) \sin\omega t \tag{7.113a}$$

The Lagrangian equation for this problem is

$$\frac{d}{dt}\left(\frac{\partial T}{\partial \dot{q}_i}\right) - \frac{\partial T}{\partial q_i} + \frac{\partial U}{\partial q_i} = \frac{\partial W}{\partial q_i} \tag{7.116}$$

$$\frac{\partial T}{\partial \dot{q}_i} = \int_0^l \rho I_p \sum_1^N g_i g_j \dot{q}_i dx$$

$$\frac{d}{dt}\left(\frac{\partial T}{\partial \dot{q}_i}\right) = \int_0^l \rho I_p \sum_1^N g_i g_j \ddot{q}_i dx \tag{c}$$

$$\frac{\partial T}{\partial q_i} = 0 \tag{d}$$

$$\frac{\partial V}{\partial q_i} = \int_0^l G I_p \sum_1^N g'_i g'_j q_i dx \tag{e}$$

$$\frac{\partial W}{\partial q_i} = g_j \,|_{x=l}\, M_0 \sin \omega t \tag{t}$$

Substituting the above equations (c) to (f) in equation (7.116), we get

$$[M]\{\ddot{q}\} + [K]\{q\} = \{Q_0\} \sin \omega t \tag{7.117}$$

where

$$[M] = \left[\rho I_p \int_0^l \sum_1^N g_i g_j dx\right]$$

$$[K] = \left[G I_p \int_0^l \sum_1^N g_i' g_j' dx\right]$$

$$\{Q_0\} = \{g_j \,|_{x=l}\, M_0\}$$

Let us include proportional damping in equation (7.117) to give

$$[M]\{\ddot{q}\} + [C]\{\dot{q}\} + [K]\{q\} = \{Q_0\} \sin \omega t \tag{7.117a}$$

where

$$[C] = \frac{C_v}{\rho}[M]$$

Making use of modal analysis, with

$$\{q\} = [U]\{\eta\} \tag{7.118}$$

where $[U]$ is the modal matrix obtained from the free vibration problem of (7.117). Premultiplying equation (7.117a) with the transpose of the modal matrix, we get

$$[\overline{M}]\{\ddot{\eta}\} + [\overline{C}]\{\dot{\eta}\} + [\overline{K}]\{\eta\} = \{\overline{Q}_0\} \sin \omega t \tag{7.119}$$

where

$$[\overline{M}] = [U]^T [M] [U]$$

$$[\overline{C}] = [U]^T [C] [U]$$

$$[\overline{K}] = [U]^T [K] [U]$$

$$[\overline{Q}_0] = [U]^T [Q_0]$$

The forced vibration response is therefore given by

$$\eta_j = \frac{\overline{Q}_{0j}}{\overline{M}_j} \frac{\sin(\omega t - \psi_j)}{\sqrt{(p_j^2 - \omega^2)^2 + 4\xi_j^2 p_j^2 \omega^2}} \tag{7.120}$$

where

$$\xi_j = \frac{\overline{C}}{2\overline{M}_j p_j}$$

$$\psi_j = \tan^{-1} \frac{2\xi_j p_j \omega}{p_j^2 - \omega^2}$$

Using (7.114) and (7.118), θ can be determined and the stress can be obtained by using (7.110).

Case 3: Reissner method

The nonzero strain and stress are given by

$$\varepsilon_{rx} = r\theta'$$

$$\tau_{rx} = Gr\theta'$$

The first term in Reissner functional is

$$\iiint_V \tau_{ij}\varepsilon_{ij} dv = \int_0^l \int_A (Gr\theta')(r\theta') dA dx$$

$$= \int_0^l T_\theta \theta' dx \tag{g}$$

where

$$T_\theta = \int_A Gr^2\theta' dA$$

$$= GI_p\theta'$$

$$= C\theta' \tag{h}$$

The second term in Reissner functional is

$$\iiint U_0^*(\tau_{ij}) dv = \iiint \int_0^\tau r\theta' d(Gr\theta') dv$$

$$= \iiint \frac{1}{2G}(Gr\theta')^2 dv$$

$$= \int_0^l \frac{T_\theta^2}{2C} dx \tag{i}$$

The traction force term is

$$\int\int_S T_i^{(v)} u_i dA = \int_0^l M_0 \theta \sin \omega t dx$$

$$= M_0 \theta |_{x=l} \sin \omega t \tag{j}$$

The body forces are zero. Therefore, Reissner's functional is

$$I_R = \int_0^l T_\theta \theta' dx - \int_0^l \frac{T_\theta^2}{2C} dx - M_0 \theta |_{x=l} \sin \omega t \tag{k}$$

The kinetic energy is given by

$$T = \frac{1}{2} \int_0^l \rho I_p \dot{\theta}^2 dx \tag{l}$$

The dynamic Reissner's functional is

$$L = T - I_R = \frac{1}{2} \int_0^l \rho I_p \dot{\theta}^2 dx - \int_0^l T_\theta \theta' dx + \int_0^l \frac{T_\theta^2}{2C} dx + M_0 \theta |_{x=l} \sin \omega t \tag{7.121}$$

We now assume

$$\theta = \sum_1^N g_i(X) q_i(t) \tag{7.114}$$

where

$$g_i(X) = X^i - \frac{i}{i+1} X^{i+1} \tag{7.115}$$

$$X = \frac{x}{l}$$

$$T_\theta = \sum_1^N h_i(X) r_i(t) \tag{7.122}$$

where

$$h_i(x) = \frac{(1-X)^i}{i+1} \tag{7.123}$$

Both g and h satisfy the boundary conditions of the cantilever rod. Substituting (7.114) and (7.122) in equation (7.121), we get

$$L = T - I_R = \frac{1}{2} \int_0^l \rho I_p (\Sigma g_i \dot{q}_i)^2 dx - \int_0^l \Sigma h_i r_i \Sigma g'_i q_i dx$$

$$+ \int_0^l \frac{1}{2C} (\Sigma h_i r_i)^2 dx + M_0 \Sigma g_i q_i |_{x=l} \sin \omega t \tag{7.124}$$

Lagrangian equations for the above are

$$\frac{d}{dt}\left(\frac{\partial L}{\partial \dot{q}_i}\right) - \frac{\partial L}{\partial q_i} = 0$$

$$\frac{d}{dt}\left(\frac{\partial L}{\partial \dot{r}_i}\right) - \frac{\partial L}{\partial r_i} = 0 \tag{7.125}$$

The above lead to

$$\begin{bmatrix} M_1 & M_2 \\ M_3 & M_4 \end{bmatrix} \begin{Bmatrix} \ddot{q} \\ \ddot{r} \end{Bmatrix} + \begin{bmatrix} C_1 & C_2 \\ C_3 & C_4 \end{bmatrix} \begin{Bmatrix} \dot{q} \\ \dot{r} \end{Bmatrix} + \begin{bmatrix} K_1 & K_2 \\ K_3 & K_4 \end{bmatrix} \begin{Bmatrix} q \\ r \end{Bmatrix} = \begin{Bmatrix} Q_0 \\ 0 \end{Bmatrix} \tag{7.126}$$

where

$$[M_1] = \int_0^l \rho I_p g_i g_j dx$$

$$[M_2] = [M_3] = [M_4] = 0$$

$$[C_1] = \frac{C_v}{\rho}[M_1] \quad \text{etc.}$$

$$[K_1] = 0$$

$$[K' = \int_0^l h_i g_j' dx$$

$$[K_3] = -\int_0^l g_i' h_j dx$$

$$[K_4] = -\int_0^l \frac{1}{C} h_i h_j dx$$

$$\{Q_0\} = M_0 g_j |_{x=l} \sin \omega t \tag{7.127}$$

From here onwards, we make use of modal analysis as in the previous case with equation (7.118) to determine the forced vibration response. The solution in this case will lead to the angle of twist θ and the twisting moment T_θ. The stress is then given by

$$\tau = \frac{T_\theta}{C} \tag{7.128}$$

The rod with the following data is considered to illustrate the results from the above three cases.

l = 10 cm
I_p = 0.68175 cm^4
r = 0.812 cm
G = 0.843 × 10^{11} Pa
ρ = 0.00786 kg cm^3
ξ_j = 0.01
ω = 100 rad/s

Computer programs are to be used to evaluate the three different cases.

(1) Natural frequencies

Table 7.4 gives the natural frequencies obtained from exact solution, Lagrange and Reissner formulations. For the first three modes, both Lagrange and Reissner methods give quick convergence with a five-term solution. For the fourth and fifth modes, Reissner method gives improved accuracy over Lagrange method.

Table 7.4
Torsional natural frequencies–Hz

Exact	*N*	*Lagrange*	*Reissner*
8108.56	1	8161.9	8161.9
	2	8109.6	8109.6
24325.69		25057.0	25057.0
	3	8108.6	8108.6
		24432.0	24432.0
40542.81		43075.0	43075.0
	4	8108.6	8108.5
		24330.0	24330.0
		41212.0	41211.0
56759.94		62880.0	62673.0
	5	8108.6	8108.5
		24326.0	24326.0
		40607.0	40607.0
		58829.0	58130.0
72977.06		85091.0	81295.0

It is obvious that the correct results for stresses and deflections would be those obtained from an infinite-term solution of the exact solution. Figures 7.14a and 7.14b illustrate the convergence patterns for tip deflections and stresses. Both Lagrange and Reissner methods yield nearly similar values; however, Lagrange method is somewhat superior for determination of deflection. For determination of stress, Reissner method gives much faster convergence to the exact solution, whereas Lagrange method has a wide variation with the first three terms.

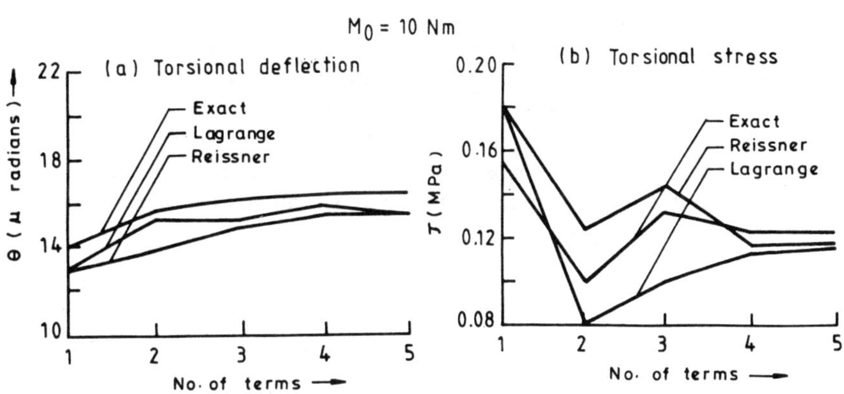

Fig. 7.14 Convergence patterns

Figures 7.15a and 7.15b show the deflection and stress distributions along the length of the rod with a five-term solution. Both the Lagrange and Reissner methods are reasonably accurate.

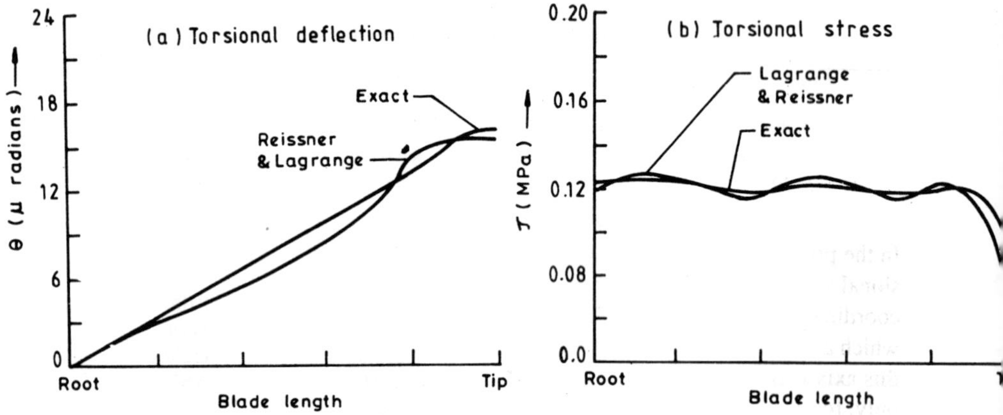

Fig. 7.15 Deflection and stress distributions

Table 7.5
Tip deflections and root stresses at resonance

	Tip deflection (m rad)			*Root stress (M Pa)*		
Mode	*Exact*	*Lagrange*	*Reissner*	*Exact*	*Lagrange*	*Reissner*
I	0.7080	0.7068	0.7068	7.55	7.54	7.54
II	0.0815	0.0800	0.0800	2.47	2.45	2.45
III	0.0305	0.0294	0.0294	1.47	1.29	1.30
IV	0.0161	0.0267	0.0267	1.02	1.84	1.62
V	0.0098	0.0006	0.0006	0.82	2.44	2.13

Table 7.5 gives the values of deflection and stress for resonant excitation frequencies with five-term solutions. Both Lagrange and Reissner methods give good results for the first three modes, the error being high for the fourth and fifth modes. Reissner method shows marginal improvement over Lagrange method. The errors in fourth and fifth modes can be expected to be high as the mode shapes here may not have converged with a five-term solution. We will discuss the same problem for bending in the next chapter.

8

Lateral Vibrations of Beams

In the previous two chapters, we considered longitudinal vibrations of bars and torsional vibrations of rods. In both the cases, the members are defined by one axial coordinate x. We will consider here one more case of structural or machine members which are one dimensional in character, executing vibrations lateral or transverse to this axis x. Such members are called beams. The vibrations can be in one direction only, for example, z axis or can be present simultaneously in two directions z and y axes.

8.1 EULER-BERNOULLI THEORY OF BEAMS

The simplest theory of lateral vibration of beams is due to Euler and Bernoulli. Consider the beam of length l shown in Fig. 8.1. The cross-sectional area of the beam

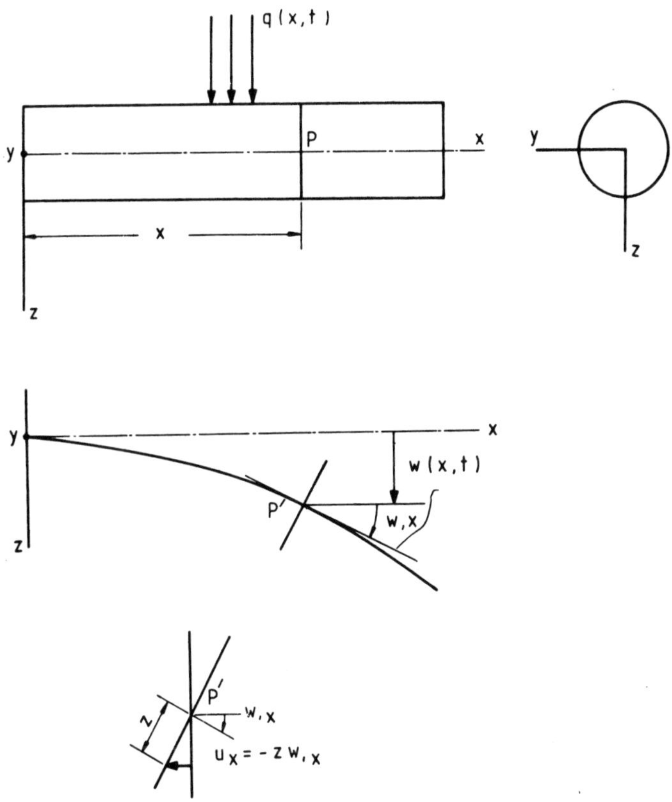

Fig. 8.1 Bending of Euler–Bernoulli beam

is A, mass density is ρ and it is acted upon by an external force $q\,(x, t)$ acting in the direction of the z axis. The beam is thus subjected to lateral vibrations, $w\,(x, t)$ in z direction. We assume here that a plane cross-section at a distance x as shown in Fig. 8.1 remains plane even after bending and has a rotation about y axis given by the slope $w_{,x}$ of the elastic curve. Thus we can define the displacement field of this beam as

$$u_x = -z w_{,x}$$

$$u_y = 0$$

$$u_z = w(x,t) \tag{8.1}$$

The corresponding strain and stress field are therefore

$$\varepsilon_{xx} = z w_{,xx} \quad \text{rest zero}$$

$$\tau_{xx} = -E z w_{,xx} \quad \text{rest zero} \tag{8.2}$$

The strain energy is

$$U = \frac{1}{2}\int_0^l \int_A E z^2 w_{,xx} dA\, dx$$

$$= \frac{1}{2}\int_0^l EI_{yy} w_{,xx}^2 dx \tag{8.3}$$

Ignoring the axial inertia due to $\dot{u}_x$ the kinetic energy of the beam is

$$T = \frac{1}{2}\int_0^l \rho A \dot{w}^2 dx \tag{8.4}$$

The work done by the external force is

$$W = \int_0^l q w dx \tag{8.5}$$

Hence, Hamilton's principle gives

$$\delta \int_{t_1}^{t_2} \left[\frac{1}{2}\int_0^l EI_{yy} w_{,xx}^2 dx - \frac{1}{2}\int_0^l \rho A \dot{w}^2 dx - \int_0^l q w dx \right] dt = 0 \tag{8.6}$$

Taking the variation of the first integral in the above equation (8.6), we have

$$I_1 = \int_{t_1}^{t_2} \int_0^l EI_{yy} \omega_{,xx} \delta w_{,xx} dx\, dt \tag{a}$$

Integrating by parts

$$I_1 = \int_{t_1}^{t_2} \left[EI_{yy} w_{,xx} \delta w_{,x} |_0^l - \int_0^l (EI_{yy} w_{,xx})_{,x} \delta w_{,x} dx \right] dt$$

i.e.,

$$I_1 = \int_{t_1}^{t_2} [EI_{yy} w_{,xx} \delta w_{,x} |_0^l - (EI_{yy} w_{,xx})_{,x} \delta w |_0^l$$

$$+ \int_0^l (EI_{yy} w_{,xx})_{,xx} \delta w \, dx] \, dt \qquad (b)$$

The second integral in equation (8.6) is

$$I_2 = \int_{t_1}^{t_2} \int_0^l \rho A \dot{w} \delta \dot{w} \, dx \, dt$$

Integrating by parts with respect to time

$$I_2 = \int_{t_1}^{t_2} - \int_0^l \rho A \ddot{w} \delta w \, dx \, dt \qquad (c)$$

The third integral is

$$I_3 = \int_{t_1}^{t_2} \int_0^l q \delta w \, dx \, dt \qquad (d)$$

Using equations (*b*) to (*d*) in equation (8.6), we get

$$\int_{t_1}^{t_2} [EI_{yy} w_{,xx} \delta w_{,x} |_0^l - (EI_{yy} w_{,xx})_{,x} \delta w |_0^l$$

$$+ \int_0^l \{(EI_{yy} w_{,xx})_{,xx} + \rho A \ddot{w} - q\} \delta w \, dx] \, dt = 0 \qquad (e)$$

Hence, we can write the governing differential equation of motion of the Euler-Bernoulli beam and the boundary conditions to be

$$(EI_{yy} w_{,xx})_{,xx} + \rho A \ddot{w} = q \qquad (8.7)$$

$$EI_{yy} w_{,xx} \delta w_{,x} |_0^l - (EI_{yy} w_{,xx})_{,x} \delta w |_0^l = 0 \qquad (8.8)$$

8.2 RAYLEIGH'S THEORY OF BEAMS

Rayleigh included the inertia term due to the velocity of an element in axial direction.

$$\dot{u}_x = -z w_{,xt} \qquad (8.9)$$

The kinetic energy due to the above component of the velocity is

$$T_x = \int_0^l \int_A \frac{1}{2} \rho z^2 \dot{w}_{,x}^2 \, dA \, dx$$

$$= \int_0^l \frac{1}{2} \rho I_{yy} \dot{w}_{,x}^2 \, dx \qquad (8.10)$$

Adding the above quantity to equation (8.4), the total kinetic energy of the system is

$$T = \frac{1}{2} \int_0^l (\rho A \dot{w}^2 + \rho I_{yy} \dot{w}_{,x}^2) dx \qquad (8.11)$$

The contribution of the second term in (8.11) above to the differential equation and boundary conditions can be obtained from the following.

$$I_2 = \delta \int_{t_1}^{t_2} \frac{1}{2} \int_0^l \rho I_{yy} \dot{w}_{,x}^2 dx\, dt$$

$$= \int_{t_1}^{t_2} \int_0^l \rho I_{yy} \dot{w}_{,x} \delta \dot{w}_{,x} dx\, dt \tag{a}$$

Integrating by parts with respect to time

$$I_2 = \int_{t_1}^{t_2} - \int_0^l \rho I_{yy} \ddot{w}_{,x} \delta w_{,x} dx\, dt$$

Integrating once again, now with respect to x, we get

$$I_2 = \int_{t_1}^{t_2} \left[-\rho I_{yy} \ddot{w}_{,x} \delta w \,|_0^l + \int_0^l (\rho I_{yy} \ddot{w}_{,x})_{,x} \delta w\, dx \right] dt \tag{b}$$

Adding equation (b) above to equations (8.7) and (8.8), Euler's theory now gets modified to

$$(EI_{yy} w_{,xx})_{,xx} - (\rho I_{yy} \ddot{w}_{,x})_{,x} + \rho A \ddot{w} = q \tag{8.12}$$

$$EI_{yy} w_{,xx} \delta w_{,x} \,|_0^l - \{(EI_{yy} w_{,xx})_{,x} - \rho I_{yy} \ddot{w}_{,x}\} \delta w \,|_0^l = 0 \tag{8.13}$$

The additional effect due to the axial inertia in the above equation is originally called rotary inertia by Rayleigh. From Fig. 8.1, it can be seen that the section at distance x, rotates about the y axis, while having the lateral motion in the z direction. In fact, Rayleigh originally derived this effect by considering the rotation of such a section, rather than by the approach we used above. Let us nondimensionalize the equations (8.12) and (8.13) by setting

$$x \leftarrow \frac{x}{l}$$

$$t \leftarrow \frac{t}{t_0}$$

$$E \leftarrow \frac{E}{E_0}$$

$$I_{yy} \leftarrow \frac{I_{yy}}{I_{yy0}}$$

$$\rho \leftarrow \frac{\rho}{\rho_0}$$

$$w \leftarrow \frac{w}{l}$$

$$q \leftarrow \frac{q}{q_0}$$

$$r_1^2 = \frac{I_{yy0}}{A_0 l^2}$$

where t_0 and q_0 are given by

$$t_0^2 = \frac{\rho_0 A_0 l^4}{E_0 I_{yy0}}$$

$$q_0 = \frac{\rho_0 A_0 l}{t_0^2}$$

$$= \frac{E_0 I_{yy0}}{l^3} \tag{8.14}$$

Equation (8.12) now becomes

$$M\ddot{w} + Kw = q \tag{8.15}$$

where the mass and stiffness operators are given by

$$M = \rho A - r_1^2 \partial_x (\rho I_{yy} \partial_x)$$

$$K = \partial_x^2 EI \partial_x^2 \tag{8.16}$$

The boundary conditions in (8.13) are

$$(EI_{yy}\partial_x^2 w)\delta\partial_x w \,|_0^1 + \{-\partial_x(EI_{yy}\partial_x^2 w) + r_1^2 \rho I_{yy}\partial_x \ddot{w}\}\,\delta w\,|_0^1 = 0 \tag{8.17}$$

For uniform beams with the reference values chosen as its original values themselves, equations (8.16) and (8.17) reduce to

$$M = 1 - r_1^2 \partial_x^2$$

$$K = \partial_x^4 \tag{8.16a}$$

$$(\partial_x^2 w)\delta\partial_x w \,|_0^1 + \{-\partial_x^3 w + r_1^2 \partial_x \ddot{w}\}\,\delta w\,|_0^1 = 0 \tag{8.17a}$$

For free vibrations, let

$$w(x,t) = W(x)e^{ipt}$$

then equation (8.15) and (8.17) become

$$(d_x^4 + r_1^2 d_x^2 p^2 - p^2)W = 0$$

$$(d_x^2 W)\delta d_x W \,|_0^1 - \{d_x^3 W + r_1^2 d_x p^2 W\}\,\delta W\,|_0^1 = 0 \tag{8.18}$$

Let

$$b = -r_1^2 p^2$$

$$c = p^2 \tag{8.19}$$

then equations (8.18) become

$$(d_x^4 - bd_x^2 - c)W = 0$$

$$(d_x^2 W)\delta d_x W \,|_0^1 + \{bd_x W - d_x^3 W\}\delta W \,|_0^1 = 0 \tag{8.18a}$$

The Laplace transform of the first equation in (8.18a) is

$$(s^4 - bs^2 - c)\overline{W} = (s^3 - bs)W(0) + s^2 W'(0)$$

$$+ sW''(0) + (W'''(0) - bW'(0)\} \tag{8.20}$$

Let us define the following positive quantities

$$s_1^2 = \frac{1}{2}(\sqrt{b^2 + 4c} + b)$$

$$s_2^2 = \frac{1}{2}(\sqrt{b^2 + 4c} - b) \tag{8.21}$$

Using (8.19), the second equation in the above gives

$$p^2 = \frac{s_2^4}{1 + r_1^2 s_2^2} \tag{8.22}$$

The quantities in (8.19) are related to the roots of the characteristic polynomial and satisfy the following relations.

$$(s^2 - s_1^2)(s^2 + s_2^2) = (s^4 - bs^2 - c) = 0$$

$$b = s_1^2 - s_2^2$$

$$c = s_1^2 s_2^2$$

$$s_1^4 - bs_1^2 - c = 0$$

$$s_2^4 + bs_2^2 - c = 0 \tag{8.23}$$

The transfer function $T(x)$ of the system is

$$T(x) = L^{-1}\frac{1}{s^4 - bs^2 - c}$$

$$= L^{-1}\frac{1}{(s^2 - s_1^2)(s^2 + s_2^2)}$$

The above can be rearranged as

$$T(x) = L^{-1}\left\{\frac{A}{s^2 - s_1^2} + \frac{B}{s^2 + s_2^2}\right\}$$

$$= \frac{1}{(s_1^2 + s_2^2)(s^2 - s_1^2)} - \frac{1}{(s_1^2 + s_2^2)(s^2 + s_2^2)}$$

i.e.,

$$T(x) = \frac{1}{s_1^2 + s_2^2}\left\{\frac{\sinh s_1 x}{s_1} - \frac{\sin s_2 x}{s_2}\right\}$$

Hence

$$T(x) = \frac{1}{s_1 s_2(s_1^2 + s_2^2)}(s_2 \sinh s_1 x - s_1 \sin s_2 x) \tag{8.24}$$

so that the mode shape and its derivatives can be written as

$$W = [(d_x^3 - b d_x)W(0) + d_x^2 W'(0) + d_x W''(0) + \{W'''(0) - bW'(0)\}]T(x)$$

$$W' = [cW(0) + d_x^3 W'(0) + d_x^2 W''(0) + d_x\{W'''(0) - bW'(0)\}]T(x)$$

$$W'' = [c d_x W(0) + (b d_x^2 + c)W'(0) + d_x^3 W''(0) + d_x^2\{W'''(0) - bW'(0)\}]T(x)$$

$$W''' - bW' = [(c d_x^2 - bc)W(0) + c d_x W'(0) + cW''(0)$$

$$+ (d_x^3 - b d_x)\{W'''(0) - bW'(0)\}]T(x) \tag{8.25}$$

Case 1: Simply supported beam

Let us consider the case of a simply supported beam with the boundary conditions

$$W(0) = W(1) = W''(0) = W''(1) = 0$$

From (8.24) we note that the conditions at $x = 0$ are satisfied. At $x = 1$, we get

$$[d_x^2 W'(0) + \{W'''(0) - bW'(0)\}]T(1) = 0$$

$$[(b d_x^2 + c)W'(0) + d_x^2\{W'''(0) - bW'(0)\}]T(1) = 0$$

The above can be written as

$$\begin{bmatrix} d_x^2 T(1) & T(1) \\ (b d_x^2 + c)T(1) & d_x^2 T(1) \end{bmatrix}\begin{Bmatrix} W'(0) \\ W'''(0) - bW'(0) \end{Bmatrix} = 0 \tag{8.26}$$

For nontrivial solution of the above characteristic equation (8.26), the determinant of the matrix should vanish, i.e.,

$$\{d_x^2 T(1)\}^2 - b d_x^2 T(1) T(1) - c\{T(1)\}^2 = 0$$

This gives

$$\sin s_2 = 0 \tag{8.27}$$

i.e.,

$$s_2 = n\pi \tag{8.28}$$

From (8.22), we therefore get the frequency expression as

$$p_n^2 = \frac{(n\pi)^4}{1 + r_1^2 (n\pi)^2} \tag{8.29}$$

To determine the mode shape, let

$$W'''(0) - bW'(0) = 1$$

Then, from equation (8.26) we have

$$W'(0) = -\frac{T(1)}{d_x^2 T(1)}$$

Therefore, from the first equation of (8.25), we have

$$W = -\frac{T(1)}{d_x^2 T(1)} d_x^2 T(x) + T(x)$$

In view of (8.27) and (8.28) the above becomes

$$W = \sin s_2 x = \sin n\pi x \cdot \quad n = 1, 2, \ldots \tag{8.30}$$

We can consider other boundary conditions and repeat the analysis as in case 1 above and obtain the following.

Case 2: Fixed-fixed beam

Boundary conditions

$$W(0) = 0; \quad W(1) = 0$$

$$W'(0) = 0; \quad W'(1) = 0$$

Frequency equation

$$\cosh s_1 \cos s_2 - \frac{1}{2}\left\{\frac{s_1}{s_2} - \frac{s_2}{s_1}\right\} \sinh s_1 \sin s_2 - 1 = 0$$

Mode shape

$$(s_2 \sinh s_1 - s_1 \sin s_2)(\cosh s_1 x - \cos s_2 x)$$

$$-(\cosh s_1 - \cos s_2)(s_2 \sinh s_1 x - s_1 \sin s_2 x)$$

Case 3 : Fixed-free cantilever beam

Boundary conditions

$$W(0) = 0; \quad W''(1) = 0$$

$$W'(0) = 0; \quad W'''(1) - bW'(1) = 0$$

Frequency equation

$$\left\{\frac{s_1^2}{s_2^2} + \frac{s_2^2}{s_1^2}\right\} \cosh s_1 \cos s_2 + \left\{\frac{s_1}{s_2} - \frac{s_2}{s_1}\right\} \sinh s_1 \sin s_2 + 2 = 0$$

Mode shape

$$s_1 s_2 (s_1 \sinh s_1 + s_2 \sin s_2)(\cosh s_1 x - \cos s_2 x)$$

$$-(s_1^2 \cosh s_1 + s_2^2 \cos s_2)(s_2 \sinh s_1 x - s_1 \sin s_2 x)$$

Case 4: Fixed-supported beam

Boundary conditions

$$W(0)=0; \quad W(1)=0$$

$$W'(0)=0; \quad W''(1)=0$$

Frequency equation

$$s_2 \tanh s_1 = s_1 \tan s_2$$

Mode shape

$$s_1 s_2 (s_1 \sinh s_1 + s_2 \sin s_2)(\cosh s_1 x - \cos s_2 x)$$

$$-(s_1^2 \cosh s_1 + s_2^2 \cos s_2)(s_2 \sinh s_1 x - s_1 \sin s_2 x)$$

Case 5: Free-free beam

Boundary conditions

$$W''(0)=0; \quad W''(1)=0$$

$$W'''(0)-bW'(0)=0; \quad W'''(1)-bW'(1)=0$$

Frequency equation

$$\cosh s_1 \cos s_2 + \frac{1}{2}\left\{\frac{s_1^3}{s_2^3} - \frac{s_2^3}{s_1^3}\right\} \sinh s_1 \sin s_2 - 1 = 0$$

Mode shape

$$s_1 s_2 (\cosh s_1 - \cos s_2)\left\{\frac{s_1^2 - s_2^2}{s_1 s_2}(s_2^2 \cosh s_1 x + s_1^2 \cos s_2 x)\right.$$

$$\left. - (s_1 \sinh s_1 x + s_2 \sin s_2 x)\right\}$$

$$+ (s_2 \sinh s_1 - s_1 \sin s_2)(s_2^2 \cosh s_1 x + s_1^2 \cos s_2 x)$$

Case 6. Simply supported-free beam

Boundary conditions

$$W(0)=0; \quad W''(1)=0$$

$$W''(0)=0; \quad W'''(1)-bW'(1)=0$$

Frequency equation

$$s_1^3 \tanh s_1 = s_2^3 \tan s_2$$

Mode shape

$$(s_1^2 + s_2^2)\cosh s_1 \cos s_2 x + 2 s_1 s_2 \cos s_2 \sinh s_1 x$$

For Euler-Bernoulli beams, without the axial inertia term

$$b = 0; \quad s_1^2 = s_2^2 = \sqrt{c} = p^2$$

and the cases 1 to 6 above reduce to the following.

Case 1: Simply-supported beam

Frequency equation

$$p = n^2\pi^2$$

Mode shape

$$W = \sin s_2 x = \sin n\pi x \quad n = 1, 2, 3, \ldots$$

Case 2: Fixed-fixed beam

Frequency equation

$$\cosh\sqrt{p}\cos\sqrt{p} - 1 = 0$$

Mode shape

$$(\sinh\sqrt{p} - \sin\sqrt{p})(\cosh\sqrt{p}\,x - \cos\sqrt{p}\,x) - (\cosh\sqrt{p} - \cos\sqrt{p})(\sinh\sqrt{p}\,x - \sin\sqrt{p}.$$

Case 3: Fixed-free cantilever beam

Frequency equation

$$\cosh\sqrt{p}\cos\sqrt{p} + 1 = 0$$

Mode shape

$$(\sinh\sqrt{p} + \sin\sqrt{p})(\cosh\sqrt{p}\,x - \cos\sqrt{p}\,x) - (\cosh\sqrt{p} + \cos\sqrt{p})(\sinh\sqrt{p}\,x - \sin\sqrt{p}\,x)$$

Case 4: Fixed-supported beam

Frequency equation

$$\tanh\sqrt{p} = \tan\sqrt{p}$$

Mode shape

$$(\sinh\sqrt{p} + \sin\sqrt{p})(\cosh\sqrt{p}\,x - \cos\sqrt{p}\,x) - (\cosh\sqrt{p} + \cos\sqrt{p})(\sinh\sqrt{p}\,x - \sin\sqrt{p}\,x)$$

Case 5: Free-free beam

Frequency equation

$$\cosh\sqrt{p}\cos\sqrt{p} - 1 = 0$$

Mode shape

$$(\cosh\sqrt{p} - \cos\sqrt{p})(\sinh\sqrt{p}\,x + \sin\sqrt{p}\,x) - (\sinh\sqrt{p} - \sin\sqrt{p})(\cosh\sqrt{p}\,x + \cos\sqrt{p}\,x)$$

Case 6: Simply supported-free beam

Frequency equation

$$\tanh\sqrt{p} = \tan\sqrt{p}$$

Mode shape

$$\cosh\sqrt{p}\cos\sqrt{p}\,x+\cos\sqrt{p}\sinh\sqrt{p}\,x$$

Except in the case of simply supported beams, in all other cases, the frequency equation is given by a transcendental equation. The natural frequencies can therefore be determined by only a numerical solution of this equation.

For the case of simply supported beam, we can obtain a simple solution. To take into account the rotary inertia effect the correction factor for the natural frequency of the beam is given from equation (8.29).

$$\frac{p_{nri}}{p_{\text{Euler}}}=\frac{1}{\sqrt{1+r_1^2(n\pi)^2}}$$

This correction is a function of mode number and r_1 and is plotted in Fig. 8.2. One can see that the correction is important for short and stubby beams i.e., beams with larger depth, and is significant for higher modes.

8.3 TIMOSHENKO BEAM THEORY

Consider a short stubby beam as shown in Fig. 8.3. We assume that the deformation in the beam is purely due to shear and that a vertical element before deformation remains vertical after deformation and moves by a distance w in the transverse z direction. The displacement field is therefore given by

$$u_x=0$$

$$u_y=0$$

$$u_z=w(x,t) \tag{8.31}$$

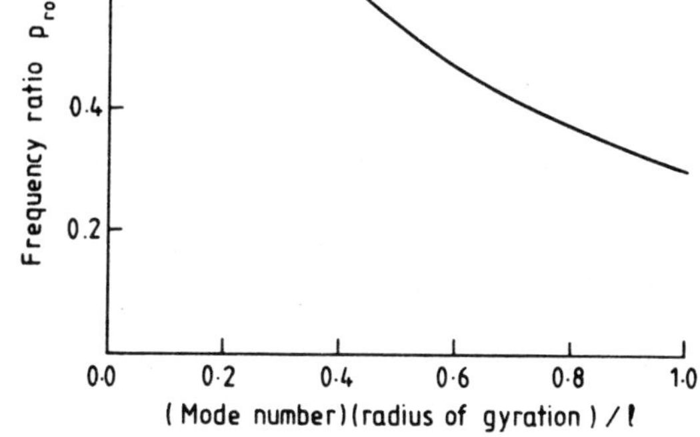

Fig. 8.2 Rayleigh correction for simply supported beam

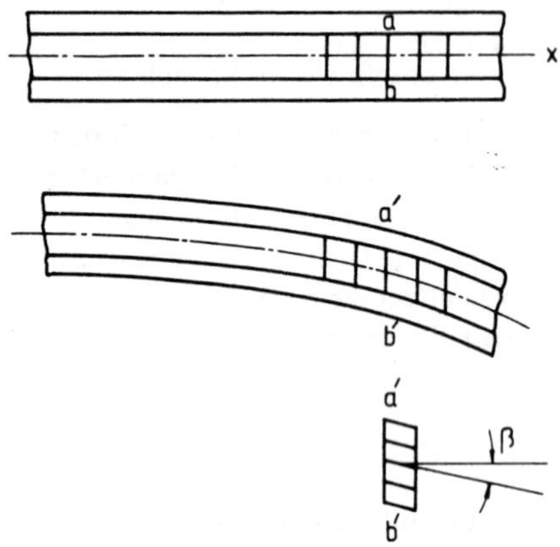

Fig. 8.3 Deflection of a shear beam

The line elements tangential to the centre line of the beam undergo a rotation $\beta\,(x, t)$, shown in Fig. 8.4, which obviously gives the shear angle γ_{xz}. The strains from (8.31) are

$$\varepsilon_{xz} = \frac{1}{2} w_{,x}$$

rest $$\varepsilon = 0 \tag{8.32}$$

The stress field again consists of only one component

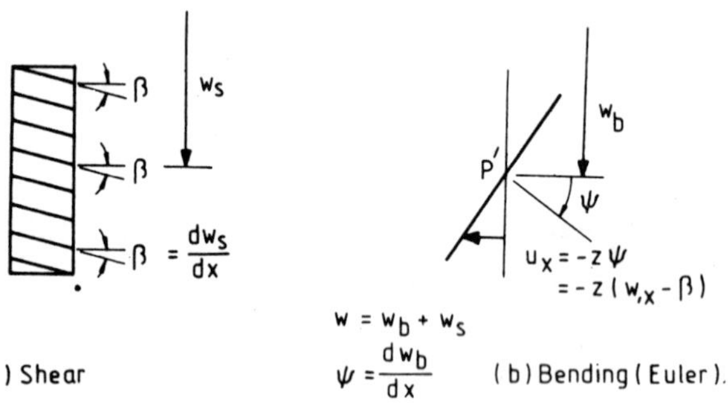

Fig. 8.4 Timoshenko beam field

$$\tau_{xz} = G w_{,x} \tag{8.33}$$

The above equation (8.33) shows that the shear stress across the cross-section of the beam is uniform and independent of z direction. This is not correct in reality and therefore we use a shear correction factor k' to equation (8.33), rather than making the theory more complicated by having one more independent parameter z in the analysis. Thus

$$\tau_{xz} = k' G w_{,x} \tag{8.33a}$$

We can continue in the usual manner to derive the equation of motion of a beam subjected purely to a shear force. This again is not a real case. We know that the significant component of deformation arises due to bending moment as we studied in the case of Euler and Rayleigh beams. We should now combine the effect of shear considered above with Euler beam theory. Thus, we assume the following displacement field.

$$u_x = -z\psi(x,t) = -z\{w_{,x} - \beta(x)\}$$

$$u_y = 0$$

$$u_z = w(x,t) \tag{8.34}$$

Note that the total slope of the beam consists of two parts, one due to bending, as in Fig. 8.1, which is $\psi(x)$ and the other due to only shear, which is $\beta(x)$. The axial displacement of a point at a distance z from the centre line is only due to the bending slope and that the shear slope has no contribution to this. Thus there are two dependent variables in this problem. The strain field from equation (8.34) is

$$\varepsilon_{xx} = -z\psi_{,x}$$

$$\varepsilon_{xz} = \frac{1}{2}(w_{,x} - \psi)$$

$$\text{rest} \quad \varepsilon = 0 \tag{8.35}$$

The corresponding stresses are

$$\tau_{xx} = -Ez\psi_{,x}$$

$$\tau_{xz} = k'G(w_{,x} - \psi)$$

$$\text{rest} \quad \tau = 0 \tag{8.36}$$

The strain energy is

$$U = \frac{1}{2}\int_0^l \{EI_{yy}\psi_{,x}^2 + k'AG(w_{,x} - \psi)^2\}\, dx \tag{8.37}$$

The kinetic energy is

$$T = \frac{1}{2}\int_0^l (\rho A\dot{w}^2 + \rho I_{yy}\dot{\psi}^2)dx \tag{8.38}$$

The work done by external force $q(x,t)$ is

$$\int_0^l qwdx \tag{8.39}$$

Hamilton's principle now leads to

$$\delta\int_{t_1}^{t_2}[\frac{1}{2}\int_0^l \{EI_{yy}\psi_{,x}^2 + k'AG(w_{,x} - \psi)^2\}\, dx$$

$$-\frac{1}{2}\int_0^l (\rho A\dot{w}^2 + \rho I_{yy}\dot{\psi}^2)dx - \int_0^l qw dx]dt = 0$$

i.e.,

$$\int_{t_1}^{t_2}\left[\int_0^l \{EI_{yy}\psi_{,x}\delta\psi_{,x} + k'AG(w_{,x}-\psi)\delta(w_{,x}-\psi)\}\,dx - \int_0^l (\rho A\dot{w}\delta\dot{w} + \rho I_{yy}\dot{\psi}\delta\dot{\psi})dx - \int_0^l q\,\delta w\,dx\right]dt = 0 \qquad (a)$$

There are six integrals in (*a*) above, which are evaluated below

$$I_1 = \int_{t1}^{t2}\int_0^l EI_{yy}\psi_{,x}\delta\psi_{,x}dx\,dt$$

Integrating by parts, we get

$$I_1 = \int_{t1}^{t2}\left[EI_{yy}\psi_{,x}\delta\psi\,|_0^l - \int_0^l (EI_{yy}\psi_{,x})_{,x}\delta\psi dx\right]dt$$

The second integral is

$$I_2 = \int_{t1}^{t2}\int_0^l k'AG\,(w_{,x}-\psi)\,\delta w_{,x}dxdt$$

Integrating by parts

$$I_2 = \int_{t1}^{t2}\left[k'AG(w_{,x}-\psi)\,\delta w\,|_0^l - \int_0^l \{kAG\,(w_{,x}-\psi)\}_{,x}\delta w dx\right]dt$$

The third integral is

$$I_3 = \int_{t1}^{t2}\int_0^l -k'AG(w_{,x}-\psi)\delta\psi dx\,dt$$

The fourth integral is

$$I_4 = \int_{t1}^{t2}\int_0^l \rho A\,\dot{w}\,\delta\dot{w}\,dx\,dt$$

Integrating by parts with respect to time

$$I_4 = \int_{t1}^{t2} - \int_0^l \rho A\,\ddot{w}\,\delta w\,dx\,dt$$

The fifth integral is

$$I_5 = \int_{t1}^{t2}\int_0^l \rho I_{yy}\dot{\psi}\delta\dot{\psi}dx\,dt$$

Integrating by parts with respect to time, we get

$$I_5 = \int_{t1}^{t2} - \int_0^l \rho I_{yy}\ddot{\psi}\delta\psi\,dx\,dt$$

The last integral in (*a*) is

$$I_6 = \int_{t1}^{t2}\int_0^l q\,\delta w\,dx\,dt$$

Substituting the above six integrals in (*a*), we get

$$\int_{t1}^{t2} [k'AG(w_{,x} - \psi)\delta w \,|_0^l + EI_{yy}\psi_{,x}\delta\psi \,|_0^l$$

$$+ \int_0^l [-\{k'AG(w_{,x} - \psi)\}_{,x} + \rho A\ddot{w} - q]\, dx\, \delta w$$

$$+ \int_0^l \{-(EI_{yy}\psi_{,x})_{,x} - k'AG(w_{,x} - \psi) + \rho I_{yy}\ddot{\psi}\}\,\delta\psi]\, dt = 0 \qquad (b)$$

Equation (*b*) leads to the following differential equations of motion and the boundary conditions.

$$-\{k'AG(w_{,x} - \psi)\}_{,x} + \rho A\ddot{w} = q$$

$$-(EI_{yy}\psi_{,x})_{,x} - k'AG(w_{,x} - \psi) + \rho I_{yy}\ddot{\psi} = 0 \qquad (8.40)$$

$$k'AG(w_{,x} - \psi)\delta w \,|_0^l = 0$$

$$EI_{yy}\psi_{,x}\delta\psi \,|_0^l = 0 \qquad (8.41)$$

Let us nondimensionalize the equations (8.12) and (8.13) by setting

$$x \leftarrow \frac{x}{l}$$

$$t \leftarrow \frac{t}{t_0}$$

$$E \leftarrow \frac{E}{E_0}$$

$$G \leftarrow \frac{G}{G_0}$$

$$A \leftarrow \frac{A}{A_0}$$

$$I_{yy} \leftarrow \frac{I_{yy}}{I_{yy0}}$$

$$\rho \leftarrow \frac{\rho}{\rho_0}$$

$$w \leftarrow \frac{w}{w_0}$$

$$\psi \leftarrow \frac{\psi}{\psi_0}$$

$$q \leftarrow \frac{q}{q_0}$$

$$M \leftarrow \frac{M}{M_0}$$

$$r_1^2 = \frac{I_{yy0}}{A_0 l^2}$$

$$a_1 = \frac{k' G_0 A_0 l^2}{E_0 I_{yy0}}$$

where t_0 ψ_0 M_0 and q_0 are given by

$$t_0 = \sqrt{\frac{\rho_0 A_0 l^4}{E_0 I_{yy0}}}$$

$$\psi_0 = \frac{w_0}{l}$$

$$M_0 = \frac{1}{8} q_0 l^2$$

$$q_0 = \frac{384 E_0 I_{yy0} w_0}{5 l^3} \tag{8.42}$$

Then

$$\rho A \partial_t^2 w - a_1 G A \partial_x^2 w + a_1 G A \partial_x \psi = \frac{384}{5} q$$

$$r_1^2 \partial_t^2 \psi - a_1 \partial_x w + a_1 \psi - \partial_x^2 \psi = 0 \tag{8.43}$$

$$(\partial_x w - \psi)\delta w \,|_0^1 = 0$$

$$\partial_x \psi \delta \psi \,|_0^1 = 0 \tag{8.44}$$

For a uniform beam, with the beam values as reference values, the above equations (8.43) and (8.44) become

$$\partial_t^2 w - a_1 \partial_x^2 w + a_1 \partial_x \psi = \frac{384}{5} q$$

$$r_1^2 \partial_t^2 \psi - a_1 \partial_x w + a_1 \psi - \partial_x^2 \psi = 0 \tag{8.43a}$$

$$(\partial_x w - \psi)\delta w \,|_0^1 = 0$$

$$\partial_x \psi \delta \psi \,|_0^1 = 0 \tag{8.44a}$$

Equations (8.43*a*) can be written in matrix form as

$$\begin{bmatrix} (\partial_t^2 - a_1 \partial_x^2) & a_1 \partial_x \\ -a_1 \partial_x & (r_1^2 \partial_t^2 + a_1 - \partial_x^2) \end{bmatrix} \begin{Bmatrix} w \\ \psi \end{Bmatrix} = \begin{Bmatrix} \frac{384}{5} q \\ 0 \end{Bmatrix} \tag{8.45}$$

Solving for w, we get

$$w = \frac{\frac{384}{5}(r_1^2\partial_t^2 + a_1 - \partial_x^2)q}{r_1^2\partial_t^4 + a_1\partial_t^2 - (1 + a_1 r_1^2)\partial_x^2\partial_t^2 + a_1\partial_x^4}$$

i.e.,

$$\{r_1^2\partial_t^4 + a_1\partial_t^2 - (1 + a_1 r_1^2)\partial_x^2\partial_t^2 + a_1\partial_x^4\}\, w = \frac{384}{5}(r_1^2\partial_t^2 + a_1 - \partial_x^2)q \tag{8.46}$$

The above equation (8.46) contains only one dependent variable, w and hence it is somewhat easier to solve the differential equation. However, the boundary conditions become very complicated, if ψ is eliminated from equations (8.44). A good compromise is to solve the equation (8.46) subject only to boundary conditions of w according to simple Euler theory or Rayleigh theory.

For free vibrations, equation (8.46) becomes

$$\{r_1^2\partial_t^4 + a_1\partial_t^2 - (1 + a_1 r_1^2)\partial_x^2\partial_t^2 + a_1\partial_x^4\}\, w = 0 \tag{8.47}$$

Let $w = W(x) \sin pt$, then

$$r_1^2 p^4 W - a_1 p^2 W + (1 + a_1 r_1^2) p^2 W_{,xx} + a_1 W_{,xxxx} = 0$$

i.e.,

$$[a_1 d_x^4 + (1 + a_1 r_1^2) p^2 d_x^2 + (r_1^2 p^4 - a_1 p^2)]\, W = 0$$

The above can be rearranged as

$$(d_x^4 - \bar{b} d_x^2 - \bar{c})\, W = 0 \tag{8.48}$$

where

$$\bar{b} = -\frac{(1 + a_1 r_1^2)p^2}{a_1}$$

$$\bar{c} = -\frac{(r_1^2 p^4 - a_1 p^2)}{a_1} \tag{8.49}$$

The Laplace transform of equation (8.48) is

$$(s^4 - \bar{b}s^2 - \bar{c})\overline{W} = (s^3 - \bar{b}s)W(0) + s^2 W'(0)$$
$$+ sW''(0) + \{W'''(0) - \bar{b}W'(0)\} \tag{8.50}$$

Let us define the following positive quantities

$$s_1^2 = \frac{1}{2}\left(\sqrt{\bar{b}^2 + 4\bar{c}} + \bar{b}\right)$$

$$s_2^2 = \frac{1}{2}\left(\sqrt{\bar{b}^2 + 4\bar{c}} - \bar{b}\right) \tag{8.51}$$

Using (8.49), the second equation in (8.51) gives

$$r_1^2 p^4 - p^2\{s_2^2(1 + a_1 r_1^2) + a_1\} + a_1 s_2^4 = 0 \tag{8.52}$$

The quantities in (8.49) are related to the roots of the characteristic polynomial and satisfy the following relations.

$$(s^2 - s_1^2)(s^2 + s_2^2) = (s^4 - \bar{b}s^2 - \bar{c}) = 0$$

$$\bar{b} = s_1^2 - s_2^2$$

$$\bar{c} = s_1^2 s_2^2$$

$$s_1^4 - \bar{b}s_1^2 - \bar{c} = 0$$

$$s_2^4 + \bar{b}s_2^2 - \bar{c} = 0 \tag{8.53}$$

The transfer function is given by (8.24) and the mode shapes and its derivatives can be written as in (8.25)

$$T(x) = \frac{1}{s_1 s_2(s_1^2 + s_2^2)}(s_2 \sinh s_1 x - s_1 \sin s_2 x) \tag{8.44}$$

so that

$$W = [(d_x^3 - \bar{b}d_x)W(0) + d_x^2 W'(0) + d_x W''(0) + \{W'''(0) - \bar{b}W'(0)\}]T(x)$$

$$W' = [\bar{c}W(0) + d_x^3 W'(0) + d_x^2 W''(0) + d_x\{W'''(0) - \bar{b}W'(0)\}]T(x)$$

$$W'' = [\bar{c}d_x W(0) + (\bar{b}d_x^2 + \bar{c})W'(0) + d_x^3 W''(0) + d_x^2\{W'''(0) - \bar{b}W'(0)\}]T(x)$$

$$W''' - \bar{b}W' = [(\bar{c}d_x^2 - \bar{b}\bar{c})W(0) + \bar{c}d_x W'(0) + \bar{c}W''(0)$$

$$+ (d_x^3 - \bar{b}d_x)\{W'''(0) - \bar{b}W'(0)\}]T(x) \tag{8.55}$$

Case 1: Simply-supported beam

Let us consider a simply supported beam with its correct kinematical boundary condition, $w = 0$, at both the ends and as an approximation, the classical natural boundary condition, $w'' = 0$ again at both the ends. Then, following the analysis in the previous section 8.2, we get

$$s_2 = n\pi \tag{8.56}$$

From (8.51), we have

$$s_2^2 = \frac{1}{2}\sqrt{\frac{(1 + a_1 r_1^2)^2}{a_1^2}p^4 - 4\frac{r_1^2 p^4 - a_1 p^2}{a_1} + \frac{1 + a_1 r_1^2}{2a_1}p^2} \tag{c}$$

Rearranging (*c*) above, we get

$$r_1^2 p^4 - p^2\{s_2^2(1 + a_1 r_1^2) + a_1\} + a_1 s_2^4 = 0 \tag{d}$$

With the help of (8.56), the above frequency equation (*d*) becomes

$$r_1^2 p^4 - p^2\{n^2\pi^2(1 + a_1 r_1^2) + a_1\} + a_1 n^4 \pi^4 = 0 \tag{8.57}$$

We note that there are two roots of the above equation (8.57). The lower one corresponds to the modified Euler bending coupled with the shear bending and the higher one belongs to the shear bending mode coupled with Euler bending. The shear bending mode is very high since the beam is quite rigid for this deformation and is not of much practical importance. The lower bending mode frequency is of importance to design engineers, which is

$$p_1^2 = \frac{n^2\pi^2(1+a_1r_1^2)+a_1-\sqrt{n^4\pi^4(1-a_1r_1^2)^2+a_1^2+2a_1n^2\pi^2(1+a_1r_1^2)}}{2r_1^2}$$

Therefore, the corrected frequency with Timoshenko shear and Rayleigh rotary inertia effects taken into account together is given by

$$\frac{p_1}{p_c} = \sqrt{\frac{n^2\pi^2(1+a_1r_1^2)+a_1-\sqrt{n^4\pi^4(1-a_1r_1^2)^2+a_1^2+2a_1n^2\pi^2(1+a_1r_1^2)}}{2r_1^2n^4\pi^4}} \tag{8.58}$$

The above equation (8.58) cannot be used directly when r_1 is zero. Then the classical Euler theory or Rayleigh's theory is to be used. For very small values of r_1, we need high precision to calculate the above correction factor. In fact for small value of r_1, the corrections are of small order and we can express

$$p^2 = p_c^2 + \delta^2 \tag{e}$$

Substituting (e) above in equation (8.57) and neglecting δ^4 term, we get

$$r_1^2(p_c^4 + p_c^2\delta^2) - (p_c^2 + \delta^2)\{n^2\pi^2(1+a_1r_1^2)+a_1\} + a_1n^4\pi^4 = 0 \tag{f}$$

From (f) above we can get

$$\delta^2 = \frac{-r_1^2p_c^4 + p_c^2\{n^2\pi^2(1+a_1r_1^2)+a_1\} - a_1n^4\pi^4}{2p_c^2r_1^2 - n^2\pi^2(1+a_1r_1^2) - a_1} \tag{g}$$

Substituting the above equation (g) in (e) and rearranging, we get

$$p^2 = \frac{r_1^2p_c^4 - a_1n^4\pi^4}{2p_c^2r_1^2 - n^2\pi^2(1+a_1r_1^2) - a_1} \tag{8.59}$$

Hence

$$\frac{p}{p_c} = \sqrt{\frac{r_1^2(n\pi)^4 - a_1}{2(n\pi)^4r_1^2 - n^2\pi^2(1+a_1r_1^2) - a_1}} \tag{8.60}$$

We find from (8.42) that equations (8.58) and (8.60) are functions of E/Gk' and (radius of gyration/l). The frequency ratios of a simply supported beam, according to both Rayleigh and Timoshenko effects are plotted in Figs. 8.5 to 8.7 for the first three modes, with the approximate formula (8.60) for very small values of (radius of gyration/l). We note that Rayleigh's rotary inertia is more significant in reducing the natural frequencies in higher modes for beams of stubby cross-section. For other boundary conditions, we can follow a similar analysis as given in the previous section and this section, with correct kinematical boundary conditions and approximate natural boundary conditions. Even then it can be seen that the analysis is quite involved. This becomes more complicated if one is interested in calculations of a beam with nonuniform cross-section. The energy methods become useful in the case of such problems. This is illustrated in case(2).

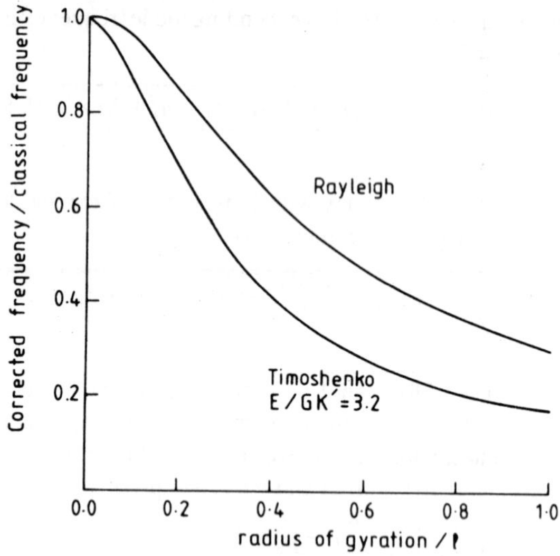

Fig. 8.5 Rayleigh and Timoshenko corrections -I mode

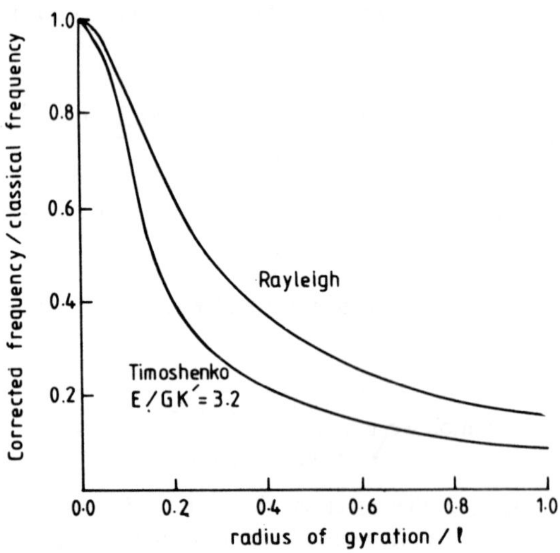

Fig. 8.6 Rayleigh and Timoshenko corrections -II mode

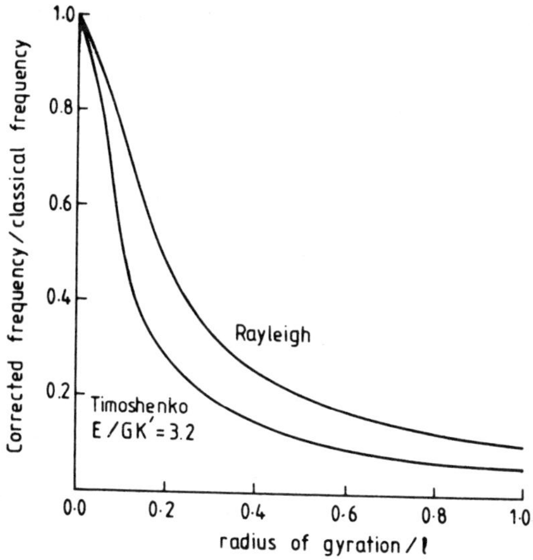

Fig. 8.7 Rayleigh and Timoshenko corrections -III mode

Case 2: Tapered cantilever beam by Ritz method

The strain and kinetic energy expressions are

$$U = \frac{1}{2}\int_0^l \{EI_{yy}\psi_{,x}^2 + k'AG(w_{,x} - \psi)^2\}\,dx \tag{8.37}$$

$$T = \frac{1}{2}\int_0^l (\rho A\dot{w}^2 + \rho I_{yy}\dot{\psi}^2)dx \tag{8.38}$$

Writing

$$w = w_b + w_s$$

$$w_{s,x} = w_{,x} - \psi \tag{8.61}$$

The Lagrangian expression for a beam in lateral vibration can be written as

$$L = \frac{1}{2}\int_0^l (EI_{yy}w_{b,xx}^2 + k'AGw_{s,x}^2)dx - \frac{1}{2}\int_0^l (\rho A\dot{w}^2 + \rho I_{yy}\dot{w}_{b,x}^2)dx \tag{8.62}$$

For free vibration we assume

$$w_b(x,t) = W_b(x)\sin pt$$

$$w_s(x,t) = W_s(x)\sin pt$$

and

$$w(x,t) = W(x)\sin pt \tag{f}$$

Then equation (8.62) becomes

$$L = \frac{1}{2}\int_0^l (EI_{yy}W_{b,xx}^2 + k'AGW_{s,x}^2)\sin^2 pt\,dx$$

$$-\frac{1}{2}\int_0^l (\rho AW^2 - \rho I_{yy}W_{b,x}^2)p^2\cos^2 pt\,dx \tag{g}$$

We integrate the above expression over a sufficiently long period of time T^* and obtain

$$L = \frac{1}{2T^*}\int_0^{T^*}\int_0^l \{(EI_{yy}W_{b,xx}^2 + k'AGW_{s,x}^2)\sin^2 pt$$

$$- (\rho AW^2 + \rho I_{yy}W_{b,x}^2)p^2\cos^2 pt\}\,dx\,dt$$

i.e.,

$$L = \frac{1}{4}\int_0^l \{EI_{yy}W_{b,xx}^2 + k'AGW_{s,x}^2 - (\rho AW^2 + \rho I_{yy}W_{b,x}^2)p^2\}\,dx \tag{h}$$

Let us assume the following shape functions

$$W_b = \sum_r A_r h_r(X)$$

$$= \sum_r A_r \{(r+2)X^{r+1} - rX^{r+2}\}$$

$$W_s = \sum_r B_r g_r(X)$$

$$= \sum_r B_r \{(r+1)X^r - rX^{r+1}\} \tag{8.63}$$

where

$$X = \frac{x}{l}$$

which satisfy the boundary conditions for a cantilever beam viz.,

$$W_s = W_b = W_{b,x} = 0 \quad x = 0$$

$$W_{s,x} = W_{b,xx} = 0 \quad x = l$$

Let us consider a cantilever of rectangular cross-section, with breadth b and thickness t as

$$b = b_0(1 - \alpha X)$$

$$t = t_0(1 - \beta X) \tag{8.64}$$

so that

$$A = b_0 t_0(1 - \alpha X)(1 - \beta X)$$

$$I_{yy} = \frac{1}{12}b_0 t_0^3(1 - \alpha X)(1 - \beta X)^3 \tag{i}$$

Substituting (8.63) and (*i*) in equation (*h*), we get

$$L = \frac{1}{4}\int_0^l \left[\frac{1}{12}Eb_0 t_0^3(1 - \alpha X)(1 - \beta X)^3\left(\sum_r A_r h_{r,xx}\right)^2\right.$$

$$+ k'Gb_0 t_0(1 - \alpha X)(1 - \beta X)\left(\sum_r B_r g_{r,x}\right)^2$$

$$-\left\{\rho b_0 t_0(1-\alpha X)(1-\beta X)\left(\sum_r A_r h_r + \sum_r B_r g_r\right)^2\right.$$

$$\left.\left.+\frac{1}{12}\rho b_0 t_0^3(1-\alpha X)(1-\beta X)^3\left(\sum_r A_r h_{r,x}\right)^2\right\}p^2\right]dx$$

We can now use the Ritz method to obtain

$$\begin{bmatrix} P-Qp^2 & R-Sp^2 \\ T-Up^2 & V-Wp^2 \end{bmatrix}\begin{Bmatrix} A \\ B \end{Bmatrix}=0 \tag{8.65}$$

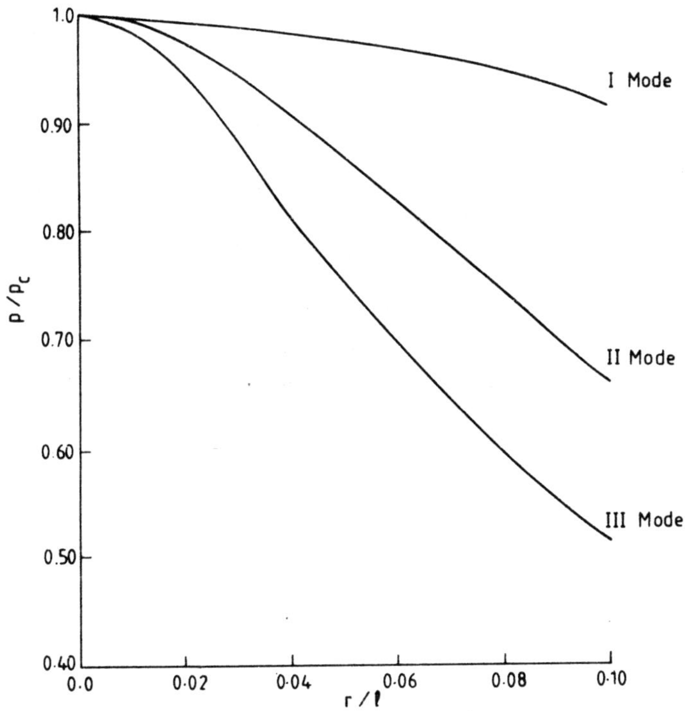

Fig. 8.8 Solution by Ritz method for a cantilever beam, E/Gk′ = 3.2

where

$$P_{rj}=\int_0^l \frac{1}{12}Eb_0 t_0^3(1-\alpha X)(1-\beta X)^3 h_{r,xx}h_{j,xx}dx$$

$$Q_{rj}=\int_0^l \left\{\rho b_0 t_0(1-\alpha X)(1-\beta X)h_r h_j+\frac{1}{12}\rho b_0 t_0^3(1-\alpha X)(1-\beta X)^3 h_{r,x}h_{j,x}\right\}dx$$

$$R_{rj} = 0$$

$$S_{rj} = \int_0^l \rho b_0 t_0 (1 - \alpha X)(1 - \beta X) g_r h_j dx$$

$$T_{rj} = 0$$

$$U_{rj} = S_{jr}$$

$$V_{rj} = \int_0^l k' G b_0 t_0 (1 - \alpha X)(1 - \beta X) g_{r,x} g_{j,x} dx$$

$$W_{rj} = \int_0^l \rho b_0 t_0 (1 - \alpha X)(1 - \beta X) g_r g_j dx \tag{8.66}$$

We can solve the above eigen value problem on a computer and determine the natural frequencies of the Timoshenko cantilever beam. The results are given for a uniform beam in Fig. 8.8 for $E/Gk' = 3.2$. The correction factors are plotted in Figs. 8.9 to 8.11 as a function of breadth taper parameter α for four different values of r/l at the root. It can be seen that the correction factors decrease as the taper parameter α increases. Also, they decrease more as α increases for larger values of r/l at the root. This is to be expected since the beam becomes relatively stubbier with a larger value of r/l. Similarly, the correction factors as a function of thickness taper β are given in Figs. 8.12 to 8.14. Here the correction factors increase as taper parameter β increases. This can be expected again as the beams become relatively slender with increase in thickness taper β. Figure 8.15 gives experimental results of several tapered cantilever beams with the same root cross-section along with Euler's theory and Timoshenko's theory. It is clear from Fig. 8.15 that Timoshenko corrections are important for higher modes

8.4 VOLTERRA'S THEORY

We recall Timoshenko's theory where the displacement and strain fields are given by

$$u_x = -z\psi(x,t) = -z\{w_{,x} - \beta(x)\}$$

$$u_y = 0$$

$$u_z = w(x,t) \tag{8.34}$$

$$\varepsilon_{xx} = -z\psi_{,x}$$

$$\varepsilon_{xz} = \frac{1}{2}(w_{,x} - \psi)$$

rest $$\varepsilon = 0 \tag{8.35}$$

We note from the above that the strain components ε_{yy} and ε_{zz} are zero in the presence of the axial strain ε_{xx}. This is not acceptable from the point of view of theory of elasticity, where an axial strain in one direction is always accompanied by Poisson strains in the other two directions. This discrepancy in Timoshenko beam theory is removed by Volterra by assuming the following field.

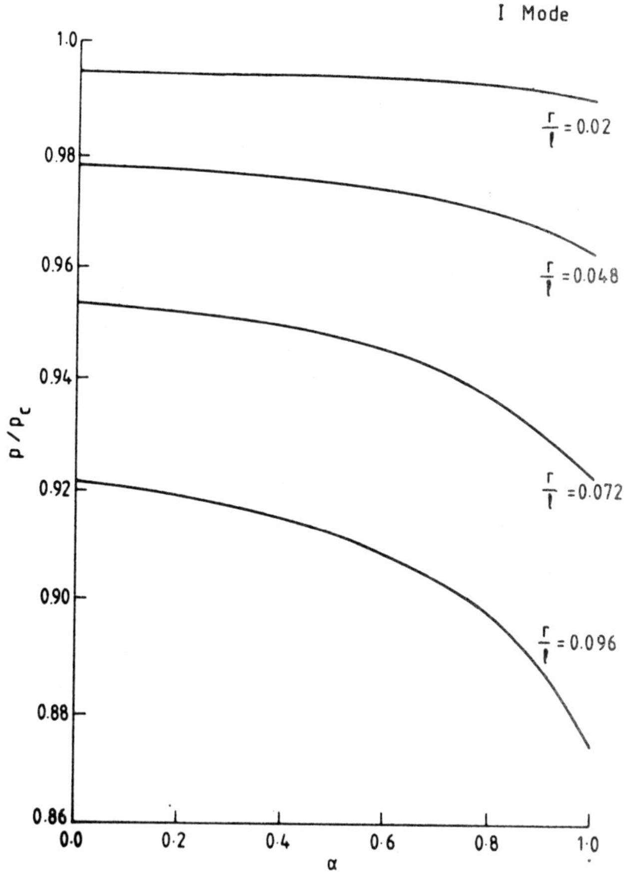

Fig. 8.9 Effect of breadth taper α on the ratio p/p_c for the I mode

$$u_x = -z\psi$$

$$u_y = \mu y z \psi_{,x}$$

$$u_z = w(x,t) + \frac{1}{2}\mu z^2 \psi_{,x} \tag{8.67}$$

The corresponding strain field is therefore

$$\varepsilon_{xx} = -z\psi_{,x}$$

$$\varepsilon_{yy} = \mu z \psi_{,x}$$

$$\varepsilon_{zz} = \mu z \psi_{,x}$$

$$\varepsilon_{xy} = \frac{1}{2}\mu y z \psi_{,xx}$$

$$\varepsilon_{yz} = \frac{1}{2}\mu y \psi_{,x}$$

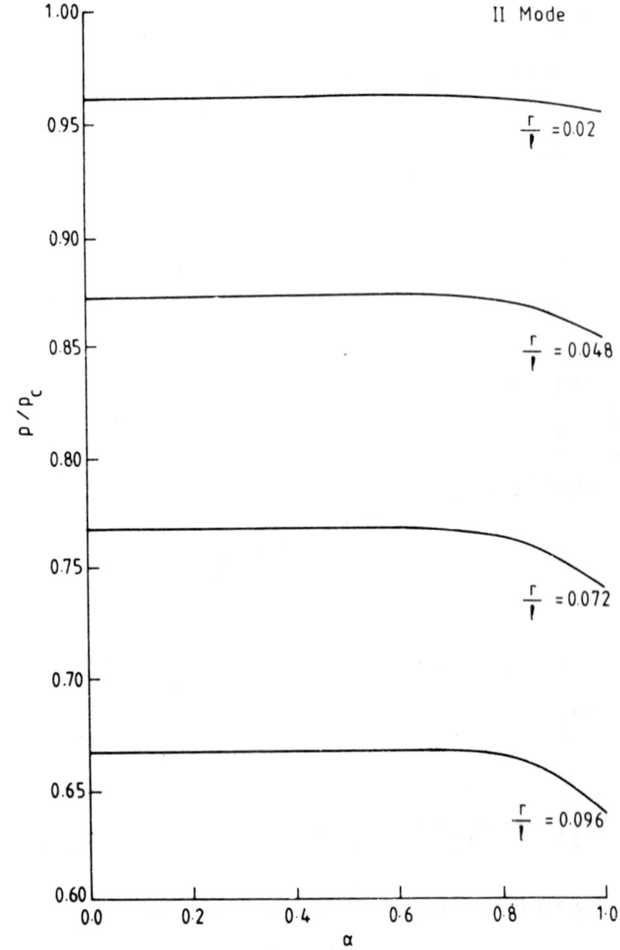

Fig. 8.10 Effect of breadth taper α on the ratio p/p_c for the II mode

$$\varepsilon_{xz} = \frac{1}{2}\left(w_{,x} - \psi + \frac{1}{2}\mu z^2 \psi_{,xx}\right) \tag{8.68}$$

The stress field can now be obtained as

$$\tau_{xx} = -Ez\psi_{,x}$$

$$\tau_{yy} = 0$$

$$\tau_{zz} = 0$$

$$\tau_{xy} = \mu G\, yz\psi_{,xx}$$

$$\tau_{yz} = \mu G\, y\psi_{,x}$$

$$\tau_{xz} = G\left(w_{,x} - \psi + \frac{1}{2}\mu z^2 \psi_{,xx}\right) \tag{8.69}$$

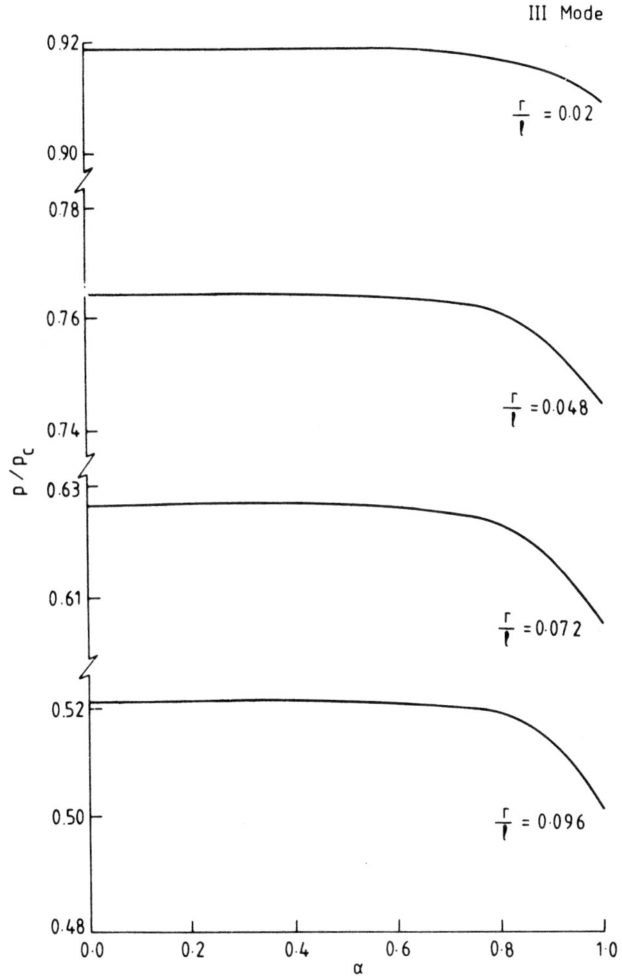

Fig. 8.11 Effect of breadth taper α on the ratio p/p_c for the III mode

We now can obtain the potential energy as

$$U = \frac{1}{2}\int_0^l [(EI_{yy} + \mu^2 GI_{zz})\psi_{,x}^2 + GAw_{,x}^2$$

$$+ GA\psi^2 - 2GAw_{,x}\psi + G\mu I_{yy}\psi_{,xx}w_{,x} - G\mu I_{yy}\psi_{,xx}\psi]\,dx \tag{8.70}$$

In the above equation (8.70) the terms of $y^2 z^2$ and z^4 are neglected. Similarly, the kinetic energy is obtained as

$$T = \frac{1}{2}\int_0^l [\rho I_p \dot{\psi}^2 + \rho A \dot{w}^2 + \mu\rho I_p \dot{w}\dot{\psi}_{,x}]\,dx \tag{8.71}$$

The work of external force q is

$$\int_0^l \left\{qw + \frac{1}{2}qa^2\psi_{,x}\right\}dx \tag{8.72}$$

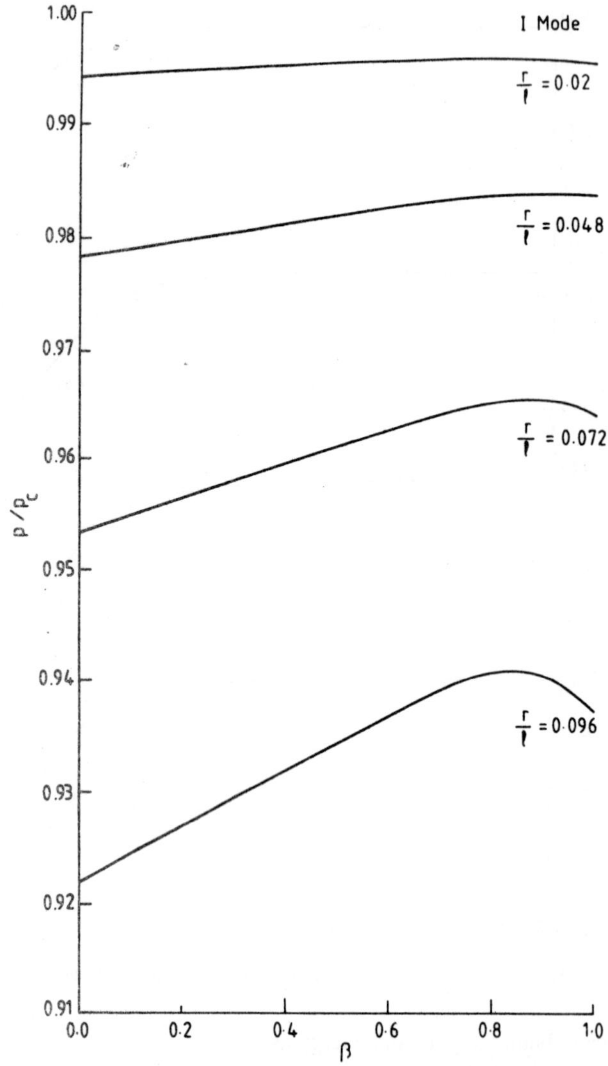

Fig. 8.12 Effect of thickness taper β on the ratio p/p_c for the I mode

where a is the z coordinate of the contact surface with the force q. Application of Hamilton's principle now gives

$$\delta \int_{t1}^{t2} \left[\frac{1}{2} \int_0^l \{(EI_{yy} + \mu^2 GI_{zz})\psi_{,x}^2 + GAw_{,x}^2\} \right.$$

$$+ GA\psi^2 - 2GAw_{,x}\psi + G\mu I_{yy}\psi_{,xx}w_{,x} - G\mu I_{yy}\psi_{,xx}\psi\} dx$$

$$\left. - \frac{1}{2}\int_0^l \{\rho I_p \dot{\psi}^2 + \rho A\dot{w}^2 + \mu\rho I_p \dot{w}\dot{\psi}_{,x}\} dx - \int_0^l \left\{ qw + \frac{1}{2} qa^2 \psi_{,x} \right\} dx \right] dt = 0 \quad (8.73)$$

i.e.,

$$\int_{t1}^{t2} \left[\left[\int_0^l \{(EI_{yy} + \mu^2 GI_{zz})\psi_{,x}\delta\psi_{,x} + GAw_{,x}\delta w_{,x}\} \right. \right.$$

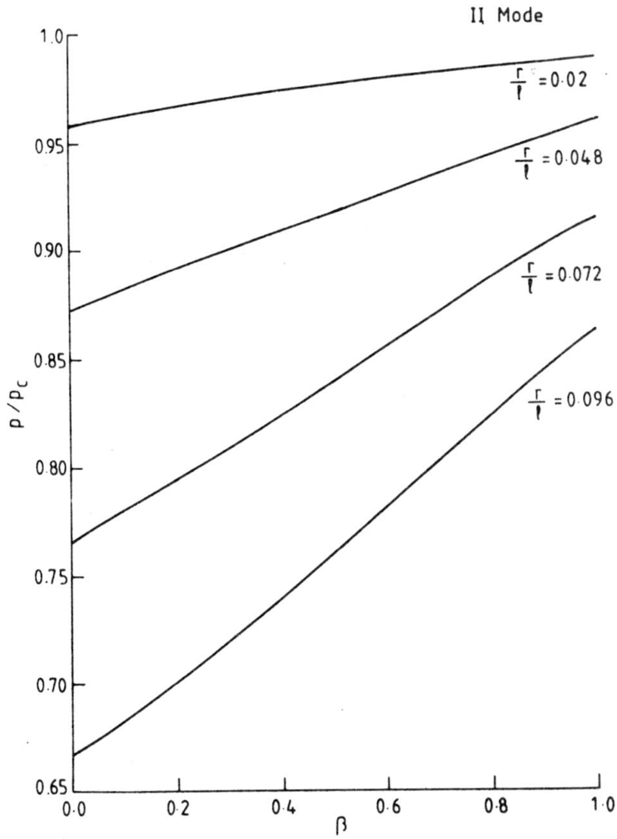

Fig. 8.13 Effect of thickness taper β on the ration p/p_c for II mode

$$+GA\psi\delta\psi - 2GA\psi\delta w_{,x} - 2GA w_{,x}\delta\psi + G\mu I_{yy}\delta\psi_{,xx}w_{,x} + G\mu I_{yy}\psi_{,xx}\delta w_{,x}$$

$$-G\mu I_{yy}\delta\psi_{,xx}\psi - G\mu I_{yy}\psi_{,xx}\delta\psi\}\,dx$$

$$-\int_0^l \{\rho I_p\dot{\psi}\delta\dot{\psi} + \rho A\dot{w}\delta\dot{w} + \mu\rho I_p\delta\dot{w}\dot{\psi}_{,x} + \mu\rho I_p\dot{w}\delta\dot{\psi}_{,x}\}\,dx$$

$$-\int_0^l\left\{q\,\delta w + \frac{1}{2}qa^2\delta\psi_{,x}\right\}dx\Bigg]dt = 0 \tag{8.74}$$

Integrating by parts and grouping the appropriate terms, we get

$$\rho A\ddot{w} - \{GA(w_{,x} - \psi)\}_{,x} - \frac{\mu}{2}(GI_{yy}\psi_{,xx})_{,x} + \frac{\mu}{2}\rho I_p\ddot{\psi}_{,x} = q \tag{8.75}$$

$$\{(EI_{yy} + \mu^2 GI_{zz})\psi_{,x}\}_{,x} + \frac{\mu}{2}(\rho I_p\ddot{w})_{,x} + GA(w_{,x} - \psi)$$

$$+\frac{\mu}{2}GI_{yy}\psi_{,xx} - \rho I_p\ddot{\psi} + \frac{\mu}{2}\{GI_{yy}(w_{,x} - \psi)\}_{,xx} = \frac{\mu}{2}(a^2q)_{,x} \tag{8.76}$$

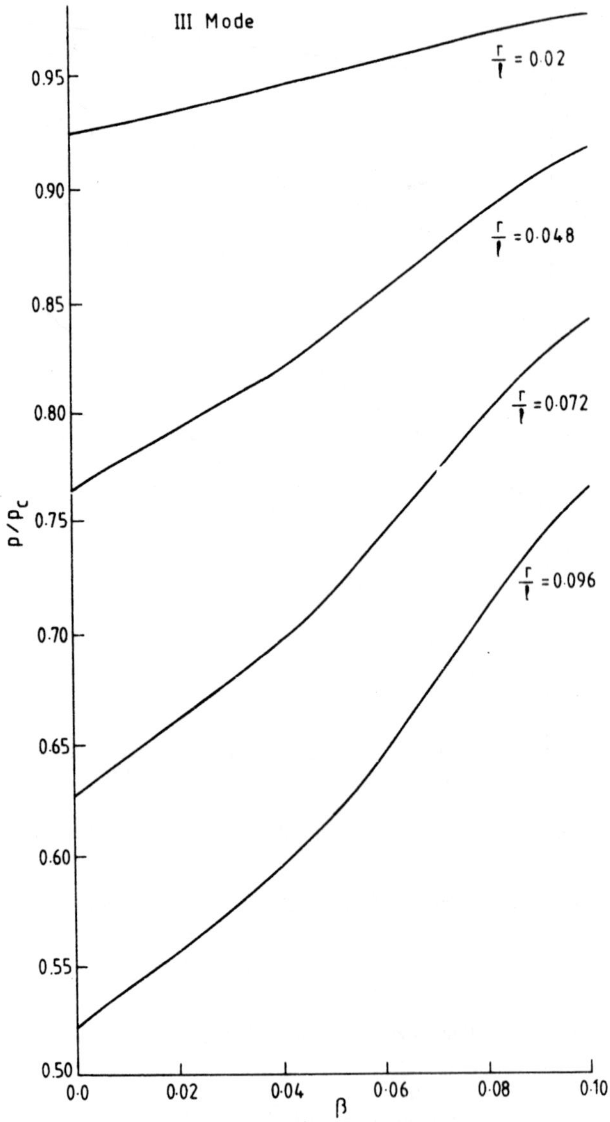

Fig. 8.14 Effect of thickness taper β on the ration p/p_c for the III mode

$$\left\{GA(w_{,x} - \psi) + \frac{\mu}{2} GI_{yy} \psi_{,xx}\right\} \delta w \, |_0^l = 0$$

$$\left\{(EI_{yy} + \mu^2 GI_{zz})\psi_{,x} + \frac{\mu}{2}(\rho I_p \ddot{w}) - \frac{\mu}{2}\{GI_{yy}(w_{,x} - \psi)\}_{,x} - \frac{\mu}{2} a^2 q\right\} \delta\psi \, |_0^l = 0$$

$$\frac{\mu}{2} GI_{yy}(w_{,x} - \psi) \delta\psi_{,x} \, |_0^l = 0 \tag{8.77}$$

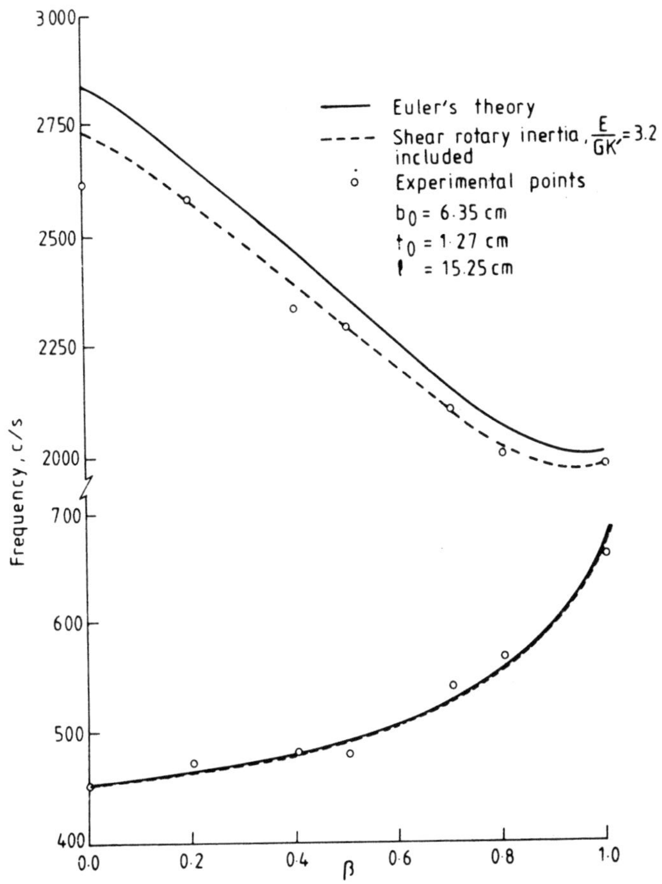

Fig. 8.15 Effect of thickness taper β on the first two modes of a cantilever beam

Volterra's theory is exact from theory of elasticity point of view. However, it is too unwieldy for a practical solution. The improvement of results that one may obtain from the solution of equations (8.75) and (8.76), against the results obtained from Timoshenko theory, does not warrant such a need for the moment. In fact, from engineering point of view, it is fairly sufficient to use Euler or Rayleigh theory for most of the problems and where needed one can perhaps use Timoshenko theory.

8.5 FREE AND FORCED VIBRATION OF BEAMS

Rayleigh's theory

The equations of motion and boundary conditions according to Rayleigh theory are given by

$$M\ddot{w} + Kw = q \tag{8.15}$$

$$(EI_{yy}\partial_x^2 w)\delta\partial_x w \,|_0^l + \{-\partial_x(EI_{yy}\partial_x^2 w) + r_1^2\rho I_{yy}\partial_x\ddot{w}\}\,\delta w \,|_0^l = 0 \tag{8.16}$$

where the mass and stiffness operators are given by

$$M = \rho A - r_1^2 \partial_x(\rho I_{yy} \partial_x)$$

$$K = \partial_x^2 EI \partial_x^2 \tag{8.17}$$

For uniform beams with the reference values chosen as its original values themselves, the operators given in (8.17) are

$$M = 1 - r_1^2 \partial_x^2$$

$$K = \partial_x^4 \tag{8.17a}$$

with the corresponding boundary conditions

$$(\partial_x^2 w)\delta\partial_x w \,|_0^1 + \{-\partial_x^3 w + r_1^2 \partial_x \ddot{w}\}\,\delta w\,|_0^1 = 0 \tag{8.16a}$$

Expressing the solution in the form

$$w = W_i(x) q_i(t) \tag{8.78}$$

equation (8.15) becomes

$$MW_i(x)\ddot{q}_i(t) + KW_i(x)q_i(t) = q \tag{8.15a}$$

Forming the scalar product with W_j, equation (8.15a) gives

$$\int_0^1 MW_i(x)W_j(x)\ddot{q}_i(t)dx + \int_0^1 KW_i(x)W_j(x)q_i(t)dx = \int_0^1 qW_j(x)dx$$

With the help of orthogonality conditions (*see* section 6.5), we have

$$m_j\ddot{q}_j + k_j q_j = f_j \quad \text{(no sum)} \tag{8.79}$$

where

$$m_j = \int_0^1 MW_i(x)W_i(x)dx \quad \text{(no sum)}$$

$$k_j = \int_0^1 KW_i(x)W_i(x)dx \quad \text{(no sum)}$$

$$f_j = \int_0^1 qW_j(x)dx \tag{8.80}$$

The complete solution of equation (8.79) can be obtained by using Duhamel's integral

$$q_j(t) = q_j(0)\cos p_j t + \frac{\dot{q}_j(0)}{p_j}\sin p_j t$$

$$+\frac{1}{m_j p_j}\int_0^t f_j(\tau)\sin p_j(t-\tau)d\tau \quad \text{(no sum)} \tag{8.81}$$

Then, the dynamic displacement is given by

$$w(x,t) = \sum_{1,2,}^{\infty} \{q_j(0)\cos p_j t + \frac{\dot{q}_j(0)}{p_j}\sin p_j t$$

$$+\frac{1}{m_j p_j}\int_0^t f_j(\tau)\sin p_j(t-\tau)d\tau\} W_j(x) \quad \text{(no sum)} \tag{8.82}$$

For the case of a free-free beam, the rigid body displacement should be superimposed over the deformation given by the above equation.

The initial conditions over the displacement are generally prescribed for the system and they should be transformed into the generalized coordinates in order to use the equation (8.81). This can be achieved by expanding the prescribed initial conditions in term of orthogonal modes. Thus if $f(x)$ and $g(x)$ are respectively the nondimensional initial displacement and velocity of the bar, then,

$$w(x,0) = f(x) = q_i(0)W_i(x)$$

$$\dot{w}(x,0) = g(x) = \dot{q}_i(0)W_i(x) \tag{8.83}$$

Taking scalar product of the above with respect to MW_j we have

$$\int_0^1 f(x)MW_j(x)dx = \int_0^1 q_i(0)MW_j(x)W_i(x)dx$$

$$\int_0^1 g(x)MW_j(x)dx = \int_0^1 \dot{q}_i(0)MW_j(x)W_i(x)dx \tag{8.83a}$$

With the help of orthogonal properties of the system, equations (8.83*a*) above give us

$$q_j(0) = \frac{1}{m_j}\int_0^1 f(x)MW_j(x)dx$$

$$\dot{q}_j(0) = \frac{1}{m_j}\int_0^1 g(x)MW_j(x)dx \quad j = 1,2,\ldots \quad \text{(no sum)} \tag{8.84}$$

The above values can be substituted in equations (8.81) and (8.82) to obtain the complete solution.

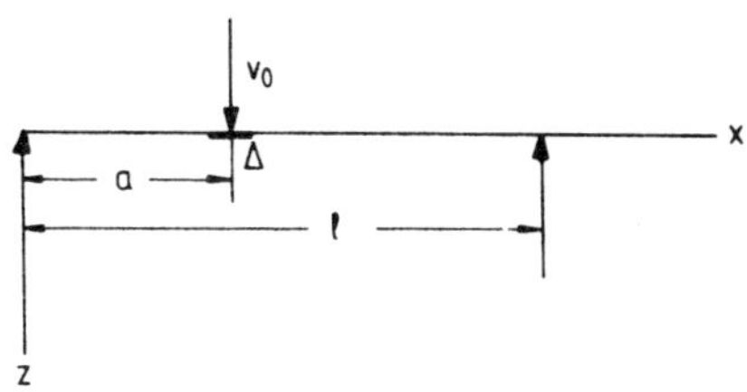

Fig. 8.16 Initial condition v_0 due to impact at $t = 0$ over a distance Δ

Example 1

Let us consider a simply supported beam shown in Fig. 8.16 subjected to an initial velocity v_0, at a distance a from the left end of the beam arising out of an impact occurring over a small distance Δ at this point. Then from equations (8.83)

$$w(x,0)=f(x)=0$$

$$\dot{w}(x,0)=g(x)=v_0\delta(x-a)$$

Then, from (8.84) we have

$$q_j(0)=0$$

$$\dot{q}_j(0)=\frac{2}{1+r_1^2(j\pi)^2}\int_0^1 v_0\Delta\delta(x-a)(1-r_1^2\partial_x^2)\sin j\pi x\,dx \quad j=1,2,\ldots \quad \text{(no sum)}$$

i.e.,

$$\dot{q}_j(0)=\frac{2v_0\Delta}{1+r_1^2(j\pi)^2}(\sin j\pi a+r_1^2(j\pi)^2\cos j\pi a) \quad j=1,2,\ldots \quad \text{(no sum)}$$

From (8.82), we get the complete solution

$$w(x,t)=\sum_{1,2,\ldots}^{\infty}\frac{\dot{q}_j(0)}{p_j}W_j(x)\sin p_j t$$

i.e.,

$$w(x,t)=\sum_{1,2,\ldots}^{\infty}\frac{2v_0\Delta}{p_j\{1+r_1^2(j\pi)^2\}}\{\sin j\pi a+r_1^2(j\pi)^2\cos j\pi a\}\sin j\pi x\sin p_j t$$

If the rotary inertia term is ignored, then

$$w(x,t)=\sum_{1,2,\ldots}^{\infty}\frac{2v_0\Delta}{p_j}\sin j\pi a\sin j\pi x\sin p_j t$$

If $a=1/2$, then the above reduces to

$$w(x,t)=\sum_{1,3,5,\ldots}^{\infty}\frac{2v_0\Delta}{p_j}\sin j\pi x\sin p_j t$$

i.e.,

$$w(x,t)=\frac{2v_0\Delta}{\pi^2}\sum_{1,3,5,\ldots}^{\infty}\frac{1}{j^2}\sin j\pi x\sin p_j t$$

It is seen that the modes which are symmetrical about the middle of the beam are only excited, and the amplitudes of the consecutive modes of vibration decrease by the factor $1/j^2$. If rotary inertia is included this symmetry is lost. In fact, rotary inertia terms give antisymmetric terms in the solution, i.e., the even terms with $j=2, 4$, etc.

Example 2

Let us consider now a force $P_0 \sin \omega t$ applied at a distance a from the left support. The forced vibration response from (8.81) is given by

$$q_j(t)=\frac{2}{p_j\{1+r_1^2(j\pi)^2\}}\int_0^t f_j(\tau)\sin p_j(t-\tau)d\tau \quad \text{(no sum)}$$

The generalised force in jth mode from (8.80) is

$$f_j = \int_0^1 \delta(x-a)P_0 \sin\omega t \sin j\pi x \, dx$$

i.e.,

$$f_j = P_0 \sin j\pi a \sin\omega t$$

Therefore

$$q_j(t) = \frac{2}{p_j\{1+r_1^2(j\pi)^2\}} \int_0^t p_0 \sin j\pi a \sin\omega\tau \sin p_j(t-\tau) d\tau \quad \text{(no sum)}$$

i.e.,

$$q_j(t) = \frac{2}{p_j^2\{1+r_1^2(j\pi)^2\}} P_0 \sin j\pi a \frac{1}{1-\left(\frac{\omega}{p_j}\right)^2}\left(\sin\omega t - \frac{\omega}{p_j}\sin p_j t\right) \quad \text{(no sum)}$$

Hence, the dynamic displacement is

$$w(x,t) = \sum_{1,2,\ldots}^{\infty} \frac{2}{p_j^2\{1+r_1^2(j\pi)^2\}} P_0 \sin j\pi a \frac{1}{1-\left(\frac{\omega}{p_j}\right)^2}\left(\sin\omega t - \frac{\omega}{p_j}\sin p_j t\right)\sin j\pi x$$

The free vibration response component eventually dies out and the steady-state solution is

$$w(x,t) = \sum_{1,2,\ldots}^{\infty} \frac{2}{p_j^2\{1+r_1^2(j\pi)^2\}} P_0 \sin j\pi a \frac{1}{1-\left(\frac{\omega}{p_j}\right)^2} \sin\omega t \sin j\pi x$$

If the rotary inertia is neglected, the response further simplifies to

$$w(x,t) = \sum_{1,2,\ldots}^{\infty} \frac{2}{j^4\pi^4} P_0 \sin j\pi a \frac{1}{1-\left(\frac{\omega}{j^2\pi^2}\right)^2} \sin\omega t \sin j\pi x$$

Timoshenko theory

Consider now a Timoshenko beam with its differential equations of motion and the boundary conditions given by

$$\rho A \ddot{w} - a_1 GA \partial_x^2 w + a_1 GA \partial_x \psi = \frac{384}{5} q$$

$$r_1^2 \ddot{\psi} - a_1 \partial_x w + a_1 \psi - \partial_x^2 \psi = 0 \tag{8.43}$$

$$(\partial_x w - \psi)\delta w \,|_0^1 = 0$$

$$\partial_x \psi \delta\psi \,|_0^1 = 0 \tag{8.44}$$

For a uniform beam, with the beam values as reference values, (8.43) become

$$\ddot{w} - a_1 \partial_x^2 w + a_1 \partial_x \psi = \frac{384}{5} q$$

$$r_1^2 \ddot{\psi} - a_1 \partial_x w + a_1 \psi - \partial_x^2 \psi = 0 \tag{8.43a}$$

where

$$a_1 = \frac{k'GAl^2}{EI_{yy}}$$

$$r_1^2 = \frac{I_{yy}}{Al^2} \quad (see\ 8.42)$$

Let us multiply the first equation in (8.43*a*) by sin $n\pi x$ and the second by cos $n\pi x$ and integrate over the length 0 to 1, to obtain

$$\int_0^1 \ddot{w} \sin n\pi x dx - \int_0^1 a_1 \partial_x^2 w \sin n\pi x dx$$

$$+ \int_0^1 a_1 \partial_x \psi \sin n\pi x dx = \int_0^1 \frac{384}{5} q \sin n\pi x dx$$

$$\int_0^1 r_1^2 \ddot{\psi} \cos n\pi x dx - \int_0^1 a_1 \partial_x w \cos n\pi x dx$$

$$+ \int_0^1 a_1 \psi \cos n\pi x dx - \int_0^1 \partial_x^2 \psi \cos n\pi x dx = 0 \tag{8.85}$$

We can substitute the following Fourier sine and cosine transforms for the integrals in the above

$$w_n = \int_0^1 w \sin n\pi x dx$$

$$\psi_n = \int_0^1 \psi \cos n\pi x dx$$

$$q_n = \int_0^1 q \sin n\pi x dx$$

$$\int_0^1 \partial_x w \cos n\pi x dx = \{w(1,t) \cos n\pi - w(0,t)\} + n\pi w_n$$

$$\int_0^1 \partial_x^2 w \sin n\pi x dx = -n\pi\{w(1,t) \cos n\pi - w(0,t)\} - n^2\pi^2 w_n$$

$$\int_0^1 \partial_x \psi \sin n\pi x dx = -n\pi\psi_n$$

$$\int_0^1 \partial_x^2 \psi \cos n\pi x dx = -\frac{48}{5}\{M(1,t) \cos n\pi - M(0,t)\} - n^2\pi^2\psi_n \tag{8.86}$$

where

$$M(0,t) = -\frac{5}{48}\partial_x \psi(0,t)$$

$$M(1,t) = -\frac{5}{48}\partial_x \psi(1,t)$$

Equation (8.85) now becomes

$$\ddot{w}_n + a_1 n\pi\{w(1,t)\cos n\pi - w(0,t)\} + a_1 n^2\pi^2 w_n - a_1 n\pi\psi_n = \frac{384}{5} q_n$$

$$r_1^2\ddot{\psi}_n - a_1\{w(1,t)\cos n\pi - w(0,t)\} - a_1 n\pi w_n + a_1\psi_n$$

$$+\frac{48}{5}\{M(1,t)\cos n\pi - M(0,t)\} + n^2\pi^2\psi_n = 0 \qquad (8.85a)$$

We can rewrite (8.85a)

$$-(d_t^2 + a_1 n^2\pi^2)w_n + a_1 n\pi\psi_n = -\frac{384}{5} q_n + a_1 n\pi\{w(1,t)\cos n\pi - w(0,t)\}$$

$$-a_1 n\pi w_n + (n^2\pi^2 + a_1 + r_1^2 d_t^2)\psi_n = a_1\{w(1,t)\cos n\pi - w(0,t)\}$$

$$-\frac{48}{5}\{M(1,t)\cos n\pi - M(0,t)\} \qquad (8.85b)$$

Expressing (8.85b) in matrix form

$$Aq_n = f \qquad (8.87)$$

i.e.,

$$\begin{bmatrix} a_{11} & a_{12} \\ a_{21} & a_{22} \end{bmatrix} \begin{Bmatrix} w_n \\ \psi_n \end{Bmatrix} = \begin{Bmatrix} f_1 \\ f_2 \end{Bmatrix}$$

where

$$a_{11} = -(d_t^2 + a_1 n^2\pi^2)$$

$$a_{12} = n\pi a_1$$

$$a_{21} = -n\pi a_1$$

$$a_{22} = (n^2\pi^2 + a_1 + r_1^2 d_t^2)$$

$$f_1 = -\frac{384}{5} q_n + a_1 n\pi\{w(1,t)\cos n\pi - w(0,t)\}$$

$$f_2 = a_1\{w(1,t)\cos n\pi - w(0,t)\} - \frac{48}{5}\{M(1,t)\cos n\pi - M(0,t)\} \qquad (8.88)$$

We can now take the Laplace transform of the above

$$\overline{A}\overline{q}_n = \overline{f} + \overline{B}q_n(0) + \overline{C}\dot{q}_n(0) \qquad (8.89)$$

where

$$\overline{A} = \begin{bmatrix} \overline{a}_{11} & \overline{a}_{12} \\ \overline{a}_{21} & \overline{a}_{22} \end{bmatrix}$$

$$\overline{B} = \begin{bmatrix} \overline{b}_{11} & \overline{b}_{12} \\ \overline{b}_{21} & \overline{b}_{22} \end{bmatrix}$$

$$\overline{C} = \begin{bmatrix} \overline{c}_{11} & \overline{c}_{12} \\ \overline{c}_{21} & \overline{c}_{22} \end{bmatrix} \tag{8.90}$$

In (8.90) above

$$\overline{a}_{11} = -(s^2 + a_1 n^2 \pi^2)$$

$$\overline{a}_{12} = n\pi a_1$$

$$\overline{a}_{21} = -n\pi a_1$$

$$\overline{a}_{22} = (n^2\pi^2 + a_1 + r_1^2 s^2)$$

$$\overline{b}_{11} = a_{11,2} + a_{11,3} s$$

$$\overline{b}_{12} = 0; \quad \overline{b}_{21} = 0$$

$$\overline{b}_{22} = a_{22,2} + a_{22,3} s$$

$$\overline{c}_{11} = a_{11,3}$$

$$\overline{c}_{12} = 0; \quad \overline{c}_{21} = 0$$

$$\overline{c}_{22} = a_{22,3} \tag{8.91}$$

In (8.91) $\overline{a}_{11}, \overline{a}_{12} \ldots$ polynomials are arranged in the form

$$\sum_{k=1}^{N+1} a_{ij,k} a^{k-1}$$

The transfer matrix of the system, being the inverse of the characteristic matrix, is given by

$$\overline{T} = [\overline{t}_{ij}] = \begin{bmatrix} \dfrac{\overline{a}_{22}}{\Delta} & -\dfrac{\overline{a}_{12}}{\Delta} \\ -\dfrac{\overline{a}_{21}}{\Delta} & \dfrac{\overline{a}_{11}}{\Delta} \end{bmatrix} \tag{8.92}$$

where

$$\Delta = \overline{a}_{11}\overline{a}_{22} - \overline{a}_{21}\overline{a}_{12}$$

$$= (n\pi a_1)^2 - (s^2 + a_1 n^2\pi^2)(n^2\pi^2 + a_1 + r_1^2 s^2) \tag{8.93}$$

The solution in the transformed plane is then given by

$$\overline{q}_n = \overline{T}\overline{f} + \overline{T}\,\overline{B} q_n(0) + \overline{T}\,\overline{C}\dot{q}_n(0) \tag{8.94}$$

The first term on the right hand side of (8.94) above gives the steady-state solution and the next two terms correspond to transient response of the system.

Example 3

Consider the following data.

$$w(0,t) = 0$$

$$w(1,t) = 0$$

$$M(0,t) = 0$$

$$M(1,t) = 0$$

$$q(x,t) = e^{i\omega t}$$

From (8.86), we have

$$q_n = \int_0^1 e^{i\omega t} \sin n\pi x \, dx$$

$$= \frac{1 - \cos n\pi}{n\pi} e^{i\omega t}$$

Also from equation (8.88), the excitations f_1 and f_2 are given by

$$f_1 = -\frac{384}{5} \frac{1 - \cos n\pi}{n\pi} e^{i\omega t}$$

$$f_2 = 0$$

The steady-state response from (8.94) is

$$\bar{q}_n = \overline{T}\bar{f}$$

We can replace d_t by $j\omega$ in the above and obtain the steady-state response. Thus, the steady-state response is characterized by

$$w_n(j\omega) = \frac{-\bar{a}_{22}(j\omega)}{\Delta(j\omega)} \frac{384}{5} \frac{1 - \cos n\pi}{n\pi} e^{i\omega t}$$

The magnitude of the above complex quantity represents the dynamic magnification factor and its modulus represents the phase angle by which the response lags behind the excitation in the nth mode.

Example 4

Let the beam be moving with an initial velocity $\dot{w}(x,0) = v$ and let the rigid body motion be arrested at time $t = 0$. The transforms of the initial conditions and excitations are therefore given by

$$\dot{w}_n(0) = \frac{v t_0}{w_0} \frac{1 - \cos n\pi}{n\pi}$$

$$w_n(0) = 0$$

$$\psi_n(0) = 0$$

$$\psi_n(0) = 0$$

$$f_1(0) = 0$$

$$f_2(0) = 0$$

Assuming $v = w_0/t_0$ and substituting the above in (8.89)

$$\overline{w}_n(s) = \frac{2}{n\pi}\frac{\overline{a}_{22}\overline{c}_{11} - \overline{a}_{12}\overline{c}_{21}}{\Delta} \quad n = 1,3,\ldots$$

$$= 0 \quad n = 2,4,\ldots$$

$$\overline{\psi}_n(s) = \frac{2}{n\pi}\frac{\overline{a}_{11}\overline{c}_{21} - \overline{a}_{21}\overline{c}_{11}}{\Delta} \quad n = 1,3,\ldots$$

$$= 0 \quad n = 2,4,\ldots$$

Because of the symmetry in the system, antisymmetric mode components are not present. The solution is obtained by inverting the Laplace transform and Fourier transform.

$$w(x,t) = \sum_{1,3,\ldots}^{\infty} w_n(t)\sin n\pi x$$

$$\psi(x,t) = \sum_{1,3,\ldots}^{\infty} \psi_n(t)\cos n\pi x$$

8.6 LATERAL VIBRATION OF BEAMS SUBJECT TO AXIAL FORCE

Consider a beam in transverse vibration subjected to an axial compressive force S, Figure 8.17 shows an element dx of the beam. The change in the elemental length is given by

$$ds - dx = \sqrt{dx^2 + (w_{,x})^2 dx^2} - dx \tag{8.95}$$

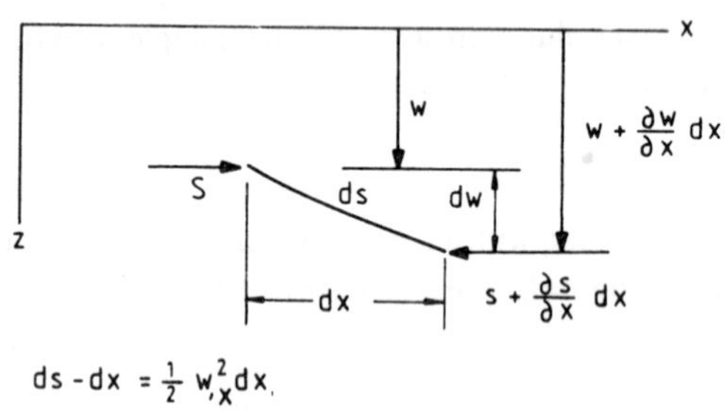

Fig. 8.17 A beam element with axial forces

Equation (8.95) above can be approximated for small amplitudes of vibration as

$$ds - dx = \frac{1}{2} w_{,x}^2 dx \tag{8.95a}$$

Hence, the work done by the axial compressive forces is given by

$$W_s = \int_0^l \frac{1}{2} S w_{,x}^2 dx \tag{8.96}$$

The corresponding term in Hamilton's principle becomes

$$I = \delta \int_{t1}^{t2} \int_0^l \frac{1}{2} S w_{,x}^2 dx\, dt$$

i.e.,

$$I = \int_{t1}^{t2} \int_0^l S w_{,x} \delta w_{,x} dx\, dt$$

Integrating by parts

$$I = \int_{t1}^{t2} \left[S w_{,x} \delta w \mid_0^l - \int_0^l (S w_{,x})_{,x} \delta w dx \right] dt$$

We will add this term to Rayleigh's theory in equations (8.12) and (8.13). Thus

$$(EI_{yy} w_{,xx})_{,xx} + (S w_{,x})_{,x} - (\rho I_{yy} \ddot{w}_{,x})_{,x} + \rho A \ddot{w} + q \quad (8.97)$$

$$EI_{yy} w_{,xx} \delta w_{,x} \mid_0^l - \{(EI_{yy} w_{,xx})_{,x} + S w_{,x} - \rho I_{yy} \ddot{w}_{,x}\} \delta w \mid_0^l = 0 \quad (8.98)$$

We can nondimensionalize (8.97) and (8.98) according to (8.14). Then

$$M\ddot{w} + Kw = q \quad (8.99)$$

where the mass and stiffness operators are given by

$$M = \rho A - r_1^2 \partial_x (\rho I_{yy} \partial_x)$$

$$K = \partial_x^2 EI \partial_x^2 + a_2 \partial_x^2 \quad (8.100)$$

and the boundary conditions are

$$(EI_{yy} \partial_x^2 w) \delta \partial_x w \mid_0^1$$

$$+ \{-\partial_x (EI_{yy} \partial_x^2 w) - a_2 \partial_x w + r_1^2 \rho I_{yy} \partial_x \ddot{w}\} \delta w \mid_0^1 = 0 \quad (8.101)$$

In the above equations (8.100) and (8.101)

$$a_2 = \frac{Sl^2}{E_0 I_{yy0}} \quad (8.102)$$

is the nondimensional axial compressive force.

For uniform beams with the reference values chosen as its original values themselves, equations (8.100) and (8.101) reduce to

$$M = 1 - r_1^2 \partial_x^2$$

$$K = \partial_x^4 + a_2 \partial_x^2 \quad (8.100a)$$

$$(\partial_x^2 w) \delta \partial_x w \mid_0^1 + \{-\partial_x^3 w - \partial_x w + r_1^2 \partial_x \ddot{w}\} \delta w \mid_0^1 = 0 \quad (8.101a)$$

For free vibrations, let

$$w(x,t) = W(x) e^{ipt}$$

Then, equations (8.100*a*) and (8.101*a*) become

$$\{d_x^4 + (a_2 + r_1^2 p^2) d_x^2 - p^2\} W = 0 \quad (8.103)$$

$$(d_x^2W)\delta d_xW\,|_0^1 - \{d_x^3W + (a_2 + r_1^2p^2)d_xW\}\,\delta W\,|_0^1 = 0 \tag{8.104}$$

Let

$$b = -(a_2 + r_1^2p^2)$$

$$c = p^2$$

Then, equations (8.103) and (8.104) become

$$(d_x^4 - bd_x^2 - c)W = 0 \tag{8.103a}$$

$$(d_x^2W)\delta d_xW\,|_0^1 + \{bd_xW - d_x^3W\}\,\delta W\,|_0^1 = 0 \tag{8.104a}$$

The Laplace transform of the first equation in (8.103a) is

$$(s^4 - bs^2 - c)\overline{W} = (s^3 - bs)W(0) + s^2W'(0)$$

$$+ sW''(0) + \{W'''(0) - bW'(0)\} \tag{8.106}$$

The following positive quantities are defined as in (8.21).

$$s_1^2 = \frac{1}{2}(\sqrt{b^2 + 4c} + b)$$

$$s_2^2 = \frac{1}{2}(\sqrt{b^2 + 4c} - b) \tag{8.107}$$

Using (8.105), the second equation in the above gives

$$p^2 = \frac{s_2^4 - s_2^2a_2}{1 + r_1^2s_2^2} \tag{8.108}$$

The transfer function, as also the mode shape and its derivatives remain the same as in (8.24) and (8.25).

$$T(x) = \frac{1}{s_1s_2(s_1^2 + s_2^2)}(s_2 \sinh s_1x - s_1 \sin s_2x) \tag{8.24}$$

Hence, the mode shape and its derivatives can be written as

$$W = [(d_x^3 - bd_x)W(0) + d_x^2W'(0) + d_xW''(0) + \{W'''(0) - bW'(0)\}]T(x)$$

$$W' = [cW(0) + d_x^3W'(0) + d_x^2W''(0) + d_x\{W'''(0) - bW'(0)\}]T(x)$$

$$W'' = [cd_xW(0) + (bd_x^2 + c)W'(0) + d_x^3W''(0) + d_x^2\{W'''(0) - bW'(0)\}]T(x)$$

$$W''' - bW' = [(cd_x^2 - bc)W(0) + cd_xW'(0) + cW''(0)$$

$$+ (d_x^3 - bd_x)\{W'''(0) - bW'(0)\}]T(x) \tag{8.25}$$

Let us consider a simply-supported beam, whose frequency equation is given by equation (8.28).

$$s_2 = n\pi \tag{8.28}$$

From (8.108) we get the natural frequency for a beam in transverse vibration, subjected to axial compressive force.

$$p^2 = \frac{(n\pi)^4 - (n\pi)^2 a_2}{1 + r_1^2 s_2^2}$$

$$= \frac{(n\pi)^4}{1 + r_1^2 s_2^2}\left(1 - \frac{a_2}{n^2\pi^2}\right) \tag{8.109}$$

We see from (8.109) above that the natural frequency decreases with an increase in the axial compressive force. The mode shape of the system remains same as in (8.30).

$$W = \sin s_2 x = \sin n\pi x \quad n = 1, 2, \ldots \tag{8.30}$$

From (8.109) we note that the frequency becomes zero when

$$a_2 = (n\pi)^2$$

i.e.,

$$S = n^2\pi^2 \frac{EI_{yy}}{l^2} \tag{8.110}$$

The beam experiences instability and buckles as a column for forces equal and beyond the value of S in (8.110). Equation (8.110), in fact, is the Euler buckling load for a simply supported beam.

If the axial force is tensile, then the natural frequency is given by

$$p^2 = \frac{(n\pi)^4}{1 + r_1^2 s_2^2}\left(1 + \frac{a_2}{n^2\pi^2}\right) \tag{8.109a}$$

That is, for a beam with tensile axial force, the natural frequency increases and there will not be any buckling.

The analysis of free and forced vibration of a beam subjected to an axial force can be carried out in the same way as in section 8.5.

A special case of the above problem is that in which large amplitudes of vibration of a beam occur in lateral motion. The total extension of the beam according to equation (8.95a) is

$$e = \int_0^l \frac{1}{2} w_{,x}^2 dx \tag{a}$$

The extension of the beam in the axial direction can be considered to be arising out of an axial tensile force N given by

$$N = \frac{1}{2}\frac{AE}{l}\int_0^l w_{,x}^2 dx \tag{b}$$

Neglecting the term of rotary inertia, the corresponding differential equation of motion of the beam for lateral vibrations can be written directly from equation (8.97).

$$EI_{yy} w_{,xxxx} - \frac{AE}{2l} w_{,xx} \int_0^l w_{,x}^2 dx + \rho A \ddot{w} = q \tag{8.97a}$$

Equation (8.97a) is nonlinear and we can assume a solution given by

$$w(x,t) = af(x)\eta(t) \tag{c}$$

where the assumed shape function can be chosen to be the mode shape of the corresponding linear system. We can use Ritz-Galerkin method to solve (8.97a) as follows.

$$\int_0^T \int_0^l \left\{ EI_{yy} a f_{,xxxx}\eta - \frac{AE}{2l} a^3\eta^3 f_{,xx} \int_0^l f_{,x}^2 dx + \rho A a f\ddot{\eta} - q \right\} f\eta \, dx\, dt = 0 \tag{d}$$

If $f(x)$ satisfies the linear equation

$$f_{,xxxx} - \lambda^4 f \tag{e}$$

where

$$\lambda^4 = \frac{\rho A p_1^2}{EI_{yy}} \tag{f}$$

$$f(x) = \sin\frac{\pi x}{l},$$

and p_1 is the natural frequency of the first mode of vibration of the linear beam. Equation (d) now becomes

$$\int_0^T \int_0^l \{EI_{yy} a\lambda^4 \sin^2\frac{\pi x}{l}\eta^2$$

$$+\frac{AE}{2l} a^3\eta^4\pi^4 l^4 \sin^2\frac{\pi x}{l} \int_0^l \cos^2\frac{\pi x}{l} dx$$

$$+\rho A a \sin^2\frac{\pi x}{l}\eta\ddot{\eta} - q \sin\frac{\pi x}{l}\eta\}\, dx\, dt = 0 \tag{g}$$

If

$$q = P_0 \sin \omega t$$

and

$$\eta = \sin \omega t$$

Then, equation (g) becomes

$$\int_0^T \int_0^l \{EI_{yy} a\lambda^4 \sin^2\frac{\pi x}{l} \sin^2 \omega t$$

$$+\frac{AE\pi^4}{4l^4} a^3 \sin^4 \omega t \sin^2\frac{\pi x}{l}$$

$$-\rho A a \omega^2 \sin^2\frac{\pi x}{l} \sin^2 \omega t - P_0 \sin\frac{\pi x}{l} \sin^2 \omega t\}\, dx\, dt = 0 \tag{h}$$

Integrating with respect to time over 0 to $2\pi/\omega$, the above becomes

$$\int_0^l \left\{ EI_{yy} a\lambda^4 \frac{\pi}{\omega} \sin^2\frac{\pi x}{l} + \frac{3AE\pi^5}{16\omega l^4} a^3 \sin^2\frac{\pi x}{l} \right\}$$

$$\left\{-\rho A a \omega \pi \sin^2\frac{\pi x}{l} - P_0\frac{\pi}{\omega}\sin\frac{\pi x}{l}\right\}dx = 0 \qquad (i)$$

Now, integrating with respect to x and rearranging, we get

$$\frac{1}{2}EI_{yy}la\lambda^4 + \frac{3}{32}\frac{AE\pi^4}{l^3}a^3 - \frac{1}{2}\rho Ala\omega^2 - \frac{2P_0l}{\pi} = 0 \qquad (j)$$

There are three roots possible for a in equation (j) above. Depending on the value of ω, they can all be real or one root real and the other complex conjugate. Let

$$\bar{a} = \frac{a}{r_g}$$

$$s = \frac{l}{r_g}$$

$$P = \frac{P_0l^3}{EI_{yy}}$$

$$r = \frac{\omega}{p}$$

$$r_g = \frac{\sqrt{I_{yy}}}{A}$$

Let us choose $Ps = 0.25$, then equation (j) becomes

$$0.1875\pi^5\bar{a}^3 + \pi^5(1 - r^2)\bar{a} - 1 = 0 \qquad (k)$$

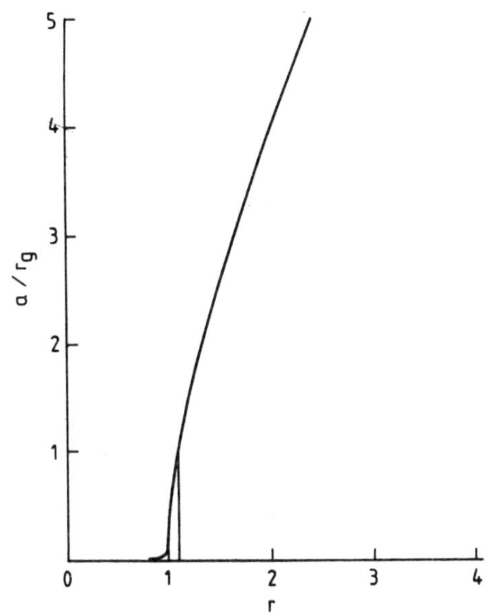

Fig. 8.18 Response of a nonlinear beam

Figure 8.18 shows the solution of (j) of the beam, which is a typical Duffing's equation response with a jump characteristic of the hardening type.

8.7 BEAM ON ELASTIC FOUNDATION

Consider now a beam resting on a Winkler elastic foundation as shown in Fig. 8.19. The property of foundation is defined by its modulus k, which is the force per unit length of the beam necessary to produce a unit compression in the foundation. It is customary to neglect the mass of the foundation. The work done by the reaction force of the foundation is given by

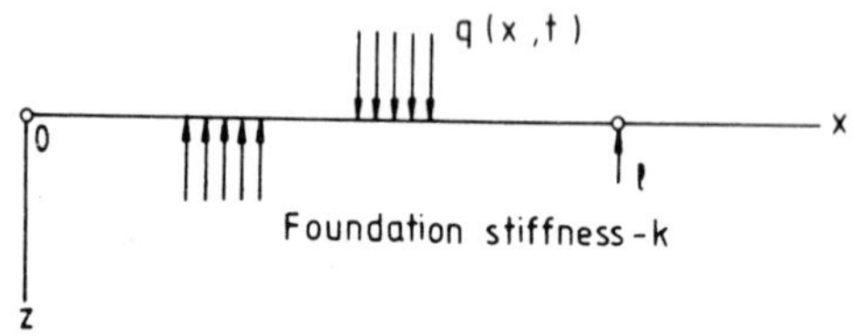

Fig. 8.19 Beam on a Winkler elastic foundation

$$W_f = -\frac{1}{2}\int_0^l kw^2 dx \tag{8.111}$$

The corresponding term that arises in the Hamilton's principle for the beam is

$$I_f = -\delta \int_{t1}^{t2} \int_0^l \frac{1}{2} kw^2 dx$$

$$= -\int_{t1}^{t2} \int_0^l kw\delta w\, dx\, dt$$

Rayleigh's beam equation (8.12) can now be modified to

$$(EI_{yy}w_{,xx})_{,xx} + kw - (\rho I_{yy}\ddot{w}_{,x})_{,x} + \rho A\ddot{w} = q \tag{8.112}$$

The boundary conditions remain unchanged as given in equation (8.13).

$$EI_{yy}w_{,xx}\delta w_{,x} |_0^l - \{(EI_{yy}w_{,xx})_{,x} - \rho I_{yy}\ddot{w}_{,x}\}\delta w |_0^l = 0 \tag{8.13}$$

Following the nondimensionalization in equation (8.14), equations (8.112) and (8.13) become

$$M\ddot{w} + Kw = q \tag{8.113}$$

where the mass and stiffness operators are given by

$$M = \rho A - r_1^2 \partial_x(\rho I_{yy}\partial_x)$$

$$K = \partial_x^2 EI \partial_x^2 + a_3 \tag{8.114}$$

$$(EI_{yy}\partial_x^2 w)\delta\partial_x w |_0^l + \{-\partial_x(EI_{yy}\partial_x^2 w) + r_1^2 \rho I_{yy}\partial_x \ddot{w}\}\delta w |_0^l = 0 \tag{8.17}$$

where the nondimensional foundation stiffness a_3 is

$$a_3 = \frac{kl^4}{EI_{yy0}} \tag{8.115}$$

For uniform beams with the reference values chosen as its original values themselves, equations (8.114) and (8.17) reduce to

$$M = 1 - r_1^2 \partial_x^2$$

$$K = \partial_x^4 + a_3 \tag{8.114a}$$

$$(\partial_x^2 w)\delta\partial_x w \,|_0^l + (-\partial_x^3 w + r_1^2 \partial_x \ddot{w})\delta w \,|_0^l = 0 \tag{8.17a}$$

For free vibrations,

$$w(x,t) = W(x)e^{ipt}$$

Then, equations (8.113) and (8.17) become

$$(d_x^4 + r_1^2 d_x^2 p^2 + a_3 - p^2)W = 0$$

$$[(d_x^2 W)\delta d_x W]_0^1 - [\{d_x^3 W + r_1^2 d_x p^2 W\}\,\delta W]_0^1 = 0 \tag{8.116}$$

Let

$$b = -r_1^2 p^2$$

$$c = p^2 - a_3 \tag{8.117}$$

Then, equations (8.116) become

$$(d_x^4 - b d_x^2 - c)W = 0$$

$$[(d_x^2 W)\delta d_x W]_0^1 + [(b d_x W - d_x^3 W)\delta W]_0^1 = 0 \tag{8.118}$$

We follow the Laplace transformation of this problem as in section 8.2, and using (8.117), equation (8.21) gives us

$$p^2 = \frac{s_2^4 + a_3}{1 + r_1^2 s_2^2} \tag{8.119}$$

The transfer function and the mode shape functions remain same as in (8.24) and (8.25).

$$T(x) = \frac{1}{s_1 s_2 (s_1^2 + s_2^2)} (s_2 \sinh s_1 x - s_1 \sin s_2 x) \tag{8.24}$$

$$W = [(d_x^3 - b d_x)W(0) + d_x^2 W'(0) + d_x W''(0) + \{W'''(0) - bW'(0)\}]T(x)$$

$$W' = [cW(0) + d_x^3 W'(0) + d_x^2 W''(0) + d_x\{W'''(0) - bW'(0)\}]T(x)$$

$$W'' = [c d_x W(0) + (b d_x^2 + c)W'(0) + d_x^3 W''(0) + d_x^2\{W'''(0) - bW'(0)\}]T(x)$$

$$W''' - bW' = [(c d_x^2 - bc)W(0) + c d_x W'(0) + cW''(0)$$

$$+ (d_x^3 - b d_x)\{W'''(0) - bW'(0)\}]T(x) \tag{8.25}$$

For the simply-supported beam, the frequency equation also remains the same as in equation (8.28), which is

$$s_2 = n\pi \tag{8.28}$$

Therefore, the natural frequency is given from (8.119) and the mode shape remains unaltered from (8.30).

$$p^2 = \frac{n\pi^4 + a_3}{1 + r_1^2 n\pi^2} \tag{8.120}$$

The study of free and forced vibration can be carried out in a similar way as in section 8.5.

8.8 VISCOELASTIC BEAM ON VISCOELASTIC FOUNDATION

In this section, we will consider the general problem of a viscoelastic Timoshenko beam subjected to an axial compressive force and resting on a viscoelastic foundation. Let the beam be uniform in cross-section and that the viscoelastic model be represented by a linear viscoelastic solid. Equation (A4.50 in Appendix 4) gives the general stress strain relation for this model.

$$P_1(\partial_t)P_3(\partial_t)\sigma_{ij} = P_2(\partial_t)P_3(\partial_t)\varepsilon_{ij} + \frac{1}{3}\delta_{ij}\{P_1(\partial_t)P_4(\partial_t) - P_2(\partial_t)P_3(\partial_t)\}e \tag{A.4.50}$$

Also, equation (A4.45) gives

$$P_3(\partial_t)s(x,t) = P_4(\partial_t)e(x,t) \tag{A4.45}$$

Hence, equation (A4.50) can be written as

$$P_1P_3\sigma_{ij} = P_2P_3\varepsilon_{ij} + \frac{1}{3}\delta_{ij}\{P_1P_4 - P_2P_3\}\frac{P_3}{P_4}s$$

From the above, we can write

$$E(\partial_t) = \frac{3P_2P_4}{2P_1P_4 + P_2P_3}$$

$$G(\partial_t) = \frac{P_2}{2P_1} \tag{8.121}$$

Let the polynomial operators be given by

$$P_1(\partial_t) = 1 + t_1\partial_t$$

$$P_2(\partial_t) = 2G(I + t_2\partial_t)$$

$$P_3(\partial_t) = \partial_t$$

$$P_4(\partial_t) = 3K\partial_t \tag{8.122}$$

where t_1 t_2 are relaxation time constants of the beam material and K is the bulk modulus of the material given by

$$K = \lambda + \frac{2}{3}G$$

With the help of equations (A3.17) and (8.122), equations (8.121) can be written as

$$E(\partial_t) = \frac{1 + t_2\partial_t}{1 + t_3\partial_t}$$

$$G(\partial_t) = \frac{1 + t_2\partial_t}{1 + t_1\partial t} \tag{8.123}$$

where

$$t_3 = \frac{2}{3}(1 + \mu)t_1 + \frac{1}{3}(1 - 2\mu)t_2$$

If the material is incompressible, $\mu \cong 0.5$, $t_3 \cong t_1$, then E and G can be taken as same. Similarly, the behaviour of the foundation can also be represented by a standard linear solid viscoelastic model

$$k(\partial_t) = \frac{1 + t_4\partial_t}{1 + t_5\partial_t} \tag{8.124}$$

We will assume that the beam and the foundation are bonded together. This means that the foundation can exert both compressive and tensile reactions on the beam, i.e., the foundation exerts a lateral reaction kw resisting the lateral motion of the beam. The foundation will also exert a reaction force $k_1 u_x(x, h)$ in the axial direction resisting the axial motion of the beam, where $2h$ is the depth of the beam. Thus, the reactive forces set by the foundation on the beam in the lateral and axial directions are

$$F_z = -kw$$

$$F_x = -k_1 h\psi \tag{8.125}$$

where, as before, ψ is the beam slope due to bending alone.

We now can derive the differential equations of motion and the boundary conditions for Timoshenko beam. The displacement field is given by (8.34).

$$u_x = -z\psi(x,t) = -z\{w_{,x} - \beta(x)\}$$

$$u_y = 0$$

$$u_z = w(x,t) \tag{8.34}$$

The strain and stress fields are given by (8.25) and (8.36).

$$\varepsilon_{xx} = -z\psi_{,x}$$

$$\varepsilon_{xz} = \frac{1}{2}(w_{,x} - \psi)$$

rest $$\varepsilon = 0 \tag{8.35}$$

$$\tau_{xx} = -Ez\psi_{,x}$$

$$\tau_{xz} = k'G(w_{,x} - \psi)$$

rest $$\tau = 0 \tag{8.36}$$

The strain and kinetic energies are

$$U = \frac{1}{2}\int_0^l (EI_{yy}\psi_{,x}^2 + k'AG(w_{,x} - \psi)^2)\, dx \tag{8.37}$$

$$T = \frac{1}{2}\int_0^l (\rho A\dot{w}^2 + \rho I_{yy}\dot{\psi}^2) dx \tag{8.38}$$

The work done by external forces is given by (8.39), (8.111) and (8.96), besides the work due to the horizontal reaction force $-k_1\, h\psi$. Therefore

$$W = \int_0^l \left(qw - \frac{1}{2}kw^2 - \frac{1}{2}k_1 h^2\psi^2 + \frac{1}{2}Sw_{,x}^2 \right) dx \tag{8.126}$$

We note that the additional terms to Timoshenko equations (8.40) come from the foundation reactions and the axial force of equation (8.126). Making use of Hamilton's principle, we arrive at the following:

$$\rho A\ddot{w} + \{(S - k'AG)\partial_x^2 + k\}\, w + k'AG\partial_x\psi = q$$

$$-k'AG\partial_x w + \rho I_{yy}\ddot{\psi} + \{k'AG + k_1 h^2 - \partial_x(\partial_x EI_{yy})\}\,\psi = 0 \tag{8.127}$$

The boundary conditions remain same as

$$k'AG(w_{,x} - \psi)\delta w \,|_0^l = 0$$

$$EI_{yy}\psi_{,x}\delta\psi \,|_0^l = 0 \tag{8.41}$$

Following the nondimensionalization of (8.42)

$$\rho A\partial_t^2 w + \{(a_2 - a_1 GA)\partial_x^2 + a_3\}\, w + a_1 GA\partial_x\psi = \frac{384}{5} q$$

$$r_1^2\ddot{\psi} - a_1\partial_x w + (a_1 + a_4 k_1 - \partial_x^2)\psi = 0 \tag{8.128}$$

with the boundary conditions

$$(\partial_x w - \psi)\delta w \,|_0^1 = 0$$

$$\partial_x\psi\delta\psi \,|_0^1 = 0 \tag{8.44}$$

where

$$r_1^2 = \frac{I_{yy0}}{A_0 l^2}$$

$$a_1 = \frac{k' G_0 A_0 l^2}{E_0 I_{yy0}}$$

$$a_2 = \frac{Sl^2}{E_0 I_{yy0}}$$

$$a_3 = \frac{kl^4}{E_0 I_{yy0}}$$

$$a_4 = \frac{k_1 h^2 l^2}{E_0 I_{yy0}}$$

For a uniform beam, with the beam values as reference values, the differential equations for the viscoelastic Timoshenko beam on a viscoelastic foundation subjected to a compressive axial force become

$$\partial_t^2 w + [\{a_2 - a_1 G(\partial_t)\}\partial_x^2 + a_3 k(\partial_t)]w + a_1 G(\partial_t)\partial_x \psi = \frac{384}{5} q$$

$$r_1^2 \ddot{\psi} - a_1 G(\partial_t)\partial_x w + \{a_1 G(\partial_t) + a_4 k_1(\partial_t) - E(\partial_t)\partial_x^2\}\psi = 0 \qquad (8.128a)$$

We will convert the above partial differential equations to ordinary differential equations by employing Fourier sine and cosine transforms. Let us multiply the first equation in (8.128a) by sin $n\pi x$ and the second by cos $n\pi x$ and integrate over the length 0 to 1, to obtain

$$\int_0^1 \ddot{w} \sin n\pi x\,dx + \int_0^1 [\{a_2 - a_1 G(\partial_t)\}\partial_x^2 + a_3 k(\partial_t)]w \sin n\pi x\,dx$$

$$+\int_0^1 a_1 G(\partial_t)\partial_x \psi \sin n\pi x\,dx = \int_0^1 \frac{384}{5} q \sin n\pi x\,dx$$

$$\int_0^1 r_1^2 \ddot{\psi} \cos n\pi x\,dx - \int_0^1 a_1 G(\partial_t)\partial_x w \cos n\pi x\,dx$$

$$+\int_0^1 \{a_1 G(\partial_t) + a_4 k_1(\partial_t) - E(\partial_t)\partial_x^2\}\psi \cos n\pi x\,dx = 0 \qquad (8.129)$$

With the help of the following Fourier sine and cosine transforms

$$w_n = \int_0^1 w \sin n\pi x\,dx$$

$$\psi_n = \int_0^1 \psi \cos n\pi x\,dx$$

$$q_n = \int_0^1 q \sin n\pi x\,dx$$

$$\int_0^1 \partial_x w \cos n\pi x\,dx = \{w(1,t)\cos n\pi - w(0,t)\} + n\pi w_n$$

$$\int_0^1 \partial_x^2 w \sin n\pi x\,dx = -n\pi\{w(1,t)\cos n\pi - w(0,t)\} - n^2\pi^2 w_n$$

$$\int_0^1 \partial_x \psi \sin n\pi x\,dx = -n\pi\psi_n$$

$$\int_0^1 \partial_x^2 \psi \cos n\pi x\,dx = -\frac{48}{5}\{M(1,t)\cos n\pi - M(0,t)\} - n^2\pi^2\psi_n \qquad (8.86)$$

where

$$M(0,t) = -\frac{5}{48}\partial_x \psi(0,t)$$

$$M(1,t) = -\frac{5}{48}\partial_x \psi(1,t)$$

Equations (8.129) now become

$$\ddot{w}_n + a_1 G(\partial_t) n\pi\{w(1,t)\cos n\pi - w(0,t)\} + [-\{a_2 - a_1 G(\partial_t)\} n^2\pi^2$$

$$+ a_3 k(\partial_t)] w_n - a_1 G(\partial_t) n\pi\psi_n = \frac{384}{5} q_n$$

$$r_1^2 \ddot{\psi}_n - a_1 G(\partial_t)\{w(1,t)\cos n\pi - w(0,t)\} - a_1 G(\partial_t) n\pi w_n + \{a_1 G(\partial_t)$$

$$+ a_4 k_1(\partial_t) - E(\partial_t) n^2\pi^2\}\psi_n + \frac{48}{5}\{M(1,t)\cos n\pi - M(0,t)\} = 0 \qquad (8.130)$$

For simply-supported beam, the boundary conditions from (8.44) can be specified by the time-dependent functions

$$w_0^*(t) = w(0,t)$$

$$w_1^*(t) = w(1,t)$$

$$M_0^*(t) = -\frac{5}{48} E(\partial_t)\partial_x \psi(0,t) = M(0,t)$$

$$M_1^*(t) = -\frac{5}{48} E(\partial_t)\partial_x \psi(1,t) = M(1,t) \qquad (8.131)$$

Introducing

$$E(\partial_t) = \frac{1 + t_2\partial_t}{1 + t_3\partial_t}$$

$$G(\partial_t) = \frac{1 + t_2\partial_t}{1 + t_1\partial_t} \qquad (8.123)$$

where

$$t_3 = \frac{2}{3}(1+\mu)t_1 + \frac{1}{3}(1-2\mu)t_2$$

$$k(\partial_t) = \frac{1 + t_4\partial_t}{1 + t_5\partial_t} \qquad (8.124)$$

and

$$k_1(\partial_t) = 1 + t_6\partial_t \qquad (8.132)$$

equations (8.131) can be written as

$$A q_n = f \qquad (8.133)$$

i.e.,

$$\begin{bmatrix} a_{11} & a_{12} \\ a_{21} & a_{22} \end{bmatrix} \begin{Bmatrix} w_n \\ \psi_n \end{Bmatrix} = \begin{Bmatrix} f_1 \\ f_2 \end{Bmatrix}$$

where

$$a_{11} = -\{(1+t_1d_t)(1+t_5d_t)d_t^2$$

$$+n^2\pi^2(a_8+t_8d_t)(1+t_5d_t)+a_3(1+t_4d_t)(1+t_1d_t)\}$$

$$a_{12} = n\pi a_1(1+t_2d_t)(1+t_5d_t)$$

$$a_{21} = -n\pi a_1(1+t_2d_t)(1+t_3d_t)$$

$$a_{22} = \{n^2\pi^2(1+t_1d_t)(1+t_2d_t)+a_1(1+t_2d_t)(1+t_5d_t)$$

$$+a_4(1+t_1d_t)(1+t_3d_t)(1+t_6d_t)+r_1^2(1+t_1d_t)(1+t_3d_t)d_t^2\}$$

$$f_1 = -\frac{384}{5}(1+t_1d_t)(1+t_5d_t)q_n+(a_8+t_8d_t)(1+t_5d_t)n\pi\{w(1,t)\cos n\pi - w(0,t)\}$$

$$f_2 = a_1(1+t_2d_t)(1+t_3d_t)\{w(1,t)\cos n\pi - w(0,t)\}$$

$$-\frac{48}{5}(1+t_1d_t)(1+t_3d_t)\{M(1,t)\cos n\pi - M(0,t)\}$$

$$a_8 = a_1 - a_2$$

$$t_8 = a_1t_2 - a_2t_1 \tag{8.134}$$

We can now take the Laplace transform of (8.133).

$$\overline{A}\overline{q}_n = \overline{f}+\overline{B}q_n(0)+\overline{C}\dot{q}_n(0)+\overline{D}\ddot{q}_n(0)+\overline{E}\dddot{q}_n(0) \tag{8.135}$$

where

$$\overline{A} = \begin{bmatrix} \overline{a}_{11} & \overline{a}_{12} \\ \overline{a}_{21} & \overline{a}_{22} \end{bmatrix}$$

$$\overline{B} = \begin{bmatrix} \overline{b}_{11} & \overline{b}_{12} \\ \overline{b}_{21} & \overline{b}_{22} \end{bmatrix}$$

$$\overline{C} = \begin{bmatrix} \overline{c}_{11} & \overline{c}_{12} \\ \overline{c}_{21} & \overline{c}_{22} \end{bmatrix}$$

$$\overline{D} = \begin{bmatrix} \overline{d}_{11} & \overline{d}_{12} \\ \overline{d}_{21} & \overline{d}_{22} \end{bmatrix}$$

$$\overline{E} = \begin{bmatrix} \overline{e}_{11} & \overline{e}_{12} \\ \overline{e}_{21} & \overline{e}_{22} \end{bmatrix} \tag{8.136}$$

In the above

$$\overline{a}_{11} = -[t_1t_5s^4+(t_1+t_5)s^3+(n^2\pi^2t_5t_8+a_3t_1t_4)s^2$$

$$+\{n^2\pi^2(t_8+a_8t_5)+a_3(t_1+t_4)\}s+a_8n^2\pi^2+a_3]$$

$$\overline{a}_{12} = n\pi a_1\{1+(t_2+t_5)s+t_2t_5s^2\}$$

$$\overline{a}_{21} = -n\pi a_1\{1+(t_2+t_3)s+t_2t_3s^2\}$$

$$\bar{a}_{22} = [r_1^2 t_1 t_3 s^4 + \{a_4 t_1 t_3 t_6 + r_1^2(t_1 + t_3)\} s^3$$

$$+\{n^2\pi^2 t_1 t_2 + a_1 t_2 t_3 + a_4(t_1 t_3 + t_3 t_6 + t_6 t_1)\} s^2$$

$$+\{n^2\pi^2(t_1 + t_2) + a_1(t_2 + t_3) + a_4(t_1 + t_3 + t_6)\} s + n^2\pi^2 + a_1 + a_4]$$

$$\bar{b}_{11} = a_{11,2} + a_{11,3}s + a_{11,4}s^2 + a_{11,5}s^3$$

$$\bar{b}_{12} = a_{12,2} + a_{12,3}s$$

$$\bar{b}_{21} = a_{21,2} + a_{21,3}s$$

$$\bar{b}_{22} = a_{22,2} + a_{22,3}s + a_{22,4}s^2 + a_{22,5}s^3$$

$$\bar{c}_{11} = a_{11,3} + a_{11,4}s + a_{11,5}s^2$$

$$\bar{c}_{12} = a_{12,3}$$

$$\bar{c}_{21} = a_{21,3}$$

$$\bar{c}_{22} = a_{22,3} + a_{22,4}s + a_{22,5}s^2$$

$$\bar{d}_{11} = a_{11,4} + a_{11,5}s$$

$$\bar{d}_{12} = 0; \quad \bar{d}_{21} = 0$$

$$\bar{d}_{22} = a_{22,4} + a_{22,5}s$$

$$\bar{e}_{11} = a_{11,5}$$

$$\bar{e}_{12} = 0; \quad \bar{e}_{21} = 0$$

$$\bar{e}_{22} = a_{22,5} \tag{8.137}$$

In the above $\bar{a}_{11}, \bar{a}_{12} \ldots$ polynomials are arranged in the form

$$\sum_{k=1}^{N+1} a_{ij,k}\, a^{k-1}$$

The transfer matrix of the system, being the inverse of the characteristic matrix, is given by

$$\bar{T} = [\bar{t}_{ij}] = \begin{bmatrix} \dfrac{\bar{a}_{22}}{\Delta} & -\dfrac{\bar{a}_{12}}{\Delta} \\ -\dfrac{\bar{a}_{21}}{\Delta} & \dfrac{\bar{a}_{11}}{\Delta} \end{bmatrix} \tag{8.138}$$

where

$$\Delta = \bar{a}_{11}\bar{a}_{22} - \bar{a}_{21}\bar{a}_{12}$$

$$= (n\pi a_1)^2 - (s^2 + a_1 n^2\pi^2)(n^2\pi^2 + a_1 + r_1^2 s^2) \tag{8.139}$$

The solution in the transformed plane is then given by

$$\bar{q}_n = \bar{T}\bar{f} + \bar{T}\bar{B}q_n(0) + \bar{T}\bar{C}\dot{q}_n(0) + \bar{T}\bar{D}\ddot{q}_n(0) + \bar{T}\bar{E}\dddot{q}(0) \tag{8.140}$$

The first term on the right hand side above gives the steady-state solution and the next four terms correspond to transient-response of the system. Let us consider some examples using the above analysis.

Example 1

Consider a viscoelastic Timoshenko beam on an elastic foundation with the following initial conditions and excitation force.

$$w(0,t) = 0$$

$$w(1,t) = 0$$

$$M(0,t) = 0$$

$$M(1,t) = 0$$

$$q(x,t) = e^{j\omega t}$$

The transform of the external force is

$$q_n = \int_0^1 e^{j\omega t} \sin n\pi x\, dx$$

$$= \frac{1-\cos n\pi}{n\pi} e^{j\omega t}$$

The excitations f_1 and f_2 are

$$f_1 = -\frac{384}{5} \frac{1-\cos n\pi}{n\pi}(1-j\omega t_1)(1-j\omega t_5)e^{j\omega t}$$

$$f_2 = 0$$

The steady-state response from (8.140) is

$$\bar{q}_n = \bar{T}\bar{f}$$

We can replace d_t by $j\omega$ in the above and obtain the steady-state response. Thus, the steady-state response is characterized by

$$w_n(j\omega) = \frac{-\bar{a}_{22}(j\omega)}{\Delta(j\omega)} \frac{384}{5} \frac{1-\cos n\pi}{n\pi}(1-j\omega t_1)(1-j\omega t_5)e^{j\omega t}$$

The magnitude of the above complex quantity represents the dynamic magnification factor and its modulus represents the phase angle by which the response lags behind the excitation in the nth mode. It will be difficult to make numerical calculations by hand for the problem above. A suitable computer program should be developed to determine the response of the system. The response obtained from such a program is plotted in Fig. 8.20 for two different values of r_1^2. The effect of r_1^2 is not appreciable, as this can be seen from the closeness of the two curves corresponding to two extreme cases of $r_1^2 = 0.0025$ and 0.0001.

Figure 8.21 shows the effect of viscoelasticity of the Timoshenko beam t_1 on its frequency response for a given value of r_0/l and t_2. It can be seen that the subcritical response is lower with lower values of t_1. As t_1 increases, the resonant frequency decreases, and the subcritical response increases while the supercritical response decreases. The reverse holds good for the parameter t_2, as can be seen from Fig. 8.22. This behaviour can be easily attributed to the fact that t_2 occurs in the numerator and t_1 occurs in the denominator of the viscoelasticity modulus operator of the beam.

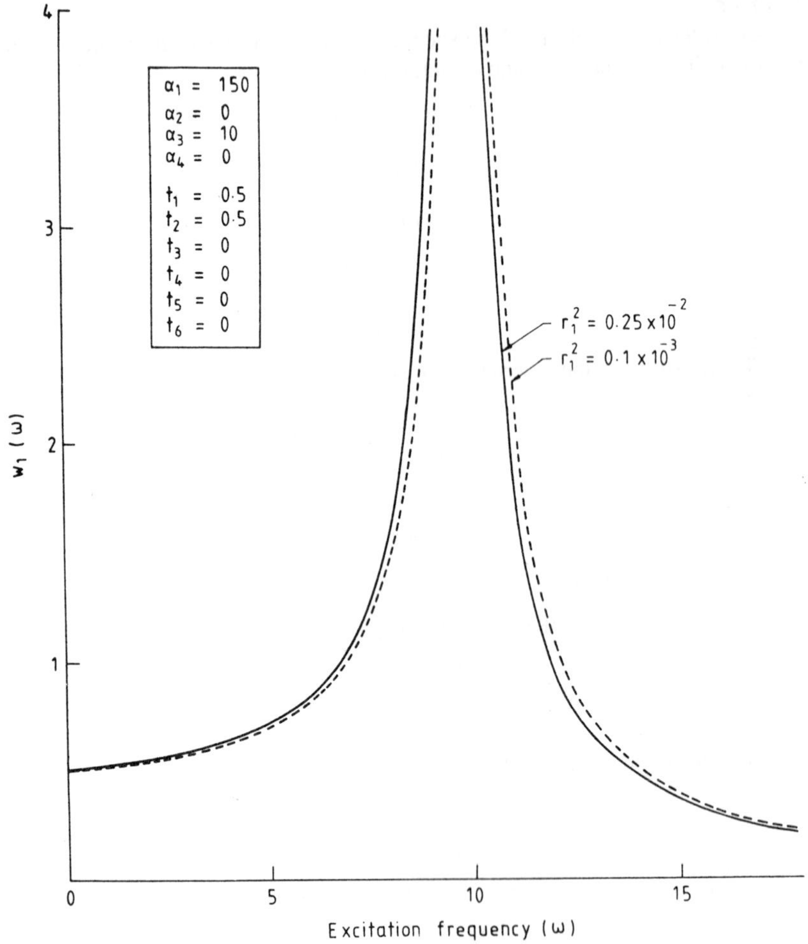

Fig. 8.20 Effect of r_0/l ratio on the frequency response of viscoelastic Timoshenko beam

Figure 8.23 shows the effect of axial force on the response of the viscoelastic Timoshenko beam. As the axial compressive force is increased the natural frequency decreases, which was also observed in section 8.6.

Example 2

Consider a force P moving on the beam from the left support $x = 0$, with a velocity v. The force can be defined as

$$q(x,t) = P\delta(x - vt) \quad 0 \le t \le \frac{l}{v}$$

$$= 0 \qquad \frac{l}{v} \le t \le \infty$$

We can nondimensionalize the force $P = q_0 l$ and $v \leftarrow vt_0/l$, then, the above becomes

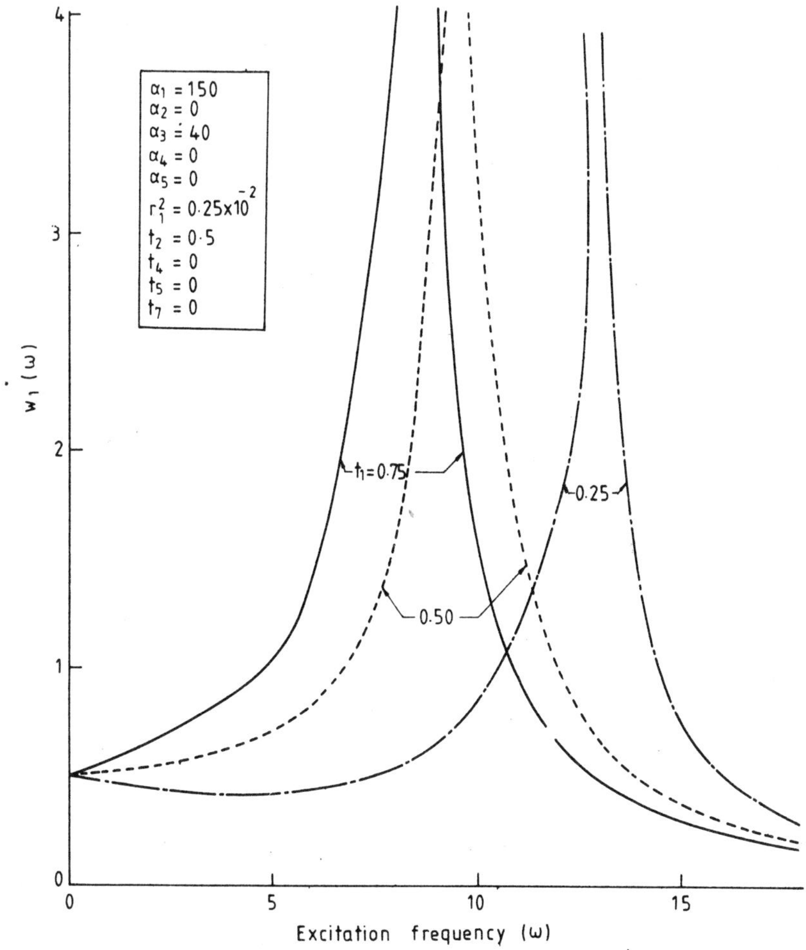

Fig. 8.21 Effect of viscoelasticity of a Timoshenko beam on its frequency response–effect of t_1

$$q(x,t) = \delta(x - vt) \quad 0 \le t \le \frac{l}{v}$$

$$= 0 \qquad \frac{l}{v} \le t \le \infty$$

The initial conditions of the problem are

$$w(x,0) = 0$$

$$\dot{w}(x,0) = 0$$

$$\psi(x,0) = 0$$

The sine transform of the external force is

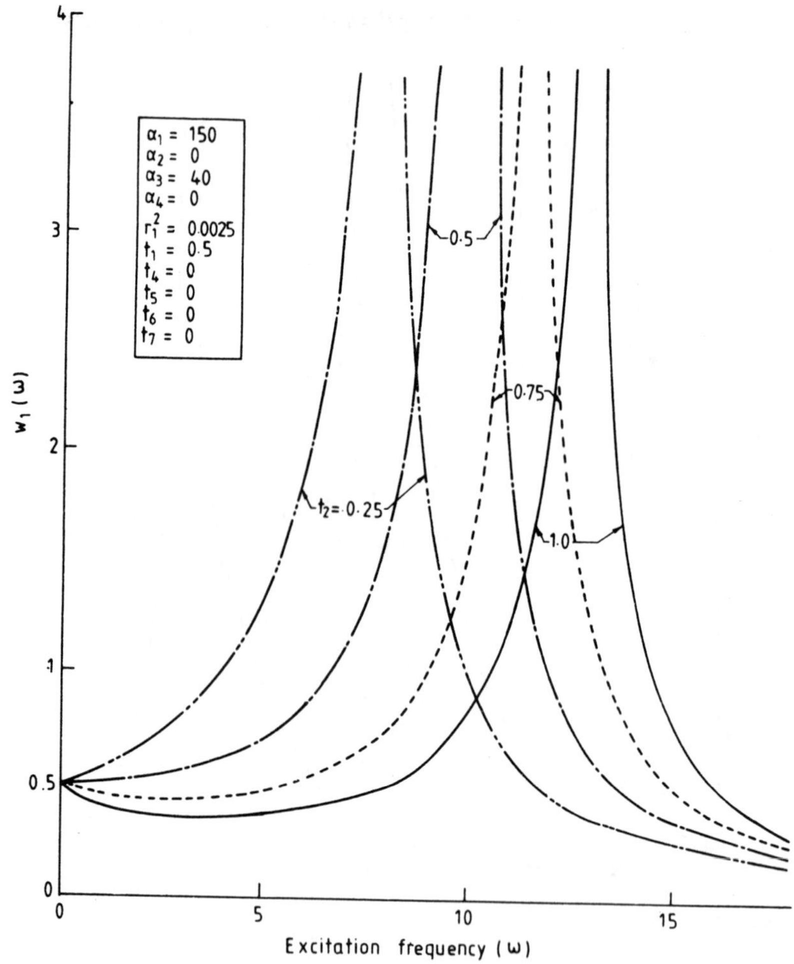

Fig. 8.22 Effect of viscoelasticity of a Timoshenko beam on its frequency response–effect of t_2

$$q_n(t) = \int_0^1 q \sin n\pi x \, dx$$

$$= H\left(\frac{1}{v} - t\right) \sin n\pi v t$$

Using the following formulae for Laplace transforms

$$L\left[H\left(\frac{1}{v} - t\right) \sin n\pi v t\right] = \frac{n\pi v}{s^2 + (n\pi v)^2}(1 - e^{-s/v} \cos n\pi)$$

$$L\left[H\left(\frac{1}{v} - t\right) \cos n\pi v t\right] = \frac{s}{s^2 + (n\pi v)^2}(1 - e^{-s/v} \cos n\pi)$$

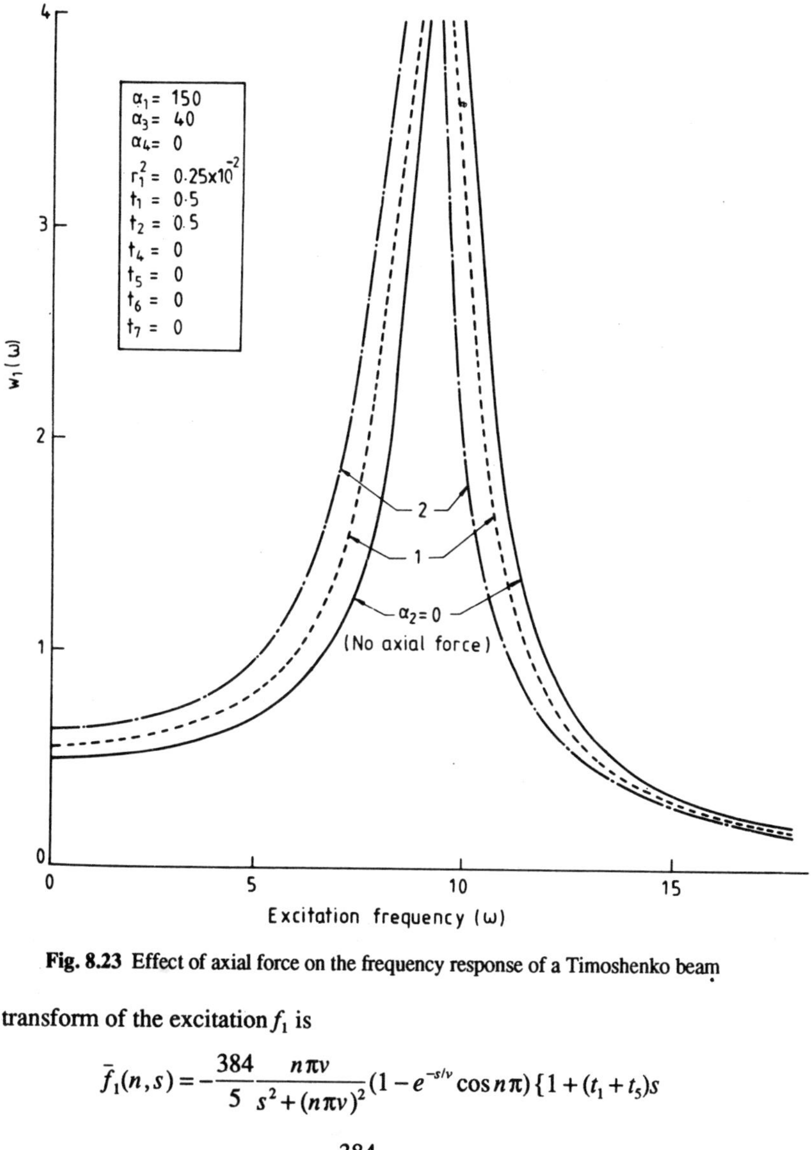

Fig. 8.23 Effect of axial force on the frequency response of a Timoshenko beam

the transform of the excitation f_1 is

$$\bar{f}_1(n,s) = -\frac{384}{5}\frac{n\pi v}{s^2+(n\pi v)^2}(1-e^{-s/v}\cos n\pi)\{1+(t_1+t_5)s$$

$$-t_1t_5(n\pi v)^2 + \frac{384}{5}t_1t_5 n\pi v e^{-s/v}\cos n\pi$$

Finally, the inversion of Laplace and Fourier transforms leads to the following solution of the beam subjected to the moving force.

$$w(x,t) = -\frac{384}{5}\sum_{1,2,\ldots} n\pi v \sin n\pi x F_2(n,t) \qquad 0 \le t \le \frac{1}{v}$$

$$w(x,t) = -\frac{384}{5}\sum n\pi v \sin n\pi x F_2(n,t)$$

$$+\frac{384}{5}\sum_{1,2,\ldots} n\pi v \sin n\pi x \cos n\pi x F_2\left(n, t-\frac{1}{v}\right)$$

$$+\frac{384}{5}\sum_{1,2,\ldots} n\pi v \sin n\pi x \cos n\pi x t_1 t_5 F_1\left(n, t-\frac{1}{v}\right) \quad \frac{1}{v}\le t\le\infty$$

where

$$F_1(n,t) = L^{-1}\frac{\overline{a}_{22}}{\Delta}$$

$$F_2(n,t) = L^{-1}\frac{\overline{a}_{22}}{\Delta}\,\frac{1+(t_1+t_5)s - t_1 t_5 (n\pi v)^2}{s^2+(n\pi v)^2}$$

The last two terms in $w(x, t)$ above represent the free vibration solution after the force has moved out of the beam. It will be difficult to make numerical calculations by hand for the problem above. A suitable computer program should be developed to determine the response of the system.

Figure 8.24 gives the forced vibration response obtained from such a program for a viscoelastic Timoshenko beam with no axial force resting on a viscoelastic foundation. As the force approaches the right support, the response reaches the maximum flat value. As the force leaves the beam, the response sharply decreases and is quickly damped out in the presence of t_4.

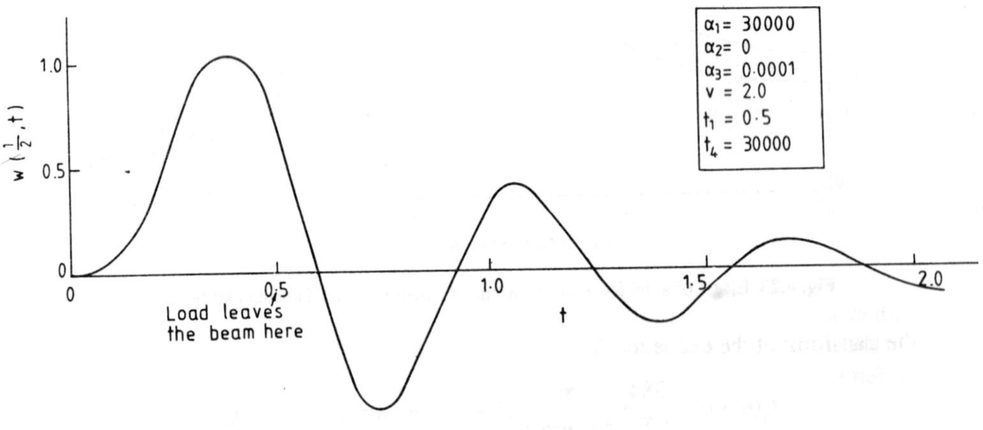

Fig. 8.24 Forced vibration of a viscoelastic Timoshenko beam subjected to a moving force

8.9 LAMINATED BEAMS

In the previous section we considered beams made of viscoelastic materials. Though viscoelastic materials have excellent damping properties, they are weak as structural members. Therefore, to provide both good damping and structural properties simultaneously, laminated beams with alternate layers of viscoelastic material have considerable advantage over the conventional single layer beam construction. We will first consider the case of a beam with n layers of thickness $2h$, with the material and cross-sectional properties E_i, G_i, A_i, and I_{yyi} as shown in Fig. 8.25a. A local coordinate system is used for each layer so that ith layer particles are located with respect to the frame of reference (x_i, y_i, z_i), with the x_i axis coinciding with the line of centroids of the ith layer as shown. The local frames are all referred to a global reference frame (x, y, z). We can refer to the elements in the ith layer either with respect to the local frame or with respect to the global frame. For convenience, we will consider the z axis of the global frame as well as local frames to coincide and let c_i be the z-coordinate of the origin of (x_i, y_i, z_i) frame. Then the transformation formulae for the coordinates of a particle P in the ith layer are

$$x = x_i$$

$$y = y_i$$

$$z = c_i + z_i \quad i = 1, 2, \ldots n \tag{8.141}$$

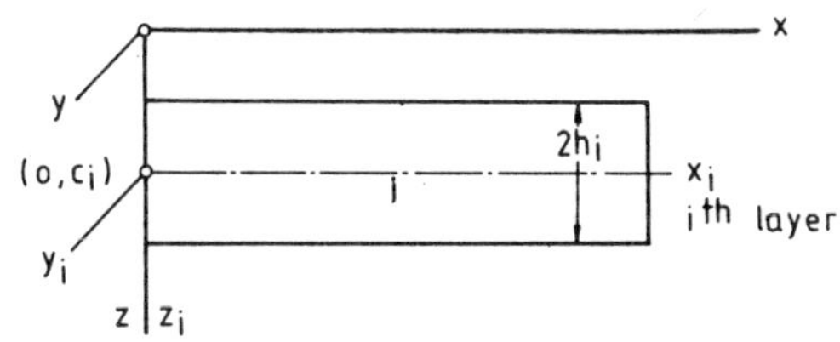

Fig. 8.25 (a) Geometry of layered beams

Let us assume Timoshenko model for each layer of the laminated beam, then the deformation field with respect to the local frame for the ith layer is

$$u_x(x_i, y_i, z_i, t) = u_i(x_i, t) - z_i\psi_i(x_i, t) \quad \text{(no sum)}$$

$$u_y = 0$$

$$u_z(x_i, y_i, z_i, t) = w_i(x_i, t) \tag{8.142}$$

where u_i, ψ_i are the axial deformation in ith centroidal layer and slope due to bending in the ith layer, respectively. We can transform the above into the global frame of reference (x, y, z) by using (8.141), i.e.,

$$u_x(x, y, z, t) = u_i(x, t) - (z - c_i)\psi_i(x, t)$$

$$u_y = 0$$

$$u_z(x, y, z, t) = w(x, t) \tag{8.142a}$$

Following Hamilton's principle we can obtain the differential equations of motion and the boundary conditions in the usual manner as given below. The strain field is

$$\varepsilon_{xx} = u_{i,x} - z_i\psi_{i,x}$$

$$\varepsilon_{xx} = \frac{1}{2}(w_{,x} - \psi_i)$$

rest $$\varepsilon = 0$$

The corresponding stresses are

$$\tau_{xx} = E(u_{i,x} - z_i \psi_{i,x})$$

$$\tau_{xx} = k'G(w_{,x} - \psi_i)$$

rest $$\tau = 0$$

The strain energy is

$$U = \frac{1}{2}\sum_{i=1}^{n} \int_0^l \{E_i I_{yyi} \psi_{i,x}^2 + k' A_i G_i (w_{,x} - \psi_i)^2 + E_i A_i u_{i,x}^2\}\, dx \tag{a}$$

The kinetic energy is

$$T = \frac{1}{2}\sum_{i=1}^{n} \int_0^l (\rho_i A_i \dot{w}^2 + \rho_i I_{yyi} \dot{\psi}_i^2 + \rho_i A_i \dot{u}_i^2) dx \tag{b}$$

The work done by external force $q\ (x, t)$ is

$$\int_0^l q w dx \tag{c}$$

Hamilton's principle now leads to

$$\Sigma \rho_i A_i \ddot{w} - \Sigma \partial_x \{k' G_i A_i (w_{,x} - \psi_i)\} = q$$

$$\rho_i I_{yyi} \ddot{\psi}_i - k' G_i A_i (w_{,x} - \psi_i) - \partial_x (E_i I_{yyi} \partial_x \psi_i) = 0$$

$$\rho_i A_i \ddot{u}_i - \partial_x (E_i A_i \partial_x u_i) = 0$$

$$\Sigma k' G_i A_i (w_{,x} - \psi_i) \delta w \,\big|_0^l = 0$$

$$E_i I_{yyi} \partial_x \psi_i \delta \psi_i \,\big|_0^l = 0$$

$$E_i A_i \partial_x u_i \delta u_i \,\big|_0^l = 0 \tag{8.143}$$

From the above we note that the lateral motion and axial motions are decoupled. This is obvious because, we have not provided any bonding between the interfaces of the alternate layers. We can use the method of Lagrangian multipliers to provide for the constraints and derive the corresponding differential equations of motion and the boundary conditions. This is illustrated first for the simple case of obtaining the Euler/Rayleigh theories from Timoshenko theory.

Euler/Rayleigh beam theories can be obtained from the fact that they are same as Timoshenko's theory with the additional nonholonomic constraint viz., the slope due to bending alone is same as the slope of the deflection curve, i.e.,

$$\psi = w_{,x}$$

However, we cannot directly obtain Rayleigh theory from Timoshenko theory by setting, the above relation as one of the boundary condition vanishes under this constraint. This difficulty is avoided by using the method of Lagrangian multipliers. Timoshenko's deformation field is

$$u_x = -z\psi(x,t)$$

$$u_y = 0$$

$$u_x = w(x,t) \tag{8.34}$$

We now provide the following constraint to the above

$$\lambda(w_{,x} - \psi) = 0 \tag{8.144}$$

where λ is an arbitrary Lagrangian multiplier function.

The above constraint contributes to the Lagrangian functional that is to be minimized. Therefore, we modify the strain energy function in equation (8.37).

$$U = \frac{1}{2}\int_0^l \{EI_{yy}\psi_{,x}^2 + k'AG(w_{,x} - \psi)^2 + \lambda(w_{,x} - \psi)\}\,dx \tag{8.145}$$

The kinetic energy remains same as in (8.38).

$$T = \frac{1}{2}\int_0^l (\rho A\dot{w}^2 + \rho I_{yy}\dot{\psi}^2)dx \tag{8.38}$$

The work done by external force $q\,(x, t)$ is given by equation (8.39).

$$\int_0^l qw\,dx \tag{8.39}$$

Hamilton's principle now leads to the following differential equations

$$\rho A\ddot{w} - \{k'AG(w_{,x} - \psi)\}_{,x} - \lambda_{,x} = q$$

$$\rho I_{yy}\ddot{\psi} - (EI_{yy}\psi_{,x})_{,x} - k'AG(w_{,x} - \psi) - \lambda = 0$$

$$\delta\lambda(w_{,x} - \psi) = 0 \tag{8.146}$$

and the boundary conditions

$$\{k'AG(w_{,x} - \psi) + \lambda\}\,\delta w\,|_0^l = 0$$

$$EI_{yy}\psi_{,x}\delta\psi\,|_0^l = 0 \tag{8.147}$$

For a uniform beam, we can express equation (8.146) in matrix form as

$$\begin{bmatrix} \rho A\partial_t^2 - k'GA\partial_x^2 & k'GA\partial_x & -\partial_x \\ -k'GA\partial_x & \rho I_{yy}\partial_t^2 + k'GA - EI_{yy}\partial_x^2 & -1 \\ \partial_x & -1 & 0 \end{bmatrix} \begin{Bmatrix} w \\ \psi \\ \lambda \end{Bmatrix} = \begin{Bmatrix} q \\ 0 \\ 0 \end{Bmatrix} \tag{8.148}$$

We can now eliminate ψ and λ to get the following differential equation containing only w.

$$(\rho A\partial_t^2 - \rho I_{yy}\partial_x^2\partial_t^2 + EI_{yy}\partial_x^4)w = q \tag{8.12a}$$

which is same as Rayleigh's equation (8.12) for a uniform beam.

The second equation of (8.148) gives

$$\lambda = (\rho I_{yy}\partial_x\partial_t^2 - EI_{yy}\partial_x^3)w \tag{d}$$

Substituting the above and the third equation of (8.148) in the first equation of (8.147), we get

$$(\rho I_{yy}\partial_x\partial_t^2 w - EI_{yy}\partial_x^3 w)\delta w\,|_0^l = 0 \tag{e}$$

The above equation (d) and the second of the equation (8.147) together give the same boundary conditions (8.13) for the Rayleigh beam of uniform cross-section. Thus, we can use the method of Lagrangian multipliers to include any constraints in the displacement field. This procedure will now be applied to include some constraints on the displacement field of a laminated beam with a specific arrangement of its layers.

Oberst's two-layer sandwiched beam theory

In this case there are only two layers in the laminated beam. In the top layer 1, we have the global reference frame (x, y, z) along the line of centroids and the layer 2 is at a distance c_2 as shown in Fig. 8.25b. It is assumed that the plane cross-sections of the beam continue to remain plane after deformation as in Euler's theory and that the displacements at the bonding surface of the two layers are continuous. Then the constraints are

$$\begin{aligned} \lambda_1(u_1 - u_2 - c_2\psi_2) &= 0 \\ \lambda_2(w_{,x} - \psi_1) &= 0 \\ \lambda_3(w_{,x} - \psi_2) &= 0 \end{aligned} \tag{8.149}$$

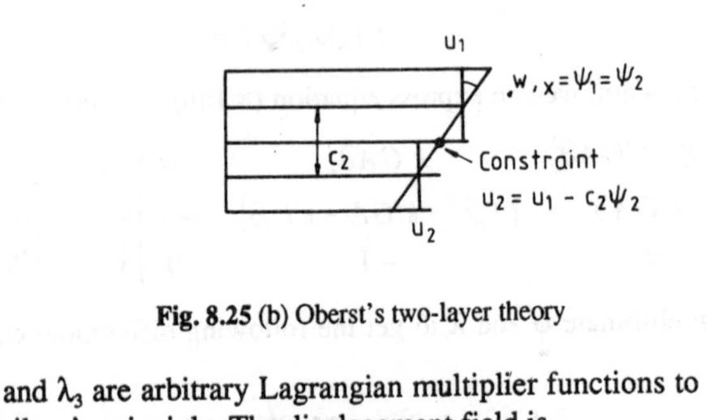

Fig. 8.25 (b) Oberst's two-layer theory

where λ_1, λ_2 and λ_3 are arbitrary Lagrangian multiplier functions to be determined by using Hamilton's principle. The displacement field is

$$u_{x1} = u_1 - z_1\psi_1$$

$$u_{x2} = u_2 - z_2\psi_2$$

$$u_y = 0$$

$$u_z = w(x,t) \tag{8.150}$$

The strain energy can be shown to be

$$U = \frac{1}{2}\sum_{i=1}^{2}\int_0^l \{E_i I_{yyi}\psi_{i,x}^2 + k'G_iA_i(w_{,x} - \psi_i)^2 + E_iA_iu_{i,x}^2\}\,dx \tag{8.151}$$

The contribution due to the constraints to the functional to be minimized is

$$D = \int_0^l \{\lambda_1(u_1 - u_2 - c_2\psi_2) + \lambda_2(w_{,x} - \psi_1) + \lambda_3(w_{,x} - \psi_2)\}\,dx \tag{8.152}$$

The kinetic energy is

$$T = \frac{1}{2}\sum_{i=1}^{2}\int_0^l (\rho_iA_i\dot{w}^2 + \rho_iI_{yyi}\dot{\psi}_i^2 + \rho_iA_i\dot{u}_i^2)dx \tag{8.153}$$

The work done by the external force is

$$W = \int_0^l qw\,dx \tag{8.154}$$

We can now form Hamilton's principle that

$$\delta\int_{t1}^{t2}(U + D - T - W)dt = 0 \tag{8.155}$$

and obtain the following differential equations and the boundary conditions.

$$\rho_1A_1\ddot{u}_1 - \partial_x(E_1A_1\partial_xu_1) + \lambda_1 = 0$$

$$E_1A_1\partial_xu_1\delta u_1\,|_0^l = 0$$

$$\rho_2A_2\ddot{u}_2 - \partial_x(E_2A_2\partial_xu_2) + \lambda_1 = 0$$

$$E_2A_2\partial_xu_2\delta u_2\,|_0^l = 0$$

$$\sum_{i=1}^{2}\rho_iA_i\ddot{w} - \sum_{i=1}^{2}k'G_iA_i(w_{,x} - \psi_i)_{,x} - (\lambda_{2,x} + \lambda_{3,x}) = q$$

$$\left\{\lambda_2 + \lambda_3 + \sum_{i=1}^{2}k'G_iA_i(w_{,x} - \psi_i)\right\}_0^l = 0$$

$$\rho_1\ddot{\psi}_1 - (E_1I_{yy1}\psi_{1,x})_{,x} - \lambda_2 = 0$$

$$E_1I_{yy1}\psi_{1,x}\delta\psi_1\,|_0^l = 0$$

$$\rho_2\ddot{\psi}_2 - (E_2I_{yy2}\psi_{2,x})_{,x} - \lambda_2 = 0$$

$$E_2I_{yy2}\psi_{2,x}\delta\psi_2\,|_0^l = 0$$

$$\delta\lambda_1(u_1 - u_2 - c_2\psi_2) = 0$$

$$\delta\lambda_2(w_{,x} - \psi_1) = 0$$

$$\delta\lambda_3(w_{,x} - \psi_2) = 0$$

We can now eliminate u_2, ψ_1, ψ_2, λ_1, λ_2 and λ_3 to obtain the following equations of motion and the boundary conditions.

$$(\rho_1 A_1 + \rho_2 A_2)\ddot{u}_1 - \partial_x(E_1 A_1 + E_2 A_2)\partial_x u_1$$

$$- \rho_2 A_2 c_2 \partial_x \ddot{w} + \partial_x\{E_2 A_2 \partial_x(c_2 \partial_x w)\} = 0$$

$$(\rho_1 A_1 + \rho_2 A_2)\ddot{w} + \partial_x^2(E_1 A_1 + E_2 A_2)\partial_x^2 w - \partial_x(\rho_1 I_{yy1} + \rho_2 I_{yy2})\partial_x \ddot{w}$$

$$+ \partial_x(c_2 \rho_1 A_1 \ddot{u}_1) - \partial_x\{c_2 \partial_x(E_1 A_1 \partial_x u_1)\} = q$$

$$E_1 A_1 \partial_x u_1 \delta u_1 \,|_0^l = 0$$

$$E_2 A_2 \partial_x u_2 \delta u_2 \,|_0^l = 0$$

$$\{(\rho_1 I_{yy1} + \rho_2 I_{yy2})\partial_x \ddot{w} - \partial_x(E_1 I_{yy1} + E_2 I_{yy2})\partial_x^2 w$$

$$- c_z \rho_1 A_1 \ddot{u}_1 + c_2 \partial_x(E_1 A_1 \partial_x u_1)\}\,\delta w \,|_0^l = 0$$

$$(E_1 I_{yy1} + E_2 I_{yy2})\partial_x^2 \delta w_{,x} \,|_0^l = 0$$

We may like to note here that the above could have been derived in a simpler way by assuming Euler field to start with.

Ditaranto's three layer sandwiched beam theory

In this theory three layers are considered, as shown in Fig. 8.26, the outer two layers elastic and the middle layer viscoelastic. The thickness of each layer is taken as $2h_i$ and the breadth of all the layers as b. The outer two layers are considered to be Euler type and the middle layer is subjected to lateral bending as well as shear in the axial plane as shown in Fig. 8.26. The figure also shows the continuity relations between the two interfaces. The force resultant in the middle viscoelastic layer can be considered negligible as its elastic modulus is very small compared to the moduli of the outer two layers. The constraining assumptions for the three-layer beam can therefore be given as

(1) Equilibrium of normal forces

$$E_1 A_1 u_{1,x} + E_3 A_3 u_{3,x} = 0$$

(2) Continuity of axial displacements at the top interface

$$u_2 - u_1 + h_1\psi_1 + h_2(\psi_2 + \phi_2) = 0$$

(3) Continuity of axial displacements at the lower interface

$$u_2 - u_3 - h_2(\psi_2 + \phi_2) + h_3\psi_3 = 0$$

(4) Euler model for top layer

$$w_{,x} - \psi_1 = 0$$

(5) Assumed shear stress-strain relation for the middle layer

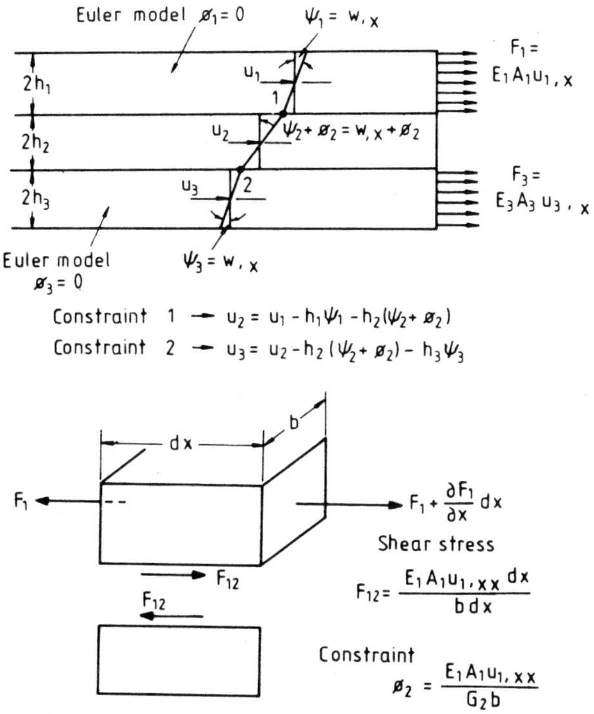

Fig. 8.26 Ditaranto's three-layer theory

$$\phi_2 = \frac{E_1A_1}{Gb}u_{1,xx} = 0$$

where G is the shear modulus of the middle layer.

(6) Euler model for bottom layer

$$w_{,x} = \psi_3 = 0 \tag{8.156}$$

From condition 1 in (8.156) above we get

$$u_3 = -\frac{K_1}{K_3}u_1 \tag{f}$$

where

$$K_i = E_iA_i \quad \text{(no sum)}$$

From conditions 2 and 3 in (8.156) and using the above relation (f), we get

$$Su_1 - \Delta w_{,x} - 2h_2\phi_2 = 0 \tag{g}$$

where

$$S = \frac{K_1 + K_3}{K_3}$$

$$\Delta = h_1 + 2h_2 + h_3$$

With the help of condition 5 of equation (8.156), the above equation (g) gives

$$\left(S - \frac{1}{R}\partial_x^2\right)u_1 = \Delta w_{,x} \tag{h}$$

where

$$R = \frac{Gb}{2h_2K_1}$$

The displacement field for the system is assumed as

$$u_{x1} = u_1 - z_1 w_{,x}$$

$$u_{x2} = u_2 - z_2(w_{,x} + \phi_2)$$

$$u_{x3} = u_3 - z_3 w_{,x}$$

$$u_y = 0$$

$$u_z = w$$

The strain field is then given by

$$\varepsilon_{xx1} = u_{1,x} - z_1 w_{,xx}$$

$$\varepsilon_{xx2} = u_{2,x} - z_2(w_{,xx} + \phi_{2,x})$$

$$\varepsilon_{xx3} = u_{3,x} - z_3 w_{,xx}$$

$$\varepsilon_{xz2} = -\frac{1}{2}\phi_2$$

The stress field then can be obtained as

$$\tau_{xx1} = E_1(u_{1,x} - z_1 w_{,xx})$$

$$\tau_{xx2} = E_2\{u_{2,x} - z_2(w_{,xx} + \phi_{2,x})\}$$

$$\approx 0$$

$$\tau_{xx3} = E_3(u_{3,x} - z_3 w_{,xx})$$

$$\tau_{xz2} = -G\phi_2$$

$$\approx 0$$

In the above τ_{xx2}, τ_{xz2} are neglected for the viscoelastic material in comparison with the corresponding quantities of the elastic outer layers for the purpose of calculation of the strain energy of the system. The strain energy of the system thus is given by

$$U = \frac{1}{2}\int_0^l K_1 S u_{,x}^2 dx + \frac{1}{2}\int_0^l (B_1 + B_3) w_{,xx}^2 dx$$

where

$$B_i = E_i I_{yyt} \quad \text{(no sum)}$$

The contribution of the constraint (h) to the strain energy is

$$D = \int_0^l \lambda\left\{\left(S - \frac{1}{R}\partial_x^2\right)u_1 - \Delta w_{,x}\right\}dx$$

The kinetic energy is

$$T = \frac{1}{2}\int_0^l \rho A \dot{w}^2 dx$$

Applying Hamilton's principle to the above, we get

$$\delta\int_{t1}^{t2}\left[\frac{1}{2}\int_0^l K_1 S u_{1,x}^2 dx + \frac{1}{2}\int_0^l (B_1 + B_3) w_{,xx}^2 dx\right.$$

$$\left. + \int_0^l \lambda\left\{\left(S - \frac{1}{R}\partial_x^2\right)u_1 - \Delta w_{,x}\right\}dx - \frac{1}{2}\int_0^l \rho A \dot{w}^2 dx\right]dt = 0$$

i.e.,

$$\int_{t1}^{t2}\left[\int_0^l K_1 S u_{1,x}\delta u_{1,x} dx + \int_0^l (B_1 + B_3)\, w_{,xx}\,\delta w_{,xx} dx - \int_0^l \rho A \dot{w}\delta\dot{w} dx\right.$$

$$+ \int_0^l \lambda S\, \delta u_1 dx - \int_0^l \frac{\lambda}{R}\delta u_{1,xx} dx - \int_0^l \lambda\Delta\delta w_{,x} dx$$

$$\left. + \int_0^l \left\{\left(S - \frac{1}{R}\partial_x^2\right)u_1 - \Delta w_{,x}\right\}\delta\lambda dx\right]dt = 0$$

The above can be evaluated by integrating each term by parts.

$$I_1 = \int_{t1}^{t2}\left[\int_0^l K_1 S u_{1,x}\delta u_{1,x} dx\right]dt$$

$$= \int_{t1}^{t2}\left[K_1 S u_{1,x}\delta u_1 \big|_0^l - \int_0^l K_1 S u_{1,xx}\delta u_1 dx\right]dt$$

$$I_2 = \int_{t1}^{t2}\left[\int_0^l (B_1 + B_3) w_{,xx}\delta w_{,xx} dx\right]dt$$

$$= \int_{t1}^{t2}\left[(B_1 + B_3) w_{,xx}\delta w_{,x}\big|_0^l - (B_1 + B_3) w_{,xxx}\delta w\big|_0^l + \int_0^l (B_1 + B_3) w_{,xxxx}\delta w dx\right]dt$$

$$I_3 = -\int_{t1}^{t2}\left[\int_0^l \rho A \dot{w}\delta\dot{w} dx\right]dt$$

$$= \int_{t1}^{t2}\left[\int_0^l \rho A \ddot{w}\delta w dx\right]dt$$

$$I_4 = \int_{t1}^{t2}\int_0^l \lambda S\delta u_1 dx\, dt$$

$$I_5 = -\int_{t1}^{t2}\int_0^l \frac{\lambda}{R}\delta u_{1,xx} dx\, dt$$

$$= \int_{t1}^{t2} \left[-\frac{\lambda}{R} \delta u_{1,x} |_0^l + \frac{\lambda_{,x}}{R} \delta u_1 |_0^l - \int_0^l \frac{\lambda_{,xx}}{R} \delta u_1 dx \right] dt$$

$$I_6 = -\int_{t1}^{t2} \int_0^l \lambda\Delta\, \delta w_{,x} dx\, dt$$

$$= \int_{t1}^{t2} \left[-\lambda\Delta\delta w |_0^l + \int_0^l \lambda_{,x} \Delta\delta w\, dx \right] dt$$

$$I_7 = \int_{t1}^{t2} \int_0^l \left\{ \left(S - \frac{1}{R} \partial_x^2 \right) u_1 - \Delta w_{,x} \right\} \delta\lambda dx] dt$$

Hence, we get the following differential equations and boundary conditions for the system.

$$-K_1 S u_{1,xx} + \lambda S - \frac{\lambda_{,xx}}{R} = 0$$

$$(B_1 + B_3) w_{,xxxx} + \rho A \ddot{w} + \Delta\lambda_{,x} = q$$

$$\left(S - \frac{1}{R} \partial_x^2 \right) u_1 - \Delta w_{,x} = 0 \tag{8.157}$$

$$\left(K_1 S u_{1,x} + \frac{\lambda_{,x}}{R} \right\} \delta u_1 |_0^l - \frac{\lambda}{R} \delta u_{1,x} |_0^l = 0$$

$$(B_1 + B_3) w_{,xx} \delta w_{,x} |_0^l - \{(B_1 + B_3) w_{,xxx} + \lambda\Delta\} \delta w |_0^l = 0 \tag{8.158}$$

Neglecting the shear term in the first equation of (8.157), we can write them in matrix form as

$$\begin{bmatrix} -K_1 S \partial_x^2 & 0 & S \\ 0 & (B_1 + B_3)\partial_x^4 + \rho A \partial_t^2 & \Delta\partial_x \\ \left(S - \frac{1}{R} \partial_x^2 \right) & -\Delta\partial_x & 0 \end{bmatrix} \begin{Bmatrix} u_1 \\ w \\ \lambda \end{Bmatrix} = \begin{Bmatrix} 0 \\ P \\ 0 \end{Bmatrix} \tag{8.159}$$

We can eliminate w, λ in (8.159) above and obtain the following differential equation.

$$u'''''' - \left(\frac{R K_1 \Delta^2}{B_1 + B_3} - RS \right) u'''' - \frac{R S \rho A}{B_1 + B_3} \ddot{u} + \frac{\rho A}{B_1 + B_3} \ddot{u}'' = 0 \tag{8.160}$$

n-layer theory

We consider a general layered beam with n layers. The displacements are taken to be continuous across the interfaces, which of course disturbs the symmetry of the system. Let u_i, u_{i+1} denote the axial displacements of the top and bottom surfaces of ith layer. We will consider the local coordinate system for each layer and a known reference frame for the n-layered beam. We choose the following field for the ith layer, see Fig. 8.27.

$$u_x = \frac{z_i + h_i}{2h_i} u_{i+1} - \frac{z_i - h_i}{2h_i} u_i$$

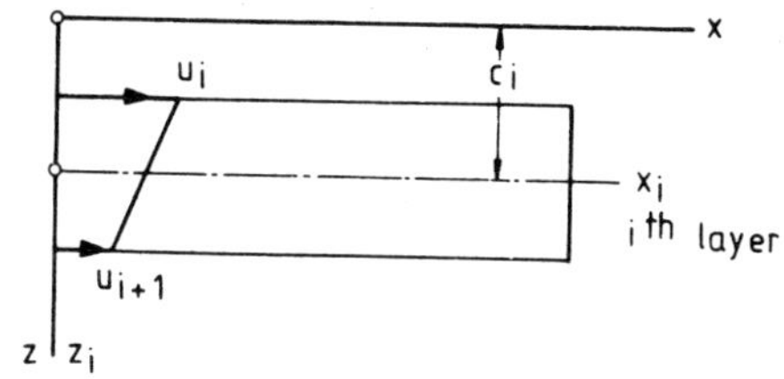

Fig. 8.27 Geometry of deformation of N–layer sandwich beams

$$u_y = 0$$

$$u_z = w \tag{8.161}$$

The strains in *i*th layer are

$$\varepsilon_{xx} = \frac{z_i + h_i}{2h_i} u_{i+1,x} - \frac{z_i - h_i}{2h_i} u_{i,x}$$

$$\varepsilon_{xz} = \frac{1}{2}\left(w_{,x} + \frac{u_{i+1} - u_i}{2h_i} \right)$$

rest

$$\varepsilon = 0 \tag{8.162}$$

Then, the stresses are

$$\tau_{xx} = E_i \left\{ \frac{z_i + h_i}{2h_i} u_{i+1,x} - \frac{z_i - h_i}{2h_i} u_{i,x} \right\}$$

$$\tau_{xz} = G_i \left(w_{,x} + \frac{u_{i+1} - u_t}{2h_i} \right)$$

rest

$$\tau = 0 \tag{8.163}$$

The energy expressions are

$$U = \frac{1}{2} \sum_{i=1}^{n} \int_0^l \left\{ E_i a_i \, (u_{i+1,x} + u_{i,x})^2 \right.$$

$$\left. - 2E_i \alpha_i u_{i+1,x} u_{i,x} + G_i A_i \left(w_{,x} + \frac{u_{i+1} - u_i}{2h_i} \right)^2 \right\} dx$$

$$T = \frac{1}{2} \sum_{i=1}^{n} \int_0^l \{ \rho_i a_i \, (\dot{u}_{i+1} + \dot{u}_l)^2 - 2\rho_i \alpha_i \dot{u}_{i+1} \dot{u}_i + \rho_i A_i \dot{w}^2 \} \, dx$$

$$W = \int_0^l q w dx \tag{8.164}$$

where

$$a_i = \frac{I_{yyi} + A_i h_i^2}{4h_i^2}$$

$$\alpha_i = \frac{I_{yyi} + A_i h_i^2}{4h_i^2}$$

We can now apply Hamilton's principle and obtain the following differential equations and boundary conditions.

$$\sum_{i=1}^{n} \rho_i A_i \ddot{w} - \sum_{i=1}^{n} \partial_x \left\{ G_i A_i \left(w_{,x} + \frac{u_{i+1} - u_i}{2h_i} \right) \right\} = q$$

$$(\rho_i a_i + \rho_{i-1} a_{i-1})\ddot{u}_i - (\rho_i \alpha_i \ddot{u}_{i+1} + \rho_{i-1}\alpha_{i-1}\ddot{u}_{i-1}$$

$$-\partial_x \{(E_i a_i + E_{i-1} a_{i-1}) u_{i,x} - E_i \alpha_i u_{i+1,x} - E_{i-1}\alpha_{i-1} u_{i-1,x}\} = 0 \qquad (8.165)$$

$$\sum_{i=1}^{n} G_i A_i \left(w_{,x} + \frac{u_{i+1} - u_i}{2h_i} \right) \delta w \,\Big|_0^l = 0$$

$$\{(E_i a_i + E_{i-1} a_{i-1}) u_{i,x} - E_i \alpha_i u_{i+1,x} - E_{i-1}\alpha_{i-1,x} u_{i-1,x}\} \,\delta u_i \,\Big|_0^l = 0 \qquad (8.166)$$

It should be noted that

$$u_0 = 0\text{:}\ \rho_0, E_0, \cdots = 0\text{:}\ \rho_{n+1}, E_{n+1}' \cdots = 0$$

Let the layers be incompressible Kelvin solid type. Then the elastic moduli are to be replaced by

$$E_i \leftarrow E_i\,(1 + t_{1i} f(i)\partial_t)$$

$$G_i \leftarrow G_i\,(1 + t_{2i} f(i)\partial_t) \qquad (8.167)$$

where $f(i)$ is a function that depends on the arrangement of elastic and viscoelastic layers. Let the first layer be elastic, the next layer viscoelastic, and that they alternate subsequently, then

$$f(i) = \frac{1 + (-1)^i}{2}$$

If the first layer is viscoelastic, then $f(i)$ is

$$f(i) = \frac{1 - (-1)^i}{\cdot 2}$$

Steady-state solution for laminated beams

The differential equations governing the lateral motion of a laminated beam from equation (8.143) can be written as

$$\Sigma \rho_i A_i \ddot{w} - \Sigma \partial x \,\{k' G_i A_i\,(w_{,x} - \psi_i)\} = q$$

$$\rho_i I_{yyi} \ddot{\psi}_i - k' G_i A_i\,(w_{,x} - \psi_i) - \partial_x\,(E_i I_{yyi} \partial_x \psi_i) = 0$$

$$\Sigma k' G_i A_i\,(w_{,x} - \psi_i)\,\delta w \,\Big|_0^l = 0$$

$$E_i I_{yyi} \partial_x \psi_i \delta\psi_i \,\Big|_0^l = 0 \qquad (8.168)$$

In the above we assume that the viscoelastic layers are almost incompressible and are alternating with elastic layers.

$$E_i = E_i \{1 + t_i f(i)\partial_t\}$$

$$G_i = G_i \{1 + t_i f(i)\partial_t\} \tag{8.169}$$

where $f(i)$ is a function that depends on the arrangement of elastic and viscoelastic layers. If the first layer is elastic, the next layer viscoelastic, and that they alternate subsequently, then

$$f(i) = \frac{1 + (-1)^i}{2} \tag{8.170a}$$

If the first layer is viscoelastic, then $f(i)$ is

$$f(i) = \frac{1 - (-1)^i}{2} \tag{8.170b}$$

We can nondimensionalize the above by setting

$$x \leftarrow \frac{x}{l}$$

$$t \leftarrow \frac{t}{t_0}$$

$$w \leftarrow \frac{w}{l}$$

$$\psi_i = \frac{\psi_i}{\psi_{i0}}$$

$$q \leftarrow \frac{q}{q_0}$$

$$a_i = \frac{k' G_i A_i t_0^2}{\Sigma \rho_j A_j l^2}$$

$$b_i = \frac{G_i t_0^2}{\rho_i l^2}$$

$$\psi_{i0} = \frac{l}{r_i}$$

$$r_i^2 = \frac{\rho_i I_{yyi}}{\Sigma \rho_j A_j}$$

$$q_0 = \Sigma \frac{\rho_i A_i l}{t_0^2} \tag{8.171}$$

In the above a_i, b_i are measures of velocity of propagation of dilatational wave and longitudinal waves in ith layer, respectively. Then

$$\ddot{w} - \Sigma E_i\,(\partial_t)\,a_i\partial_x^2 w + \Sigma E_i\,(\partial_t)a_i\psi_{i0}\partial_x\psi_i = q$$

$$\ddot{\psi}_i - E_i\,(\partial_t)\,a_i\psi_{i0}\partial_x w + E_i\,(\partial_t)\,(a_i\psi_{i0}^2 - b_i\partial_x^2)\,\psi_i = 0$$

$$i = 1, 2, \cdots n \tag{8.172}$$

Properties of the elastic layered system

The above $n + 1$ equations are in the form of Timoshenko beam equations. First we examine the properties of the corresponding elastic layers system. We can express these equations of (8.172) in a matrix form.

$$M\ddot{u} + Ku = f \tag{8.173}$$

where

$$u = \begin{Bmatrix} w \\ \psi_1 \\ \cdot \\ \cdot \\ \psi_n \end{Bmatrix} \quad f = \begin{Bmatrix} q \\ 0 \\ \cdot \\ \cdot \\ 0 \end{Bmatrix}$$

$M = I$ the identity matrix operator and

$$K = \begin{bmatrix} -a_i\partial_x^2 & a_1\psi_{10}\partial x & a_2\psi_{20}\partial_x & \cdots & \cdots \\ a_1\psi_{10}\partial_x & a_1\psi_{10}^2 - b_1\partial_x^2 & 0 & 0 & \cdots \\ a_2\psi_{20}\partial_x & 0 & a_2\psi_{20}^2 - b_2\partial_x^2 & 0 & \cdots \\ a_3\psi_{30}\partial x & 0 & 0 & a_3\psi_{30}^2 - b_3\partial_x^2 & \cdots \\ \cdots & \cdots & \cdots & \cdots & \cdots \end{bmatrix} \tag{8.174}$$

We can show that the stiffness operator is symmetric with respect to a class of functions that satisfy the boundary conditions

$$(\Sigma G_i A_i \partial_x w - \Sigma G_i A_i \psi_{i0}\psi_i)\delta w \,|_0^1 = 0 \tag{8.175}$$

Consider the scalar product of Ku with respect to v, where $u = u\,(x)$ and $v = v\,(x)$ are arbitrary functions satisfying the boundary conditions above. Then

$$\int_0^1 (Ku)\,v dx = \int_0^1 (k_{ij}u_j)\,v_i dx$$

$$= \int_0^1 -\Sigma a_i\,(\partial_x^2 u_1)\,v_1 dx + \int_0^1 a_1\psi_{10}\,(\partial_x u_2)v_1 dx + \int_0^1 a_2\psi_{20}\,(\partial_x u_2)\,v_1 dx$$

$$+ \int_0^1 -a_1\psi_{10}\,(\partial_x u_1)\,v_2 dx + \int_0^1 \{(a_1\psi_{10}^2 - b_1\partial_x^2)\,u_2\}\,v_1 dx + \cdots$$

$$+ \int_0^1 -a_2\psi_{20}\,(\partial_x u_1)\,v_3 dx + \cdots + \int_0^1 \{(a_2\psi_{20}^2 - b_2\partial_x^2)\,u_3\}\,v_3 dx$$

$$+ \cdots \tag{8.176}$$

Integrating the above by parts, we get

$$\int_0^1 (Ku)\, v\, dx =$$

$$= -[\Sigma a_i\,(\partial_x u_1 + v_1 + u_1\partial_x v_1)] + [a_1\psi_{10}u_2v_1] +$$

$$-[a_1\psi_{10}u_1v_2] + [-b_1\,(\partial_x u_2 v_2) + b_1\,(u_2\partial_x v_2)] +$$

$$+\cdots$$

$$+\cdots$$

$$\int_0^1 \Sigma a_i\,(\partial_x^2 v_1)v_1 dx - \int_0^1 a_1\psi_{10}\,(\partial_x v_1)\,u_2 dx -$$

$$+\int_0^1 a_1\psi_{10}\,(\partial_x v_2)\,u_1 dx + \int_0^1 \{(a_1\psi_{10}^2 - b_1\partial_x^2)v_2\}\; u_2 dx$$

$$+\cdots \tag{8.177}$$

The first series of terms in the above vanish in view of the boundary conditions, while the second series represents $\int uKvdx$. Hence

$$\int_0^1 (Ku)\, v\, dx = \int_0^1 u\,(Kv)\, dx$$

The symmetry of Timoshenko beam operator is a special case of the above. Now we can solve the forced vibration problem of a laminated beam with alternate viscoelastic layers by using the eigen function expansion process. Let us consider the case of a simply-supported beam. For steady-state solution, we assume

$$q\,(x,t) = q\,(x)\,e^{j\omega t}$$

$$w\,(x,t) = w\,(x)\,e^{j\omega t}$$

$$\psi_i\,(x,t) = \psi_i\,(x)\,e^{j\omega t}$$

Then, equation (8.172) becomes

$$-\omega^2 w - \Sigma E_t\,(j\omega)\,a_i d_x^2 w - \Sigma E_i\,(j\omega)\,a_i\psi_{i0} d_x\psi_i = q$$

$$-\omega^2\psi_i - E_t\,(j\omega)\,a_i\psi_{i0} d_x w + E_i\,(j\omega)\,(a_i\psi_{i0}^2 - b_i d_x^2)\,\psi_i = 0$$

$$i = 1, 2, \cdots n \tag{8.178}$$

The mode shapes of simply supported layered beams are given by

$$w_k\,(x) = \sin k\pi x$$

$$\psi_{ik}\,(x) = \cos k\pi x$$

and the eigen function expansion solution is same as using finite Fourier transforms of equations (8.178). Thus, multiplying the first equation by $\sin k\pi x$ and the second equation by $\cos k\pi x$ of (8.178), and integrating over the region 0 to 1, we get

$$\{\Sigma E_i\,(j\omega)\,a_i\,(k\pi)^2 - \omega^2\}\,w_k - \Sigma E_t\,(j\omega)a_i\psi_{i0}k\pi\psi_{ik} = q_k$$

$$-E_i\,(j\omega)\,a_i\psi_{i0}k\pi w_k - \{E_i\,(j\omega)\,(a_i\psi_{i0}^2 + b_i\,(k\pi)^2) - \omega^2\}\,\psi_{ik} = 0 \tag{8.179}$$

where

$$w_k = \int_0^1 w \sin k\pi x \, dx$$

$$\psi_{ik} = \int_0^1 \psi_i \cos k\pi x \, dx$$

$$q_k = \int_0^1 q \sin k\pi x \, dx$$

$$\int_0^1 d_x^2 w \sin k\pi x \, dx = -k^2\pi^2 w_k$$

$$\int_0^1 d_x \psi_i \sin k\pi x \, dx = -k\pi\psi_{ik}$$

$$\int_0^1 d_x^2 \psi_i \cos k\pi x \, dx = -k^2\pi^2 \psi_{ik}$$

Solving the system of the $n + 1$ equations in (8.179), we get

$$w_k = \frac{q_k}{\{\Sigma E_i\,(j\omega)\,a_i\,(k\pi)^2 - \omega^2\} - \sum_{i=1}^{n} \frac{E_i^2\,(j\omega)\,a_i^2\,\psi_{i0}^2\,(k\pi)^2}{\left(E_i\,(j\omega)\left(a_i\psi_{i0}^2 + b_i\,(k\pi)^2\right) - \omega^2\right)}}$$

$$\psi_{ik} = \frac{E_i(j\omega)\,a_i\psi_{i0}k\pi w_k}{E_i\,(j\omega)\,(a_i\psi_{i0}^2 + b_i\,(k\pi)^2) - \omega^2} \tag{8.189}$$

Using the above equations and the properties of the Fourier transforms, the steady state solution can be expressed in the following final form.

$$w\,(x,t) = \sum_0^{\infty} w_k \sin k\pi x\, e^{j\omega t}$$

$$\psi_i\,(x,t) = \sum_0^{\infty} \left(\psi_{ik} \cos k\pi x + \frac{1}{2}\psi_0\right) e^{j\omega t} \tag{8.190}$$

It is thus noted that $w_k \sin k\pi x$ denotes the contribution of kth mode to the amplitude of vibration, and w_k represents the dynamic magnification factor in kth mode vibration and contains information on the effectiveness of damping of the system.

The dynamic magnification factor of a three-layered beam with a viscoelastic layer in the middle and surrounded by elastic layers of uniform thickness, is given in Fig. 8.28. The maximum magnification in the beam can be considerably reduced by increasing the viscoelasticity of the middle layer.

8.10 COUPLED BENDING VIBRATIONS OF BEAMS

In all the earlier sections, we considered bending vibration of beams in one plane. They occur when the beam is straight, and the direction of external force coincides with one of the principal directions of the cross-section, along the entire length of the beam. If the xyz axes system is such that the y and z directions are principal axes of the beam along its whole length, the beam can vibrate only in z direction with displacement w, or the beam can have vibrations only in y direction with displacement v, depending on the direction of the external force in the z or y directions, respectively. Such vibrations are uncoupled vibrations. In examples such as pre-twisted beams, such an uncoupled motion is not possible, since the directions of principal axes change in their orientation with respect to y and z axes along the x axis of the beam. We then have coupled bending-bending vibrations. The bending motion in xy and xz planes get coupled together and they occur simultaneously.

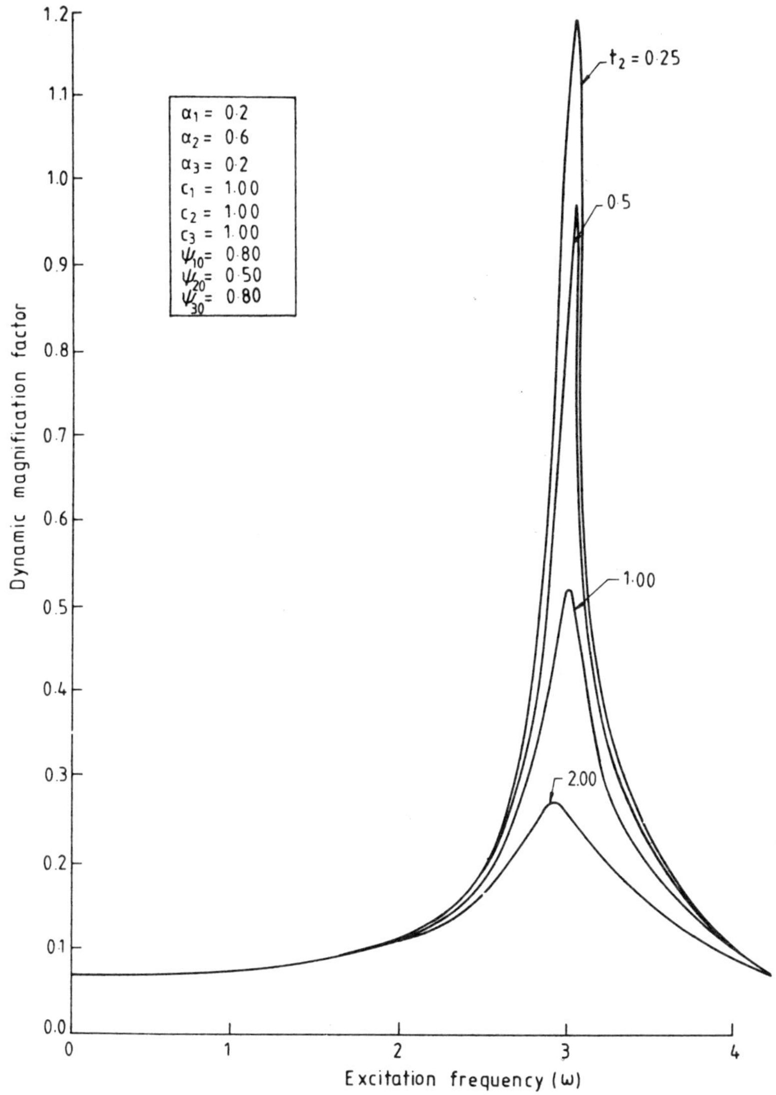

Fig. 8.28 Frequency response of laminated beams

Consider the coupled bending-bending motion shown in Fig. 8.29. In the xz plane, the motion is similar as we considered before, with the displacement w and the slope $w_{,x}$. In addition to this, we now account for the displacement v, occurring simultaneously in the xy plane. Hence the displacement field is given by

$$u_x = -zw' - yv'$$

$$u_y = v$$

$$u_x = w \tag{8.191}$$

In (8.191) above, we assumed Euler field for simplicity. The corresponding strain field is

$$\varepsilon_{xx} = -zw'' - yv''$$

rest $$\varepsilon = 0 \tag{8.192}$$

The stress field is

$$\tau_{xx} = -Ezw'' - Eyv''$$

rest $$\tau = 0 \tag{8.192}$$

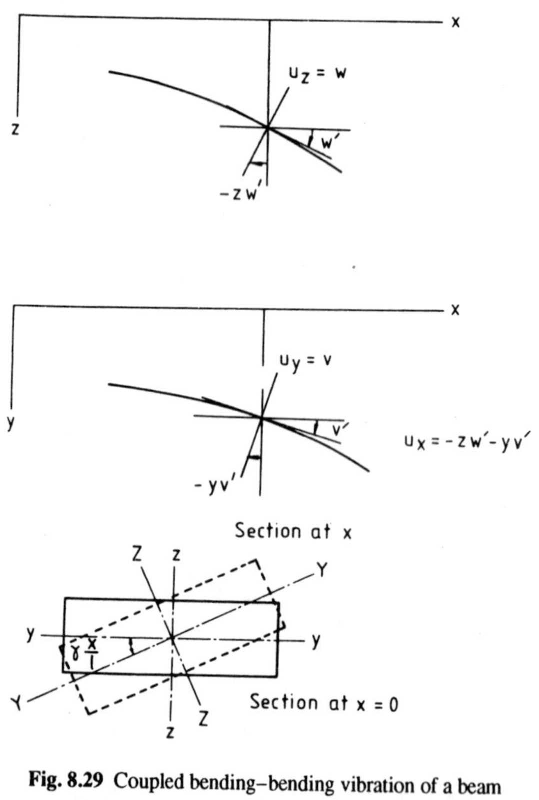

Fig. 8.29 Coupled bending–bending vibration of a beam

Then, the strain energy of the beam is given by

$$U = \frac{1}{2}\int_0^l \int_A \{Ez^2 w''^2 + Ey^2 v''^2 + Ezyv''w''\}\, dx\,dA$$

i.e.,

$$U = \frac{1}{2}\int_0^l EI_{yy} w''^2 + dx + \frac{1}{2}\int_0^l EI_{zz} v''^2 dx + \int_0^l EI_{yz} v''w''\, dx \tag{8.193}$$

If we neglect the axial inertia, the kinetic energy is given by

$$T = \frac{1}{2}\int_0^l \rho A\,(\dot{v}^2 + \dot{w}^2)\, dx \tag{8.194}$$

Let us consider free vibrations of the beam in coupled bending-bending vibration. Then, Hamilton's principle leads to

$$\delta \int_{t1}^{t2} \int_0^l \left\{ \frac{1}{2} EI_{yy} w''^2 + \frac{1}{2} EI_{zz} v''^2 + EI_{yz} v''w'' - \frac{1}{2}\rho A\,(\dot{v}^2 + \dot{w}^2) \right\} dx\,dt = 0 \tag{8.195}$$

i.e.,

$$\int_{t1}^{t2} \int_0^l \{EI_{yy} w''\delta w'' + EI_{zz} v''\delta v''$$

$$+ EI_{yz} v''\delta w'' + EI_{yz} w''\delta v'' - \rho A\,(\dot{v}\delta\dot{v} + \dot{w}\delta\dot{w})\}\, dx\,dt = 0$$

Integrating by parts, we get

$$I_1 = \int_{t1}^{t2} \int_0^l EI_{yy} w''\delta w''\, dx\,dt$$

i.e.,

$$I_1 = \int_{t1}^{t2} \left[EI_{yy} w''\delta w' \Big|_0^l - \int_0^l (EI_{yy} w'')'\delta w'\, dx \right] dt$$

$$= \int_{t1}^{t2} \left[EI_{yy} w''\delta w' \Big|_0^l - (EI_{yy} w'')'\delta w \Big|_0^l + \int_0^l (EI_{yy} w'')''\delta w\, dx \right] dt$$

$$I_2 = \int_{t1}^{t2} \int_0^l EI_{zz} v''\delta v''\, dx\,dt$$

i.e.,

$$I_2 = \int_{t1}^{t2} \left[EI_{zz} v''\delta v' \Big|_0^l - \int_0^l (EI_{zz} v'')'\delta v'\, dx \right] dt$$

$$= \int_{t1}^{t2} \left[EI_{zz} v''\delta v' \Big|_0^l - (EI_{zz} v'')'\delta v \Big|_0^l + \int_0^l (EI_{zz} v'')''\delta v\, dx \right] dt$$

$$I_3 = \int_{t1}^{t2} \int_0^l EI_{yz} w''\delta v''\, dx\,dt$$

i.e.,

$$I_3 = \int_{t1}^{t2}\left[EI_{yz}w''\delta v'\,|_0^l - \int_0^l (EI_{yz}w'')'\delta v' dx\right]dt$$

$$= \int_{t1}^{t2}\left[EI_{yz}w''\delta v'\,|_0^l - (EI_{yz}w'')'\delta v\,|_0^l + \int_0^l (EI_{yz}w'')''\delta v\, dx\right]dt$$

$$I_4 = \int_{t1}^{t2}\int_0^l EI_{yz}v''\delta w'' dxdt$$

i.e.,

$$I_4 = \int_{t1}^{t2}\left[EI_{yz}v''\delta w'\,|_0^l - \int_0^l (EI_{yz}v'')'\delta w' dx\right]dt$$

$$= \int_{t1}^{t2}\left[EI_{yz}v''\delta w'\,|_0^l - (EI_{yz}v'')'\delta w\,|_0^l + \int_0^l (EI_{yz}v'')''\delta w\, dx\right]dt$$

$$I_5 = \int_{t1}^{t2}\int_0^l -(\rho A\ddot{w}\delta w + \rho A\ddot{v}\delta v)\, dxdt$$

Grouping the terms of δw and δv, we get the following differential equations and boundary conditions.

$$(EI_{yy}w'')'' + (EI_{yz}v'')'' + \rho A\ddot{w} = 0$$

$$(EI_{zz}v'')'' + (EI_{yz}w'')'' + \rho A\ddot{v} = 0 \qquad (8.196)$$

$$\{EI_{yy}w'' + EI_{yz}v''\}\,\delta w'\,|_0^l - \{(EI_{yy}w'')' + (EI_{yz}v'')'\}\,\delta w\,|_0^l = 0$$

$$\{EI_{zz}v'' + EI_{yz}w''\}\,\delta v'\,|_0^l - \{(EI_{zz}v'')' + (EI_{yz}w'')'\}\,\delta v\,|_0^l = 0 \qquad (8.197)$$

Let us consider a rectangular cross-section beam of breadth b and thickness t, pre-twisted uniformly along its length by an angle γ. We can then express the second moments and the product moment of area of cross-section of the beam in terms of the principal moments of area of the cross-section as follows:

$$I_{yy} = I_{ZZ}\sin^2\gamma X + I_{YY}\cos^2\gamma X$$

$$I_{zz} = I_{ZZ}\cos^2\gamma X + I_{YY}\sin^2\gamma X$$

$$I_{yz} = \frac{1}{2}(I_{ZZ} - I_{YY})\sin 2\gamma X \qquad (8.198)$$

where

$$X = \frac{x}{l}$$

For simple harmonic motion with frequency p, the differential equations of the beam are

$$(EI_{yy}w'')'' + (EI_{yz}v'')'' - \rho A p^2 w = 0$$

$$(EI_{zz}v'')'' + (EI_{yz}w'')'' - \rho A p^2 v = 0 \qquad (8.196a)$$

The problem is not easily amenable for a closed form solution and therefore, we can adopt any approximate method of solution. Let us choose Galerkin method for this purpose and consider the specific case of a cantilever beam fixed at the left end $x = 0$ and pre-twisted through an angle γ till the free end at $x = l$. For the cantilever we can assume the following the mode shapes.

$$w = \sum_{r=1}^{n} A_r f_r (X)$$

$$= \sum_{r=1}^{n} A_r \left\{ \frac{1}{6}(r+2)(r+3)X^{r+1} - \frac{1}{3} r (r+3) X^{r+2} + \frac{1}{6} r (r+1) X^{r+3} \right\}$$

$$v = \sum_{r=1}^{n} B_r f_r (X) \tag{8.199}$$

The second moments of area of cross-section of the beam can be expressed as

$$I_{yy} = I_{YY} (\bar{r} \sin^2 \gamma X + \cos^2 \gamma X)$$

$$I_{zz} = I_{YY} (\bar{r} \cos^2 \gamma X + \sin^2 \gamma X)$$

$$I_{yz} = \frac{1}{2} I_{YY} (\bar{r} - 1) \sin 2\gamma X \tag{8.198a}$$

$$\bar{r} = \frac{I_{ZZ}}{I_{YY}}$$

Substituting the above relations in the differential equations (8.196*a*) we get

$$(\bar{r} \sin^2 \gamma X + \cos^2 \gamma X) w'''' + \frac{1}{2} (\bar{r} - 1) \sin 2\gamma X v'''' - \frac{\rho A l^4 p^2}{E I_{YY}} w = 0$$

$$(\bar{r} \cos^2 \gamma X + \sin^2 \gamma X) v'''' + \frac{1}{2} (\bar{r} - 1) \sin 2\gamma X w'''' - \frac{\rho A l^4 p^2}{E I_{YY}} v = 0 \tag{8.196b}$$

where differentiation in the above is now given by

$$' = \frac{d}{dX}$$

We can now substitute the shape functions of (8.199) in the above to obtain the error functions as

$$\epsilon_1 = (\bar{r} \sin^2 \gamma X + \cos^2 \gamma X) \Sigma A_r f_r''''$$
$$+ \frac{1}{2} (\bar{r} - 1) \sin 2\gamma X \, \Sigma B_r f_r'''' - \frac{\rho A l^4 p^2}{E I_{YY}} \Sigma A_r f_r$$

$$\epsilon_2 = (\bar{r} \cos^2 \gamma X + \sin^2 \gamma X) \Sigma B_r f_r''''$$
$$+ \frac{1}{2} (\bar{r} - 1) \sin 2\gamma X \, \Sigma A_r f_r'''' - \frac{\rho A l^4 p^2}{E I_{YY}} \Sigma B_r f_r \tag{8.200}$$

We can now apply Galerkin procedure as follows:

$$\int_0^1 \in_1 f_j(X)\,dX = 0$$

$$\int_0^1 \in_2 f_j(X)\,dX = 0 \quad j = 1, 2, \cdots \tag{8.201}$$

i.e.,

$$\int_0^1 (\bar{r}\sin^2\gamma X + \cos^2\gamma X)\,\Sigma A_r f_r'''' f_j\,dX$$

$$+\frac{1}{2}\int_0^1 (\bar{r}-1)\sin 2\gamma X\,\Sigma B_r f_r'''' f_j dX - \int_0^1 \frac{\rho A l^4 p^2}{EI_{YY}} \Sigma A_r f_r f_j dX = 0$$

$$\int_0^1 (\bar{r}\cos^2\gamma X + \sin^2\gamma X)\,\Sigma B_r f_r'''' f_j\,dX$$

$$+\frac{1}{2}\int_0^1 (\bar{r}-1)\sin 2\gamma X\,\Sigma A_r f_r'''' f_j dX - \int_0^1 \frac{\rho A l^4 p^2}{EI_{YY}} \Sigma B_r f_r f_j dX = 0$$

For a finite n-term solution, the above eigen value problem can be written as

$$\begin{bmatrix} P-\lambda^2 Q & R-\lambda^2 S \\ T-\lambda^2 U & V-\lambda^2 W \end{bmatrix} = 0 \tag{8.202}$$

where

$$\lambda^2 = \frac{\rho A l^4 p^2}{EI_{YY}}$$

$$P_{rj} = \int_0^1 (\bar{r}\sin^2\gamma X + \cos^2\gamma X) f_r'''' f_j dX$$

$$Q_{rj} = \int_0^1 f_r f_j dX$$

$$R_{rj} = \frac{1}{2}\int_0^1 (\bar{r}-1)\sin 2\gamma X f_r'''' f_j dX$$

$$S_{rj} = 0$$

$$T_{rj} = \frac{1}{2}\int_0^1 (\bar{r}-1)\sin 2\gamma X f_r'''' f_j dX$$

$$U_{rj} = 0$$

$$V_{rj} = \int_0^1 (\bar{r}\cos^2\gamma X + \sin^2\gamma X) f_r'''' f_j dX$$

$$W_{rj} = \int_0^1 f_r f_j dX$$

$r = 1, 2 \ldots n$
$j = 1, 2 \ldots n$

We note from the above that the natural frequencies of a pre-twisted beam are a function of $\bar{r}$ and γ only. We can develop a computer program to solve the eigen value problem of (8.202) and obtain these values. A five-term solution will be sufficient to obtain the coupled modes of the first three modes of vibration in z and y directions. We can study the frequency ratios p/p_{z1} as a function of the ratio of principal second moments of area of the cross-section, r, and the different pre-twist angles. These results are shown in Fig. 8.30, which shows the coupling effects between the z and y modes clearly.

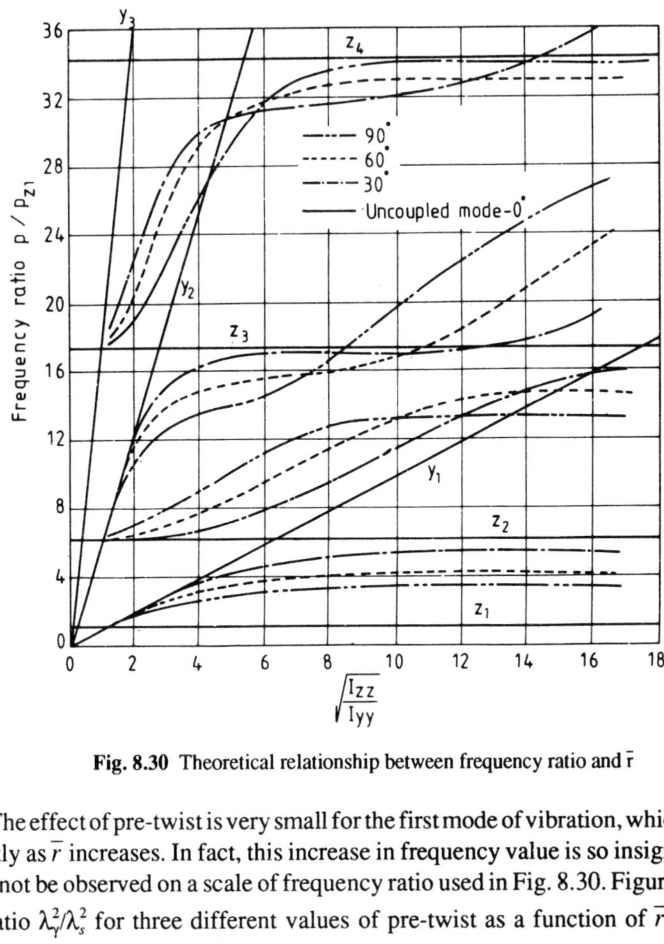

Fig. 8.30 Theoretical relationship between frequency ratio and $\bar{r}$

The effect of pre-twist is very small for the first mode of vibration, which increases slightly as $\bar{r}$ increases. In fact, this increase in frequency value is so insignificant that it cannot be observed on a scale of frequency ratio used in Fig. 8.30. Figure 8.31 gives the ratio $\lambda_\gamma^2/\lambda_s^2$ for three different values of pre-twist as a function of $\bar{r}$. The eigen values of the pre-twisted blade and the straight uniform blade are represented by λ_γ, λ_s respectively in the z or y directions as the case may be.

The second coupled mode belongs to the first uncoupled y_1 mode for $\bar{r} = 39.3$ and for $\bar{r} > 39.3$, it belongs to the second uncoupled z_2 mode as can be seen from Fig. 8.30. Figure 8.32 shows a more detailed picture of this variation with a discontinuity at $\bar{r} = 39.3$. The second coupled mode—whether it belongs to first mode in xy plane or to the second mode in xz plane, as the case may be depending on $\bar{r}$ — is less than the corresponding uncoupled modes. The y_1 mode increases as $\bar{r}$ increases upto 39.3, the effect being more predominant for larger pre-twists. For $\bar{r}$ greater than 39.3, the

z_2 mode increases with $\bar{r}$ for a given pre-twist and the frequency decreases rapidly as pre-twist increases for a given value of $\bar{r}$.

The third coupled mode increases with pre-twist over uncoupled frequencies, z_2 modes for $\bar{r} < 39.3$ and y_1 modes for $\bar{r} > 39.3$, as can be seen from Figs. 8.30 and 8.33. For a given pre-twist, the coupled z_2 mode increases with $\bar{r}$ ($\bar{r} < 39.3$) and the uncoupled y_1 mode decreases with $\bar{r}$ ($\bar{r} > 39.3$).

The transition point for the fourth uncoupled mode occurs at

$$\bar{r} = 7.84\sqrt{\frac{I_{ZZ}}{I_{YY}}} = \left(\frac{7.855}{4.694}\right)^2 = 2.8$$

as shown in Figs. 8.30 and 8.34. For a given $\bar{r}$ the uncoupled frequency decreases with pre-twist (only z_3 modes are shown), and for a given pre-twist, the coupled frequency increases with $\bar{r}$.

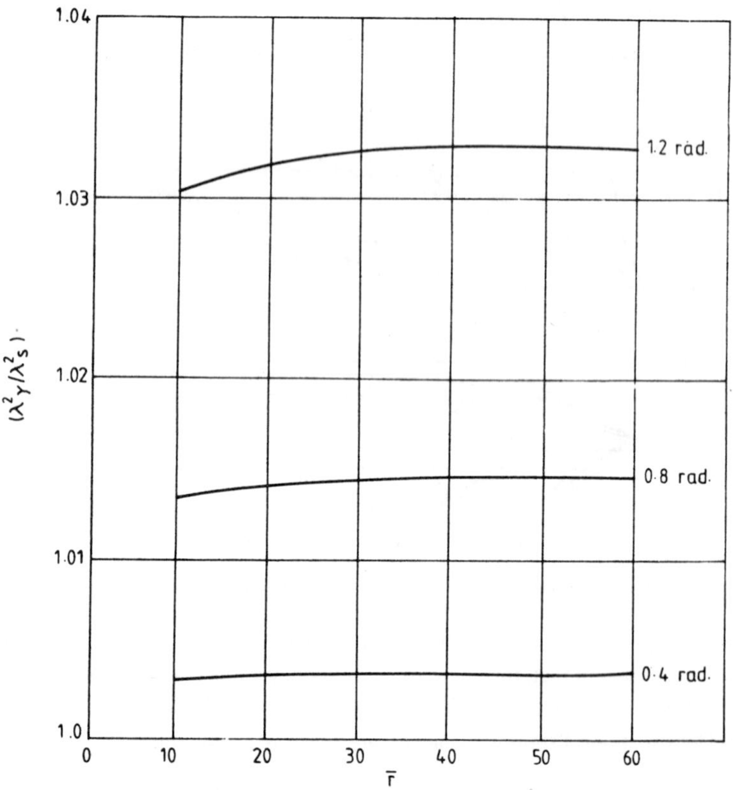

Fig. 8.31 Effect of $\bar{r}$ on frequency parameter ratio in first coupled mode for different pre-twists

8.11 COUPLED BENDING-TORSION VIBRATION OF BEAMS

When the cross-section of a beam is symmetric, its centroid and centre of flexure coincide and we have either pure torsion or pure lateral motion, as we discussed in the previous chapter and so far in this chapter, respectively. If the cross-section is asymmetric, the centroid and centre of flexure do not coincide and we have coupled bending torsion motion. This is the case that we consider in this section.

Let the cross-section of a beam be asymmetric as shown in Fig. 8.35, where the centre of flexure is at a distance r_z, r_y from the principal axes system located at the centroid, in the directions of z and y axes, respectively. The beam is considered to be straight and therefore there is no coupled bending-bending motion arising out of pre-twist. It may be, however, noted that there is a simultaneous bending-bending motion occurring through coupling of bending in each direction with torsion as will be illustrated below.

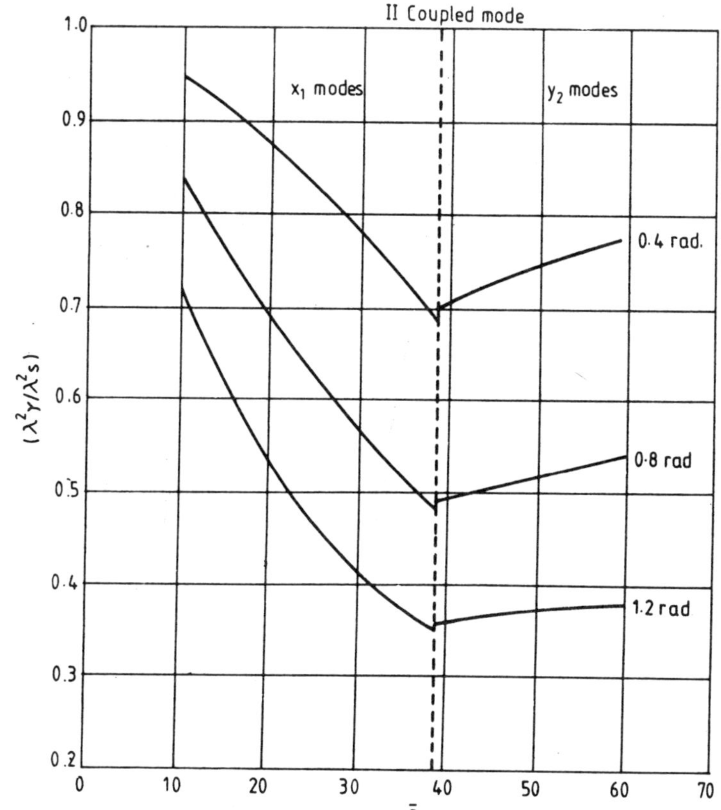

Fig. 8.32 Effect of $\bar{r}$ on frequency parameter ratio in second coupled mode for different pre-twists

A point P in the cross-section undergoes a rotation about the centre of flexure through an angle θ to the point P_1 and a bending displacement v, parallel to y axis from P_1 to P_2 and further a bending displacement w, parallel to z axis from P_2 to P' The point P finally undergoes a displacement parallel to x axis, i.e., perpendicular to the plane of Fig. 8.35, due to bending in both the planes. This field is represented by

$$u_x = -zw' - yv'$$

$$u_y = v - z_c\theta$$

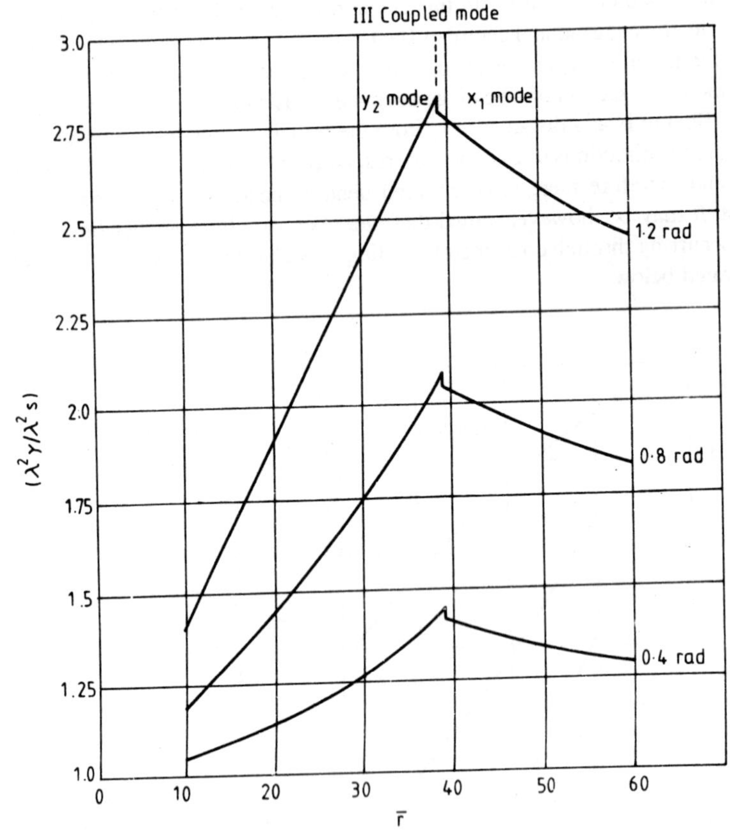

Fig. 8.33 Effect of $\bar{r}$ on frequency parameter ratio in third coupled mode for different pre-twists

$$= v - (z - r_z)\theta$$

$$u_z = w + y_c\theta$$

$$= w + (y - r_y)\theta \tag{8.203}$$

The strain field is given by

$$\varepsilon_{xx} = -zw'' - yv''$$

$$\varepsilon_{xy} = -\frac{1}{2}(z - r_z)\,\theta'$$

$$\varepsilon_{xz} = +\frac{1}{2}(y - r_y)\,\theta'$$

rest
$$\varepsilon = 0 \tag{8.204}$$

Hence, the stress field is

$$\tau_{xx} = -Ezw'' - Eyv''$$

$$\tau_{xy} = -G\,(z - r_z)\,\theta'$$

$$\tau_{xz} = G\,(y - r_y)\,\theta'$$

rest $$\tau = 0 \tag{8.205}$$

Then, the strain energy of the beam is given by

$$U = \frac{1}{2}\int_0^l \int_A \{Ez^2 w''^2 + Ey^2 v''^2 + Ezyv''w'' + \frac{1}{2}G\,(z - r_z)^2\theta'^2 + G\,(y - r_y)^2\theta'^2\}\; dx\,dA$$

Letting y and z to be the principal axes, the above becomes

$$U = \frac{1}{2}\int_0^l EI_{yy} w''^2 dx + \frac{1}{2}\int_0^l EI_{zz} v''^2 dx + \frac{1}{2}\int_0^l \int_A G\,(z_c^2 + y_c^2)\,\theta'^2 dA\,dx \tag{8.206}$$

i.e.,

$$U = \frac{1}{2}\int_0^l EI_{yy} w''^2 dx + \frac{1}{2}\int_0^l EI_{zz} v''^2 dx + \frac{1}{2}\int_0^l C\theta'^2 dx \tag{8.206a}$$

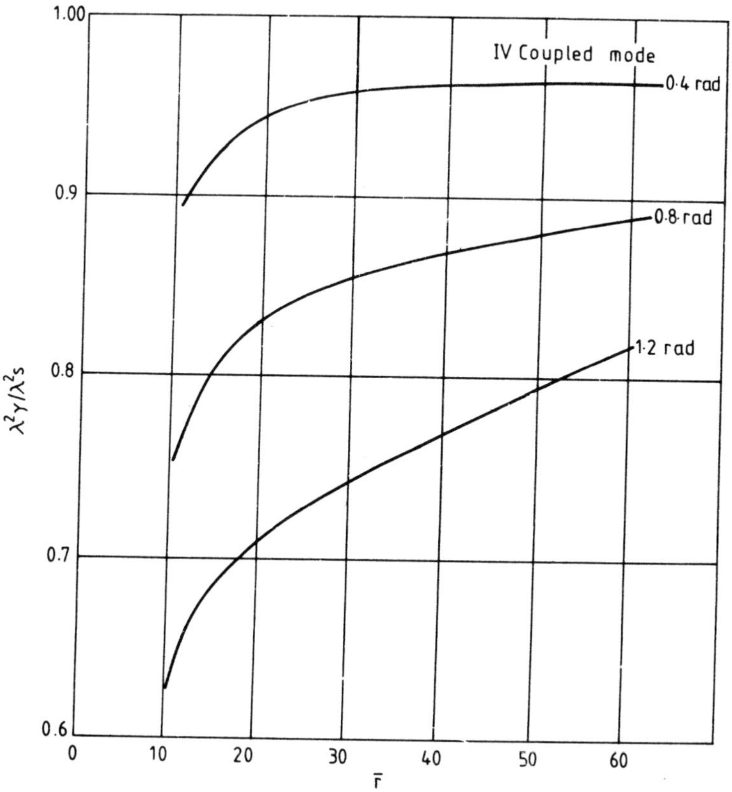

Fig. 8.34 Effect of $\bar{r}$ on frequency parameter ratio in fourth coupled mode for different pre-twists

where

$$C = GI_{cf}$$

Neglecting the axial velocity components, the velocities of the beam element are

$$\dot{u}_x = 0$$

$$\dot{u}_y = \dot{v} - z_c\dot{\theta}$$

$$\dot{u}_z = \dot{w} + y_c\dot{\theta} \tag{8.207}$$

If we neglect the axial inertia, the kinetic energy is given by

$$T = \frac{1}{2}\int_0^l \rho A\,(\dot{v}^2 + \dot{w}^2)\,dx + \frac{1}{2}\int_0^l \int_A \rho\,(z_c^2 + y_c^2)\,\dot{\theta}^2 dA\,dx$$

$$+ \int_0^l \int_A (-z_c\dot{v} + y_c\dot{w})\dot{\theta} dA\,dx \tag{8.208}$$

i.e.,

$$T = \frac{1}{2}\int_0^l \rho A\,(\dot{v}^2 + \dot{w}^2)\,dx + \frac{1}{2}\int_0^l I_{cf}\dot{\theta}^2 dx + \int_0^l \rho A\,(r_z\dot{v} - r_y\dot{w})\,\dot{\theta} dx \tag{8.208}$$

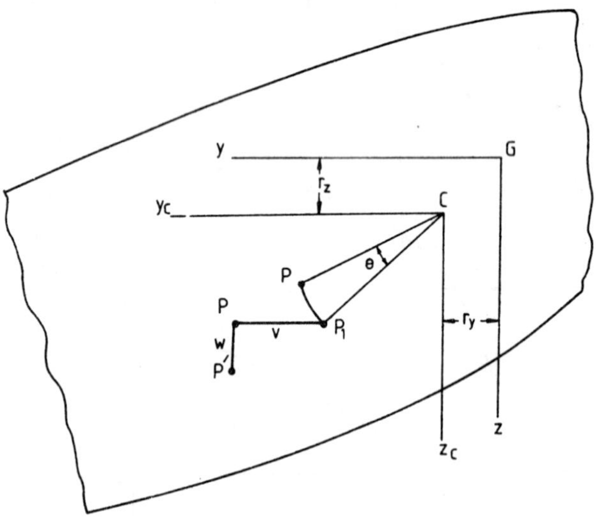

Fig. 8.35 Asymmetric cross–section of a beam

Let us consider free vibrations of the beam in coupled bending-torsion vibration. Then, Hamilton's principle leads to

$$\delta\int_{t1}^{t2}\int_0^l \left\{\frac{1}{2}EI_{yy}w''^2 + \frac{1}{2}EI_{zz}v''^2 + \frac{1}{2}C\theta'^2\right.$$

$$\left. - \frac{1}{2}\rho A\,(\dot{v}^2 + \dot{w}^2) - \frac{1}{2}I_{cf}\dot{\theta}^2 - \rho A\,(r_z\dot{v} - r_y\dot{w})\dot{\theta}\right\} dx\,dt = 0 \tag{8.209}$$

i.e.,

$$\int_{t1}^{t2}\int_0^l \{EI_{yy}w''\delta w'' + EI_{zz}v''\delta v'' + EI_{yz}v''\delta w'' + C\theta'\delta\theta'$$

$$-\rho A\,(\dot{v}\delta\dot{v} + \dot{w}\delta\dot{w}) - I_{cf}\dot{\theta}\delta\dot{\theta} - \rho A\,(r_z\dot{v} - r_y\dot{w})\,\delta\dot{\theta}$$

$$-\rho A\,(r_z\delta\dot{v} - r_y\delta\dot{w})\dot{\theta}\}\;dx\,dt = 0 \tag{8.209a}$$

The first two integrals are considered in the previous section (*see* equations 8.196 and 8.197). The third integral was considered in chapter 7 (*see* equation 7.8). The fourth integral is also considered in (8.196), while the fifth integral was taken into account in equation (7.8). The last two integrals give

$$I_7 = \int_{t1}^{t2}\int_0^l \rho A\,(r_z\dot{v} - r_y\dot{w})\,\delta\dot{\theta}dx\,dt$$

Integrating by parts, we get

$$I_7 = \int_{t1}^{t2}\int_0^l -\rho A\,(r_z\ddot{v} - r_y\ddot{w})\,\delta\theta dx\,dt$$

$$I_8 = \int_{t1}^{t2}\int_0^l \rho A\,(r_z\delta\dot{v} - r_y\delta\dot{w})\,\dot{\theta}dx\,dt$$

Integrating by parts, we have

$$I_8 = \int_{t1}^{t2}\int_0^l -\rho A\,(r_z\delta v - r_y\delta w)\,\ddot{\theta}dx\,dt$$

i.e.,

$$I_8 = \int_{t1}^{t2}\int_0^l \{-\rho A r_z\ddot{\theta}\delta v + \rho A r_y\ddot{\theta}\delta w)\,dx\,dt$$

We can combine now the appropriate integrals and obtain the following differential equations and the boundary conditions for the beam in coupled bending torsion motion.

$$(EI_{yy}w'')'' + \rho A\ddot{w} = \rho A r_y\ddot{\theta}$$

$$(EI_{zz}v'')'' + \rho A\ddot{v} = -\,\rho A r_z\ddot{\theta}$$

$$-(C\theta')' = I_{cf}\ddot{\theta} = \rho A r_y\ddot{w} - \rho A r_z\ddot{v} \tag{8.210}$$

$$\{EI_{yy}w''\}\,\delta w'\,|_0^l - \{(EI_{yy}w'')'\}\,\delta w\,|_0^l = 0$$

$$\{EI_{zz}v''\}\,\delta v'\,|_0^l - \{(EI_{zz}v'')'\}\,\delta v\,|_0^l = 0$$

$$C\theta'\delta\theta\,|_0^l = 0 \tag{8.211}$$

For free vibrations with frequency p, the above differential equations (8.210) reduce to

$$(EI_{yy}w'')'' - \rho A p^2 w = -\,\rho A p^2 r_y\theta$$

$$(EI_{zz}v'')'' - \rho A p^2 v = \rho A p^2 r_z \theta$$

$$-(C\theta')' - I_{cf}p^2\theta = -\rho A p^2 r_y w + \rho A p^2 r_z v \qquad (8.210a)$$

Let us set

$$w \leftarrow \frac{w}{l}$$

$$v \leftarrow \frac{v}{l}$$

$$x \leftarrow \frac{x}{l}$$

$$r_y \leftarrow \frac{r_y}{l}$$

$$r_z \leftarrow \frac{r_z}{l} \qquad (8.212)$$

Let us consider a uniform beam, then equations (8.210*a*) become

$$w'''' - \frac{\rho A l^4}{EI_{yy}} p^2 w = -\frac{\rho A l^4}{EI_{yy}} p^2 r_y \theta$$

$$v'''' - \frac{\rho A l^4}{EI_{zz}} p^2 v = -\frac{\rho A l^4}{EI_{zz}} p^2 r_z \theta$$

$$-\theta'' - \frac{I_{cf} l^2}{C} p^2 \theta = -\frac{\rho A l^4}{C} p^2 r_y w + \frac{\rho A l^4}{C} p^2 r_z v \qquad (8.213)$$

i.e.,

$$w'''' - \frac{p^2}{\lambda_z^2} w = -\frac{p^2}{\lambda_z^2} r_y \theta$$

$$v'''' - \frac{p^2}{\lambda_y^2} v = \frac{p^2}{\lambda_y^2} r_z \theta$$

$$-\theta'' - \frac{p^2}{\lambda_\theta^2}\theta = -\mu_y \frac{p^2}{\lambda_z^2} r_y w + \mu_z \frac{p^2}{\lambda_y^2} r_z v \qquad (8.213a)$$

where

$$\mu_y = \frac{EI_{yy}}{C}$$

$$\mu_z = \frac{EI_{zz}}{C}$$

$$\lambda_z^2 = \frac{EI_{yy}}{\rho A l^4}$$

$$\lambda_y^2 = \frac{EI_{zz}}{\rho A l^4}$$

$$\lambda_\theta^2 = \frac{C}{I_{cf} l^2} \tag{8.214}$$

From equations (8.213) above we note that bending displacement w and θ are coupled because of the distance of centre of flexure from the centre of gravity in y direction r_y. Similarly, the bending displacement v is coupled with θ because of the distance of the centre of flexure from the centre of gravity in z direction r_z. In other words, if the cross-section is symmetric about yy axis, i.e., $r_z = 0$, then the coupling takes place between the bending displacement w and θ.

As in the previous section, we find it convenient to adopt any approximate method of solution. Let us choose Galerkin method and consider again the case of a uniform cantilever beam fixed at the left end $x = 0$ and free at the right end $x = 1$. For the cantilever we can assume the following mode shapes.

$$w = \sum_{r=1}^{n} A_r f_r(x)$$

$$= \sum_{r=1}^{n} A_r \left\{ \frac{1}{6}(r+2)(r+3)x^{r+1} - \frac{1}{3} r\,(r+3) x^{r+2} + \frac{1}{6} r\,(r+1) x^{r+3} \right\}$$

$$v = \sum_{r=1}^{n} B_r f_r(x)$$

$$\theta = \sum_{r=1}^{n} C_r F_r(x)$$

$$= \sum_{r=1}^{n} C_r x^r \left\{ 1 - \frac{r}{r+1} x \right\} \tag{8.215}$$

Substituting the assumed shape functions (8.215) in the differential equations (8.213a), we get

$$\epsilon_1 = \Sigma A_r f_r'''' - \frac{p^2}{\lambda_z^2} \Sigma A_r f_r + \frac{p^2}{\lambda_z^2} r_y \Sigma C_r F_r$$

$$\epsilon_2 = \Sigma B_r f_r'''' - \frac{p^2}{\lambda_y^2} \Sigma B_r f_r - \frac{p^2}{\lambda_y^2} r_z \Sigma C_r F_r$$

$$\epsilon_3 = -\Sigma C_r F_r'' - \frac{p^2}{\lambda_\theta^2} \Sigma C_r F_r + \mu_y \frac{p^2}{\lambda_z^2} r_y \Sigma A_r f_r - \mu_z \frac{p^2}{\lambda_y^2} r_z \Sigma B_r f_r \tag{8.216}$$

Now, we can apply Galerkin method to the above and obtain

$$\int_0^1 \epsilon_1 f_j(x)\, dx = 0$$

$$\int_0^1 \epsilon_2 f_j(x)\, dx = 0$$

$$\int_0^1 \in_3 F_j(x)\,dx = 0 \quad j = 1,2,\cdots \tag{8.217}$$

i.e.,

$$\int_0^1 \left\{\Sigma A_r f_r'''' - \frac{p^2}{\lambda_z^2}\Sigma A_r f_r + \frac{p^2}{\lambda_z^2} r_y \Sigma C_r F_r\right\} f_j dx = 0$$

$$\int_0^1 \left\{\Sigma B_r f_r'''' - \frac{p^2}{\lambda_y^2}\Sigma B_r f_r - \frac{p^2}{\lambda_y^2} r_z \Sigma C_r F_r\right\} f_j dx = 0$$

$$\int_0^1 \left\{\Sigma C_r F_r'' + \frac{p^2}{\lambda_\theta^2}\Sigma C_r F_r \right.$$

$$\left. -\mu_y \frac{p^2}{\lambda_z^2} r_y \Sigma A_r f_r + \mu_z \frac{p^2}{\lambda_y^2} r_z \Sigma B_r f_r\right\} F_j dx = 0 \tag{8.217a}$$

We can express the above in the following eigen value problem

$$\begin{bmatrix} D - p^2E & F - p^2G & H - p^2I \\ J - p^2K & L - p^2M & N - p^2O \\ P - p^2Q & R - p^2S & T - p^2U \end{bmatrix} \begin{Bmatrix} A \\ B \\ C \end{Bmatrix} = 0 \tag{8.218}$$

where

$$D_{rj} = \int_o^1 f_r'''' f_j dx$$

$$E_{rj} = \int_0^1 \frac{1}{\lambda_z^2} f_r f_j dx$$

$$F_{rj} = 0$$

$$G_{rj} = 0$$

$$H_{rj} = 0$$

$$I_{rj} = \int_0^1 \frac{r_z}{\lambda_y^2} F_r f_j dx$$

$$J_{rj} = 0$$

$$K_{rj} = 0$$

$$L_{rj} = \int_0^1 f_r'''' f_j dx$$

$$M_{rj} = \int_0^1 \frac{1}{\lambda_y^2} f_r f_j dx$$

$$N_{rj} = 0$$

$$O_{rj} = \int_0^1 \frac{r_z}{\lambda_y^2} F_r f_j dx$$

$$P_{rj} = 0$$

$$Q_{rj} = \int_0^1 \frac{\mu_y}{\lambda_z^2} r_y f_r F_j dx$$

$$R_{rj} = 0$$

$$S_{rj} = -\int_0^1 \frac{\mu_z}{\lambda_y^2} r_z f_r F_j dx$$

$$T_{rj} = \int_0^1 F_r'' F_j dx$$

$$U_{rj} = -\int_0^1 \frac{1}{\lambda_\theta^2} F_r F_j dx \tag{8.219}$$

One can write a computer program based on the above to formulate and solve the eigen value problem of an asymmetric straight uniform beam to determine its coupled bending-bending torsion natural frequencies.

The effect of asymmetry of cross-section on the coupled bending and torsion modes is shown in Figs. 8.36 through 8.39. The bending displacement w and torsion

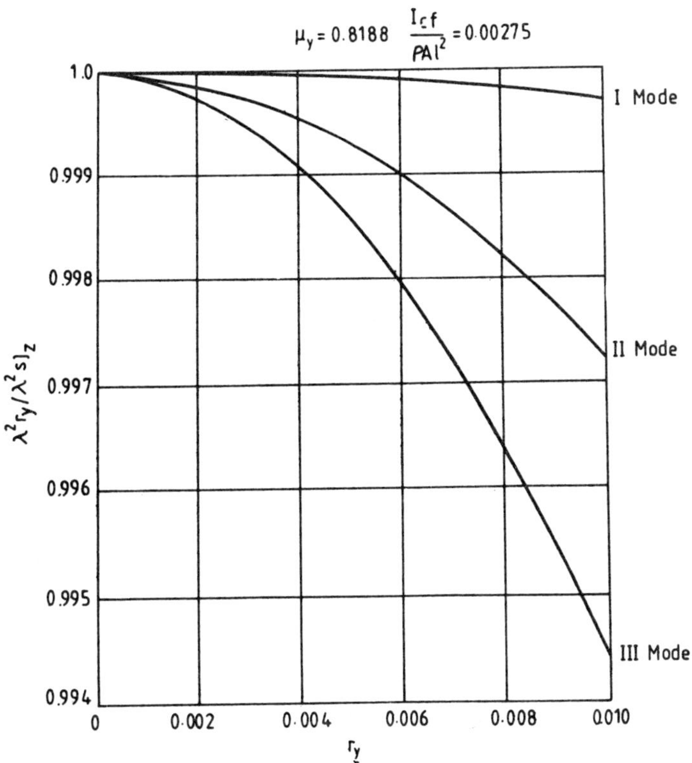

Fig. 8.36 Effect of Asymmetry r_y on frequency parameter ratio in flapwise bending

displacement θ are coupled because of r_y. This effect is shown in Figs. 8.36 and 8.37. The effect of r_z on the coupling of bending modes v and torsion modes θ is shown in Figs. 8.38 and 8.39. We note that the bending frequencies decrease with asymmetry, while the torsion frequency increases with asymmetry of the cross-section. The effect of asymmetry is negligible in the fundamental bending as well as torsion modes. Its importance is not very significant even in higher modes, though asymmetry effects the higher modes more than the fundamental modes.

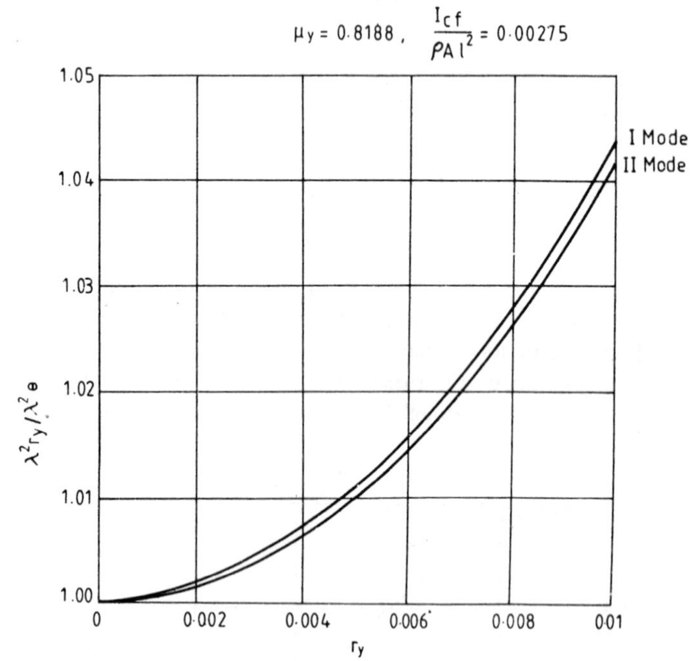

Fig. 8.37 Effect of Asymmetry r_y on frequency parameter ratio in torsion

8.12 VIBRATION OF ROTATING BEAMS

In this section, we will consider the case of a beam mounted on a rotating disk at a stagger angle. The disk is of radius R and rotates at an angular velocity ω. For simplicity, the beam is taken to be rectangular cross-section as shown in Fig. 8.40. The axis at the root of the beam where it is located on the periphery of the disk and parallel to the axis of rotation, is denoted ξ and the η is perpendicular to this as shown. The beam is mounted at an angle φ to the axis of rotation, which is called stagger angle. The setting angle of the beam is given by $\phi' = 90 - \phi$, which is measured from the η axis. Again we will use simple Euler theory for this problem. The displacement field for a point P is given by

$$u_x = -zw' + u$$

$$u_y = 0$$

$$u_z = w \tag{8.220}$$

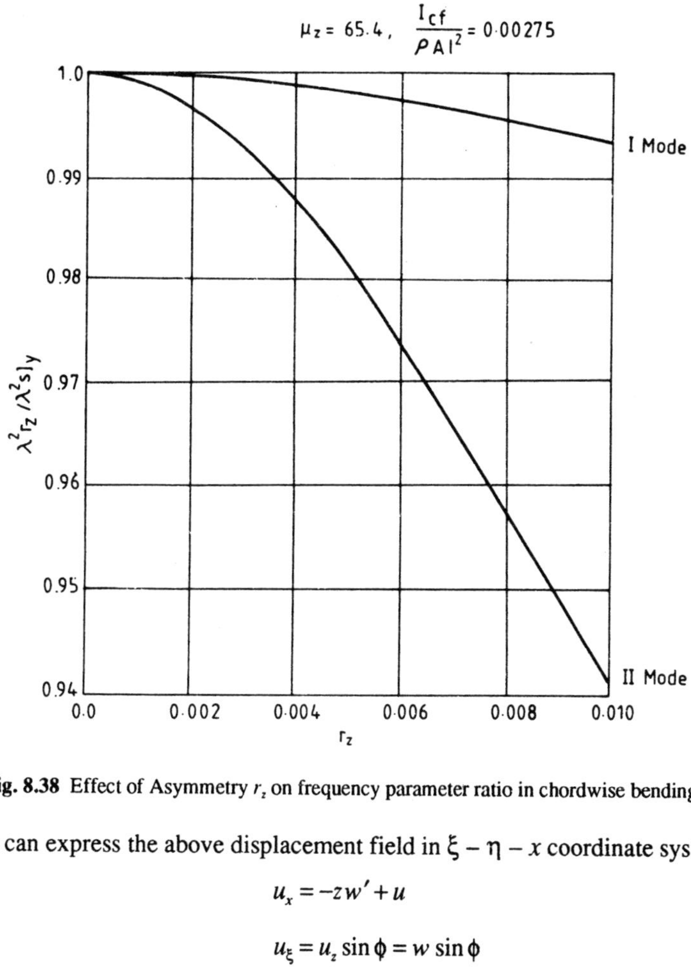

Fig. 8.38 Effect of Asymmetry r_z on frequency parameter ratio in chordwise bending

We can express the above displacement field in $\xi - \eta - x$ coordinate system as

$$u_x = -zw' + u$$

$$u_\xi = u_z \sin\phi = w \sin\phi$$

$$u_\eta = w\cos\phi \qquad (8.220a)$$

The position vector of point P' can be written as

$$CP' = CG + GP + PP'$$

$$= Re_x + \xi e_\xi + \eta e_\eta + xe_x + u_\xi e_\xi + u_\eta e_\eta + u_x e_x$$

$$= (R + x + u_x)e_x + (u_\xi + \xi)e_\xi + (u_\eta + \eta)e_\eta \qquad (8.221)$$

where e are unit vectors.

The velocity of the particle at P' is given by

$$\bar{r} = \dot{r}_x e_x + r_x \dot{e}_x + r_\xi \dot{e}_\xi + \dot{r}_\xi e_\xi + r_\eta \dot{e}_\eta + \dot{r}_\eta e_\eta \qquad (a)$$

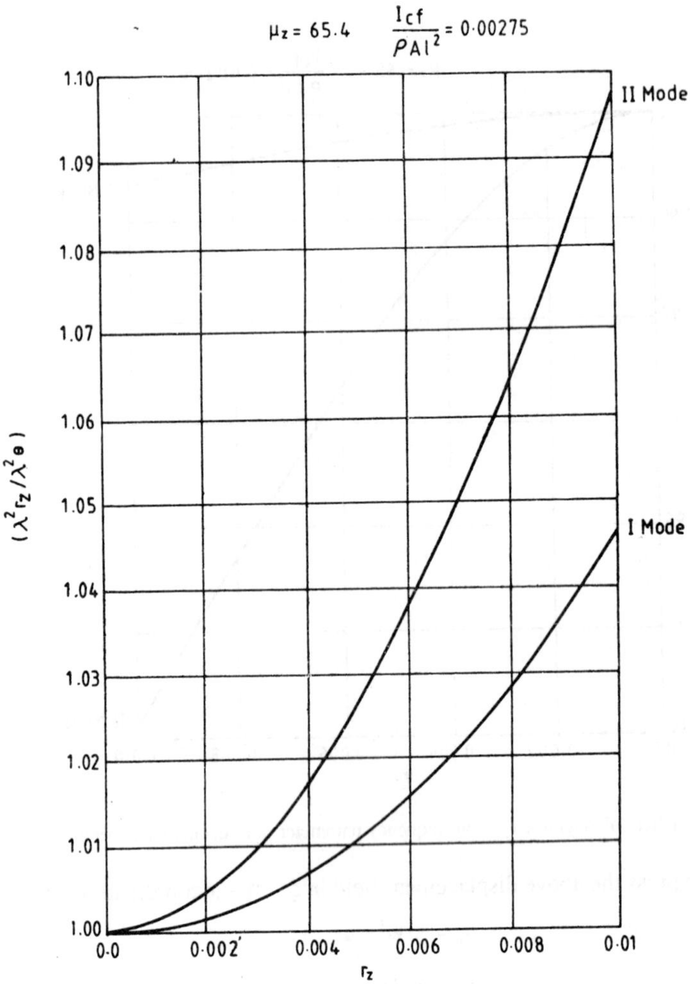

Fig. 8.39 Effect of Asymmetry r_z on frequency parameter ratio

We note that

$$\dot{e}_x = \omega e_\eta$$

$$\dot{e}_\eta = -\omega e_x$$

$$\dot{e}_\xi = 0$$

Using the above and (8.221) in (a) and neglecting axial velocity terms, we get

$$\bar{r} = (R + x + u_x)\omega e_\eta + \dot{w}\sin\phi e_\xi - (\eta + \mu_\eta)\omega e_x + \dot{w}\cos\phi e_\eta \qquad (b)$$

i.e.,

$$\bar{r} = \{(R + x + u_x)\omega + \dot{w}\cos\phi\} e_\eta + \dot{w}\sin\phi e_\xi - (z\cos\phi + w\cos\phi)\omega e_x$$

The square of the velocity is given by

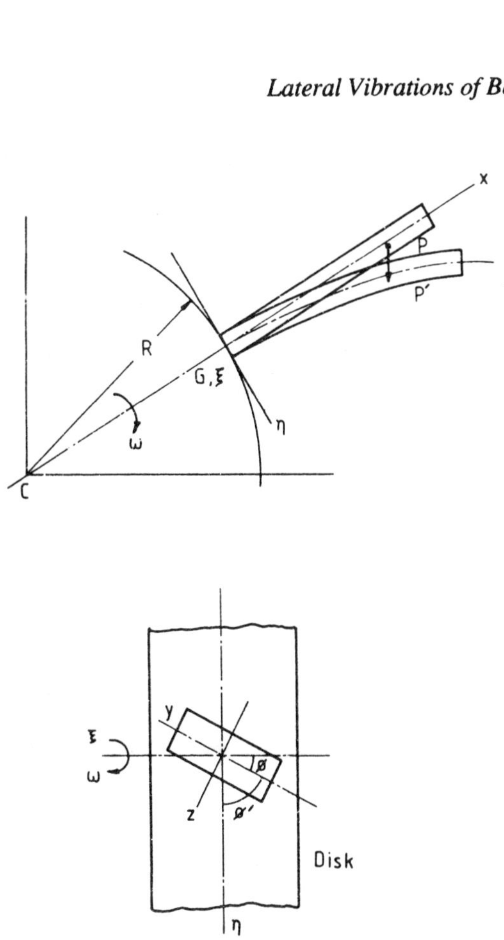

Fig. 8.40 A beam mounted on a rotating disk at a stagger angle

$$\bar{r}\bar{r} = \{(R + x + u - zw')^2 + z^2\cos^2\phi + w^2\cos^2\phi\}w^2$$

$$+ \dot{w}^2 + 2(R + x + u - zw')\omega\dot{w}\cos\phi - 2z\omega w\cos^2\phi \qquad (c)$$

In the centrifugal components of the above, we will account displacements of the particles on the line of centroid only, and neglect the Coriolis components. Then

$$\bar{r}\bar{r} = \{2(R + x)u + \dot{w}^2\cos^2\phi\}\,\omega^2 + \dot{w}^2 \qquad (8.222)$$

Hence, the kinetic energy is given by

$$T = \frac{1}{2}\int_0^l \rho A\,[\dot{w}^2 + \{2(R + x)u + w^2\cos^2\phi\}\,\omega^2]\,dx \qquad (d)$$

The inward displacement u, due to bending of the rotating beam, can be obtained from Fig. 8.41, as

$$u = -\int_0^x \frac{1}{2} w'^2 dx \qquad (8.223)$$

Therefore, equation (d) can now be written as

$$T = \frac{1}{2}\int_0^l \rho A\left[\dot{w}^2 + w^2\omega^2\cos^2\phi - (R+x)\omega^2\int_0^x w'^2 dx\right]dx \qquad (8.224)$$

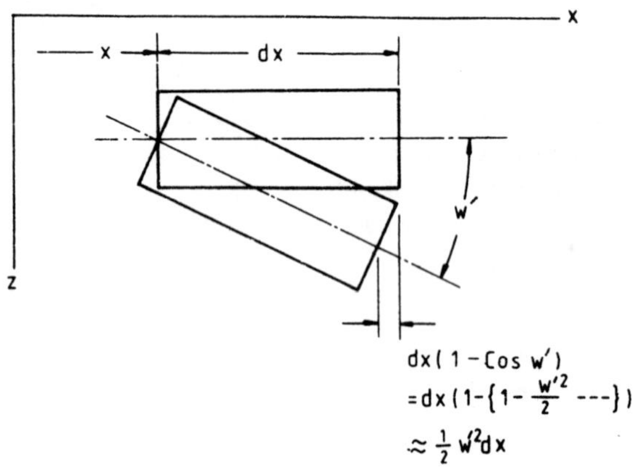

Fig. 8.41 Inward displacement of a beam in lateral bending

From (8.220), we have derived earlier the potential energy as

$$U = \frac{1}{2}\int_0^l EI_{zz} w''^2 dx \qquad (e)$$

The kinetic energy term of (8.224) has been considered in equation (5.58) and therefore the governing differential equation of the rotating beam can be obtained as

$$EI_{zz}w'''' - \rho A\omega^2\cos^2\phi w - \rho A\omega^2\left\{w'\int_x^l (R+x)dx\right\}' + \rho A\ddot{w} = 0 \qquad (8.225)$$

Let

$$w \leftarrow \frac{w}{l}$$

$$R \leftarrow \frac{R}{l}$$

$$x \leftarrow \frac{x}{l}$$

$$\alpha^2 = \frac{\rho A l^4 \omega^2}{EI_{zz}}$$

$$\lambda^2 = \frac{\rho A l^4 p^2}{EI_{zz}}$$

then

$$w'''' - \alpha^2\cos^2\phi w - \alpha^2\left\{w'\int_x^1 (R+x)dx\right\}' - \lambda^2 w = 0 \qquad (8.225a)$$

Integrating the above, we have

$$w'''' - \alpha^2 \cos^2 \phi w -$$

$$-\frac{1}{2}\alpha^2[\{2R(1-x)+(1-x^2)\}w'' - 2(R+x)w'] - \lambda^2 w = 0 \qquad (8.226)$$

We note that the boundary conditions remain same as that of stationary beam. It will be convenient to use Galerkin's process to determine the eigen values of (8.226). Let us consider again the case of a uniform cantilever beam fixed at the left end $x = 0$ and free at the right end $x = 1$, and assume the following mode shape.

$$w = \sum_r A_r f_r(x)$$

$$= \sum_{r=1}^{n} A_r \left\{ \frac{1}{6}(r+2)(r+3)x^{r+1} - \frac{1}{3}r(r+3)x^{r+2} + \frac{1}{6}r(r+1)x^{r+3} \right\} \qquad (8.227)$$

Then the error function for the differential equation (8.226) is

$$\epsilon = \sum_r A_r f_r'''' - \alpha^2 \cos^2 \phi \sum_r A_r f_r$$

$$-\frac{1}{2}\alpha^2 \left[\{2R(1-x)+(1-x^2)\} \sum_r A_r f_r'' - 2(R+x) \sum_r A_r f_r' \right] - \lambda^2 \sum_r A_r f_r \qquad (8.228)$$

According to Galerkin process

$$\int_0^1 \left[\sum_r A_r f_r'''' - \alpha^2 \cos^2 \phi \sum_r A_r f_r \right.$$

$$\left. -\frac{1}{2}\alpha^2 \left[\{2R(1-x)+(1-x^2)\} \sum_r A_r f_r'' - 2(R+x) \sum_r A_r f_r' \right] - \lambda^2 \sum_r A_r f_r \right] f_j dx = 0$$

$$j = 1, 2, \ldots n \qquad (8.229)$$

i.e.,

$$[P - \lambda^2 Q]\{A\} = 0 \qquad (8.229a)$$

where

$$P_{rj} = \int_0^1 \{f_r'''' - \alpha^2 \cos^2 \phi f_r$$

$$-\frac{1}{2}\alpha^2[\{2R(1-x)+(1-x^2)\}f_r'' - 2(R+x)f_r']\} f_j dx$$

$$Q_{rj} = \int_0^1 f_r f_j dx \qquad (8.230)$$

As before a computer program can be developed to solve the eigen value problem above and determine the natural frequencies of a beam mounted on a rotating disk at a given stagger angle.

In Fig. 8.42, the nondimensional frequency ratio $\lambda_\alpha^2/\lambda_s^2$ is plotted as a function of the rotational parameter α^2 for the first three modes. We note that the frequency ratio varies linearly, increasing with α^2. The effect of rotation is predominant in the first mode and is of less significance for higher modes. The effect of disk radius on the frequency ratio is shown in Fig. 8.43 for the first three modes and three different values of α^2. We note once again that the frequency ratio increases linearly with R, the effect being more predominant for the lower modes and higher speeds of rotation. The influence of setting angle ϕ' is shown in Fig. 8.44 for the first mode for three different values of α^2 and for two different values of α^2 of the second mode. The frequencies decrease with the setting angle linearly, the effect being predominant for the fundamental mode and of very negligible influence for higher modes.

8.13 NONLINEAR VIBRATIONS DUE TO CORIOLIS EFFECT

In the previous section, we considered the case of a beam mounted on a rotating disk at a stagger angle. We accounted for all linear terms and neglected higher order terms and Coriolis terms. In this section the additional effect due to Coriolis forces is considered. For simplicity, let us consider the case of a beam vibrating in the plane of the disk, i.e., $\phi = 0$. We recall the position vector given in equation (8.221).

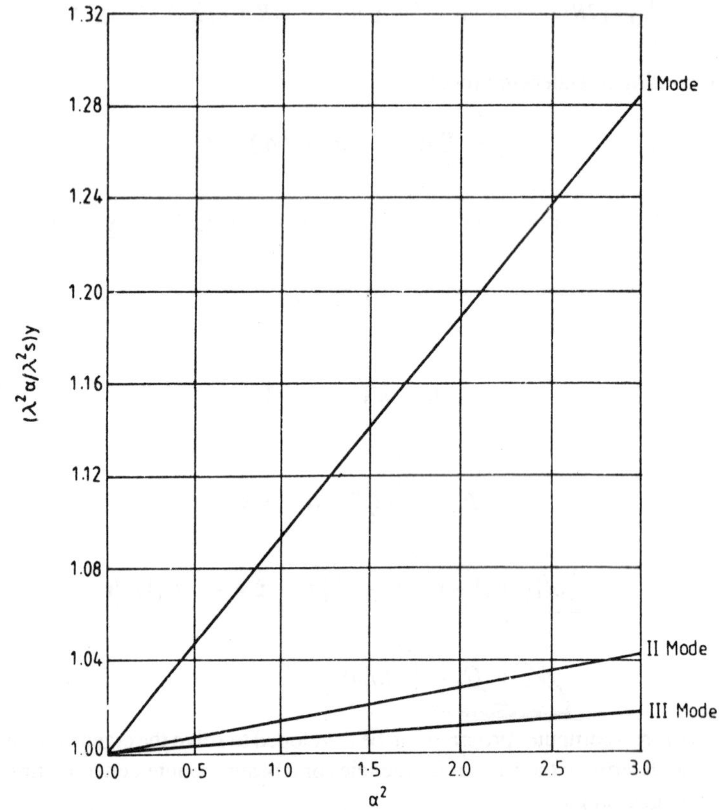

Fig. 8.42 Effect of speed of rotation on frequency parameter in flapwise bending mode at setting angle 0°

$$\bar{r} = (R + x + u_x)e_x + (u_\xi + \xi)e_\xi + (u_\eta + \eta)e_\eta \tag{8.221}$$

The velocity of the particle at P' is given by

$$\dot{\bar{r}} = \dot{r}_x e_x + r_x \dot{e}_x + r_\xi \dot{e}_\xi + \dot{r}_\xi e_\xi + r_\eta \dot{e}_\eta + \dot{r}_\eta e_\eta \tag{a}$$

We note that

$$\dot{e}_x = \omega e_\eta$$

$$\dot{e}_\eta = -\omega e_x$$

$$\dot{e}_\xi = 0$$

Using the above and (8.221) in (a), we get

$$\dot{\bar{r}} = (R + x + u_x)\omega e_\eta - \{(\eta + u_\eta)\omega - \dot{u} + z\dot{w}'\}e_x + \dot{w}e_\eta \tag{b}$$

i.e.,

$$\dot{\bar{r}} = \{(R + x + u_x)\omega + \dot{w}\}e_\eta - \{(z + w)\omega - \dot{u} + z\dot{w}'\}e_x$$

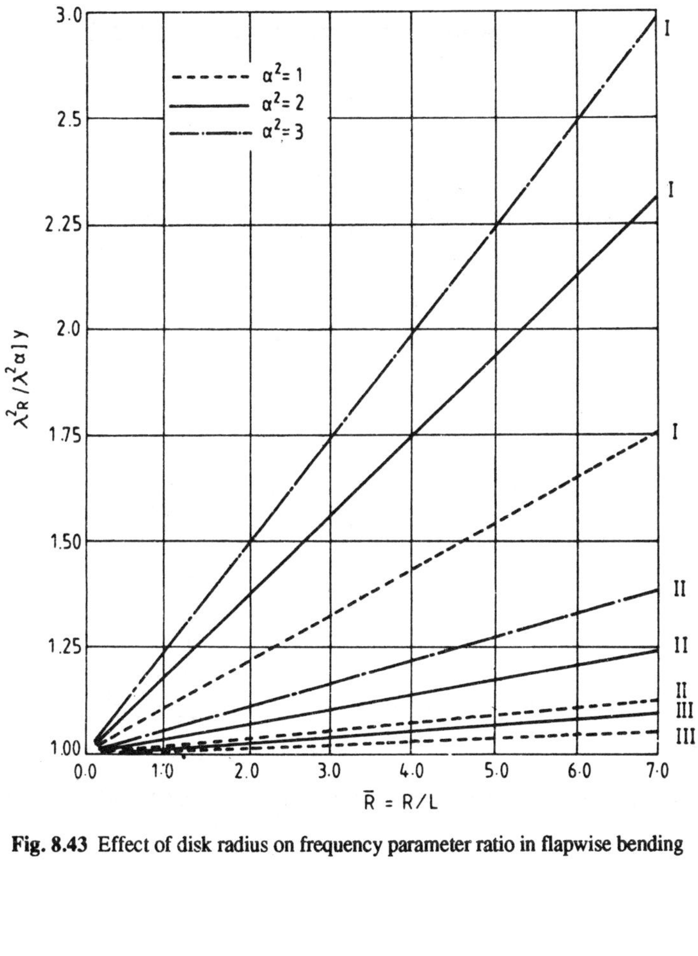

Fig. 8.43 Effect of disk radius on frequency parameter ratio in flapwise bending

The square of the velocity is given by

$$\dot{\bar{r}} \cdot \dot{\bar{r}} = (R + x + u - zw')^2\omega^2 + \{(z + w)\omega - \dot{u} + z\dot{w}'\}^2$$

$$+ \dot{w}^2 + 2(R + x + u - zw')\omega\dot{w} - 2z\omega w \qquad (c)$$

In the centrifugal components of the above, we will account displacements of the particles on the line of centroid only, including the Coriolis terms. Then

$$\dot{\bar{r}} \cdot \dot{\bar{r}} = \{2(R + x)u + w^2\}\omega^2 + \dot{w}^2 - 2\omega\dot{u}w + 2u\omega\dot{w} \qquad (8.231)$$

Hence, the kinetic energy is given by

$$T = \frac{1}{2}\int_0^l \rho A[\dot{w}^2 + \{2(R + x)u + w^2\}\omega^2 - 2\omega\dot{u}w + 2u\omega\dot{w}]dx \qquad (d)$$

The inward displacement u, due to bending of the rotating beam, is given in Fig. 8.41.

$$u = -\int_0^x \frac{1}{2}w'^2 dx \qquad (8.223)$$

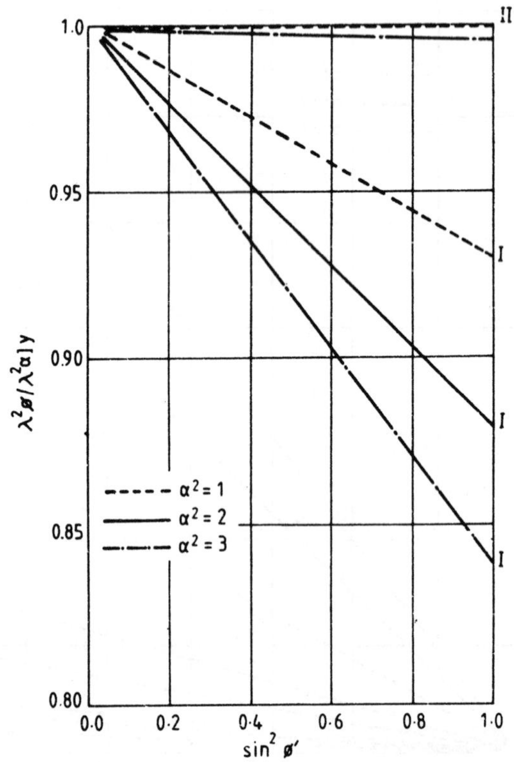

Fig. 8.44 Effect of setting angle on frequency parameter ratio in flapwise bending

Therefore, equation (*d*) becomes

$$T = \frac{1}{2}\int_0^l \rho A\left[\dot{w}^2 + w^2\omega^2 - (R+x)\omega^2 \int_0^x w'^2 dx \right.$$

$$\left. + \omega w \frac{\partial}{\partial t}\int_0^x w'^2 dx - \omega \dot{w}\int_0^x w'^2 dx\right] dx \tag{8.232}$$

The above has been assumed in equation (5.138) while illustrating Ritz method in Chapter 5. The last two integrals are Coriolis terms which are nonlinear. The potential energy remains same as in (e) in the previous section.

$$U = \frac{1}{2}\int_0^l EI_{zz} w''^2 dx \tag{e}$$

We have only two new terms in the kinetic energy expression (8.232) which are Coriolis terms. The first of these integrals in Hamilton's equation will give

$$I_1 = \frac{1}{2}\rho A\,\delta \int_{t1}^{t2}\int_0^l \omega w \frac{\partial}{\partial t}\int_0^x w'^2 dx\,dx\,dt \tag{f}$$

In (*f*) above, we replace one of the x variables by θ. Then

$$I_1 = \frac{1}{2}\rho A\,\delta \int_{t1}^{t2}\int_0^l \omega w \frac{\partial}{\partial t}\left\{\int_0^\theta w'^2 d\theta\right\} dx\,dt$$

Taking the variation

$$I_1 = \int_{t1}^{t2}\int_0^l \frac{1}{2}\rho A\,\omega\delta w \frac{\partial}{\partial t}\left\{\int_0^\theta w'^2 d\theta\right\} dx\,dt$$

$$+ \int_{t1}^{t2}\int_0^l \rho A\,\omega w \frac{\partial}{\partial t}\left\{\int_0^\theta w'\delta w' d\theta\right\} dx\,dt \tag{g}$$

In (*g*) above we have two integrals, the first one can be written as

$$I_{11} = \int_{t1}^{t2}\int_0^l \frac{1}{2}\rho A\,\omega \frac{\partial}{\partial t}\left\{\int_0^x w'^2 dx\right\} dx\,\delta w\,dt \tag{h}$$

Integrating the second integral in (g) by parts with respect to time, we get

$$I_{12} = \int_0^l \rho A\,\omega w\left\{\int_0^\theta w'\delta w' d\theta\right\}_{t1}^{t2} - \int_{t1}^{t2}\int_0^l \rho A\,\omega \dot{w}\left\{\int_0^\theta w'\delta w' d\theta\right\} dx\,dt \tag{i}$$

In view of

$$\delta w(x, t_1) = 0$$

$$\delta w(x, t_2) = 0$$

the above integral in (i) simplifies to

$$I_{12} = -\int_{t1}^{t2}\int_0^l \rho A\,\omega \dot{w}\left\{\int_0^\theta w'\delta w' d\theta\right\} dx\,dt$$

Integrating by parts once again, and reverting θ to x, we get

$$I_{12} = -\int_{t1}^{t2}\int_0^l \rho A\omega\dot{w}\left\{w'\delta w\,|_0^\theta - \int_0^\theta w''\delta w\,d\theta\right\}dx\,dt$$

i.e.,

$$I_{12} = \int_{t1}^{t2}\int_0^l \rho A\omega\dot{w}\frac{\partial}{\partial x}\int_0^x w'dx\,\delta w\,dx\,dt \qquad (j)$$

Using the result of equation (5.58), equation (j) can be written as

$$I_{12} = \int_{t1}^{t2}\int_0^l \rho A\omega\frac{\partial}{\partial x}\left\{w'\int_x^l \dot{w}\,dx\right\}\delta w\,dx\,dt \qquad (k)$$

The second Coriolis term in (8.232) gives

$$I_2 = \frac{1}{2}\delta\int_{t1}^{t2}\int_0^l \rho A\omega\dot{w}\int_0^x w'^2dx\,dx\,dt$$

i.e.,

$$I_2 = \frac{1}{2}\int_{t1}^{t2}\int_0^l \rho A\omega\delta\dot{w}\int_0^x w'^2dx\,dx\,dt + \int_{t1}^{t2}\int_0^l \rho A\omega\dot{w}\int_0^x w'\delta w'dx\,dx\,dt \qquad (l)$$

Integrating by parts with respect to time, the first integral in (l) gives

$$I_{21} = \frac{1}{2}\int_0^l\left[\rho A\omega\delta w\int_0^x w'^2dx\,\Big|_{t2}^{t2}\right]dx - \frac{1}{2}\int_{t1}^{t2}\int_0^l \rho A\omega\delta w\frac{\partial}{\partial t}\int_0^\theta w'^2d\theta\,dx\,dt$$

i.e.,

$$I_{21} = -I_{11}$$

The second integral in (1) gives

$$I_{22} = I_{12}$$

Taking the appropriate signs into account and adding the Coriolis effects to the linear rotating beam problem of (8.225), we get

$$EI_{zz}w'''' - \rho A\omega^2 w - \rho A\omega^2\left\{w'\int_x^l (R+x)dx\right\} -$$

$$-\rho A\omega\frac{\partial}{\partial t}\left\{\int_0^x w'^2dx\right\} - 2\rho A\omega\frac{\partial}{\partial x}\left\{w'\int_x^l \dot{w}\,dx\right\} + \rho A\ddot{w} = 0 \qquad (8.233)$$

Letting $m = \rho A$, carrying out the integration of the third term in the above, and adding a linear viscous dashpot at the free end of the rotating beam along with a sinusoidal excitation force, we get

$$EI_{zz}w'''' - w''\left\{m\omega^2R(l-x) + \frac{1}{2}m\omega^2(l^2-x^2)\right\} + w'(m\omega^2R + m\omega^2x)$$

$$-2m\omega\frac{\partial}{\partial x}\left\{w'\int_x^l \dot{w}\,dx\right\} - m\omega\frac{\partial}{\partial t}\left\{\int_0^x w'^2dx\right\}$$

$$-m\omega^2w + m\ddot{w} + c\dot{w}_{x=l} = F\sin pt \qquad (8.234)$$

The steady-state solution of the rotating beam above can be obtained by using Galerkin method. Let the response be represented as

$$w(x,t) = f(x)(A_1 \sin pt + A_2 \cos pt)$$

$$+ g(x)(B_1 \sin 2pt + B_2 \cos 2pt) \tag{8.235}$$

We choose the shape functions in the above to be the first two terms from (5.109) as

$$f(X) = 6X^2 - 4X^3 + X^4$$

$$g(X) = 10X^3 - 10X^4 + 3X^5 \tag{8.236}$$

where

$$X = \frac{x}{l}$$

Substituting the above in (8.234), we get the error in the differential equation as follows:

$$\in = \in_1 + \in_2 \tag{8.237}$$

where

$$\in_1 = EI_{zz}\{f''''(A_1 \sin pt + A_2 \cos pt)$$

$$+ g''''(B_1 \sin 2pt + B_2 \cos 2pt)\} - \{f''(A_1 \sin pt + A_2 \cos pt)$$

$$+ g''(B_1 \sin 2pt + B_2 \cos 2pt)\}\left\{m\omega^2 R(l - x) + \frac{1}{2} m\omega^2 (l^2 - x^2)\right\}$$

$$+ \{f'(A_1 \sin pt + A_2 \cos pt) + g'(B_1 \sin 2pt + B_2 \cos 2pt)\}\,(m\omega^2 R + m\omega^2 x)$$

$$- 2m\omega \frac{\partial}{\partial x}[\{f'(A_1 \sin pt + A_2 \cos pt)$$

$$+ g'(B_1 \sin 2pt + B_2 \cos 2pt)\} \int_x^l \{pf(A_1 \cos pt - A_2 \sin pt)$$

$$+ 2pg(B_1 \cos 2pt - B_2 \sin 2pt)\}\,dx] - m\omega \frac{\partial}{\partial t}[\{\int_0^x \{f'(A_1 \sin pt + A_2 \cos pt)$$

$$+ g'(B_1 \sin 2pt + B_2 \cos 2pt)\}^2 dx] - m(\omega^2 + p^2)\{f(A_1 \sin pt + A_2 \cos pt)$$

$$- m(\omega^2 + 4p^2)\{g(B_1 \sin 2pt + B_2 \cos 2pt)\}$$

$$\in_2 = c\{fp(A_1 \cos pt - A_2 \sin pt)$$

$$+ 2gp(B_1 \cos 2pt - B_2 \sin 2pt)\}_{x=l} - F \sin pt$$

We can use Ritz averaging process to give

$$\int_0^{\frac{2\pi}{p}} \int_0^l \in_1 f \sin pt\,dx\,dt + \int_0^{\frac{2\pi}{p}} \in_2 f_{x=l} \sin pt\,dt = 0$$

$$\int_0^{\frac{2\pi}{p}} \int_0^l \in_1 f \cos pt\, dx\, dt + \int_0^{\frac{2\pi}{p}} \in_2 f_{x=1} \cos pt\, dt = 0$$

$$\int_0^{\frac{2\pi}{p}} \int_0^l \in_1 g \sin 2pt\, dx\, dt + \int_0^{\frac{2\pi}{p}} \in_2 g_{x=l} \sin 2pt\, dt = 0$$

$$\int_0^{\frac{2\pi}{p}} \int_0^l \in_1 g \cos 2pt\, dx\, dt + \int_0^{\frac{2\pi}{p}} \in_2 g_{x=1} \cos 2pt\, dt = 0 \tag{8.238}$$

Evaluating the above integrals, we get the following four nonlinear algebraic equations for the four unknowns A_1, A_2, B_1, B_2.

$$PA_1 + \frac{64}{65}\frac{\omega p}{l}(A_1B_1 + A_2B_2) - \frac{9cp}{ml}A_2 = \frac{3F}{ml}$$

$$PA_2 + \frac{64}{65}\frac{\omega p}{l}(A_1B_2 - A_2B_1) - \frac{9cp}{ml}A_1 = 0$$

$$QB_1 + \frac{32}{65}\frac{\omega p}{l}(A_1^2 - A_2^2) - \frac{18cp}{ml}B_2 = 0$$

$$QB_2 + \frac{32}{65}\frac{\omega p}{l}2A_1A_2 + \frac{18cp}{ml}B_1 = 0 \tag{8.239}$$

where

$$P = \frac{144EI_{zz}}{5ml^4} + \omega^2\left(\frac{18}{5}\frac{R}{l} + \frac{122}{45}\right) - \frac{104}{45}(\omega^2 + p^2)$$

$$Q = \frac{240EI_{zz}}{7ml^4} + \omega^2\left(\frac{10}{3}\frac{R}{l} + \frac{1850}{693}\right) - \frac{1304}{693}(\omega^2 + 4p^2)$$

Let

$$A_1 = a \sin\phi$$

$$A_2 = a \cos\phi$$

$$B_1 = b \sin\psi$$

$$B_2 = b \cos\psi$$

then equations (8.239) reduce to

$$C_0 + C_1a^2 + C_2a^4 + C_3a^6 = 0$$

$$b = Sa^2 \tag{8.240}$$

where

$$C_0 = \frac{9F^2}{m^2l^2}$$

$$C_1 = \frac{81c^2p^2}{m^2l^2} + p^2$$

$$C_2 = \frac{648S^2c^2p^2}{m^2l^2} - 4PQS^2$$

$$C_4 = \frac{1296S^4c^2p^2}{m^2l^2} + 4Q^2S^4$$

and

$$S = \frac{32m\omega p}{65\sqrt{324c^2p^2 + Q^2l^2m^2}}$$

We can write a suitable computer program and obtain the amplitudes of the first and second harmonics a and b of a rotating cantilever beam. The following data is used to illustrate the response of a typical rotating beam.

Length cf the blade = 25.4 cm
Disk radius = 15.24 cm
Breadth of the beam = 2.54 cm
Thickness of the beam = 0.3175 cm
Density = 0.00784 kg/cm^3
Young's modulus = 2.068×10^{11} Pa
Magnitude of force at the tip = 2224 N

The amplitude ratios a/A_{st}, b/A_{st} (A_{st} being the static deflection of the beam by setting $p = 0$) can be studied as a function of the frequency ratio p/p_{lr} (p_{lr} being the natural frequency of the linear beam obtained by setting c_1 and F equal to zero). The response is given in Figs. 8.45 through 8.48 for three different damping ratios $\xi = c/(2mlp_{lr}) = 0.0, 0.001$, and 0.01. The undamped results show that the first and second harmonic responses both have a hardening type of resonance near the linear resonant frequency and a softening type of resonance near the frequency ratio 0.6. When damping is included, the first harmonic response in Fig. 8.47 comprises of resonance as in the classical linear case, with a hardening characteristic. The second harmonic has, however, resonance in both the regions of the frequency ratio 0.6 and 1.0.

8.14 REISSNER'S DYNAMIC FUNCTIONAL AND STRESS ANALYSIS

In this section, we will study the stresses induced in a beam due to periodic excitation. The dynamic stress analysis is feasible in an exact manner only in a very few cases. As in the case of dynamic analysis for free and forced vibration, we may have to resort to approximate methods of solution. We should make sure that the results obtained by a chosen method are appropriate and within reasonable accuracy. It will be shown in this section that Reissner's dynamic functional is superior over the conventional Lagrangian functional for determining the dynamic stresses.

Consider a simple cantilever fixed at $x = 0$ and acted on by a sinusoidal force $P_0 \sin \omega t$ at the free end at $x = l$ as shown in Fig. 8.49. The depth of the beam cross-section is taken as $2b$. The dynamic stresses due to the force at the free end can be determined by the exact solution of an Euler-Bernoulli beam without any approximation. We will examine solutions of this problem using a five term solution by Lagrangian as well as Reissner's dynamic functional in the following.

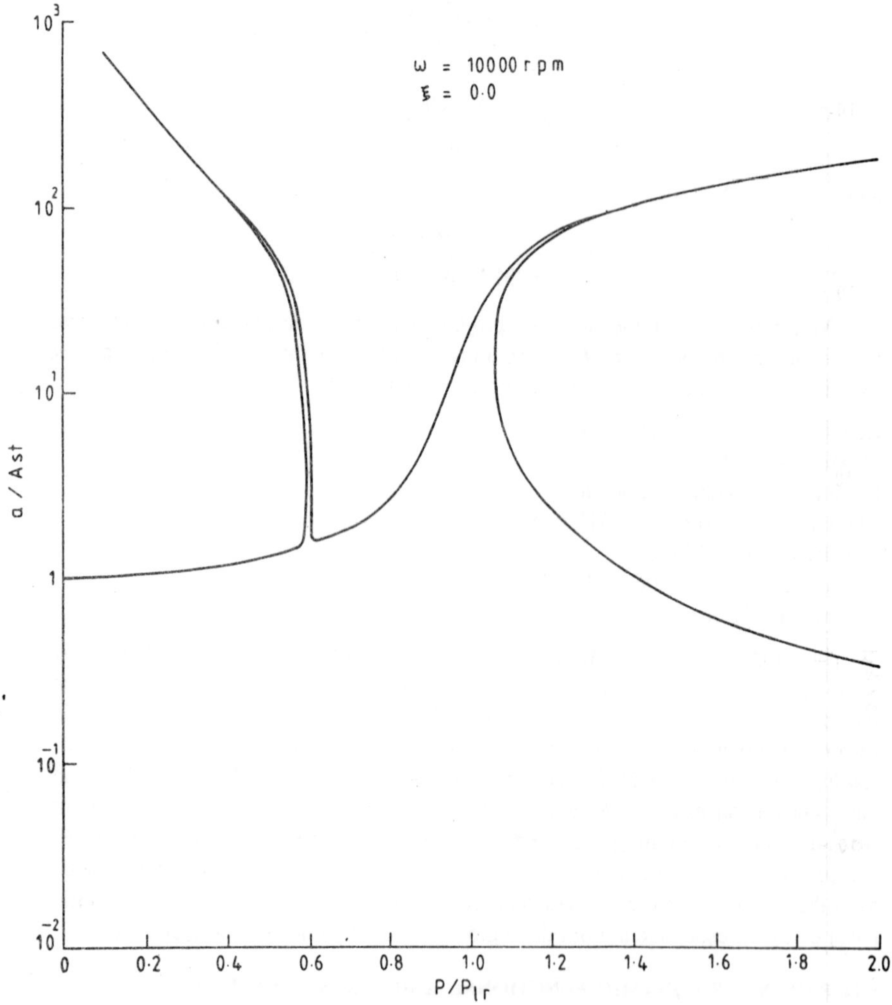

Fig. 8.45 Amplitude ratio a/A_{st} against frequency ratio p/p_{1r}

Case 1 Exact solution

The differential equation of an Euler beam with the exciting force at the free end is given by

$$a^2 \frac{\partial^4 w}{\partial x^4} + \frac{\partial^2 w}{\partial t^2} = P_0 \left.\right|_{x=l} \sin \omega t \tag{8.241}$$

where

$$a^2 = \frac{EI_{yy}}{\rho A} \tag{8.242}$$

The natural frequencies and mode shapes have been determined earlier (case 3, section 8.2).

Frequency equation

$$\cosh\sqrt{\frac{p_i}{a}}\,l\cos\sqrt{\frac{p_i}{a}}\,l+1=0$$

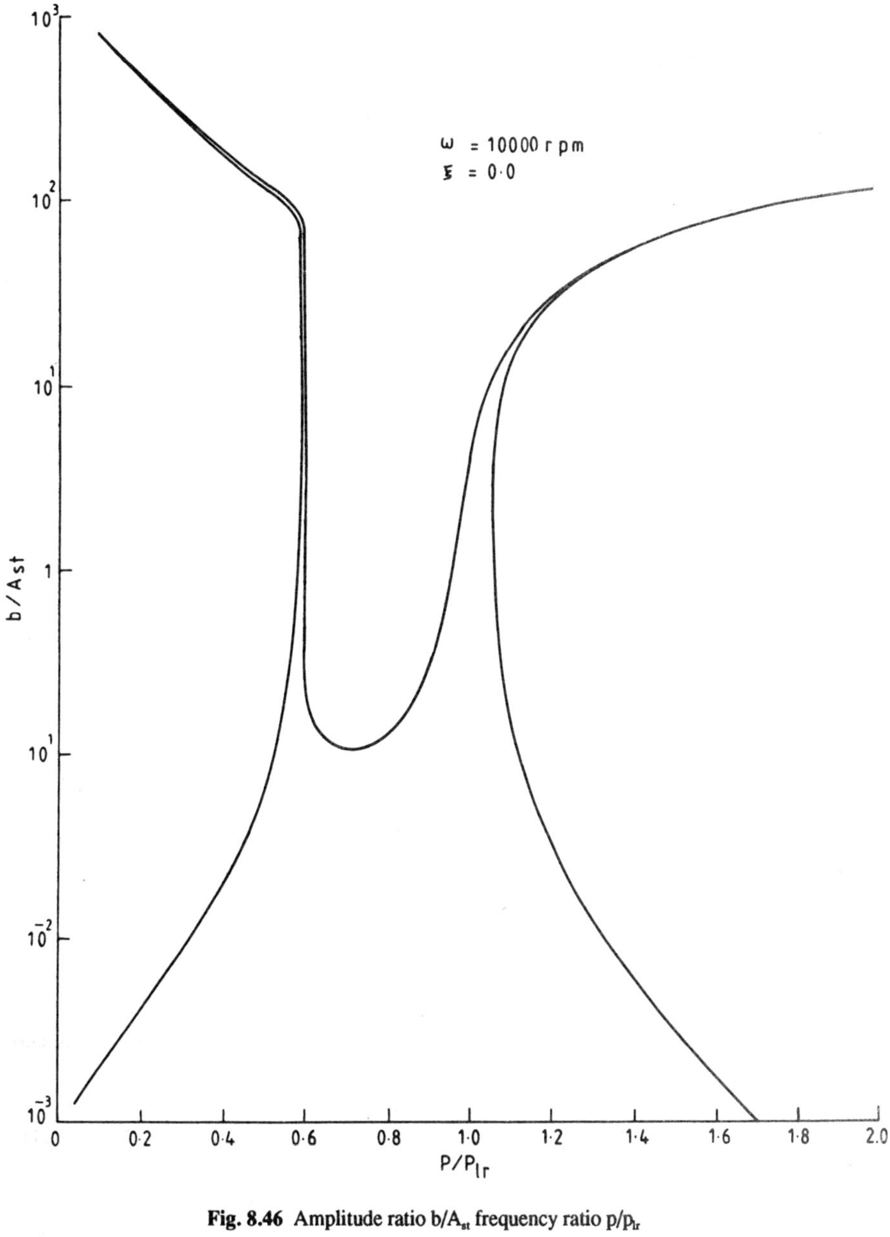

Fig. 8.46 Amplitude ratio b/A_{st} frequency ratio p/p_{lr}

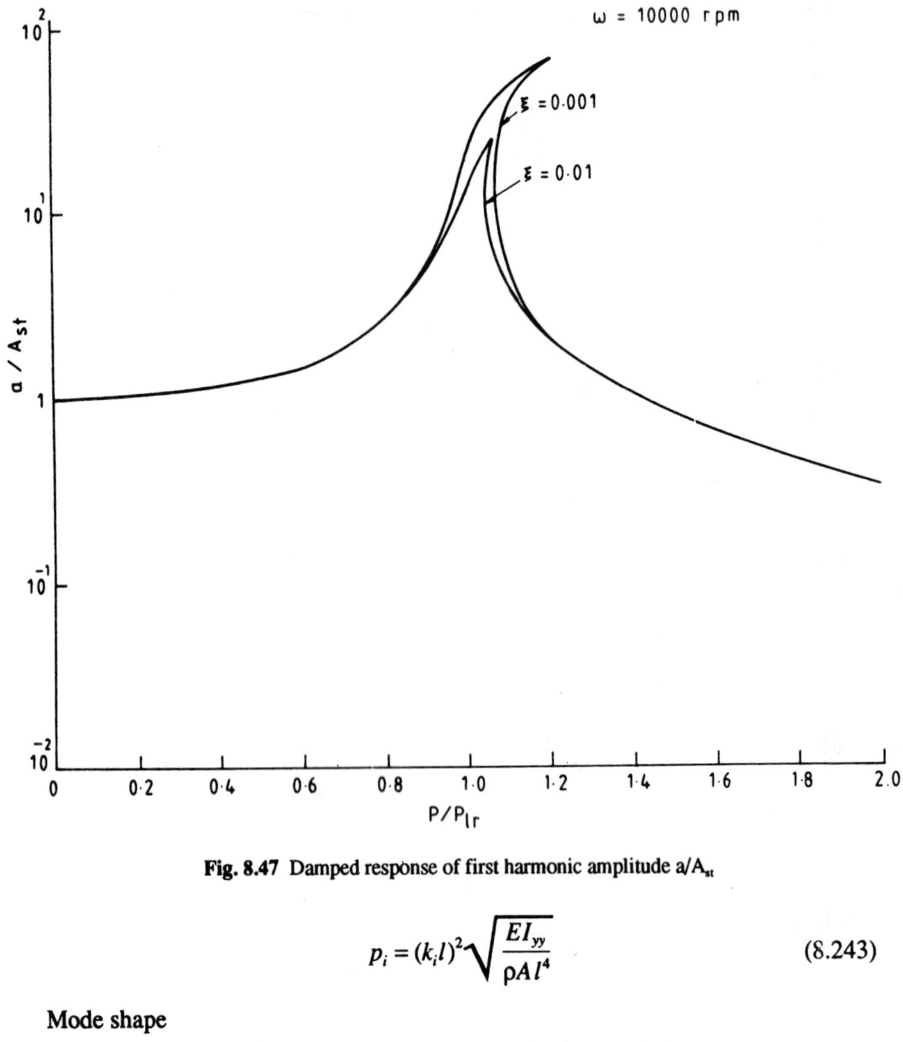

Fig. 8.47 Damped response of first harmonic amplitude a/A_{st}

$$p_i = (k_i l)^2 \sqrt{\frac{EI_{yy}}{\rho A l^4}} \tag{8.243}$$

Mode shape

$$\left(\sinh\sqrt{\frac{p_i}{a}}\, l + \sin\sqrt{\frac{p_i}{a}}\, l\right)\left(\cosh\sqrt{\frac{p_i}{a}}\, x\right.$$

$$\left.-\cos\sqrt{\frac{p_i}{a}}\, x\right) - \left(\cosh\sqrt{\frac{p_i}{a}}\, l + \cos\sqrt{\frac{p_i}{a}}\, l\right)\left(\sinh\sqrt{\frac{p_i}{a}}\, x - \sin\sqrt{\frac{p_i}{a}}\, x\right)$$

Using eigen expansion process the solution for steady-state vibration is obtained as

$$w = \sum_j \phi_j(t) X_j(x)$$

$$\ddot{\phi}_i + \frac{a^2 (k_i l)^4}{l^4} \phi_t = \frac{P_0}{\rho A l} \{X_i\}_{x=l} \sin \omega t$$

$$\phi_i = \frac{P_0 l^3}{\rho A} \frac{\{X_i\}_{x=l}}{(kl)^4 a^2 - \omega^2 l^4} \sin \omega t$$

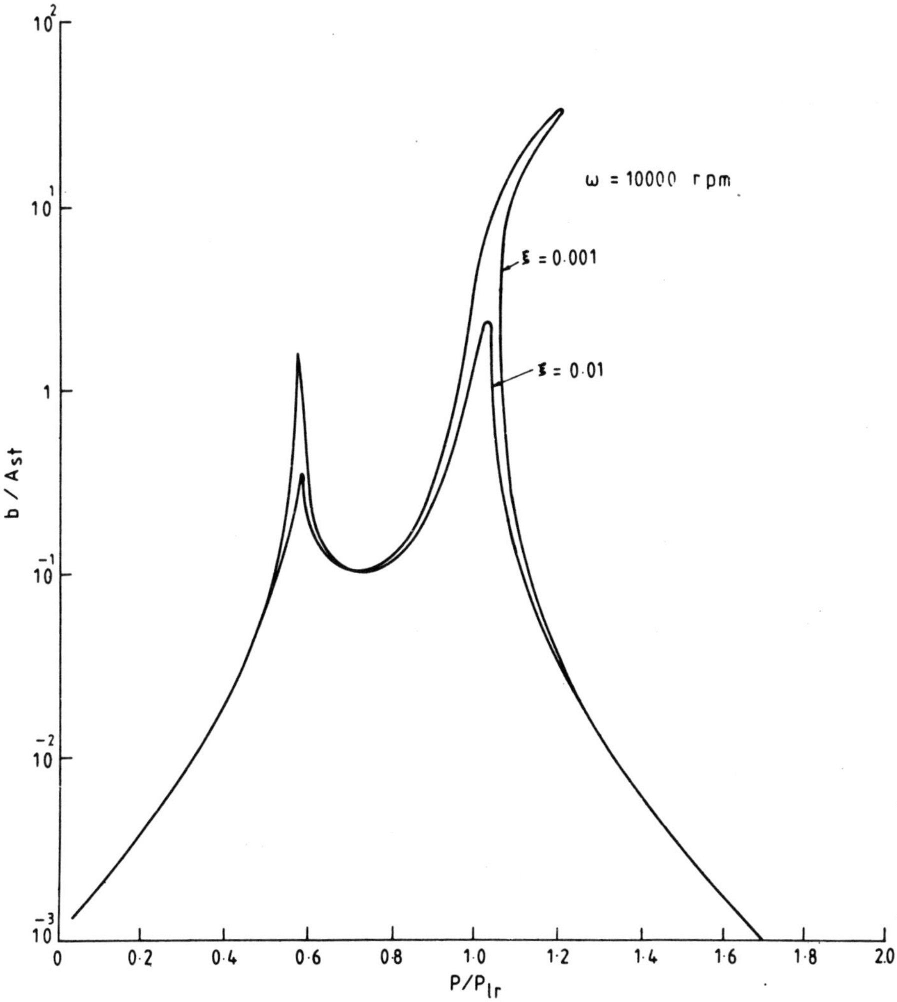

Fig. 8.48 Damped response of second harmonic amplitude b/A_{st}

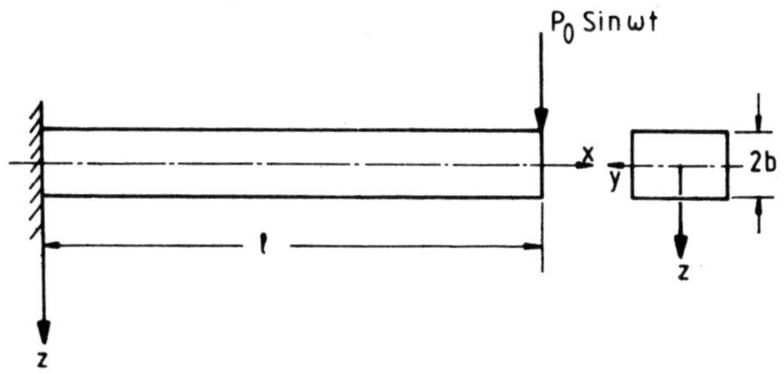

Fig. 8.49 A cantilever excited by sinusoidal force at free end

Therefore

$$w = \frac{P_0}{\rho A l} \sum_i \frac{X_i \{X_i\}_{x=l}}{p_i^2 - \omega^2} \sin \omega t$$

If we include damping in the analysis, and ξ_i is the modal damping ratio in ith mode, the above gets modified to

$$w = \frac{P_0}{\rho A l} \sum_i \frac{X_i \{X_i\}_{x=l}}{\sqrt{(p_i^2 - \omega^2)^2 + 4\xi_i^2 p_i^2 \omega^2}} \sin(\omega t - \psi_i) \tag{8.244}$$

The bending stress is then given by

$$\sigma = Ew''b$$

$$= \frac{EP_0 b}{\rho A l} \sum_i \frac{X_i'' \{X_i\}_{x=l}}{\sqrt{(p_i^2 - \omega^2)^2 + 4\xi_i^2 p_i^2 \omega^2}} \sin(\omega t - \psi_i) \tag{8.245}$$

Case 2 Lagrangian method

The kinetic energy is

$$T = \frac{1}{2} \rho A \int_0^l \dot{w}^2 dx$$

The strain energy is

$$U = \frac{1}{2} EI_{yy} \int_0^l w''^2 dx$$

Let

$$w = \sum_{i=1}^{n} f_i(x) q_i(t) \tag{8.246}$$

where

$$f_i(x) = \frac{1}{6}(i+2)(i+3)X^{i+1} - \frac{1}{3} i(i+3)X^{i+2} + \frac{1}{6} i(i+1)X^{i+3} \qquad X = \frac{x}{l}$$

Using the Lagrangian equation

$$\frac{d}{dt}\left\{\frac{\partial T}{\partial \dot{q}}\right\} - \frac{\partial T}{\partial q} + \frac{\partial U}{\partial q} = 0$$

we get

$$[M]\{\ddot{q}\} + [K]\{q\} = 0$$

where

$$[M] = \left[\rho A \int_0^l f_i f_j dx\right]$$

$$[K] = \left[EI_{yy} \int_0^l f_i'' f_j'' dx\right]$$

$$\{q\} = \begin{Bmatrix} q_1 \\ q_2 \\ \dots \\ q_n \end{Bmatrix}$$

For the forced vibration analysis

$$[M]\{\ddot{q}\} + [K]\{q\} = \{Q_0\}\sin\omega t$$

where

$$\{Q_0\} = \{P_0 f_j|_{x=l}\}$$

Using modal analysis, we get

$$\{q\} = [U]\{\eta\} \qquad (a)$$

$$[\overline{M}]\{\ddot{\eta}\} + [\overline{K}]\{\eta\} = \{\overline{Q}_0\}\sin\omega t$$

where

$$[\overline{M}] = [U]^T[M][U]$$
$$[\overline{K}] = [U]^T[K][U]$$

Including modal damping, the forced vibration response is

$$\eta_k = \frac{\overline{Q}_{0k}}{M_k\sqrt{(p_i^2-\omega^2)^2 + 4\xi_i^2 p_i^2\omega^2}}\sin(\omega t - \psi_i) \qquad (8.247)$$

Making use of (8.246) and (a), the bending stress can be determined as in equation (8.245).

Case 3 Reissner method

The Reissner's functional is

$$I_R = \iiint_V \{\tau_{ij}\varepsilon_{ij} - U_0^*(\tau_{ij})\}\,dv$$

$$-\iiint_V B_i u_i dv - \iint_s T_i^{(v)} u_i dA \qquad (5.88)$$

The nonzero strain and stress are

$$\varepsilon_{xx} = -zw''$$

$$\tau_{xx} = -Ezw''$$

and the bending moment is

$$M_y = -EI_{yy}w''$$

$$\iiint_V \tau_{ij}\varepsilon_{ij}dv = -\int_0^l M_x w'' dx$$

$$\iiint_V U_0{}^*(\tau_{ij})dv = \iiint\int_0^{\tau_{xx}} \frac{\tau}{E}\,d(\tau_{xx})dv$$

$$= \iiint_V \frac{\tau_{xx}^2}{2E}dv$$

$$= \int_0^l \frac{M_y^2}{2EI_{yy}}dx$$

Neglecting the body forces, the Reissner's dynamic functional for free vibration problem is

$$L = T - I_R = \int_0^l \left[\frac{1}{2}\rho A \dot{w}^2 + M_y w'' + \frac{M_y^2}{2EI_{yy}}\right] dx \tag{8.248}$$

Let

$$w = \sum_{i=1}^{n} B_i(t) f_i(X)$$

$$M_y = \sum_{i=1}^{n} D_i(t) h_i(X)$$

where

$$f_i(X) = \frac{1}{6}(i+2)(i+3)X^{i+1} - \frac{1}{3}i(i+3)X^{i+2} + \frac{1}{6}i(i+1)X^{i+3} \quad X = \frac{x}{l}$$

$$h_i(x) = \frac{(1-X)^i}{i+1}$$

Substituting the above in (8.248), we get

$$L = T - I_R = \int_0^l \Bigg[\frac{1}{2}\rho A \{\Sigma \dot{B}_i(t) f_i(X)\}^2$$

$$+ \{\Sigma D_i(t) h_i(X)\}\{\Sigma B_i(t) f_i''(X)\} + \frac{\{\Sigma D_i(t) h_i(X)\}^2}{2EI_{yy}}\Bigg] dx$$

Lagrangian equations for the above are

$$\frac{d}{dt}\left(\frac{\partial L}{\partial \dot{B}_i}\right) - \frac{\partial L}{\partial B_i} = 0$$

$$\frac{d}{dt}\left(\frac{\partial L}{\partial \dot{D}_i}\right) - \frac{\partial L}{\partial D_i} = 0$$

The above lead to

$$\begin{bmatrix} M_1 & M_2 \\ M_3 & M_4 \end{bmatrix}\begin{Bmatrix} \ddot{B} \\ \ddot{D} \end{Bmatrix} + \begin{bmatrix} K_1 & K_2 \\ K_3 & K_4 \end{bmatrix}\begin{Bmatrix} B \\ D \end{Bmatrix} = 0$$

$$[M_1] = \left[\rho A \int_0^l f_i f_j dx\right]$$

$$[M_2] = [M_3] = [M_4] = 0$$

$$[K_1] = 0$$

$$[K_2] = \left[-\int_0^l h_i f_j'' dx\right]$$

$$[K_3] = \left[-\int_0^l f_i'' h_j dx\right]$$

$$[K_4] = \left[-\frac{1}{EI_{yy}}\int_0^l h_i h_j dx\right]$$

The traction force for the forced vibration problem is

$$\int\int_s T_i^{(v)} u_i dA = \int_0^l F(x,t) w(x) dx$$

We can incorporate the external force acting on the beam and including modal damping, we get

$$\begin{bmatrix} M_1 & M_2 \\ M_3 & M_4 \end{bmatrix} \begin{Bmatrix} \ddot{B} \\ \ddot{D} \end{Bmatrix} + \begin{bmatrix} C_1 & C_2 \\ C_3 & C_4 \end{bmatrix} \begin{Bmatrix} \dot{B} \\ \dot{D} \end{Bmatrix} + \begin{bmatrix} K_1 & K_2 \\ K_3 & K_4 \end{bmatrix} \begin{Bmatrix} B \\ D \end{Bmatrix} = \begin{Bmatrix} Q_0 \\ 0 \end{Bmatrix} \quad (8.249)$$

where

$$C_1 = \frac{C_v}{\rho M_1}$$

$$Q_0 = F_0 \{f_i\}_{x=l}$$

We can follow the procedure from here onwards as in Lagrangian method and determine the stress from

$$\sigma = \frac{M_y}{I_{yy}} b \quad (8.250)$$

It is obvious a computer program will be useful to carry out the analysis. For the purpose of illustration, let us consider the following data.

$$l = 10\text{cm}$$

$$I_{yy} = 0.00675\text{cm}^4$$

$$E = 2.11 \times 10^{11}\text{Pa}$$

$$A = 0.9\text{cm}^2$$

$$b = 0.15\text{cm}$$

$$\rho = 0.00786 \text{ kg/cm}^3$$

$$\omega = 100 \text{ rad/s}$$

$$\xi_k = 0.01$$

Figures 8.50a and 8.50b show the convergency of the deflections and stresses at the free end, respectively. In the exact method, as many number of terms as possible should be taken into account. We find from these figures that Reissner's method gives correct results even with one-term solution for both deflection and stress. With Lagrange method, one needs at least four terms to get convergency in the deflection, but no convergence is obtained for stress even with a five-term solution. Figures 8.50c and 8.50d give the deflection and stress distributions along the length of the beam with a five-term solution. It can be seen that with Lagrange method, the estimates of stress are in considerable error in comparison to the exact method, particularly at the root where the stress values are large.

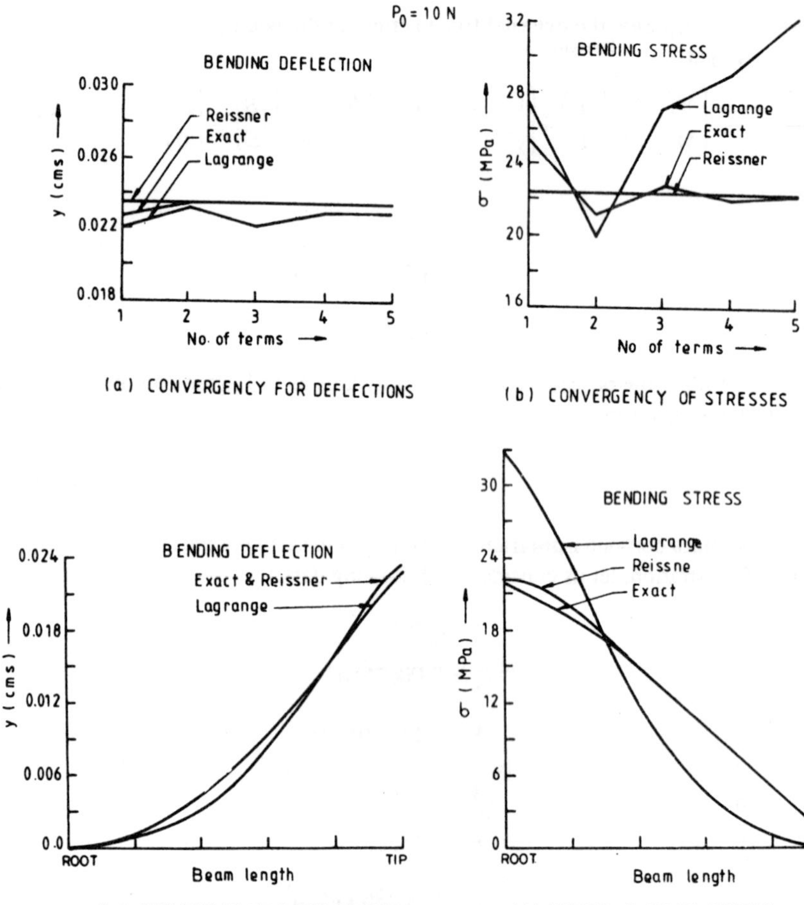

Fig. 8.50 Bending deflection and stresses of a cantilever beam

Table 8.1 gives the convergency pattern of the natural frequencies by the Lagrangian and Reissner methods in comparison with the exact values of the cantilever beam. Both these methods give quick convergency and near-exact results for the first three modes. For the fourth and fifth modes, Reissner method gives better results over Lagrangian method.

The tip deflections and stresses for resonant excitation of the beam in the first five modes obtained by a five-term solution are given in Tables 8.2 and 8.3, respectively. Since a five-term solution is used, we should expect good results for the first three modes. Reissner method is in very good agreement with the exact values of stresses in comparison with Lagrange method. In the fourth and fifth modes, Reissner method predicts stresses much closer to the exact values, in comparison with

Lagrange method. The above discussion shows that Reissner method formulation may be preferred over Lagrangian method when stresses are of importance in the analysis.

Table 8.1
Natural frequencies (Hz) in bending by Lagrangian and Reissner methods compared with exact values

n	*Exact value*	*Lagrange method*	*Reissner method*
1	248.65	249.67	241.70
2		248.68	248.48
	1558.36	1606.40	1473.40
3		248.68	248.67
		1558.50	1541.00
	4363.89	4686.10	4104.40
4		248.68	248.68
		1558.50	1558.50
		4368.60	4187.80
	8551.67	9638.70	8251.40
5		248.68	248.68
		1558.40	1558.40
		4366.60	4365.00
		8599.70	8563.30
	14135.01	16901.00	14468.00

Table 8.2
Resonant response–tip deflection (cm)

Resonance	*Exact value*	*Lagrange method*	*Reissner method*
I	1.1368	1.1352	1.1366
II	0.0296	0.0286	0.0296
III	0.0039	0.0037	0.0039
IV	0.0010	0.0011	0.0011
V	0.0004	0.0001	0.0001

Table 8.3
Resonant response–tip stress (MPa)

Resonance	*Exact value*	*Lagrange method*	*Reissner method*
I	1261.30	1270.89	1261.95
II	199.73	205.36	199.27
III	71.00	73.76	71.02
IV	35.63	41.20	39.50
V	21.78	39.04	27.35

Tensors

A1.1 *Index Notation*

Range convention: Index notation is commonly used to denote the three cartesian coordinates of a point. Let x_1, x_2, x_3 represent the three coordinate axes as in Fig. A1.1. The coordinates of point P (a_1, a_2, a_3) can be written in the index notation as a_i. Two indices are used to denote 9 quantities, e.g., in a_{ij} both i and j can take the values 1, 2 and 3 representing $a_{11}, a_{12}, \ldots a_{33}$. Similarly, a_{ijk} will represent a set of 27 quantities and a_{ijkl} a set of 81 quantities. In the range convention, each subscript can appear only once in a term.

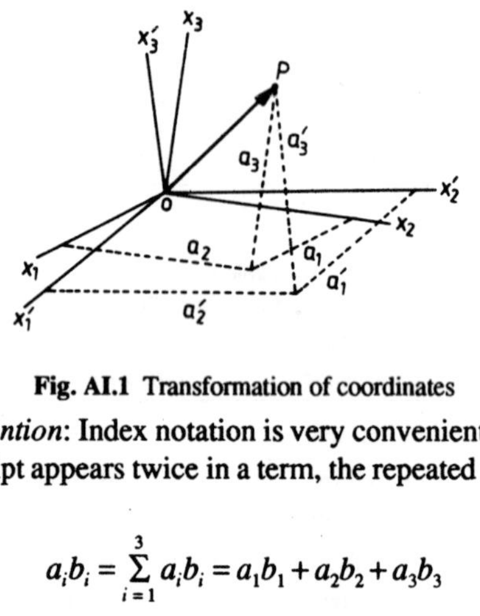

Fig. AI.1 Transformation of coordinates

Summation convention: Index notation is very convenient in writing summation quantities. If a subscript appears twice in a term, the repeated index is summed from 1 to 3. e.g.,

$$a_i b_i = \sum_{i=1}^{3} a_i b_i = a_1 b_1 + a_2 b_2 + a_3 b_3 \tag{A 1.1}$$

$$p_{ij} q_i r_j = \sum_{i=1}^{3} \sum_{j=1}^{3} p_{ij} q_i r_j = p_{11} q_1 r_1 + p_{12} q_1 r_2 + p_{13} q_1 r_3$$

$$+ p_{21} q_2 r_1 + p_{22} q_2 r_2 + P_{23} q_2 r_3 + p_{31} q_3 r_1 + p_{32} q_3 r_2 + p_{33} q_3 r_3 \tag{A 1.2}$$

An unrepeated index remains free, as in the following:

$$a_{ij} b_i = \sum_{i=1}^{3} a_{ij} b_i = a_{1j} b_1 + a_{2j} b_2 + a_{3j} b_3 \tag{A 1.3}$$

Comma convention: This is used to denote partial differentiation. A subscript comma followed by an index, say i, indicates partial differentiation with respect to each x_i, e.g.,

$$u_{i,j} = \frac{\partial u_i}{\partial x_j} \tag{A 1.4}$$

The above represents 9 quantities viz.,

$$\frac{\partial u_1}{\partial x_1}, \frac{\partial u_1}{\partial x_2}, \quad \cdots \quad \frac{\partial u_3}{\partial x_3}$$

The second partial derivative is represented by

$$u_{i,jk} = \frac{\partial^2 u_i}{\partial x_j \partial x_k} \tag{A1.5}$$

The range and summation conventions apply to indices following the comma also. i.e.,

$$a_{i,i} = \frac{\partial a_i}{\partial x_i} = \frac{\partial a_1}{\partial x_1} + \frac{\partial a_2}{\partial x_2} + \frac{\partial a_3}{\partial x_3} \tag{A1.6}$$

A1.2 *Tensors*

Consider two cartesian coordinate systems with the common origin O, as shown in Fig. A1.1. Let the point P have coordinates a_1, a_2, a_3 in x_1, x_2, x_3 axis system. The coordinates of the same point in x_1', x_2', x_3' can be obtained from the following linear transformation.

$$a'_1 = \alpha_{11} a_1 + \alpha_{12} a_2 + \alpha_{13} a_3$$

$$a'_2 = \alpha_{21} a_1 + \alpha_{22} a_2 + \alpha_{23} a_3$$

$$a'_3 = \alpha_{31} a_1 + \alpha_{32} a_2 + \alpha_{33} a_3 \tag{A1.7}$$

In the above the quantities α_{ij} are the direction cosines.

$$\alpha_{ij} = \cos(x'_i, x_j) \tag{A1.8}$$

Equations (A1.7) can be written in the index notation as

$$a'_i = \alpha_{ij} a_j \tag{A1.9}$$

The above defines the first order tensor, in which a_i are its components in the x_i coordinate system and a_i' are the components of the same tensor in the x_i' system.

A second order tensor a_{ij} referred to x_i coordinate system transforms to another set of x_i' coordinate system by the following.

$$a'_{ij} = \alpha_{ik} \alpha_{jl} a_{kl} \tag{A1.10}$$

Kronecker delta is a special second order tensor defined as

$$\begin{aligned} \delta_{ij} &= 1 \quad i = j \\ &= 0 \quad i \neq j \end{aligned} \tag{A1.11}$$

The above is useful as a substitution tensor, e.g.,

$$\delta_{ij} a_j = a_i$$

$$\delta_{ij} a_{jk} = a_{ik} \tag{A1.12}$$

Gauss' Theorem

Consider a second order tensor τ_{ij} defined at each point in space in the domain V (*see* Fig. A2.1). In the direction x_1, lines parallel to the x_1 axis intersect the boundary only twice. An infinitesimal prism of volume δV parallel to the x_1 axis with sides dx_2, dx_3 is shown in the figure. Several such prisms make up the whole volume of the domain V. Over the infinitesimally small prism consider the integral as follows:

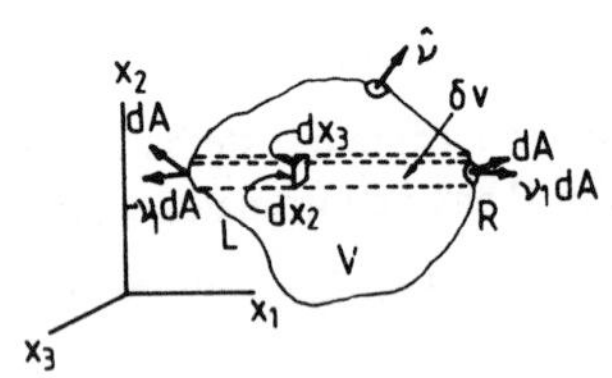

Fig. A2.1 Conversion of volume integral to surface integral

$$\iiint_{\delta V} \frac{\partial \tau_{ij}}{\partial x_1} dx_1 dx_2 dx_3$$

Integrating with respect to x_1, we get

$$\iiint_{\delta V} \frac{\partial \tau_{ij}}{\delta x_1} dx_1 dx_2 dx_3 = \iint \{\tau_{ij} dx_2 dx_3\}_R - \iint \{\tau_{ij} dx_2 dx_3\}_L \tag{A2.1}$$

Let ν be the unit normal to the boundary surface of V, with ν_1, ν_2, ν_3 its direction cosines. Then, the area on the right hand side is $\nu_1\, dA$ and that on the left hand side of the prism is $-\nu_1\, dA$. Then, the above equation can be written as

$$\iiint_{\delta V} \frac{\partial \tau_{ij}}{\partial x_1} dx_1 dx_2 dx_3 = \iint \{\tau_{ij} \nu_1 dA\}_R + \iint \{\tau_{ij} \nu_1 dA\}_L \tag{A2.2}$$

Integrating over all the prisms, we get

$$\iiint_{V} \frac{\partial \tau_{ij}}{\partial x_1} dx_1 dx_2 dx_3 = \iint_R \tau_{ij} \nu_1 dA + \iint_L \tau_{ij} \nu_1 dA \tag{A2.3}$$

i.e.,

$$\iiint_{V} \frac{\partial \tau_{ij}}{\partial x_1} dx_1 dx_2 dx_3 = \iint_s \tau_{ij} \nu_1 dA \tag{A2.4}$$

In general (A2.4) is

$$\iiint_{V} \frac{\partial \tau_{ij..}}{\partial x_i} dV = \iint_s \tau_{ij..} \nu_i dA \tag{A2.5}$$

If the tensor is of first order, (A2.5) becomes, divergence theorem.

Linear Elasticity Principles

A3.1 *Force Distribution*

Body force: The body force distribution acts directly on the distributed matter of the body in its domain V. This is represented by $B\ (x, y, z, t)$ in vector form or in index notation $B_i\ (x_1, x_2, x_3)$. This is a force intensity function expressed per unit mass or per unit volume of the matter of the body.

Surface traction: If the force applied is restricted to the surface of the body as in the case of a beam or plate, such force distributions are termed surface tractions, denoted as $T\ (x, y, z, t)$ or $T_i\ (x_1, x_2, x_3)$. This force need not be normal to the surface element (*see* Fig. A3.1) and therefore we can denote the unit normal to the surface element by the notation

$$T^{(v)}(x, y, z, t) \quad \text{or} \quad T_i^{(v)}(x_1, x_2, x_3, t)$$

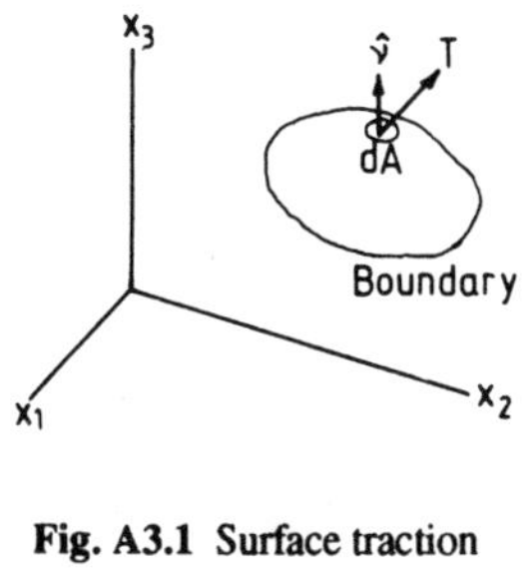

Fig. A3.1 Surface traction

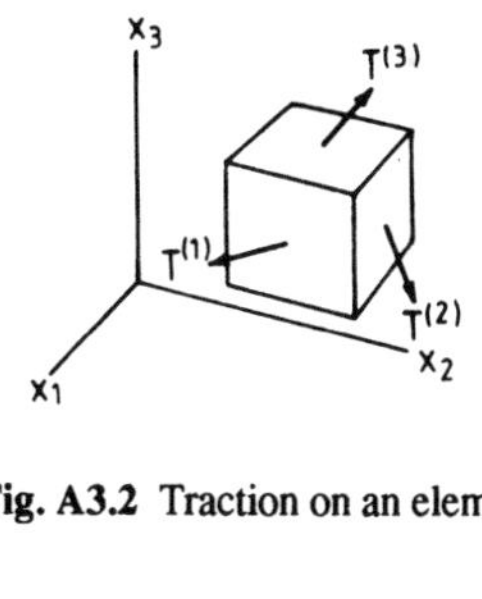

Fig. A3.2 Traction on an element

A3.2 *Cauchy's Formula*

Stress: Stress is defined at a point in the body, which is force per unit area. Consider an infinitely small rectangular parallelepiped of the body at a point, shown in Fig. A3.2. The traction forces on the surface of the element at this point are

$$T_i^{(1)} = \tau_{11}, \tau_{12}, \tau_{13}$$

$$T_i^{(2)} = \tau_{21}, \tau_{22}, \tau_{23}$$

$$T_i^{(3)} = \tau_{31}, \tau_{32}, \tau_{33}$$

This forms the stress tensor

$$T_j^{(i)} = \tau_{ij} \tag{A3.1}$$

i.e.,

$$T_j^{(i)} = \begin{bmatrix} \tau_{11} & \tau_{12} & \tau_{13} \\ \tau_{21} & \tau_{22} & \tau_{23} \\ \tau_{31} & \tau_{32} & \tau_{33} \end{bmatrix} \tag{A3.2}$$

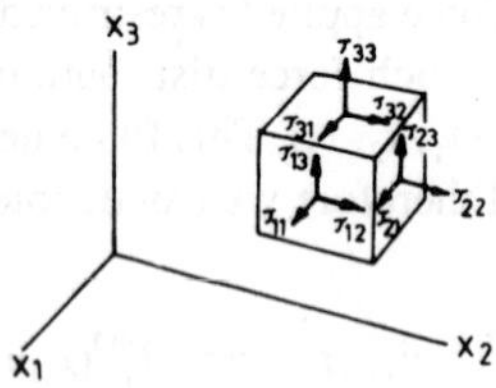

Fig. A3.3 Notation for stress components

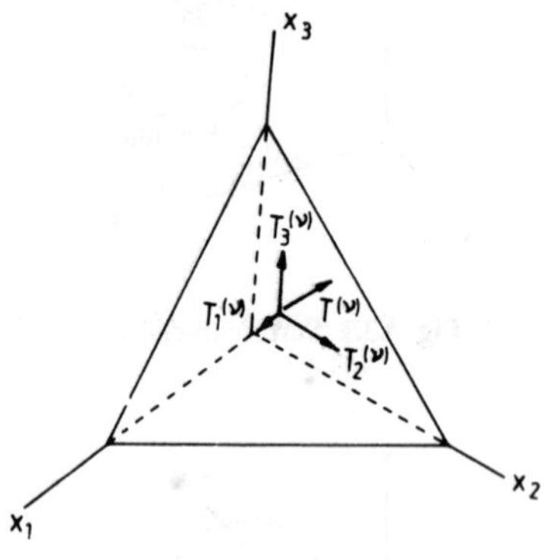

Fig. A3.4 Traction vector on an interface

The notation for these stress components is given in Fig. A3.3. The main diagonal elements in (A3.2) are normal stresses and the off diagonal elements are the shear stresses. Note that the stress tensor is symmetric.

The traction vector for any other interface as in Fig. A3.4 can be obtained by simple equilibrium conditions and is given by Cauchy's formula, which is

$$T_i^{(\nu)} = \tau_{ij} \nu_j \tag{A3.3}$$

Principle stresses: We can use equation (A1.10) to transform the stress components above into another system of axes x'_1, x'_2, x'_3 using the direction cosines α_{ij} as

$$\tau'_{ij} = \alpha_{ik}\alpha_{jl}\tau_{kl} \tag{A3.4}$$

If the stress vector on any interface is normal to the surface itself, we have principal planes, and the stresses on these planes which are normal, the principal stresses σ. The shear stresses on these planes are zero.

$$\sigma\nu_i = \tau_{ij}\nu_j$$

i.e.,

$$\{\tau_{ij} - \sigma\delta_{ij}\}\nu_j = 0$$

with the condition

$$\nu_i\nu_i = 1$$

The above lead to the following cubic equation from which the principal stresses can be determined.

$$\sigma^3 - I_\tau\sigma^2 + II_\tau\sigma - III_\tau = 0 \tag{A3.5}$$

where

$$I_\tau = \tau_{11} + \tau_{22} + \tau_{33}$$

$$II_\tau = \tau_{11}\tau_{22} + \tau_{22}\tau_{33} + \tau_{11}\tau_{33} - \tau_{12}^2 - \tau_{13}^2 - \tau_{23}^2$$

$$III_\tau = \tau_{11}\tau_{22}\tau_{33} - \tau_{11}\tau_{23}^2 - \tau_{22}\tau_{13}^2 - \tau_{33}\tau_{12}^2 + 2\tau_{12}\tau_{13}\tau_{23}$$

are the tensor invariants, as they are independent of the reference at a point.

A3.3 *Equations of Motion*

Let the domain of the body be V and its surface be represented by S. Using Newton's third law, we conclude that the tractions on the bounding surface only do not cancel and therefore

$$\iint_S T_i^{(\nu)} dA + \iiint_V B_i dV = \iiint_{V_i} \dot{V}_i \rho dV$$

where ρ is the mass density and V is the velocity vector at any point in the body. With the help of (A3.3), the above becomes

$$\iint_S \tau_{ji}\nu_j dA + \iiint_V B_i dV = \iiint_V \dot{V}_i \rho dV$$

We now use equation (A2.5) to transform the surface integral in the above and obtain

$$\iiint_V \{\tau_{ji,j} + B_i - \dot{V}_i\rho\} dV = 0$$

i.e.,

$$\tau_{ji,j} + B_i = \dot{V}_i \rho \qquad (A3.6)$$

A3.4 *Strain*

Consider a body shown in Fig. A3.5 with two points p and q at positions x_i and $x_i + dx_i$. The distance between these two points is given by

$$(ds)^2 = dx_i dx_i \qquad (A3.7)$$

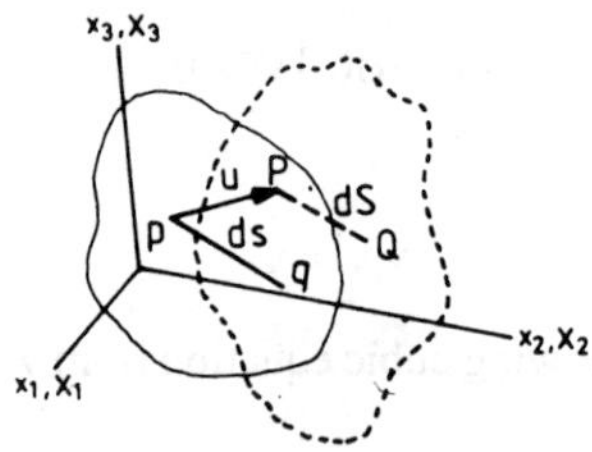

Fig. A3.5 Deformations in a body

When the body deforms, the points p and q move to P and Q, respectively. For the deformed state it is convenient to rename the axis system X_i. That is, the deformation can be depicted by a mapping of each point from coordinate x_i to coordinate X_i. Then

$$X_i = X_i(x_1, x_2, x_3)$$

$$x_i = x_i(X_1, X_2, X_3)$$

From the above we have

$$dX_i = \left\{\frac{\partial X_i}{\partial x_j}\right\} dx_j$$

$$dx_i = \left\{\frac{\partial x_i}{\partial X_j}\right\} dX_j$$

Equation (A3.7) can now be written as

$$(ds)^2 = dx_i dx_i = \frac{\partial x_i}{\partial X_m}\frac{\partial x_i}{\partial X_k} dX_m dX_k$$

Similarly, for the deformed state

$$(dS)^2 + dX_i dX_i = \frac{\partial X_i}{\partial x_m}\frac{\partial X_i}{\partial x_k} dx_m dx_k$$

From the above two equations, we get

$$(dS)^2 - (ds)^2 = \left[\frac{\partial X_k}{\partial x_i}\frac{\partial X_k}{\partial x_i} - \delta_{ij}\right] dx_i dx_j$$

Writing the above

$$(dS)^2 - (ds)^2 = 2\varepsilon_{ij} dx_i dx_j$$

where

$$\varepsilon_{ij} = \frac{1}{2}\left[\frac{\partial X_k}{\partial x_i}\frac{\partial X_k}{\partial x_i} - \delta_{ij}\right] \qquad (A3.8)$$

In the above ε_{ij} are expressed as functions of the coordinates in the undeformed state, also called Lagrangian coordinates, and it is known as Green's strain tensor.

The displacement field is given by

$$u_i = X_i - x_i$$

Therefore

$$\frac{\partial X_i}{\partial x_j} = \frac{\partial u_i}{\partial x_j} + \delta_{ij}$$

Substituting the above in equation (A3.8), we have

$$\varepsilon_{ij} = \frac{1}{2}\left[\frac{\partial u_i}{\partial x_j} + \frac{\partial u_j}{\partial x_i} + \frac{\partial u_k}{\partial x_i}\frac{\partial u_k}{\partial x_j}\right]$$

For infinitesimal strain, i.e.,

$$\frac{\partial u_i}{\partial x_j} \ll 1$$

the above ε_{ij} becomes

$$\varepsilon_{ij} = \frac{1}{2}\left(\frac{\partial u_i}{\partial x_j} + \frac{\partial u_j}{\partial x_i}\right) = \frac{1}{2}(u_{i,j} + u_{j,i}) \qquad (A3.9)$$

The above are strain displacement relations and the components are given by

Normal strains:

$$\varepsilon_{xx} = \frac{\partial u_x}{\partial x}$$

$$\varepsilon_{yy} = \frac{\partial u_y}{\partial y}$$

$$\varepsilon_{zz} = \frac{\partial u_z}{\partial z}$$

Shear strains:

$$\varepsilon_{xy} = \frac{1}{2}\left\{\frac{\partial u_x}{\partial y} + \frac{\partial u_y}{\partial x}\right\} = \frac{1}{2}\gamma_{xy}$$

$$\varepsilon_{yz} = \frac{1}{2}\left\{\frac{\partial u_y}{\partial z} + \frac{\partial u_z}{\partial y}\right\} = \frac{1}{2}\gamma_{yz}$$

$$\varepsilon_{zx} = \frac{1}{2}\left\{\frac{\partial u_z}{\partial x} + \frac{\partial u_x}{\partial z}\right\} = \frac{1}{2}\gamma_{zx}$$

where γ_{ij} are engineering strains.

The normal strains are simply the elongation per unit original length of a vanishingly small line segment in the respective directions. The shear strain γ_{ij} is the change in the right angle of vanishingly small line segments originally in the i and j directions at a point. The transformation equations for the strain tensor are obtained from (A1.10).

A3.5 *Rotation Tensor*

Consider the tensor $u_{i,j}$ which can be written as follows

$$u_{i,j} = \frac{1}{2}(u_{i,j} + u_{j,i}) - \frac{1}{2}(u_{i,j} - u_{j,i})$$

i.e.,

$$u_{i,j} = \varepsilon_{ij} + \omega_{ij}$$

ω_{ij} in the above is called rotation tensor, which is

$$\omega_{ij} = \frac{1}{2}(u_{i,j} - u_{j,i}) \qquad (A3.10)$$

The rotation tensor is skew-symmetric, its diagonal elements are zero and the off diagonal elements are

$$\omega_{32} = -\omega_{23}$$

$$\omega_{13} = -\omega_{31}$$

$$\omega_{21} = -\omega_{12}$$

It can be shown that the above represent the rigid body rotation of an element of volume about x_1, x_2 and x_3 axes, respectively.

A3.6 *Compatibility Conditions*

From the strain displacement relations

$$\varepsilon_{ij} = \frac{1}{2}(u_{i,j} + u_{j,i}) \qquad (A3.9)$$

the strains can be easily determined given the displacement field. However, it is not an easy matter to determine the displacements given the strain components. The displacement field consists of only three functions, viz., u_i and are to be determined from six partial differential equations in (A3.9). Therefore, to ensure single valued continuous solution for the displacement field, we impose certain restrictions on the strain field. This results in compatibility conditions. Taking derivatives of (A3.9)

$$\varepsilon_{ij,kl} = \frac{1}{2}(u_{i,jkl} + u_{j,ikl})$$

$$\varepsilon_{kl,ij} = \frac{1}{2}(u_{k,lij} + u_{l,kij})$$

$$\varepsilon_{lj,ki} = \frac{1}{2}(u_{j,lik} + u_{l,jik})$$

$$\varepsilon_{ki,lj} = \frac{1}{2}(u_{k,ijl} + u_{i,kjl})$$

From the above we find

$$\varepsilon_{ij,kl} + \varepsilon_{kl,ij} - \varepsilon_{lj,ki} - \varepsilon_{ki,lj} = 0 \tag{A3.11}$$

The above is independent of the displacements and they specify the relations between the strains that should be satisfied for obtaining a single valued continuous solution of displacement field from a strain field. They are called compatibility conditions, which are 81 in number. Actually only six of them are independent which are

$$\frac{\partial^2 \varepsilon_{xx}}{\partial y^2} + \frac{\partial^2 \varepsilon_{yy}}{\partial x^2} = \frac{\partial^2 \gamma_{xy}}{\partial x \partial y}$$

$$\frac{\partial^2 \varepsilon_{yy}}{\partial z^2} + \frac{\partial^2 \varepsilon_{zz}}{\partial y^2} = \frac{\partial^2 \gamma_{yz}}{\partial y \partial z}$$

$$\frac{\partial^2 \varepsilon_{zz}}{\partial x^2} + \frac{\partial^2 \varepsilon_{xx}}{\partial z^2} = \frac{\partial^2 \gamma_{zx}}{\partial z \partial x}$$

$$2\frac{\partial^2 \varepsilon_{xx}}{\partial y \partial z} = \frac{\partial}{\partial x}\left\{-\frac{\partial \gamma_{yz}}{\partial x} + \frac{\partial \gamma_{xz}}{\partial y} + \frac{\partial \gamma_{xy}}{\partial z}\right\}$$

$$2\frac{\partial^2 \varepsilon_{yy}}{\partial z \partial x} = \frac{\partial}{\partial y}\left\{-\frac{\partial \gamma_{zx}}{\partial y} + \frac{\partial \gamma_{yx}}{\partial z} + \frac{\partial \gamma_{yz}}{\partial x}\right\}$$

$$2\frac{\partial^2 \varepsilon_{zz}}{\partial z \partial x} = \frac{\partial}{\partial z}\left\{-\frac{\partial \gamma_{xy}}{\partial z} + \frac{\partial \gamma_{zy}}{\partial x} + \frac{\partial \gamma_{zx}}{\partial y}\right\} \tag{A3.12}$$

A3.7 *Strain Energy*

The stress tensor arising out of equilibrium considerations and the strain tensor derived out of kinematical considerations are related to each other through constitutive laws. These laws relate stress and strain directly and uniquely, i.e.,

$$\tau_{ij} = \tau_{ij}(\varepsilon_{11}, \varepsilon_{12}, \ldots, \varepsilon_{33})$$

Consider an integral

$$U_0 = \int_0^{\varepsilon_{ij}} \tau_{ij} d\varepsilon_{ij} \tag{A3.13}$$

where U_0 is a point function, i.e., it is independent of the path. Hence, the integrand in the above must be a perfect differential. So

$$dU_0 = \tau_{ij} d\varepsilon_{ij}$$

Hence

$$\frac{\partial U_0}{\partial \varepsilon_{ij}} = \tau_{ij} \qquad (A3.14)$$

Then U_0 is called the strain energy density function and it can be shown that equation (A3.13) gives the mechanical work performed at a point on an element per unit volume during the deformation of a body. The total strain energy of the body is

$$U = \iiint_V U_0 dV = \iiint_V \left\{ \int_0^{\varepsilon_{ij}} \tau_{ij} d\varepsilon_{ij} \right\} dV \qquad (A.315)$$

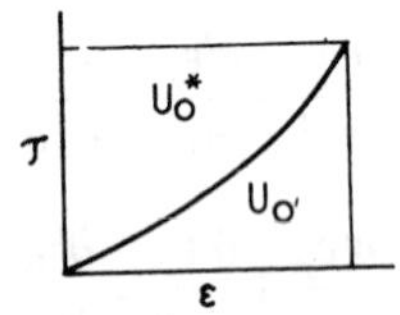

Fig. A3.6 Strain energy and complementary strain energy

If the stress strain diagram from a tensile test is given by Fig. A3.6, U_0 is simply the area under the curve for a given strain. We can also consider the area between the curve and the stress axis and use a complementary energy density function denoted by U^*_0. Then

$$dU_0^* = \varepsilon_{ij} d\tau_{ij}$$

and

$$U_0^* = \int_0^{\tau_{ij}} \varepsilon_{ij} d\tau_{ij} \qquad (A3.13a)$$

Therefore

$$\frac{\partial U_0^*}{\partial \tau_{ij}} = \varepsilon_{ij} \qquad (A3.14a)$$

For linear elastic behaviour, it is obvious that

$$U_0 = U_0^*$$

A3.8 *Hooke's Law*

For linear elastic behaviour, the most general constitutive law relating the stress and strain tensors can be written as

$$\tau_{ij} = C_{ijkl} \varepsilon_{kl} \qquad (A3.16)$$

where C_{ijkl} has 81 terms. Since τ_{ij} is symmetric

$$C_{ijkl} = C_{jikl}$$

Similarly, since ε_{ij} is symmetric

$$C_{ijkl} = C_{ijlk}$$

The strain energy density function from (A3.13) is

$$U_0 = \frac{1}{2} C_{ijkl} \varepsilon_{ij} \varepsilon_{kl}$$

where the tensor C_{ijkl} should be symmetric, therefore

$$C_{ijkl} = C_{klij}$$

Using the above three symmetric properties, it can be shown that there are only 21 independent terms in C_{ijkl}. We restrict ourselves to a discussion on isotropic materials, i.e., the tensor C_{ijkl} must have the same components for all references. It can be shown that a tensor is fourth order isotropic with the following property.

$$C_{ijkl} = \lambda \delta_{ij} \delta_{kl} + \beta \delta_{ik} \delta_{jl} + \gamma \delta_{il} \delta_{jk}$$

where, λ, β, γ are scalar constants. β, γ can be expressed as

$$\beta = G + B$$

$$\gamma = G - B$$

Then

$$C_{ijkl} = \lambda \delta_{ij} \delta_{kl} + G(\delta_{ik} \delta_{jl} + \delta_{il} \delta_{jk}) + B(\delta_{ik} \delta_{jl} - \delta_{il} \delta_{jk})$$

The tensor associated with B in the above is skew-symmetric and therefore we set $B = 0$ to obtain

$$C_{ijkl} = \lambda \delta_{ij} \delta_{kl} + G(\delta_{ik} \delta_{jl} + \delta_{il} \delta_{jk})$$

Substituting the above in equation (A3.16) and simplifying, we get the most general Hooke's law for isotropic bodies

$$\tau_{ij} = \lambda \delta_{ij} \varepsilon_{ll} + 2G \varepsilon_{ij} \qquad (A3.17)$$

In the above λ and G are Lame constants and they can be expressed in terms of Young's modulus E and Poisson's ratio μ as

$$\lambda = \frac{E\mu}{(1+\mu)(1-2\mu)}$$

$$G = \frac{E}{2(1+\mu)}$$

A3.9 *Problems in Linear Elasticity*

From the above sections, we conclude that the complete system of governing equations for homogeneous isotropic bodies under linear elastic behaviour are given by

$$\tau_{ji,j} + B_i = \dot{V}_i \rho \qquad (A3.6)$$

$$\varepsilon_{ij} = \frac{1}{2}(u_{i,j} + u_{j,i}) \qquad (A3.9)$$

$$\tau_{ij} = \lambda \delta_{ij} \varepsilon_{ll} + 2G\varepsilon_{ij} \qquad (A3.17)$$

There are fifteen equations and fifteen unknowns. When explicit use of the displacement field is not made, we must ensure that the compatibility conditions (A3.12) are satisfied.

Linear Viscoelasticity

A4.1 *Viscoelasticity*

Viscoelasticity theory is a natural extension of classical theory of elasticity, to take into account the energy absorption in a continuous system. Just as we developed various mechanisms of damping for discrete systems, e.g., equivalent viscous damping, Coulomb damping, structural damping etc., several models can be developed to consider the energy dissipation in continuous systems to model the behaviour of deformable solids. We discuss these models briefly in this appendix.

In classical theory of elasticity, the stress deformation relations are considered to be constant, i.e. independent of time. This is not true for many rubber-like materials, whose deformations increase with time, even though the force applied is constant. After the removal of force, the deformations do not vanish altogether. To account for the time variations of the mechanical properties of the body, we consider them to be elastic as well as viscous simultaneously, i.e., they are considered viscoelastic. The elastic property follows Hooke's law and the viscous property follows Newtons' law. We will restrict any relations to be linear and that the deformations are small. There are two important functions when we discuss viscoelasticity, which are creep and relaxation.

Consider a viscoelastic bar under simple axial tension. The bar is loaded at time $t = 0$ by a constant force, which produces in the bar a stress σ_0. The resulting strain is now considered to be a function of time, $\varepsilon(t)$. We define the following creep function.

$$\phi(t;\sigma_0) = \frac{\varepsilon(t)}{\sigma_0} \qquad (A4.1)$$

As soon as the force is applied, the strain registered will be

$$\varepsilon(t = 0^+) = \frac{\sigma_0}{E}$$

The deformation gradually increases in a form similar to that shown in Fig. A4.1, from point P to R. At any instant of time t, the total deformation is given by

$$\begin{aligned}\varepsilon(t) &= \varepsilon(0^+) + \varepsilon_t(t) \\ &= \frac{\sigma_0}{E} + \sigma_0\alpha(t) \qquad (A4.2)\end{aligned}$$

From (A4.1), we get the creep function

$$\begin{aligned}\phi(t;\sigma_0) &= \frac{1}{E} + \alpha(t) \\ &= \phi(0;\sigma_0) + \alpha(t) \qquad (A4.3)\end{aligned}$$

When $\alpha(t) = 0$, we have classical elasticity. The creep function increases with time t, but its derivative decreases with time. If we remove the force at any instant t_0, the deformation decreases. However, there may be a residual value ε_r, left behind permanently. The recovery curve from time t_0 is also shown in Fig. A4.1 as QS.

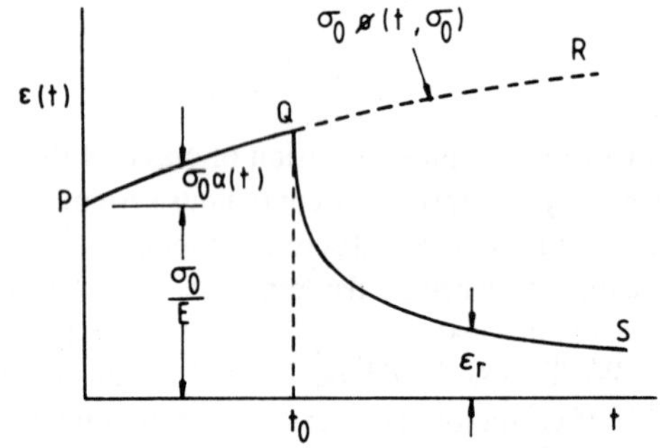

Fig. A4.1 Creep function of a viscoelastic body

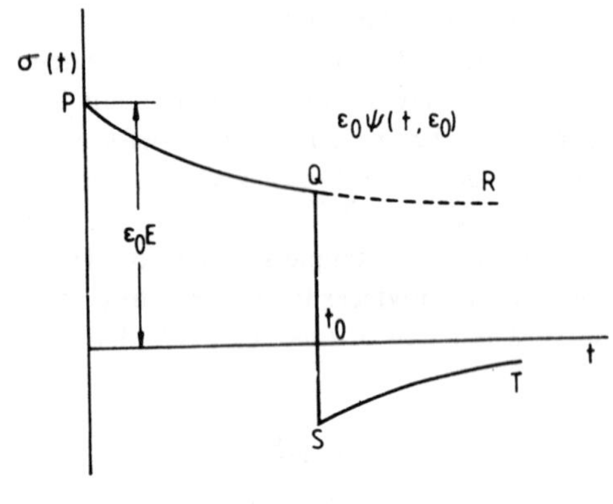

Fig. A4.2 Relaxation function of a viscoelastic body

The relaxation function for the material is defined as

$$\psi(t;\sigma_0) = \frac{\sigma(t)}{\varepsilon_0} \tag{A4.4}$$

For defining the above a specimen is first deformed to ε_0, at time $t = 0$, when the stress becomes

$$\sigma(0^+) = E\,\varepsilon_0$$

If we now let the specimen relax, the stress decreases gradually according to the curve PQR in Fig. A4.2. Then

$$\sigma(t) = \sigma(0^+) + \varepsilon_0\beta(t)$$

$$= E\varepsilon_0 + \varepsilon_0\beta(t) \tag{A4.5}$$

From (A4.4), we can now obtain the relaxation function

$$\psi(t;\varepsilon_0) = E + \beta(t) \tag{A4.6}$$

If the deformation is suddenly removed at time t_0, the change of stress is described by *PQST* in Fig. A4.2.

Let us now consider the stress σ (t) to be a continuous function of time in the interval $-\infty < \tau < t$ in (A4.1). Then, we can define

$$\varepsilon(t) = \int_{-\infty}^{t} \frac{\partial \sigma(\tau)}{\partial \tau} \phi(t-\tau) d\tau \tag{A4.7}$$

An analogous expression for (A4.4) is

$$\sigma(t) = \int_{-\infty}^{t} \frac{\partial \varepsilon(\tau)}{\partial \tau} \psi(t-\tau) d\tau \tag{A4.8}$$

The above two relations (A4.7) and (A4.8) are the basic stress deformation relations in viscoelasticity, involving the creep and relaxation functions. They are the counterparts of Hooke's law, which is the case of functions ϕ, ψ becoming independent of time.

As a special case, if σ $(t) = \sigma_0 H$ (t) at time $t = 0^+$, equation (A4.7) gives

$$\frac{\varepsilon(t)}{\sigma_0} = \int_{0}^{t} \frac{dH(\tau)}{d\tau} \phi(t-\tau) d\tau$$

which is

$$\frac{\varepsilon(t)}{\sigma_0} = \int_{0}^{t} \delta(\tau)\phi(t-\tau) d\tau$$

$$= \phi(t) \tag{A4.1a}$$

Similarly, if ε $(t) = \varepsilon_0 H$ (t) at time $t = 0^+$, equation (A4.8) gives

$$\frac{\sigma(t)}{\varepsilon_0} = \psi(t) \tag{A4.4a}$$

A more general case is the application of stress σ_0 at $t = 0^+$, which is removed at time $t = t_0$, i.e.,

$$\sigma(t) = \sigma_0\{H(t) - H(t - t_0)\}$$

Substituting the above in equation (A4.7), we obtain

$$\varepsilon_r(t) = \sigma_0\{\phi(t)H(t) - \phi(t-t_0)H(t-t_0)\}$$

$$= \varepsilon(t) - \sigma_0\phi(t-t_0)H(t-t_0) \tag{A4.9}$$

The above value is shown in Fig. A4.1.

A4.2 *Viscoelastic Models*

In practice, we would like to use simple models consisting of springs and dashpots, that will appropriately describe the viscoelastic behaviour of a system.

Case 1 Kelvin model

This is the simplest model, consisting of a spring and a dashpot connected in parallel as shown in Fig. A4.3. The extensions in the spring and the dasphot ε_s and ε_d are equal. The stress σ of the system is, however, the sum of the individual stresses of both the elements σ_s and σ_d, i.e.,

$$\varepsilon = \varepsilon_s = \varepsilon_d$$

$$\sigma = \sigma_s + \sigma_d \tag{A4.10}$$

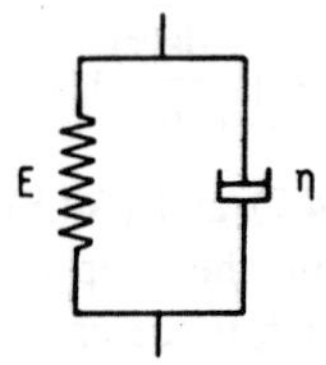

Fig. A4.3 Kelvin model

Hooke's law gives

$$\sigma_s = E\varepsilon_s = E\varepsilon(t) \tag{A4.11}$$

Newton's law gives

$$\sigma_d = \eta\dot{\varepsilon}_d$$

$$= \eta\frac{\partial\varepsilon_d}{\partial t}$$

$$= \eta\frac{\partial\varepsilon(t)}{\partial t} \tag{A4.12}$$

Substituting (A4.11) and (A4.12) in (A4.10), we get

$$\sigma(t) = E(1 + t_1\partial_t)\varepsilon(t) \tag{A4.13}$$

where

$$t_1 = \frac{\eta}{E} \tag{A4.14}$$

We can solve for ε (t) in (A4.13), assuming that the body is in free state for time $t \leq 0$, to give

$$\varepsilon(t) = \frac{H(t)}{\eta}\int_0^t \sigma(\tau)e^{-(t-\tau)/t_1}d\tau \tag{A4.15}$$

We can now calculate the creep function for the Kelvin model, by substituting σ $(t) = \sigma_0 H(t)$ in the above equation, to give

$$\frac{\varepsilon(t)}{\sigma_0} = \phi(t) = \frac{1}{E}\left(1 - e^{-t/t_1}\right) \tag{A4.16}$$

In the above t_1 is the retardation time.

Case 2 Maxwell model

This model consists of a spring and a dashpot in series, as shown in Fig. A4.4. Here

$$\varepsilon = \varepsilon_s + \varepsilon_d$$

$$\sigma = E\varepsilon_s = \eta\varepsilon_d \qquad (A4.17)$$

Fig. A4.4 Maxwell model

From the above we get

$$\frac{\partial\varepsilon(t)}{\partial t} = \frac{1}{E}\left(\frac{1}{t_1} + \partial_t\right)\sigma(t) \qquad (A4.18)$$

We can solve the above, assuming that the body is in its natural state for time $t \leq 0$, to get

$$\sigma(t) = H(t)E\int_0^t \frac{\partial\varepsilon(\tau)}{\partial\tau} e^{-(t-\tau)/t_1} d\tau \qquad (A4.19)$$

For $\varepsilon(t) = \varepsilon_0 H(t)$, in the above, we get the relaxation function for the Maxwell model, which is

$$\psi(t) = \frac{\sigma(t)}{\varepsilon_0} = Ee^{-t/t_1} \qquad (A4.20)$$

In the above t_1 is the relaxation time for the Maxwell model and is same as the retardation time for the Kelvin model.

Case 3 Generalised Maxwell model

Consider now n-number Maxwell models connected in parallel as shown in Fig. A4.5. Each of the individual models have different relaxation times

$$t_i = \frac{\eta_i}{E_i} \quad i = 1, 2, \ldots n \qquad (A4.21)$$

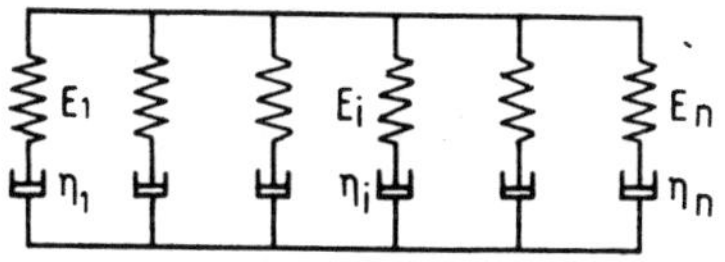

Fig. A4.5 Generalised Maxwell model

The deformations in all the n elements are equal, i.e.,

$$\varepsilon(t) = \varepsilon_i(t) \tag{A4.22}$$

The stress σ is however sum of the stresses of all individual elements

$$\sigma(t) = \sum_{i=1}^{n} \sigma_i(t) \tag{A4.23}$$

For the ith element, the following relation can be written from (A4.18).

$$\frac{\partial \varepsilon(t)}{\partial t} = \frac{1}{E_i}\left(\frac{1}{t_i} + \partial_t\right)\sigma_i(t) \qquad t_i = \frac{\eta_i}{E_i} \tag{A4.18a}$$

i.e.,

$$\sigma_i(t) = E_i \frac{\partial \varepsilon(t)/\partial t}{\left(\frac{1}{t_i} + \partial_t\right)} \tag{A4.24}$$

From (A4.23), we now have

$$\sigma(t) = \sum_{i=1}^{n} E_i \frac{\partial \varepsilon(t)/\partial t}{\left(\frac{1}{t_i} + \partial_t\right)} \tag{A4.25}$$

Let the body be in its natural state at time $t \le 0$, then applying Laplace transform to the above, we get

$$\bar{\sigma}(s) = \bar{\varepsilon}(s) \sum_{i=1}^{n} \frac{E_i}{t_i^{-1} + s} \tag{A4.26}$$

Inverting the transform, we obtain

$$\sigma(t) = H(t) \int_0^t \frac{\partial \varepsilon(t)}{\partial t} \psi(t - \tau) d\tau \tag{A4.27}$$

where the relaxation function is

$$\psi(t) = \sum_{i=1}^{n} E_i e^{-t/t_i} \tag{A4.28}$$

In general, when we choose a certain combination of springs and dashpots in a suitable viscoelastic model, the stress deformation relationship of (*A*4.25) can be expressed as

$$P(\partial_t)\sigma(t) = Q(\partial_t)\varepsilon(t) \tag{A4.29}$$

where $P(\partial_t)$ and $Q(\partial_t)$ are linear differential operators of the form

$$P(\partial_t) = \sum_{i=0}^{m} a_i(\partial_t)_i$$

$$Q(\partial_t) = \sum_{i=0}^{m} b_i(\partial_t)_i \tag{A4.30}$$

where a_i, b_i are constants. Applying Laplace transform to (*A*4.29), we obtain

$$\frac{\bar{\sigma}(s)}{\bar{\varepsilon}(s)}=\frac{Q(s)}{P(s)}$$

$$=\frac{\sum_{i=0}^{m} b_i s_i}{\sum_{i=0}^{m} a_i s_i} \tag{A4.31}$$

The Laplace transforms of the creep and relaxation functions are

$$\bar{\phi}(s)=\frac{1}{s}\frac{\bar{\varepsilon}(s)}{\bar{\sigma}(s)}$$

$$=\frac{1}{s}\frac{P(s)}{Q(s)}$$

$$\bar{\psi}(s)=\frac{1}{s}\frac{\bar{\sigma}(s)}{\bar{\varepsilon}(s)}$$

$$=\frac{1}{s}\frac{Q(s)}{P(s)} \tag{A4.32}$$

Let the order m of the operators P and Q be same. Then, we can expand the transforms in the above to yield the following:

$$\bar{\phi}(s)=\frac{1}{s}\frac{P(s)}{Q(s)}$$

$$=\frac{1}{s}\frac{P(0)}{Q(0)}+\sum_{i=1}^{m}\frac{P(\gamma_i)}{\gamma_i Q'(\gamma_i)}\frac{1}{s+\gamma_i}$$

$$\bar{\psi}(s)=\frac{1}{s}\frac{Q(s)}{P(s)}$$

$$=\frac{1}{s}\frac{Q(0)}{P(0)}+\sum_{i=1}^{m}\frac{Q(\mu_i)}{\mu_i P'(\mu_i)}\frac{1}{s+\mu_i} \tag{A4.33}$$

where γ_i are the roots of the polynomial

$$P(s)=\sum_{i=0}^{m} a_i s_i$$

$$=\prod_{i=1}^{m}(s+\gamma_i)$$

and μ_i are the roots of the polynomial

$$Q(s)=\sum_{i=0}^{m} b_i s_i$$

$$=\prod_{i=1}^{m}(s+\mu_i)$$

and

$$Q'(\gamma_i) = \frac{dQ}{ds}\}_{s=\gamma_i}$$

$$P'(\mu_i) = \frac{dP}{ds}\}_{s=\mu_i}$$

Inverting the Laplace transforms in equation (*A*4.33), we get

$$\phi(t) = \frac{P(0)}{Q(0)} + \sum_{i=1}^{m} \frac{P(\gamma_i)}{\gamma_i Q'(\gamma_i)} e^{-\gamma_i t}$$

$$\psi(t) = \frac{Q(0)}{P(0)} + \sum_{i=1}^{m} \frac{Q(\mu_i)}{\mu_i P'(\mu_i)} e^{-\mu_i t} \tag{A4.34}$$

A4.3 *Stress Deformation Relations*

For an elastic body, Hooke's law (A3.17) is

$$\sigma_{ij} = \delta_{ij}\lambda\varepsilon_{kk} + 2G\varepsilon_{ij} \tag{A4.35}$$

Subtracting $\delta_{ij}\ \sigma_{kk}/3$ from both the sides in the above equation, we have

$$s_{ij} = \sigma_{ij} - \frac{1}{3}\delta_{ij}\sigma_{kk}$$

$$= \delta_{ij}\left(\lambda\varepsilon_{kk} - \frac{1}{3}\sigma_{kk}\right) + 2G\varepsilon_{ij} \tag{A4.36}$$

From (A4.35), we also have

$$\sigma_{kk} = (3\lambda + 2G)\varepsilon_{kk} \tag{A4.37}$$

Then

$$s_{ij} = 2Ge_{ij} \tag{A4.38}$$

$$s = 3Ke \tag{A4.39}$$

where

$$e_{ij} = \varepsilon_{ij} - \frac{1}{3}\delta_{ij}\varepsilon_{kk}$$

$$s = \sigma_{kk}$$

$$e = \varepsilon_{kk}$$

$$K = \lambda + \frac{2}{3}G$$

s_{ij}, e_{ij} are stress and deformation deviators and K is the bulk modulus.

Equation (A4.38) is now generalised for a Kelvin viscoelastic body by adding to the right hand side a term representing the Newtonian viscosity, i.e., $\eta\dot{e}_{ij}$

$$s_{ij} = 2G(1 + t_1\partial_t)e_{ij} \tag{A4.40}$$

$$s = 3Ke \tag{A4.41}$$

$$t_1 = \frac{\eta}{G} \tag{A4.42}$$

For a Maxwell body, we have

$$\dot{s}_{ij} + \frac{s_{ij}}{t_1} = 2Ge_{ij}$$

$$s = 3Ke \tag{A4.43}$$

For a parallel connection of the Hooke and Maxwell models, we have

$$(1 + t_1\partial_t)s_{ij} = 2G(1 + t_2\partial_t)e_{ij}$$

$$s = 3Ke \tag{A4.44}$$

Most general equations of the above for linear viscoelastic bodies can be written as

$$P_1(\partial_t)s_{ij}(x,t) = P_2(\partial_t)e_{ij}(x,t)$$

$$P_3(\partial_t)s(x,t) = P_4(\partial_t)e(x,t) \tag{A4.45}$$

where

$$P_i(\partial_t) = \sum_{n=0}^{N_i} a_i^{(n)}\partial_t^n \qquad a_t^{(N_i)} \neq 0 \quad i = 1,2,3,4$$

where $a_i^{(n)}$ are constant coefficients.

Some special cases of the above are given below.

Case 1 Elastic body

$$a_1^{(0)} = 1$$

$$a_2^{(0)} = 2G$$

$$a_3^{(0)} = 1$$

$$a_4^{(0)} = 3K \tag{A4.46}$$

Rest of a's are zero.

Case 2 Kelvin body

$$P_1(\partial_t) = 1$$

$$P_2(\partial_t) = 2G(1 + t_1\partial_t)$$

$$P_3(\partial_t) = \partial_t$$

$$P_4(\partial_t) = 3K\partial_t \tag{A4.47}$$

Case 3 Maxwell body

$$P_1(\partial_t) = \frac{1}{t_1}\partial_t$$

$$P_2(\partial_t) = 2G\partial_t$$

$$P_3(\partial_t) = \partial_t$$

$$P_4(\partial_t) = 3K\partial_t \tag{A4.48}$$

Case 4 Standard linear viscoelastic body

This is the parallel connection of the Hooke and Maxwell bodies.

$$P_1(\partial_t) = 1 + t_2\partial_t$$

$$P_2(\partial_t) = 2G(1 + t_1\partial_t)$$

$$P_3(\partial_t) = \partial_t$$

$$P_4(\partial_t) = 3K\partial_t \tag{A4.49}$$

Sometimes, the viscoelastic body is assumed incompressible, then $P_3(\partial_t) = 0$.

Equations (A4.45) can be written also in an explicit form.

$$P_1(\partial_t)P_3(\partial_t)\sigma_{ij} = P_2(\partial_t)P_3(\partial_t)\varepsilon_{ij} + \frac{1}{3}\delta_{ij}\{P_1(\partial_t)P_4(\partial_t) - P_2(\partial_t)P_3(\partial_t)\}e \tag{A4.50}$$

Laplace Transform

A5.1 *Operations Transforms*

$f(t)=\frac{1}{2\pi i}\int_{a-i\infty}^{a+i\infty}F(s)e^{st}ds$	$F(s)=\int_0^{\infty}e^{-st}f(t)dt$
$f(t)$	$F(s)$
$f'(t)$	$sF(s)-f(0)$
$f^n(t)$	$s^nF(s)-s^{n-1}f(0)-s^{n-2}f'(0)-\ldots$ $-sf^{(n-2)}(0)-f^{(n-1)}(0)$
$\sum_{i=1}^{n}C_if_i(t)$	$\sum_{i=1}^{n}C_iF_i(s)$
$e^{at}f(t)$	$F(s-a)$
$f(t-b)$ when $t>b$ 0 when $t<b$	$e^{-bs}F(s)$
$H(t)-H(t-\tau)$ Rectangular pulse	$1-\frac{e^{-s\tau}}{s}$
$\sin\frac{\pi}{\tau}t \quad 0<t<\tau$ Half sine pulse	$\frac{\frac{\pi}{\tau}}{s^2+\frac{\pi^2}{\tau^2}}(1+e^{-s\tau})$

A5.2 *Function transforms*

1	$\frac{1}{s}$
e^{-at}	$\frac{1}{s+a}$
$\frac{t^{n-1}}{(n-1)!}$	$\frac{1}{s^n}$
$\sin at$	$\frac{a}{s^2+a^2}$
$\sinh at$	$\frac{a}{s^2-a^2}$

$\dfrac{e^{at}-e^{bt}}{a-b}$	$\dfrac{1}{(s-a)(s-b)}$
$\dfrac{1}{\omega}\sin\omega t$	$\dfrac{1}{s^2+\omega^2}$
$\dfrac{e^{-d\omega t}\sin(\omega t\sqrt{1-d^2})}{\omega\sqrt{1-d^2}} \quad d^2<1$	$\dfrac{1}{s^2+2d\omega s+\omega^2}$
$\dfrac{e^{-d\omega t}\sinh(\omega t\sqrt{d^2-1})}{\omega\sqrt{d^2-1}} \quad d^2>1$	$\dfrac{1}{s^2+2d\omega s+\omega^2}$
$te^{-\omega t}$	$\dfrac{1}{s^2+2\omega s+\omega^2}$
$\cosh at$	$\dfrac{s}{s^2-a^2}$
$\cos at$	$\dfrac{s}{s^2+a^2}$
$\dfrac{1}{\sqrt{1-d^2}}e^{-d\omega t}\sin\left(\omega t\sqrt{1-d^2}+\tau_1\right)$	$\dfrac{s}{s^2+2d\omega s+\omega^2}$
$\tau_1=\tan^{-1}\omega\dfrac{\sqrt{1-d^2}}{-d\omega} \quad d^2<1$	
$\dfrac{1}{\sqrt{d^2-1}}e^{-d\omega t}\sinh(\omega t\sqrt{d^2-1}+\tau_2)$	$\dfrac{s}{s^2+2d\omega s+\omega^2}$
$\tau_2=\tanh^{-1}\dfrac{\omega\sqrt{d^2-1}}{-d\omega}d^2>1$	
$(1-\omega t)e^{-\omega t}$	$\dfrac{s}{s^2+2\omega s+\omega^2}$
$\dfrac{a(b-c)e^{at}+b(c-a)e^{bt}+c(a-b)e^{ct}}{(a-b)(b-c)(a-c)}$	$\dfrac{s}{(s-a)(s-b)(s-c)}$
$\dfrac{(b-c)e^{at}+(c-a)e^{bt}+(a-b)e^{ct}}{(a-b)(b-c)(a-c)}$	$\dfrac{1}{(s-a)(s-b)(s-c)}$
$\dfrac{\cosh\frac{at}{\sqrt{2}}\sin\frac{at}{\sqrt{2}}-\sinh\frac{at}{\sqrt{2}}\cos\frac{at}{\sqrt{2}}}{a^3\sqrt{2}}$	$\dfrac{1}{s^4+a^4}$
$\dfrac{1}{a^2}\sin\dfrac{at}{\sqrt{2}}\sinh\dfrac{at}{\sqrt{2}}$	$\dfrac{s}{s^4+a^4}$

$\dfrac{\cos\frac{at}{\sqrt{2}}\sinh\frac{at}{\sqrt{2}}+\sin\frac{at}{\sqrt{2}}\cosh\frac{at}{\sqrt{2}}}{a\sqrt{2}}$	$\dfrac{s^2}{s^4+a^4}$
$\cos\dfrac{at}{\sqrt{2}}\cosh\dfrac{at}{\sqrt{2}}$	$\dfrac{s^3}{s^4+a^4}$
$\dfrac{1}{2a^3}(\sinh at-\sin at)$	$\dfrac{1}{s^4-a^4}$
$\dfrac{1}{2a^2}(\cosh at-\cos at)$	$\dfrac{s}{s^4-a^4}$
$\dfrac{1}{2a}(\sinh at+\sin at)$	$\dfrac{s^2}{s^4-a^4}$
$\dfrac{1}{2}(\cosh at+\cos at)$	$\dfrac{s^3}{s^4-a^4}$
$\dfrac{1}{n!}\left(1-e^{-\frac{t}{a}}\right)^n$	$\dfrac{1}{s(as+1)\ldots(as+n)}$
$\dfrac{2}{t}\sinh at$	$\ln\dfrac{s+a}{s-a}$
$\dfrac{1}{\sqrt{\pi t}}$	$\dfrac{1}{\sqrt{s}}$
$2\sqrt{\dfrac{t}{\pi}}$	$\dfrac{1}{s\sqrt{s}}$
$\dfrac{e^{-at}}{\sqrt{\pi t}}$	$\dfrac{1}{\sqrt{s+a}}$
$\dfrac{(e^{bt}-e^{at})}{2t\sqrt{\pi t}}$	$\sqrt{s-a}-\sqrt{s-b}$
$\sqrt{\dfrac{2}{\pi t}}\sin at$	$\sqrt{\dfrac{\sqrt{s^2+a^2}-s}{s^2+a^2}}$
$\sqrt{\dfrac{2}{\pi t}}\cos at$	$\sqrt{\dfrac{\sqrt{s^2+a^2}+s}{s^2+a^2}}$
$\sqrt{\dfrac{2}{\pi t}}\sinh at$	$\sqrt{\dfrac{\sqrt{s^2+a^2}-s}{s^2-a^2}}$
$\sqrt{\dfrac{2}{\pi t}}\cosh at$	$\sqrt{\dfrac{\sqrt{s^2+a^2}+s}{s^2-a^2}}$

$\dfrac{\sinh\sqrt{2at}\,\sin\sqrt{2at}}{\sqrt{\pi t}}$	$\dfrac{1}{\sqrt{s}}\sin\dfrac{a}{s}$
$\dfrac{\cosh\sqrt{2at}\,\cos\sqrt{2at}}{\sqrt{\pi t}}$	$\dfrac{1}{\sqrt{s}}\cos\dfrac{a}{s}$
$\dfrac{\cosh 2\sqrt{t}}{\sqrt{\pi t}}$	$\dfrac{e^{1/s}}{\sqrt{s}}$
$\dfrac{\sinh 2\sqrt{t}}{\sqrt{\pi}}$	$\dfrac{e^{1/s}}{s\sqrt{s}}$
$\dfrac{\sin at}{t}$	$\tan^{-1}\dfrac{a}{s}$

Nomenclature

a	area of cross-section
	also, amplitude
	also, Mathieu equation parameter
	also, crank length
	also, wave equation parameter (*see* equation 7.102)
	also, z coordinate of the contact surface with the force
	also, distance from the left end on a beam
	also, arbitrary parameter
	also, $(I_{yy} + Ah^2)/4h^2$
	also, beam parameter (*see* equation 8.242)
a, b	constants of integration
	also, functions of time
	also, *see* equation 8.240
a_1	nondimensional beam parameter (*see* equation 8.42)
a_2	nondimensional axial compressive force, (*see* equation 8.102)
a_3	nondimensional foundation stiffness in vertical direction
a_4	nondimensional foundation stiffness in horizontal direction
a_0, b_0	steady-state solution
a_1, b_1	first harmonic solution
a_3, b_3	third harmonic solution
a_1, a_2	coefficients of linear terms
b_1, b_2	coefficients of linear terms
$a_{11}, a_{12}, a_{21}, a_{22}$	partial derivative values at a singular point
	also, coefficients in the fundamental solutions
a_i	*see* equation 8.171
	also, constants in equation A4.30
a_{ni}	direction cosines
$\overline{a}$	nondimensional amplitude
A	initial amplitude
	also, function of frequency and amplitude
	also, arbitrary parameter
	also, area of cross-section
A_1, A_2	first harmonic solution of a rotating beam with Coriolis forces
A_1, A_3	functions of first and third harmonics
A_i	arbitrary parameters
A, B	coefficients in the characteristics equation for m
b	breadth
	also, coefficient in Bishop's bar equation, *see* equation 6.25
	also, coefficient in Rayleigh's beam equation, *see* equation

	8.19
	also, *see* equation 8.105
	also, *see* equation 8.117
$\overline{b}$	*see* equation 8.49
b_i	*see* equation 8.171
	also, constants in equation (A4.30)
B_1, B_2	second harmonic solution of a rotating beam with Coriolis forces
B_i	body forces
	also, arbitrary parameters
	also, $E_i\, I_{yyi}$
	also, generalised coordinates
c	damping coefficient
	also, stability parameter, *see* equation 3.160
	also, coefficient in Bishop's bar equation, *see* equation 6.25
	also, coefficient in Rayleigh's beam equation, *see* equation 8.19
	also, *see* equation (8.117)
$\overline{c}$	*see* equation 8.49
$C_0, \ldots C_4$	*see* equation 8.240
c_1, c_2	damping coefficients
	also, constants of integration
c_i	distance of z coordinate of ith layer frame from the origin
C	isocline value
	also, torsional rigidity
$[C]$	damping matrix
$C_1, - C_5$	integrals associated with torsion, *see* equation 7.99
C_r	arbitrary parameters
C_U	proportional damping factor
C_γ	additional torsional rigidity due to pre-twist
d	*see* equation 7.34
d_R	integral (*see* equation 7.85)
dF	force along the fibre and normal to the blade axis
d_x	d/dx
D	functional due to constraints
D_i	generalised coordinates
$D, E, \ldots U$	matrices
e	dilatation strain (*see* equation A4.39)
e_x, e_ξ, e_η	unit vectors in x, ξ, η directions
E	Young's modulus
$E\,(\partial_t)$	nondimensional Young's modulus (*see* equation 6.75)
$E\,(1 + t_1\, \partial_t)$	Young's modulus for Kelvin model
f	direct stress in the fibre
	also, initial displacement along the beam
$f\,(i)$	function depending on arrangement of elastic and viscoelastic layers
$f_i\,(Z)$	shape functions
f_j	generalised force in jth mode

F	excitation force, force per unit volume in chapter 6
	also, functional
F_d	damping force function
$F(t), G(t)$	time dependent coefficients (*see* equation 3.86)
$F_r(x)$	shape functions
g	acceleration due to gravity
	also, differential operator
	also, integral (*see* equation 7.19)
	also, shape function
	also, initial velocity along the beam
g_R	integral (*see* equation 7.84)
G	function of displacement and velocity
	also, shear modulus
$G(\partial_t)$	nondimensional shear modulus (*see* equation 6.75)
h	time step
	also, differential operator
	also, shape function
	also, semi depth of the beam
$H_i(\omega)$	dynamic magnification factor in *i*th mode
$H(t)$	Heaviside unit step function
i_1	Rayleigh indicator
i_2	Bishop indicator
I	Mass moment of inertia
	also, integral
	also, functional
	also, second moment of area
I_{cf}	mass moment of inertia about centre of flexure per unit length
I_p	second polar moment of cross-sectional area
I_R	Reissner's functional
I_{yy}, I_{zz}	second moments of area of cross-section
I_{yz}	product moment of area of cross-section
I_{YY}, I_{ZZ}	principal moments of area of cross-section
$I_{\phi\phi}$	*see* equation 7.22
j	$\sqrt{-1}$
k	spring stiffness
	also, lateral foundation stiffness
k_1	axial foundation stiffness
k_i	coefficient in frequency equation of a cantilever (*see* equation 8.243)
k'	shear correction factor
k_j	generalised stiffness in *j*th mode
k_{ij}	*i*th row and *j*th column element of stiffness matrix
K	torsional stiffness
	also, stiffness operator
	also, integral (*see* equation 7.82)
	also, bulk modulus
$[K]$	stiffness matrix

K_1, to K_5	Runge-Kutta parameters
K_i	$E_i A_i$
l	length
L	Lagrangian function
	also, length
	also, differential operator
	also, integral (*see* equation 7.83)
L^{-1}	Laplace inverse
m	mass
	also, stability parameter in Floquet's theory
	also, mass per unit length
	also, order of operators P, Q in (A4.30)
m_j	generalised mass in jth mode
m_{ij}	ith row and jth column element of mass matrix
M	mass of reciprocating parts
	also, mass of sleeve
	also, mass operator
	also, external moment
	also, bending moment
$[M]$	mass matrix
M_t	couple
N	number of terms in the solution
	also, axial tensile force
	also, number of layers
p	natural frequency
	also, circular frequency
p_d	damped natural frequency
$p(t)$, $q(t)$	periodic coefficients
$p(x)$	distributed force
P	intensity of external force
	also *see* equation 8.239
$P, Q, \ldots W$	matrices
$P(\partial_t)$, $Q(\partial_t)$	linear differential operator associated with general viscoelastic models (*see* equation A4.29)
P_0	intensity of external force
$[P]$, $[Q]$	matrices in eigen value problem
q	Mathieu equation parameter
	also, generalised coordinate
	also, external force per unit length
q_0	*see* equation 8.171
$q_i(t)$	generalised coordinate
Q	*see* equation 8.239
$\{Q_0\}$	forcing vector
r	frequency ratio
	also, inner radius
	also, *see* equation 7.25
	also, radius
	also, generalised coordinate

r_0	radius of gyration (*see* equation 6.15)
r_1	nondimensional parameter (*see* equation 6.15)
	also, distance of unstressed fibre from centroid
	also, *see* equation 8.14
r_2	nondimensional parameter for bars (*see* equation 6.15)
	also, nondimensional parameter for rods, *see* equation 7.36
	also, distance of line of centre of torsion from the pre-twisted fibre
r_g	$\sqrt{I_{yy}/A}$
r_i	perturbation parameters of r
	also, *see* equation 8.171
r_y, r_z	distances of the centre of flexure from centroid in y and z directions
r_x, r_ξ, r_η	components of position vector in x, ξ, η directions
$\bar{r}$	I_{zz}/I_{yy}
	also, position vector
R	Disc radius
	also, outer radius
	also, $Gb/2h_2 K_1$
s	distance
	also, Laplace parameter
	also, $\sqrt{A l^2/I_{yy}}$
	also, dilatation stress, *see* (A4.39)
s_1, s_2	characteristic values
	also, *see* equation 6.27
	also, *see* equation 8.21
	also, *see* equation 8.51
S	axial force
	also, $(K_1 + K_3)/K_3$
	also, *see* equation 8.240
S_1	surface on which traction forces act
S_2	surface on which displacements are prescribed
t	time
	also, thickness
t_0	*see* equation 6.15
	also, *see* equation 7.36
	also, *see* equation 8.14
t_1	arbitrary time
	also, relaxation time for Maxwell model
	also, retardation time for Kelvin model
t_2	arbitrary time
	also, relaxation time constant
t_3, t_4	see equation 6.97
	also, *see* equations 8.123 and 8.124
t_5	*see* equation 8.124
t_1^*	*see* equation 6.92
t_1^{**}	*see* equation 6.94

T	nondimensional time
	also, beating period
	also, time period
	also, kinetic energy
	also, average value of time in a step
$T(x)$	transfer function
$\overline{T}$	average kinetic energy
$T_i^{(v)}$	traction forces
T_θ	$C\theta'$
u	displacement in x direction
	inward displacement due to bending of a rotating beam
u_i	displacement field
u_x, u_y, y_z	displacement field in x, y, z directions
u_x, u_ξ, u_η	displacement field in x, ξ, η directions
$\overline{u}$	vector of displacements
U	potential energy
$[U]$	modal matrix
U_0	strain energy density function
$U_0{}^*$	complementary strain energy density function
v	phase plane coordinate
	also, displacement in y direction
	also, initial velocity
V	strain energy
	also, speed
	also, work done by body and traction forces
	also, volume
V^*	complementary work done by body and traction forces
$\overline{V}$	average strain energy
$V(x, y)$	Liapunov function
w	displacement
	also, displacement in z direction
W	Work
W_f	work done by foundation reaction force
$\overline{W}$	average work
W^*	complementary work
w_1, w_2	fundamental solutions
x	response
	also, piston displacement
x_i	perturbed response functions
	also, functions of time
x, y, z	global reference frame
x_i, y_i, z_i	local reference frame of ith layer
X	amplitude
	also, x/l
	also, mode shapes
X_1, X_3	first and third harmonic amplitudes
X_{st}	static deflection

X, Y	functions of x, y
y	velocity in the phase plane
	also, function of x
z	transformed response x
Z	z/l
α	linear stiffness coefficient
	also, angle made by arms from vertical
	also, breadth parameter
	also, nonuniform function, see (7.77)
	also, $(I_{yy} - Ah^2)/4h^2$
	also, rotational frequency parameter
	also, creep function
$\alpha_1, \alpha_2, \alpha_3$	taper parameters
α_{ij}	direction cosines
β	nonlinear stiffness coefficient
	also, thickness parameter
	also, slope due to shear
	also, relaxation function
β_i	damping in ith mode
β_p	helix angle due to pre-twist
γ	angular oscillation
	pre-twist angle
γ_i	roots of polynomial $P(s)$
γ_{ij}	engineering shear strains
γ_l	relative pre-twist over the length l of the rod
δ	half constant term in $F(t)$
	also, phase plane parameter
	also, variational operator
	also, denotes virtual displacement, virtual work etc.
	also, Dirac's delta function
	also, correction factor for Timoshenko beam natural frequencies
δ_{ij}	Kronecker delta
δ^t	total variation
δ^1	first variation
Δ	small distance over which impact occurs
	also $h_1 + 2h_2 + h_3$
∂_x	$\partial/\partial x$
∇^2	$\partial^2/\partial x^2 + \partial^2/\partial y^2$
ε	beating frequency
	also, phase angle
	also, axial strain
ε_0	initial strain
ε_d	extension in dashpot
ε_r	residual strain
ε_s	extension in spring
ε_{ij}	strain tensor

$\in$	small parameter
	also, error in differential equation
	also, phase angle
ζ	variation to a singular point
η	variation to a singular point
	also, transformation parameter
	also, function of x
	also, viscosity
	also, generalised coordinate
η_i	functions of x
θ	angular oscillation
	also, crank angle
	also, phase plane angle
	also, elastic twist
θ_0, θ_k	coefficients in Hill's equation
θ_{ic}, θ_{is}	cosine and sine parameters in Hills' equation
Θ	twist per unit length
κ	frequency ratio
λ	Lame's constant, *see* equation (A3.17)
	also, change in the length of the fibre
	also, frequency parameter
	also, Lagrangian multiplier function
λ_1, λ_2	characteristic values
	also, Lagrangian multiplier functions
λ_s	frequency parameter of a straight uniform beam
λ_{ry}, λ_{rz}	coupled bending-torsion frequency parameters
λ_y, λ_z, λ_θ	frequency parameters (*see* equation 8.214)
λ_γ	frequency parameter of a pre-twisted beam
λ_θ	torsional frequency parameter of a uniform beam
μ	nonlinear stiffness coefficient parameter
	also, stability parameter in Floquet's theory
	also, mass moment of inertia ratio
	also, Poisson's ratio
	also, ratio of bending rigidity to torsional rigidity, see equation (8.124)
μ_i	roots of polynomial $Q\ (s)$
ν	outward normal
	also, directional derivative
	also, Poisson's ratio
ν_i	direction cosines
ξ	damping ratio
	also, variation to a singular point
ξ, η, x	coordinate system on the rotating blade
π	total potential energy
ρ	radius of curvature
	also, mass density

σ	phase angle
	also, bending stress
	also, principal stresses
σ_0	initial stress
σ_d	stress in dashpot
σ_s	stress in spring
τ	nondimensional time
	also, stress
	also, dummy time used in Duhamel's integral
τ_{ij}	stress tensor
ϕ	phase angle
	also, unstable regime function in Hill's equation
	also, displaced position from mean equilibrium
	also, warping function
	also, stress function
	also, slope due to shear
	also, stagger angle
	also, generalized coordinate
	also, creep function
ϕ'	setting angle
ψ	slope due to bending
	also, phase angle
	also, relaxation function
ω	function, *see* equation (2.15)
	also, circular frequency
	also, angular velocity
	also, rotational speed
ω_{ij}	rotational tensor
Ω	rotational speed

Superscripts

dot denotes differentiation with time, thus, $\dot{x}$
prime denotes differentiation with x, thus, y'
tilde denotes varied path, thus, $\tilde{y}$
bar denotes Laplace transform, thus, $\overline{X} = X(s)$
bar also denotes modal matrices, thus, $[\overline{M}]$

Subscripts

suffix 0 denotes reference values, thus A_0
suffix $_{,x}$ denotes differentiation with x, thus, $u_{,x}$
suffix b indicates deflection due to pure bending, thus, w_b
suffix i indicates layer quantities, thus, w_i
suffix s indicates deflection due to shear, thus, w_s

Select Bibliography

BOOKS ON THEORY OF VIBRATION

Anderson, R.A., *Fundamentals of Vibration*, Macmillan Co., 1967.

Andronow, A.A. and Chaikin, C.E., *Theory of Oscillations*, Princeton University Press, 1949.

Andronov, A.A., Vitt, A.A. and Khaikin, S.E., *Theory of Oscillators*, Translated from Russian by F. Immerzi, Pergammon Press Ltd., 1966.

Argyris, J.H. and Kelsey, S., *Energy Theorems and Structural Analysis*, Butterworth, 1960.

Bathe, K.J., *Finite Element Procedures in Engineering Analysis*, Prentice Hall, 1982.

Beards, C.F. *Vibrations and Control Systems*,John Wiley & Sons, 1988.

Becker, M., *The Principles and Application of Variational Methods*, MIT Press, 1964.

Bendat, J.S., *Principles and Application of Random Noise Theory*, John Wiley & Sons, 1958.

Bendat, J.S. and Piersol, A.G., *Measurement and Analysis of Random Vibration Data*, John Wiley & Sons, 1965.

Beranek, L.L., *Acoustic Measurement*, John Wiley & Sons, 1949.

Beranek, L.L., *Noise and Vibration Control*, McGraw Hill Book Co., 1971.

Bernhard, R.K., *Mechanical Vibrations*, Pitman Publishing Corp, 1943.

BICERA, *A Handbook of Torsional Vibrations*, Cambridge University Press, 1958.

Biezeno, C.B. and Grammel, R., *Engineering Dynamics*, Vol 3, Steam Turbines, Blackie & Sons, 1954.

Biezeno, C.B. and Grammel, R., *Engineering Dynamics*, Vol 4, Internal Combustion Engines, Blackie & Sons, 1954.

Biezeno, C.B. and Grammel, R., *Technische Dynamik*, Springer Verlag, 1939.

Biggs, J.M., *Introduction to Structural Dynamics*, McGraw Hill Book Co., 1964.

Bishop, R.E.D., Gladwell, G.M.L. and Michaelson, S., *The Matrix Analysis of Vibration*, Cambridge University Press, 1965.

Bishop, R.E.D. and Johnson, D.C., *Mechanics of Vibration*, Cambridge University Press, 1960.

Bishop, R.E.D. and Johnson, D.C., *Vibration Analysis Tables*, Cambridge University Press, 1956.

Bisplinghoff, R.L., Ashley, H. and Halfman, R.L., *Aeroelasticity*, Addison-Wesley, 1955.

Bland, D.R., *The Theory of Linear Viscoelasticity*, Pergammon Press, 1960.

Blevins, R.D., *Flow-Induced Vibration*, Van Nostrand Co., 1977.

Bogoliubov, N.N. and Mitropolsky, J.A., *Asymptotical Methods in Theory of Nonlinear Oscillations*, Hindustan Publishing Co., 1963.

Boresi, A.P., *Theory of Elasticity*, Prentice Hall, 1965.

Broch, J.T., *Mechanical Vibration and Shock Measurement*, B & K Publ, 1984.

Brommundt, E., *Vibration of Continuous Systems*, CISM, Udine, Italy, 1969.

Burton, R., *Vibration and Impact*, Addison-Wesley, 1958.

Buzdugan, G., Mihailescu, E. and Rades, M., *Vibration Measurement*, Martinus Nijhoff Publ., 1986.

Bykhovsky, I., *Fundamentals of Vibration Engineering*, Mir Publishers, 1972.

Cain, B.S., *Vibration of Rail and Road Vehicles*, Pitman Publishing Corp, 1940.

Cheung, Y.K and Leung, A.Y.T., *Fine Element Methods in Dynamics*, Kluwer Academic Publishers, 1991.

Christensen, R.M., *Theory of Viscoelasticity–An Introduction*, Academic Press, 1971.

Church, A.H., *Mechanical Vibration*, John Wiley & Sons, 1963.

Clough, R.W. and Penzien, J., *Dynamics of Structures*, McGraw Hill Book Co., 1989.

Cole, E.B., *The Theory of Vibrations for Engineers*, Crosby Lockwood & Son Ltd., 1950.

Collacott, R.A., *Mechanical Fault Diagnosis*, Chapman and Hall, 1977.

Collatz, L., *Eigenwertprobleme und ihre Numerische Behandlung*, Chelsea Pub. Co., 1948.

Coulson, C.A., *Waves*,Oliver and Boyd, 1961.

Crafton, P.A., *Shock and Vibration in Linear Systems*, Harper Publishers, 1961.

Craig, R.R., *Structural Dynamics*, John Wiley & Sons, 1981.

Crandall, S.H., *Random Vibrations*, John Wiley & Sons, 1958.
Crandall, S.H., *Engineering Analysis*, McGraw Hill Book Co., 1956.
Crandall, S.H., Karnopp, D.C., Kurtz, E.F. Jr. and Pridmore-Brown D.C., *Dynamics of Mechanical and Electromechanical Systems*, Krieger Publ, 1982.
Crede, C.E., *Vibration and Shock Isolation*, John Wiley & Sons, 1952.
Crossley, F.R.E., *Dynamics in Machines*, Ronald Press Co., 1954.
Cunningham, W.J., *Introduction to Nonlinear Analysis*, McGraw Hill Book Co., 1958.
Del Pedro, M, Pahud, P, *Vibration Mechanics*, Kluwer Academic Publishers, 1991.
Den Hartog, J.P., *Mechanical Vibration*, McGraw Hill Book Co., 1956.
Desai, C.S. and Abel, J.F., *Introduction to the Finite Element Method–A Numerical Method for Engineering Analysis*, Van Nostrand Co., 1972.
Dimarogonas, A.D. and Paipetis, S.A., *Analytical Methods in Rotor Dynamics, Applied Science*, 1983.
Dimintberg, F.M., *Flexural Vibrations of Rotating Shafts*, Butterworth Co., 1961.
Dinca, F. and Theodosin, C., *Nonlinear and Random Vibration*, Academic Press, 1973.
Doyle, J.K. *Static and Dynamic Analysis of Structures*, Kluwer Academic Publishers, 1991
Duffing, G., *Erzwungene Schwingungen bei veranderlicher Eigenfrequenz*, F. Viewegu. Sohn, Braunschweig, 1918.
Dym, C.L. and Shames, I.H., *Solid Mechanics, A Variational Approach*, McGraw Hill Book Co., 1973.
Elishakoff, I. and Irretier, H., *Refined Dynamical Theories of Beams Plates and Shells and Their Applications*, Springer Verlag, 1986.
Elsgolc, L.E., *Calculus of Variations*, Addison-Wesley, 1961.
Evan-Iwanowski, R.M., *Resonance Oscillations in Mechanical Systems*, Elsevier, 1976.
Ewins, D.J., *Modal Testing: Theory and Practice*, Research Studies Press, 1986.
Federn, K., *Auswuchttechnik*, Vol. 1, Springer Verlag, 1977.
Fertis, D.G., *Dynamics and Vibration of Structures*, John Wiley & Sons, 1973.
Flugge, W., *Viscoelasticity*, Blaisdell Pub. Co., 1967.
Forray, M., *Variational Calculus in Science and Engineering*, McGraw Hill Book Co., 1968.
Fox, C., *An Introduction to the Calculus of Variations*, Oxford University Press, 1950.
Freberg, C.R. and Kemler, E.N., *Elements of Mechanical Vibration*, John Wiley & Sons, 1949.
Fung, Y.C., *Foundations of Solid Mechanics*, Prentice Hall, 1965.
Fung, Y.C., *An Introduction to Theory of Aeroelasticity*, John Wiley & Sons, 1955.
Gardner, M.F. and Barnes, J.L., *Transients in Linear Systems*, John Wiley & Sons, 1942
Garg, V.K. and Dukkipati, R.V., *Dynamics of Railway Vehicle Systems*, Academic Press, 1984.
Gasch, R. and Pfutzner, H., *Rotordynamik*, Spinger Verlag, 1975.
Gelfard, I.M. and Fomin, S.V., *Calculus of Variations*, Prentice Hall, 1963.
Ghosh, A, Mallik, A.K., *Theory of Mechanisms and Machines*, Afiiliated East West Press, 1976.
Gorman, D.J., *Free Vibration Analysis of Beams and Shafts*, John Wiley & Sons, 1975.
Gorman, D.J., *Free Vibration Analysis of Rectangular Plates*, Elsevier, 1982.
Gould, S.H., *Variational Methods for Eigen Value Problems*, University of Toronto, 1957.
Gross, E.E., *Measurement of Vibration*, General Radio, 1955.
Grover, G.K., *Mechanical Vibration*, Nem Chand and Bros, Roorkee, 1972.
Haag, J., *Les Mouvements Vibratoires*, Presses Universitaires, 1955.
Hagedorn, P., *Nonlinear Oscillations*, Clarendon Press, 1988.
Ham, C.W., Crane, E.J. and Rogers, W.L., *Mechanics of Machines*, McGraw Hill Book Co., 1958.
Hansen, H.M. and Chenea, P.F., *Mechanics of Vibrations*, John Wiley & Sons, 1952.
Harris, C.M. and Crede, C.E., *Shock and Vibration Handbook*, McGraw Hill Book Co., 1976.
Hatter, D.H., *Matrix Computer Methods of Vibration Analysis*, John Wiley & Sons, 1973.
Hayashi, C., *Forced Oscillations in Nonlinear Systems*, Nippon Printing Co., 1953.
Hayashi, C., *Nonlinear Oscillations in Physical Systems*, McGraw Hill Book Co., 1965.
Hayashi, C., *Nonlinear Oscillations in Physical Systems*, Princeton University, 1985.
Hodge, P.G., *Continuum Mechanics*, McGraw Hill Book Co., 1970.
Hoff, N.J., *The Analysis of Structures*, John Wiley & Sons, 1956.
Holowenko, A.R., *Dynamics of Machinery*, John Wiley & Sons, 1955.
Holzer, H., *Die Berechnung der Drehsochwingungen*, Springer Verlag, 1921.
Hort, W., *Technische Schwingungslehre*, Springer Verlag, 1922.
Huebener, K.H., *The Finite Element Method for Engineers*, John Wiley & Sons, 1975.
Hurty, W.C. and Rubenstein, M.F., *Dynamics of Structures*, Prentice Hall, 1964.

Jacobsen, L.S. and Ayre, R.S., *Engineering Vibrations*, McGraw Hill Book Co., 1963.

Kabulov, V.K., *Integral Equations of Balance Type and Their Application to Dynamic Analysis of Bars and Beams*, Izd Akad Nauk Uz SSR, 1965.

Kantorovich, L.V. and Krylov, V.I., *Approximate Methods of High Analysis*, Inter-Science Publ., 1964.

Karman, T. and Biot. M.A., *Mathematical Methods in Engineering*, McGraw Hill Book Co., 1940.

Kearton, W.J., *Steam Turbine Theory and Practice*, Pitman Publishing Corp, 1958.

Kimball, A.L., *Vibration Prevention in Engineering*, John Wiley & Sons, 1932.

Klotter, K., *Technische Schwingungslehre*, Springer Verlag, 1951.

Klotter, K., *Messung Mechanischer Schwingungen*, Springer Verlag, 1943.

Klyuev, V.V., *In Structures and Systems for Measurement of Vibration, Noise and Shock*, Mashinostroyenye, (In Russian), Moscow, 1978.

Kolsky, H., *Stress Waves in Solids*, Oxford University Press, 1953.

Kramer, E., *Maschinendynamik*, Springer Verlag, 1984.

Krylov, A.N. and Bogoliubov, N., *Introduction to Nonlinear Mechanics*, Princeton University Press, 1943.

Lalanne, M. and Berthier, P and Der Hagopian, J., *Mechanical Vibration for Engineers*, John Wiley & Sons, 1973.

Lalanne, M. and Ferraris. G., *Rotordynamics Prediction in Engineering*, John Wiley & Sons, 1990.

Lanczos, C., *The Variational Principle of Mechanics*, University of Toronto, 1949.

Langhaar, H.L., *Energy Methods in Applied Mechanics*, John Wiley & Sons, 1962.

Lazan, B.J., *Damping of Materials and Members in Structural Mechanics*, Pergammon Press, 1968.

Love, A.E.H., *Mathematical Theory of Elasticity*, Dover, 1944.

Macduff, J.N. and Curreri, J.R., *Vibration Control*, McGraw Hill Book Co., 1958.

Mallik, A.K., *Principles of Vibration Control*, Affiliated East West Press, 1990.

Manley, R.G., *Wave Form Analysis*, John Wilay & Sons, 1945.

McCallion, H., *Vibration of Linear Mechanical Systems*, Longman Publication, 1973.

McLachlan, N.W., *Theory of Vibration*, Dover Publ., 1951.

McLachlan, N.W., *Theory and Application of Mathieu Functions*, Oxford University Press, New York, 1947.

McLachlan, N.W., *Ordinary Non-Linear Differential Equations in Engineering and Physical Sciences*, Oxford University Press, New York, 1950.

Meirovitch, L., *Elements of Vibration Analysis*, McGraw Hill Book Co., 1986.

Meirovitch, L., *Analytical Methods in Vibration*, Macmillan Co., 1967.

Mickens, R.E., *An Introduction to Nonlinear Oscillations*, Cambridge University Press, 1981.

Mikhlin, S.G., *Variational Methods in Mathematical Physics*, Pergammon Press, 1964.

Minorsky, N., *Introduction to Nonlinear Mechanics*, J.W. Edwards Publisher Inc., Ann Arbor, Michigan, 1947.

Minorsky, N., *Nonlinear Oscillations*, D Van Nostrand Co. Inc., 1962

Moon, F.C., *Chaotic Vibration* John Wiley & Sons, 1987.

Morrill, B., *Mechanical Vibration*, The Ronald Press Co., 1937.

Morris, J., *The Escalator Method in Engineering Vibration Problems*, John Wiley & Sons, 1947.

Morrow, C.T., *Shock and Vibration Engineering*, John Wiley and Sons, 1963.

Morse, P.M., *Vibration and Sound*, McGraw Hill Book Co., 1948.

Mukhopadhyay, *M., Vibrations, Dynamics and Structural Systems*, Oxford and IBH, 1989.

Muller, P.C. and Schiehlen, W.O., *Linear Vibrations*, Martinus Nijhoff, 1985.

Myklestad, N.O., *Vibration Analysis*, McGraw Hill Book Co., 1944.

Myklestad, N.O., *Fundamentals of Vibration Analysis*, McGraw Hill Book Co., 1956.

Nashif, A.D., Jones, D.I.G. and Henderson, J.P., *Vibration Damping*, John Wiley & Sons, 1985.

Nayfeh, A.H. and Mook, D.T., *Nonlinear Oscillations*, John Wiley & Sons, 1979.

Newland, D.E., *An Introduction to Random Vibration and Spectral Analysis*, Longman, 1984.

Newland, D.E., *Mechanical Vibration Analysis and Computation*, John Wiley & Sons, 1989.

Nigam, N.C., *Introduction to Random Vibration*, MIT Press, 1983.

Novacki, W., *Dynamics of Elastic Systems*, Chapman and Hall, 1963.

Novozhilov, V.V., *Theory of Elasticity*, Pergammon Press, 1961.

Panovko, Ya., *Elements of the Applied Theory of Elastic Vibration*, Mir Publishers, 1971.

Pestel, E.C. and Leckie, F.A., *Matrix Methods in Elastomechanics*, McGraw Hill Book Co., 1963.

Prescott, J., *Applied Elasticity*, Dover, 1946.

Prezemieniecki, J.S., *Theory of Matrix Structural Analysis*, McGraw Hill Book Co., 1968.
Pujara, K.K., *Mechanical Vibration*, Dhanpat Rai, 1980.
Ragulskis, K.M. Jurkauskas, A.Yu., Atstupenas, V.V., Vitkute, A.Yu. and Kulvec, A.P., *Vibration of Bearings*, Oxonian Press, New Delhi, 1979.
Rao, J.S., *Rotor Dynamics*, John Wiley & Sons, 1991.
Rao, J.S., *Turbomachine Blade Vibration*, John Wiley & Sons, 1991.
Rao, J.S. and Dukkipati, *R.V. Mechanism and Machine Theory*, John Wiley & Sons, 1991.
Rao, J.S. and Gupta, K., *An Introductory Course on Theory and Practice of Mechanical Vibrations*, John Wiley & Sons, 1984..
Rao, S.S. *Mechanical Vibration*, Addison-Wesley, 1986.
Rauscher, M., *Introduction to Aeronautical Dynamics*, John Wiley & Sons, 1953.
Rayleigh, J.W.S., *The Theory of Sound*, Dover, 1945.
Rieger, N.F., *Vibrations of Rotating Machinery*, Part 1: Rotor Bearing Dynamics, Vibration Institute, 1981.
Rieger, N.F. and Rao, J.S., *Vibrations of Rotating Machinery*, Part 2: Blading and Torsional Vibrations, Vibration Institute, 1981.
Ritz, W., *Gesammelte Werke*, Gauthier-Villars, 1911.
Rogers, G.L., *Dynamics of Framed Structures*, John Wiley & Sons, 1959.
Rothbart, H.A., Cams–Design, *Dynamics and Accuracy*, John Wiley & Sons, 1956.
Routh, E.J., *Dynamics of a System of Rigid Bodies*, Dover, 1955.
Samoyovich, G.S., *Vibration Problems in Rotating Turbomachines*, Machines-stroya, 1975.
Scanlan, R.H. and Rosenbaum, R., *Introduction to the Study of Aircraft Vibration and Flutter*, Dover Publication, 1988.
Schecheter, R.S., *The Variational Method in Engineering*, McGraw Hill Book Co., 1967.
Schmidt, G. and Tondl. A., *Nonlinear Vibration*, Cambridge University Press, 1986.
Shames, I.A and Dym, C.L., *Energy and Finite Element Methods in Structural Mechanics*, Hemisphere Publ. Corp., 1985
Shigley, J.E., *Dynamics Analysis of Machines*, McGraw Hill Book Co., 1961.
Snowdon, J.C., *Vibration and Shock in Damped Mechanical Systems*, John Wiley & Sons, 1968.
Sokolnikoff, I.S., *Mathematical Theory of Elasticity*, McGraw Hill Book Co., 1956.
Srinivasan, P., *Mechanical Vibration Analysis*, Tata McGraw-Hill Book Co., 1982.
Steidel, R.F., *An Introduction to Mechanical Vibration*, John Wiley & Sons, 1971.
Stodola, A., *Steam and Gas Turbines*, McGraw Hill Book Co., 1927.
Stoker, J.J., *Nonlinear Vibrations in Mechanical and Electrical Systems*, Interscience, 1950.
Sweeney, G., *Vibration of Machine Tools*, The Machinery Publishing Co. Ltd.
Tauchert, T.R., *Energy Principles in Structural Mechanics*, McGraw Hill Book Co., 1974.
Temple, G. and Bickley, W.G., *Rayleigh's Principle and Its Applications in Engineering*, Dover, 1956.
Thomson, J.M.T and Stewart, *Nonlinear Dynamics and Chaos*, John Wiley & Sons, 1986.
Thomson, W.T., *Theory of Vibrations with Applications*, Prentice Hall, 1972.
Thornton, D.L., *Mechanics Applied to Vibrations and Balancing*, John Wiley & Sons, 1940.
Timoshenko, S.P. and Goodier, J.N., *Theory of Elasticity*, McGraw Hill Book Co., 1951.
Timoshenko, S. and Young, D.H. *Vibration Problems in Engineering*, Van Nostrand Co., 1954.
Timoshenko, S.P. and Young, D.H. *Advanced Dynamics*, McGraw Hill Book Co., 1948.
Tobias, S.A., *Machine Tool Vibration*, Blacke and Son, 1965.
Todd, F.H., *Ship Hull Vibration*, Edward Arnold, 1961.
Tondl, A., *Some Problems of Rotordynamics*, Chapman and Hall, 1965.
Tondl, A., Zvyseni Hranice, *Stability Rotorovych Systemic* Czech Academy Ved, 1985
Tong, K.N., *Theory of Mechanical Vibration*, John Wiley & Sons, 1960.
Trimmer, J.D., *Response of Physical Systems*, John Wiley & Sons, 1950.
Tse, F.S., Morse, I.E. and Hinkle, R.T., *Mechanical Vibrations*, Prentice Hall, 1963.
Tuplin, W.A., *Torsional Vibration*, John Wiley & Sons, 1934.
Vance, J.M., *Rotordynamics in Turbomachines*, John Wiley & Sons, 1988.
Van Santen, G.W., *Mechanical Vibration*, Macmillan Co., 1958.
Vernon, J.B. *Linear Vibration Theory*, John Wiley & Sons, 1967.
Vladislavlev, L.A., *Vibration of Hydro Units in Hydroelectric Power Plants*, Amerind Publishing Co., 1972.

Vlasov, V.Z. and Leontev, N.N., *Beams and Plates and Shells on Elastic Foundations*, NASA Technical Translation F-357 from Russian.

Volterra, E. and Zachmanoglou, E.C., *Dynamics of Vibrations*, Merrill, 1965.

Warburton, G.W., *The Dynamical Behaviour of Structures*, Pergammon Press, 1976.

Washizu, K., *Variational Methods in Elasticity and Plasticity*, Pergammon Press, 1982.

Weinstock, R., *Calculus of Variations with Applications to Physics and Engineering*, McGraw Hill Book Co., 1952.

Wilson, W.K., *Practical Solution of Torsional Vibration Problems*, Vol. 1, John Wiley & Sons, 1949.

Wilson, W.K., *Practical Solution of Torsional Vibration Problems*, Vol. 2, John Wiley & Sons, 1949.

Wilson, W.K., *Practical Solution of Torsional Vibration Problems*, Vol. 3, Chapman and Hall, 1965.

Weaver, W. and Johnston, P.R., *Structural Dynamics by Finite Elements*, Prentice Hall, 1987.

Wowk, V. Machinery Vibration, McGraw Hill, 1991.

Yang, S.J. And Ellison, A.J., *Machinery Noise and Measurement*, Clarendon, 1985.

Zeinkiewicz, O.C., *The Finite Element Method*, McGraw Hill Book Co, 1977.

NONLINEAR VIBRATIONS

Abramson, H.N., Response Curves for a System with Softening Restoring Force, *J Appl Mech*, v. 22, #3, 1955, p. 434.

Appleton, E.V. and Van der Pol, B., On a Type of Oscillation–Hysteresis in a Simple Triode Generator, *Phil Mag*, v. 43, 1922. p. 177.

Arnold, R.C. and Haft, E.E. Stability of Unsymmetrical Rotating Cantilever Shaft Carrying an Unsymmetrical Rotor, *J Engg for Indus, Trans ASME*, 1972, p. 243.

Ayre, R.S., Jacobsen, L.S. and Phillips, A., Steady Forced Vibration of a Nonconservative System with Variable Mass; a Pumping System, J Franklin Inst, v. 250, #4, 1950, p. 315.

Baker, J.G., Self Induced Vibrations, *Trans ASME*, APM-55-2, 1933.

Baker, J.G. and Mikina, S.J., Calculation of Dampers for Systems Subject to Self Induced Vibration, *Trans ASME*, v. 58, 1936, p. A121.

Banerji, A.K., Forced Oscillations in a Stick Slip System, *Romaninan J of Tech Sci., Appl Mech*, v. 16, #1, 1971, p. 219.

Blair, K.W. and Loud, W.S., Periodic Solutions of $x'' + cx' + g(x) = Ef(t)$ under Variation of Certain Parameters, *J Soc Indus Appl Math*, v. 8, 1960, p. 74.

Bowers, N.A., Tacoma Narrows Bridge Wrecked by Wind, Engg News Record, Nov 14, 1940, p. 647, 656.

Brock, J.E., An Iterative Numerical Method for Non-linear Vibrations, *J Appl Mech, Trans ASME*, v. 18, #1, 1951, p. 1.

Brosens, P.J., Vibrations of Unsymmetrical Rotors, *J Appl Mech Trans ASME*, 1961, p. 355.

Buland, R.N., Analysis of Nonlinear Servos by Phase Plane Delta Method, J Franklin Inst, v. 257, 1954, p. 37.

Carnegie, W.D., Rao, J.S. and Pasricha, M.S., A Theoretical Study of the Effects of Variable Inertia on the Torsional Vibrations of a Single Cylinder Engine Mechanism, Instn of Marine Engrs, London, August 1971.

Carnegie, W., Reif, Z. and Wee, R.S.S., A Numerical Technique for the Solution of Some Non-linear Systems, *Aero J of Roy Aero Soc.*, v. 73, #704, 1969, p. 680.

Carrier, G.F. and Lewis, J.A. The Relaxation Oscillations of the Van der Pol Oscillator, *Advance in Appl Mech*, Academic Press, v. 3, 1953, p. 12.

Cartwright, M.L., Forced Oscillations in Nearly Sinusoidal Systems, *J Instn Elect Engrs*, v. 95, 1948, (III), p. 88.

Clauser, F.H., The Behaviour of Nonlinear Systems, *J Aero Sci.*, v. 23, #5, 1956, p. 411.

Crandall, S.H. and Brosens, P.J., On the Stability of Rotation of a Rotor with Rotationally Unsymmetric Inertia and Stiffness Properties, *J Appl Mech Trans ASME*, 1961, p. 567.

Crossley, F.R.E., Forced Oscillations of the Centrifugal Pendulum with Wide Angles, *J Appl Mech*, v. 20, #1, 1953, p. 41.

Den Hartog, J.P., Transmission Line Vibration due to Sleet, *Trans AIEE*, v. 51, 1932, p. 1074.

Den Hartog, J.P. and Heiles, R.M., Forced Vibrations in Nonlinear Systems, *J Appl Mech, Trans ASME*, v. 58, 1936, p. A127.

Den Hartog, J.P. and Mikina, S.J., Forced Vibrations with Non-linear Spring Constants, *Trans ASME*, APM–54–15–157, 1932

Den Hartog, J.P. and Mikina, S.J., The Amplitudes of Non-harmonic Vibrations, *J Franklin Inst*, v. 216, #4, 1933, p. 459.

Evaldson, R.L., Ayre, R.S. and Jacobsen, L.S. Response of an Elastically Nonlinear System to Transient Disturbances, *J Franklin Inst*, v. 248, #6, 1949, p. 473.

Friedrichs, K.O. and Stoker, J.J., Forced Vibrations of Systems with Nonlinear Restoring Force, *Quart Appl Math*, v. 1, #2, 1943, p. 97.

Gillies, A.W., On the Transformations of Singularities and Limit Cycles of the Variational Equations of Van der Pol, *Quart J Mech and Appl Math*, v. 7, Pt 2, 1954.

Gillies, A.W., The Periodic Solutions of A Nonlinear Differential Equation of the Second Order with Unsymmetrical Damping, and A Forcing Term, *Quart J Mech and Appl Math*, v. 8, Pt. 1, 1955.

Gillies, A.W., The Periodic Solutions of the Differential Equation of A Resistance Capacitance Oscillator, *Quart J Mech and Appl Math*, v. 10, Pt. 1 1957.

Gillies, A.W., On A Class of Differential Equations Governing Nonlinear Vibrations, *Quart J Mech and Appl Math*, v. 10, Pt 3, 1957.

Goldsbrough, G.R., Torsional Vibrations in Reciprocating Engine Shafts, *Proc Royal Soc.*, A, v. 109, 1925, p. 99.

Goldsbrough, G.R., The Properties of Torsional Vibrations in Reciprocating Engine Shafts - Part I, Proc Royal soc., A, v. 113, 1926-7, p. 259.

Goldsbrough, G.R. and Baker, H., The Properties of Torsional Vibrations in Reciprocating Engine Shafts-Part II, Proc Royal soc., A, v. 113, 1926-7, p. 272.

Hayashi, C., Forced Oscillations with Nonlinear Restoring Force, *J Appl Physics*, v. 24, #2, 1953, p. 198.

Hayashi, C., Stability Investigation of the Nonlinear Periodic Oscillations, *J Appl Physics*, v. 24, #3, 1953, p. 344.

Hayashi, C., Subharmonic Oscillations in Nonlinear Systems, *J Appl Physics*, v. 24, #5, 1953, p. 521.

Hayashi, C., Initial Conditions for Certain Types of Nonlinear Oscillations, Symp Nonlinear Circuit Analysis, Polytechnic Inst of Brooklyn, Ny., v. 6, 1956, p. 63.

Hayashi, C., Higher Harmonic Oscillations in Nonlinear Forced Systems, Col Int C N R S, #148, 1964, p. 267.

Hayashi, C., Nishikawa, Y. and Abe, M. Subharmonic Oscillations of Order One Half, *IRE Trans of Professional Group on Circuit Theory*, v. CT-7, 1960, p. 102.

Hayashi, C., Ueda, Y. and Kawakami, H., Solution of Duffing's Equation using Mapping Concepts, Fourth Conf. on Nonlinear Oscillations, Prague, 1967.

Hayashi, C., Ueda, Y. and Kawakami, H., Transformation Theory as Applied to the Solutions of Nonlinear Differential Equations of the Second Order, *Int J Non-Linear Mech*, v. 4, 1969, p. 235.

Inagaki, T., Kanki, H. and Shiraki, K., Response Analysis of a General Asymmetric Rotor-Bearing System, *ASME 79-DET*-84.

Iwatsubo, T., Vibration of An Asymmetric Shaft, *JSME*, v. 37, 1971. p. 1503.

Jacobsen, L.S., On a General Method of Solving Second Order Ordinary Differential Equations by Phase Plane Displacements, *J Appl Mech*, v. 19, #4, 1952, p. 543.

Jacobsen, L.S. and Jesperson, H.J., Steady Forced Vibrations of Single Mass Systems with Symmetrical as well as Unsymmetrical Nonlinear Restoring Elements, *J Franklin Inst*, v. 220, #4, 1935, p. 467.

Keller, J.B. and Ting, L., Periodic Vibrations of Systems Governed by Nonlinear Partial Differential Equations, *Comm Pure and Appld Math*, v. 19, #4, 1966, p. 371.

Klotter, K., *Ing Arch*, v. 7, 1936, p. 87.

Klotter, K., Non-linear Vibration Problems Treated by the Averaging Method of W. Ritz, Proc. First U.S. Natl Cong Appl Mech, 1951, p. 125.

Klotter, K. and Pinney, E., Stability Criteria for Forced Vibrations in Nonlinear Systems, *J Appl Mech*, v. 20, #1, 1953, p. 9.

Krogdall, W.S., Numerical Solutions of Van der Pol's Equation, *ZAMP*, v. 11, 1960, p. 59.

Kung, G.C. and Pao, Y.H., Nonlinear Flexural Vibrations of a Clamped Circular Plate, *J Appl Mech*, v. 39, 1972, p. 1050.

Kutta, W., *Zeitscrift fur Math and Phys*, v. 46, 1901.

La Salle, J., Relaxation Oscillations, *Quart Appl Math*, April 1949, p. 1.

Li, W.H., Graphical Solution of Equations of Vibrations, *Proc ASCE*, v. 83, #EM4, 1957.

Lienard, A., Etude Des Oscillations entretenues, *Rev Gen de l'Electricite*, v. 23, 1928, p. 901.

Lourie, A. and Tchekmarev, A., On the Approximate Solution of Several Non-linear Problems of Forced Vibrations, *Russ Acad Sci. Pub, Sect Tech Mech*, v. 1, #3, 1938, p. 307.

Ludeke, C.A., An Experimental Investigation of Forced Vibration in A Mechanical System Having a Nonlinear Restoring Force, *J Appl Phys*, v. 17, 1946, p. 603.

Ludeke, C.A. Electromechanical Device for Solving Nonlinear Differential Equations, *J Appl Phy*, v. 20, 1949, p. 600.

Ludeke, C.A., A Method of Equivalent Linearization for Nonlinear Oscillatory Systems with Large Nonlinearity, *J Appl Physics*, v. 20, 1949, p. 694.

Ludeke, C.A., Predominantly Subharmonic Oscillations, *J Appl Physics*, v. 22, #11, 1951, 1321.

Ludeke, C.A. Nonlinear Phenomena, *Trans ASME*, 1957, p. 439.

Meuser, R.B. and Weibel, E.E., Vibration of a Nonlinear System during Acceleration through Resonance, *J Appl Mech*, v. 15, #1, 1948, p. 21.

Minorsky, N., Self Excited Mechanical Oscillations, *J Appl Phy*, v. 19, 1948, p. 332.

Minorsky, N., On Interaction of Nonlinear Oscillations, *J Franklin Inst*, v. 256, 1953, p. 147.

Mote, C.D., Dynamic Stability of an Axially Moving Band, *J Franklin Inst*, v. 285, #5, 1968, p. 329.

Nag, B.R., Resonance Curves of A Synchronized Oscillator, *Indian J Physics*, v. 33, #1, p. 23.

Ota, H. and Kono, K., Unstable Vibrations Induced by Rotationally Unsymmetric Inertia and Stiffness Properties, *Bull JSME*, v. 14, 1971, p. 29.

Pasricha, M.S. and Carnegie, W.D., Effects of Variable Inertia on the Damped Torsional Vibrations of Diesel Engine Systems, *J Sound Vib.*, v. 46, 1976, p. 339.

Pedersen, P.O., Subharmonics in Forced Oscillations in Dissipative Systems, Part I, *J A S A*, v. VI, 1935, p. 227.

Raucher, M., Steady Oscillations of System with Nonlinear and Unsymmetrical Elasticity, *J Appl Mech, Trans ASME*, v. 60, A169, 1938.

Runge, C., *Mathematische Annalen*, v. 46, 1895.

Schwesinger, G., On One-Term Approximations of Forced Nonharmonic Vibrations, *J Appl Mech*, v. 17, #2, 1950, p. 202.

Sethna, P.R., Steady State Motion of the One and Two-degree-of-freedom Vibrating Systems with a Nonlinear Restoring Force, *Proc. Second U.S. Natl Cong Appl Mech*, 1954, p. 69.

Soroka, W.W. Free Periodic Motions of an Undamped Two Degree of Freedom Oscillatory System with Non-linear Unsymmetrical Elasticity, *J Appl Mech, Trans ASME*, v. 18, 1950, p. 185.

Srinivasan, A.V., Nonlinear Vibrations of Beams and Plates, *Intl J Nonlinear Mechanics*, v. 1, 1966, p. 179.

Steele, C.R., Nonlinear Effects in the Problems of the Beam on a Foundation with a Moving Load, *Intl J Solids and Structures*, v. 3, #4, 1967, 565.

Tondl, A., Vibrations of Rigid Rotors with Vertical Shafts Mounted in Aerostatic Bearings, Monographs and Memoranda #14, *Natl Res Inst for Machine Design*, Bechovice, Prague, 1973.

Tondl, A., Some Problems of Self Excited Vibrations of Rotors, Monographs, and Memoranda #17, *Natl Res Inst for Machine Design*, Bechovice, 1974.

Tondl, A., On the Interaction between Self Excited and Parametric Vibrations, Monographs and Memoranda #25, *Natl Res Inst for Machine Design*, Bechovice, Prague, 1978.

Tondl, A, Analysis of the Transient Processes in Models of Pump Turbines after Sudden Loading, Monographs and Memoranda #37, *Natl Res Inst for Machine Design*, Bechovice, Prague, 1986.

Van der Pol, B., On Relaxation Oscillations, *Phil Mag*, 1926, p. 978.

Van der Pol, B., Forced Oscillations in a Circuit with Nonlinear Resistance, *Phil Mag.*, Sereis, 7, v. 3, #13, 1927, p. 65.

Van der Pol, B. and Strutt, M.J.O., On the Stability of the Solutions of Mathieu's Equations, *Phil Mag.*, Series 7, v. 5, #27, 1928, p. 18.

Van der Pol, B., The Nonlinear Theory of Electric Oscillations, *Proc IRE*, v. 22, #9, 1934, p. 1051.

Wu, M. H.L., Subharmonic Resonance of System Having Non-linear Spring with Variable Coefficient, *Proc. First U.S. Natl Cong Appl Mech*, 1951, p. 147.

Yamamoto, T. and Kono, K., Forced Vibrations of a Rotor with Rotating Inequality, *Bull JSME*, v. 14, 1971, p. 1059.

Yamamoto, T. and Ota, H., On the Unstable Vibrations of A Shaft Carrying an Unsymmetrical Rotor, *JSME*, v. 31, 1964, p. 525.

Yamamoto, T. et al., On the Forced Vibrations of the Shaft Carrying an Unsymmetrical Rotor: Forced Vibrations having the Circular Frequencies Differing from the Rotating Angular Velocity of the Shaft, *Bull JSME*, v. 9, 1966, p. 58.

Yamamoto, T. et al., On the Unstable Vibrations of A Shaft with Unsymmetrical Stiffness Carrying an Unsymmetrical Rotor, *J Appl Mech, Trans ASME*, 1968, p. 313.

Young, D., Forced Vibration of Systems with Nonlinear Restoring Force, *Proc First Midwest Conf. Solid Mechanics*, Univ of Illinois, 1953, p. 164.

Young, D. and Hess, P.N., On the Stability of Harmonic Solutions of a Modified Form of Duffing's Equation, *Proc. Second U.S. Natl Cong Appl Mech*, 1954, p. 79.

LONGITUDINAL VIBRATION OF BARS

Abramson, H.N., Plass, H.J. and Ripperger, E.A., Stress Wave Propagation in Rods and Beams, *Advances in Appld Mech*, v. 5, 1958, p. 111.

Bailey, J.J. and Finnie, I., An Analytical Study of Drill String Vibrations, *Trans ASME J Engg for Indus*, v. 82B, #2, 1960, p. 122.

Bies, D.A., Tapering A Bar for Uniform Stress in Longitudinal Oscillation, *J Aco Soc America*, v 34, #10, 1962, p. 1567.

Bishop, R.E.D., Longitudinal Waves in Beams, *Aero Qly*, v. 3, #2, 1952, p. 280.

Chang, C.S., Energy Dissipation in Longitudinal Vibration, *Proc 3rd. U.S. Natl Cong Appld Mech*, 1958, p. 108.

Chehil, D.S. and Heaps, H.S., Effect of Lateral Motion on the Longitudinal Vibrations of Tapered Bars, *J Aco Soc America*, v. 43, #1, p. 540.

Ghosh, S.K., Dynamics of Longitudinal Propagation of Elastic Disturbance Through a Medium Exhibiting Gradient of Elasticity, *Ind. J Phy*, v. 35, #1, 1961, p. 22.

Ghosh, S.K., Note on the Longitudinal Vibrations of a Paraboloidal Bar, *Ind. J Theo Phy*, v. 14, #4, 1966, p. 81.

Ghosh. S.K. and Banerjee, S.K., Dynamics of Extensional Vibrations of Free-Free Bar, *Ind J Phy*, v. 40, #3, 1966, p. 135.

Hertelendy, P., Displacement and Strain Energy Distribution in a Longitudinally Vibrating Cylindrical Rod with Viscoelastic Coating, *J Appld Mech*. v. 29, #1, 1962, p. 47.

Hertelendy, P., An Approximate Theory Governing Symmetrical Motion of Elastic Rods of Rectangular or Square Cross-section, *J Appld Mech*, v. 35, #2, 1968, p. 333.

Kabulov, V.K., The Integral Equations of the Longitudinal Vibrations of Bars, Dokaldi Akad Nauk Uz SSR, v. 3, 1956, p. 7.

Kane, J.R. and McGoldrick, R. J., Longitudinal Vibrations of Marine Propulsion Shafting Systems, *Trans Soc Nav Arch Marine Engrs*, v. 2, 1949, p. 40.

Kasin, P.A., Determination of the Fundamental Frequency of the Natural Longitudinal and Torsional Oscillations of Tapered Bars, *Tr Unta Druzhdy Naradov Patrika Humumber*, v. 28, 1967, p. 117.

Kondratov, A.S. and Chaikina, Z.S. Longitudinal Oscillations of Prismatic Rods of Variable Cross-section taking Hysteresis into account, Rasseyanie Energie Pri Kolebaniyakh, Uprug Sistem, Kiev Nauk Dumka, 1966, p. 38.

Kynch, G.J. and Green, W.A., Vibrations of Beams. I Longitudinal Modes, *Qly J Mech Appld Math*, v. 10, #1, 1951, p. 6373.

Lubkin, J.L. and Luke, Y.L., Frequencies of Longitudinal Vibrations for a Slender Rod of Variable Cross-section, *J Appld Mech*, v. 20, #2, 1953, p. 1723.

Medick, M.A., One Dimensional Theories of Wave Propagation of Vibration in Elastic Bars of Rectangular Cross-sections, *J Appld Mech*, v. 33E, 1966, p. 489.

Mindlin, R.D. and Herrman, G., A One Dimensional Theory of Compressional Waves in Elastic Rods, *Proc 1st U.S. Natl Cong on Appld Mech*, 1951, p. 187 and *Proc 2nd U.S. Natl Cong on Appld Mech*, 1954, p. 233.

Mindlin, R.D. and Mcniven, H.D., Axially Symmetric Waves in Elastic Rods, *J Appld Mech*, v. 27, #1, 1960, p. 145.

Plass, H.J. and Steyer, C.C., Studies on Longitudinal and Bending Waves in Long Elastic Rods, Univ of Texas Def Res Lab Report No. DRL-376, 1956.

Pochammer, L., Uber Fort Pflanzungsgeschwindigkeiten Kleiner Schwingungen in Einern unte Grenzten Isotropen Kreiszlinder, *J Pur Reine u. Angew Math* (Crelle), v. 81, 1876, p. 321.

Rao, D.K., *Solutions of Vibration Problems of Rods and Beams using Refined Theories*, Ph D Thesis, IIT, Kharagpur, 1972.

Rapoport, I.M., On the Oscillations of Elastic Rods, Soviet Physics Dokaldy, v. 8, #12, p. 1248.

Rozovskii, M.I., Longitudinal Oscillations of an Elastic Hereditary Rod with Singular Reaction Function, Dopoviodi Akad Nauk Ukr SSR, v. 6, 1959, p. 587.

Sarkar, S.K., Note on the Longitudinal Vibrations of a Thin Viscoelastic Rod of Variable Cross-section, Buletenul Tasi Inst Polytechnic, v. 13, 1967, p. 143.

Shestakov, V. Ya. Lengthwise Blow on a Conical Viscoelastic Bar, Studies of Theory of Structures, v. 12, 1963, Moscow (Editor I.M. Rabinovich).

Suhara, T., A Study on the Damping of Longitudinal Vibrations of an Elastic Cylinder, Report of Res Inst Elastic Engg, Kyushi Univ, v. 7, #3, 1951, p. 25.

Volterra, E., A One Dimensional Theory of Wave Propagation in Elastic Rods Based on the Method of Internal Constraints, *Ing Arch*, v. 23, #6, 1955, p. 410.

Volterra, E., Some Applications of the Method of Internal Constraints to Dynamic Problems, Deformation and Flow of Solids, Colloquium, Madrid, 1956, p. 236.

Volterra, E., Method of Internal Constraints and its Applications, *Proc ASCE*, *J Engng Mech*, v. 87, EM4, 1961, p. 103.

Volterra, E. and Zachamanoglou, E.C., Free and Forced Vibrations of Straight Elastic Bars According to the Method of Internal Constraints, *Ing Arch*, v. 25, 1957, p. 424.

Wieckowski, J., The Influence of Material Damping on Nonconservative Reactions of Elastic Transducing Torsional and Longitudinal Vibrations, *Arch Mech Stos*, v. 10, #4, 1958, p. 479.

VIBRATIONS OF RODS IN TORSION

Barr, A.D.S., Torsional Waves in Uniform Rods of Non Circular Cross-section, *J Mech Engng Sci.*, v. 4, 1962, p. 127.

Bogdonoff, J.L. and Horner, J.T., Torsional Vibration of Rotating Twisted Bars, *J Aero Sci.*, v. 23, 1956, p. 393.

Brody, W.G. and Targoff, W.P., Uncoupled Torsional Vibrations of Rotating Twisted Bars, *WADC, TR* 56-501, 1957.

Carnegie, W., Experimental Determination of Center of Flexure and Center of Torsion Coordinates of Asymmetrical Aerofoil Cross Section, *JMES*, v. 1, 1959, p. 241.

Carnegie, W. Vibrations of Pretwisted Cantilever Blading: An Additional Effect due to Torsion, *Proc Inst Mech Engrs.*, v. 176, 1962, p. 315.

Carnegie, W. and Dawson, B. Vibration Characteristics of Straight Blades of Asymmetric Airfoil Cross Section, *Aero Qly* v. 20, 1969, p. 178.

Chuvikovskii, U.S., Flexo-torsional Vibrations of Non-Prismatic Bars, Taking into Account the Shear Deformation Caused by Transverse Shear Forces and Energy Dissipation, *Izv Akad Nauk SSSR, Otd Tekh Nauk*, v. 3, 1959, p. 72.

Garland, C.F., Normal Modes of Vibration of Beams Having Noncollinear Elastic and Mass Axes, *Trans ASME*, v. 62, 1940, p. A97.

Gere, J.M., Torsional Vibrations of Beams of Thin Walled Open Section, *J. Applied Mech*, v. 21, #3, 1954, p. 381.

Lo, H., and Goulard, M., Torsion with Warping Restraint form Hamilton's Principle, *Proc 2nd Midwestern Conf on Sold Mechanics*, 1955, p. 68.

Pickett, G., Equations for Computing Elastic Constants from Flexural and Torsional Resonant Frequencies of Vibrations of Prismatic Cylinders, *Proc American Soc of Testing Materials*, v. 45, 1945, p. 846.

Rao, J.S., Fundamental Torsion Vibration of a Cantilever Beam of Triangular Cross Section with Uniform Taper, *Proc 10th Cong ISTAM*, 1965, p. 66.

Rao, J.S., Vibration of Pretwisted Tapered Cantilever Beams in Torsion, Archiwum Budowy Maszyn, Tom 18 (3), 1971, p. 443.

Rao, J.S., Coupled Bending Bending Torsion Vibration of Cantilever Beams, *J Aero Soc India*, v. 24, 1972, p. 265.

Rao, J.S., Torsional Vibration of Pretwisted Cantilever Beams, *J Inst Engrs India*, CE Div, v. 52, 1972, p. 211.

Rao, J.S., A Refined Theory of Torsional Vibration of Bars, Proc. Shock and Vibration Conf, Melbourne, 1974, p. 171.

Rao, J.S., Belgaumkar, B.M. and Carnegie, W., Torsional Vibration of a Cantilever Beam of Rectangular Cross Section with Uniform Taper, *BMEE* v. 9, 1970, p. 61.

Rao, J.S. and Carnegie, W., Solution of the Equations of Motion Coupled Bending Bending Torsion Vibrations of Turbine Blades by the Method of Ritz-Galerkin, *Intl J Mech Sci* v. 12, 1970, p. 875.

Rao, J.S. and Carnegie W., Torsional Vibration of Pretwisted Tapered Cantilever Beams Treated By Collocation Method, *Ind J Pure and Appl Physics*, v. 10, 1972, p. 566.

Reissener, E., On Nonuniform Torsion of Cylindrical Rods, *J Math Phy*, v. 31, 1952, p. 214.

Timoshenko, S.P., Theory of Bending, Torsion and Buckling of Thin walled member of open Cross-section, *J. Franklin Inst*, v., 239, 1945, p. 201, 249, 343.

Vet, M., Torsional Vibration of Beams Having Rectangular Cross Sections, *J. Acoust Soc America*, v. 34, 1962, p. 1570.

Wieckoswki, J., The Influence of Material Damping on Nonconservative Reactions of Elastic Transducing Torsional and Longitudinal Vibrations, *Arch Mech Stos*, v. 10, #4, 1958, p. 479.

TIMOSHENKO BEAMS

Anderson, R.A., Flexural Vibrations in Uniform Beam According to Timoshenko Theory, *J Appld Mech*, v. 20, #4, 1953, p. 504.

Anderson, R.A., Wave Groups in the Flexural Motion of Beams Predicted by Timoshenko Beam Theory, *J Appld Mech*, v. 21, #4, 1954, p. 388.

Barr, A.D.S., Some Notes on the Resonance of Timoshenko Beams and Effects of Lateral Inertia on Flexural Vibrations, *9th Cong Intl Mech and Appl.*, University of Brussels, v. 7, 1957, p. 448.

Barr, A.D.S., Distortion and Timoshenko Beam Equation, *J Appld Mech*, v. 26E, #1, 1959, p. 143.

Boley, B.A., On the Use of Sine Transforms in Timoshenko Beam Impact Problems, *J Appld Mech*, v. 24, #1, 1957, p. 152.

Boley, B.A. and Chao, C.C., Some Solutions of Timoshenko Beam Equation, *J Appld Mech*, v. 22, #4, 1955, p. 579.

Boley, B.A. and Chao, C.C., An Approximate Analysis of Timoshenko Beam under Dynamic Loads, *J Appld Mech*, v. 25, #1, 1958, p. 31.

Carnegie, W., A Note on the Use of Variational Method to Derive the Equations of Motion of Vibrating Beam of Uniform Cross-section taking Account of Shear Deflection and Rotary Inertia. *Bull Mech Engg Educ.*, v. 2, 1963, p. 35.

Carnegie, W., Vibrations of Pretwisted Cantilever Blading Allowing for Rotary Inertia and Shear Deflection, *J Mech Engng Sci.*, v. 6, #2, 1964, p. 105.

Carnegie, W. and Thomas, J., The Effect of Shear Deformation and Rotary Inertia on the Lateral Frequencies of Cantilever Beams in Bending, *J Engr for Indus. Trans ASME* v. 94, 1972, p. 267.

Carr, J.B., The Effect of Shear Flexibility and Rotary Inertia on the Natural Frequencies of Uniform Beams, *Aero Qly*, v. 21, #1, 1970, p. 79.

Cheng, F.Y., Vibrations of Timoshenko Beams and Frames, *J Strs Div, Proc ASCE*, v. 96, 1970, p. 551.

Chuvikovskii, U.S., Flexo-torsional Vibrations of Non-prismatic Bars, taking into account the Shear Deformation caused by Transverse Shear Forces and Energy Dissipation, *Izv Akad Nauk SSSR, Otd Tekh Nauk*, v. 3, 1959, p. 72.

Cowper, G.R., The Shear Coefficient in Timoshenko's Beam Theory, *J Appld Mech*, v. 33, #3, 1966, p. 335.

Cowper, G.R., On the Accuracy of Timoshenko's Beam Theory, *J Engng Mech*, Proc ASCE, v. 94, 1969, p. 1447.

Dangler, M.A. and Goland, M., Transverse Impact of Long Beams, Including Rotatory Inertia and Shear Effects, *Proc 1st U.S. Natl Cong Appld Mech.* 1952, p. 179.

Davies, R.M., The Frequency of Transverse Vibration of a Loaded Fixed-Free Bar III. The effect of Rotary Inertia of the Bar, *Phi Mag*, v. 23, #7, 1937, p. 563.

Davies, R.M., The Frequency of Transverse Vibration of a Loaded Fixed-Free Bar IV. The Effect of Shearing Force of the Bar, *Phi Mag*, v. 23, #7, 1937, p. 1129.

Dawson, B., Rotary Inertia and Shear in Beam Vibration Treated by Ritz Method, *J Roy Aero Soc.*, v. 72, 1968, p. 341.

Dawson, B., Ghosh, N.G. and Carnegie, W., Modal Curves of Pretwisted Beams Allowing for Shear Deformation and Rotary Inertia, *Proc 15th Cong Ind. Soc Theo and Appld Mech*, 1970.

Dolph C. L., On the Timoshenko Theory of the Transverse Vibration, *Qly J Math*, v. 12, #2, 1954, p. 175.

Ebner, A.M. and Billington, D.P., Steady State Vibrations of Damped Timoshenko Beam, *J Str Div, Proc ASCE*, v. 94, ST3, 1968, p. 736.

Fettis, G., Effect of Rotary Inertia on Higher Modes of Vibration, *J Aero Sci.*, 1949, p. 445.

Flaherty, F.T., *Viscoelastic Timoshenko Beam Vibrations*, Ph D Thesis, 1963, New York University.

Flekser, M. Sh., On the Effect of Rotary Inertia and Shear on the Transverse Vibrations of a Bar of Finite Length, *Inzheneryi Sbornik*, v. 23, 1956, p. 138.

Florence, A.L., Travelling Force on a Timoshenko Beam, *J Appld Mech*, v. 32, 1965, p. 351.

Flugge, W., Die Ausbreitung von Biegungswellen in Staben, *ZAMM*, v. 22, 1942, p. 312.

Hearmon, R.F.S., The Influence of Shear and Rotary Inertia on the Free Flexural Vibrations of Wooden Beams, *British J Appld Phys*, v. 9, #8, 1958, p. 381.

Herrman, G., Forced Motions of Timoshenko Beams, *J Appld Mech*, v. 22, 1955, p. 53.

Howe, C.E. and Howe, R.M., Application of Electronic Differential Analyzer to the Oscillation of Beams, Including Shear and Rotary Inertia, *J Appld Mech*, v. 22, #1, p. 13.

Huang, T.C., Effect of Rotary Inertia and Shear Deformation of Beam Treated by Method of Ritz and Galerkin, *Proc 3rd. U.S. Natl Cong on Appld Mech*, 1958, p. 189.

Huang, T.C., The Effect of Rotary Inertia and Shear Deformation on Frequency and Normal Modes of Vibrations with Simple End Conditions, *J Appld Mech*, v. 28, #4, 1961, p. 579, Discussion 1962, p. 596.

Huang, T.C., Eigenvalues and Modifying Quotients of Vibrations of Beams, Univ. Wis., Engr, Expt. Stn. Rept. No. 25, 1964.

Huang, T.C., Eigen functions of Vibrations of Beams, Report #26, Engng Expt Station, University of Wisconsin, 1964.

Huang, T.C., Orthogonality and Normalizing Conditions of Normal Modes of Forced Vibrations of Timoshenko Beams, Developments in Mechanics, v. 3, #2, 1965, p. 63.

Huang, T.C. and King, C.S., New Tables of Eigen Functions Representing the Normal Modes of Vibrations of Timoshenko Beams, Developments in Theoretical and Applied Mechanics,. v. 6, 1963.

Huang, T.C. and Wu, N.C., Approximate Analysis of Flexural Vibrations of Beams, *Proc 7th Mid-western Conf. on Mechanics*, 1961.

Hurty, W.C. and Rubenstein, M.F., On the Effect of Rotary Inertia and Shear in Beams, *J Franklin Inst*, v. 278, #2, 1964, p. 124.

Khachiyan, E.E., Question of the Effect of Shear Strain in the Presence of Free and Forced Oscillations of Flexible Structures, *Dokl Akad Nauk SSSR*, v. 37, #3, 1963, p. 113.

Kruszewski, E.J., Effect of Transverse Shear and Rotary Inertia on the Natural Frequencies of a Uniform Beam, *NACA TN* 1909, 1949.

Kumai, T., Effects of Shear Deflection and Rotary Inertia on the Damping of Flexural Vibrations of Ship Hulls, Rep Res Inst Appld Mech, Kyushi Univ, v. 13, #46, 1965, p. 61.

Kumai, T., Shearing Vibrations of Ships, European Ship Building, v. 5, #2, 1956, p. 32.

Lee, H.C., A Generalized Minimum Principle and its Application to the Vibrations of a Wedge with Rotary Inertia and Shear, *J Appld Mech*, v. 30, E2, 1963, p. 176.

Lee, H.C. and Bishop, R.E.D., Applications of Integral Equations to Flexural Vibrations of a Wedge with Rotary Inertia and Shear, *J Franklin Inst*, v. 277, 1964, p. 327.

Leibowitz, R.C. and Kennard, E.H., Shear and Rotary Inertia Effects on beam Vibrations, David Taylor Model Basin Report No. 1822, 1961.

Leissa, A.W. and Hwee, M., Forced Vibrations of Timoshenko Beam with Viscous Damping, Developments in Mechanics, v. 3, #2, 1965, p. 71.

Lizararev, A.D., On the Influence of Rotational Inertia and Shear on the Transverse Vibrations of Elastically Clamped Column, Izv Akad Nauk SSSR, *Otd Tekh Nauk Mekh i Mahinostr*, v. 3, 1963, p. 134.

Mindlin, R.D. and Deresiewicz, H., Timoshenko Shear Coefficient for Flexural Vibrations of Beams, *Proc 2nd U.S. Natl Cong on Appld Mech.*, 1954, p. 175.

Nederveen, C.J. and Schwarz, F.R. Correction for Shear and Rotatory Inertia on Flexural Vibrations of Beams, *British J Appld Phy*, v. 15, #3, 1964, p. 323.

Newman, M.K., Effect of Rotatory Inertia and Shear on Maximum Strain in Cantilever Impact Excitation, *J Aero Soc*, v. 22, #5, 1955, p. 313 and 348.

Novotorzov, V.I., Method of Successive Approximations Applied to Study of Vibration of Structure, taking into account the effect of Shearing Force, Rotatory Inertia and Longitudinal Vibrations of Rigid Bents, Bull Acad Des Sci. de t' Union Des Rep Sovietiques Sociatiste, 1935.

Pan. H.H., Vibrations of a Viscoelastic Timoshenko Beam, *Proc ASCE*, v. 92, EM2, 1966, p. 213.

Plass, H.J., Some Solutions of the Timoshenko Beam for Short Pulse Type Loading, *J Appld Mech*, v. 25, #3, p. 379.

Rao, D.K., *Solutions of Vibration Problems of Rods and Beams using Refined Theories*, Ph D Thesis, I I T, Kharagpur, 1972.

Rao, J.S., Kulkarni, S.V. and Subrahmanyam, K.B., Applications of the Reissner Method to a Timoshenko Beam, *Trans. ASME, J. Appld. Mech.* v. 48, 1981, p. 672.

Raskovic, D., The Properties of Eigen function of Transverse Vibrations of Homogeneous Beams, taking into consideration the Influence of Shearing and Rotary Inertia, Rozprawy Inz, v. 6, #2, 1958, p. 205.

Ravera, R.J., Response of Simply Supported Timoshenko Beams, Bell Communications Inc., Washington, D.C. 1970.

Simmonds, J.G., Modifications of the Timoshenko Beam Necessary for Thin Walled Circular Tubes, *Intl J Mech Sci.*, v. 9, #5, 1967, p. 237.

Steele, C.R., The Timoshenko Beam with a Moving Load, *J Appld Mech*, v. 35E, #3, 1968, p. 481.

Stozek, Z., On the Application of Hamilton's Principle to the Derivation of Bending Equation taking Shear into Account, Rozprawy Inz, v. 8, #2, 1960, p. 203.

Subrahmanyam, K.B., Kulkarni, S.V. and Rao, J.S., Coupled Bending-Bending Vibrations of Pre Twisted Cantilever Blading Allowing for Shear Deflection and Rotary Inertia by the Reissner Method, *Int J Mech Sci*, 23, #9, 1981, p. 517.

Subrahmanyam, K.B., Kulkarni, S.V. and Rao, J.S., Coupled Bending-Torsion Vibrations of Rotating Blades of Asymmetric Aerofoil Cross Section with Allowance for Shear Deflection and Rotary Inertia by Use of the Reissner Method, *J Sound Vib.*, v. 75, #1, 1981, p. 17.

Subrahmanyam, K.B., Kulkarni, S.V. and Rao, J.S., Analysis of Lateral Vibrations of Rotating Cantilever Blades Allowing for Shear Deflection and Rotary Inertia by Reissner and Potential Energy Methods, *Mechanism and Machine Theory*, v. 17, #4, 1982, p. 235.

Sutherland, R.L. and Goodman, L.E., Vibration of Prismatic Bars Including Rotary Inertia and Shear Deflections, Univ. of Illinois, 1954.

Sutherland, J.G. and Goodman, L.E., Vibrations of Pismatic Bars Including Rotary Inertia and Shear Correction, Rept N6-OR1-71, T.O. VI, Project NR-064-183, 1951, also see *J Appld Mech*, v. 18, 1951, p. 217.

Szidarovszky, J., Natural Vibrations of a Bar under Axial Force, Taking into Consideration the Effect of Shearing Force and Rotary Inertia, *Acta Techn*, v. 39, 1/2, 1962, p. 29.

Tabarrok, B. and Kannopp, B.H., Analysis of the Oscillations of the Timoshenko Beams, *ZAMP* v. 18, #4, 1967, p. 580.

Tang, S.C., A Solution to the Timoshenko Beam under a Moving Force, *AIAA Journal*, v. 4, #4, 1966, p. 711.

Terazawa, K. and Matsura, K., Transverse Vibrations of Higher Frequencies of Beams of Uniform Cross-section taking into account the Effect of Shear, Osaka Univ Report 8, 1958, p. 281.

Timoshenko, S.P., On the Correction for Shear of the Differential Equation for the Transverse Vibrations of Prismatic Bars, *Phi Mag*, Ser 6, v. 41, 1921, p. 744.

Timoshenko, S.P., On the Transverse Vibrations of Bars of Uniform Cross-section, *Phil Mag*, Ser 6, v. 43, 1922, p. 125.

Trail Nash, R.W. and Collar, A.R., The Effects of Shear Flexibility and Rotatory Inertia on Bending Vibrations of Beams, *Qly J Mech and Appld Math*, v. 6, #2, 1953, p. 186.

Trent, H.M. An Equivalent Circuit for Vibrating Beam which Includes Shear Motion, *J Aco Soc America*, v. 22, #3, 1950, p. 531.

Tseitlin, A., On the Effect of Shear-Deformation and Rotary Inertia in Vibrations of Beams on Elastic Foundation, *Appld Math Mech*, v. 25, #2, 1961, p. 531.

VIBRATION OF UNIFORM AND TAPERED BEAMS

Akimasa, O., Lateral Vibrations of Tapered Bars, *JSME*, v. 28, 1925, p. 429.

Ayre, R.S. and Jacobsen, L.S., Natural Frequencies of Continuous Beams of Uniform Span Length, *J Appld Mech*, v. 17, #4, 1950, p. 391.

Belgaumkar, B.M., Lakshminarayana, K. and Nakra, B.C., Vibration Characteristics of Axial Flow Compressor Blades, *J Sci Engng Res*, v. 6, Pt 2, 1962, p. 153.

Bellin, A.I., Determination of the Natural Frequencies of the Bending Vibrations of Beams, *J Appld Mech*, v. 14, #1, 1947, p. A1.

Boley, B.A., An Approximate Theory of Lateral Impact on Beams, *J Appld Mech*, v. 22, #1, 1955, p. 69.

Born, M. and Pugh, H.L.D., Vibration of a Thin Vertical Cantilever Caused by a Damped Harmonic Disturbance of the Ground, *J Inst Civil Engrs*, v. 18 #7, 1942, p. 279.

Burgreen, D., Free Vibrations of a Pin Ended Column with Constant Distance between Pin Ends, *J Appld Mechs*, v. 18, #2, 1951, p. 135.

Chandrasekharan, A.R., Dynamic Behaviour of Cantilever Beams, *J Ind. Soc Earthquake Tech*, v. 6, #2, 1969, p. 77.

Conway, H.D., Calculations of Frequencies of Truncated Pyramids, *Aircraft Engr* v. 18, 1946, p. 235.

Eisner, E., Resonant Vibrations of a Cone, *J Aco Soc America*, v. 36, 1964, p. 309.

Euler, L., Methodus Inveniendi Lineas Curvas Maximi Minimive Proprietate Gaudentes, Additamentum, De Curvis Elasticis, Lausanne, 1744.

Flekser, M.Sh., On Transverse Vibrations of Bars, *Appl Math Mech* v. 22, #5, p. 980.

Fromme, J.A. *Vibrations of the Rectangular Parallelepiped with Traction Free Boundary*, Ph D Thesis, 1967, Ohio State University.

Fromme, J.A. and Leissa, A.W., Free Vibrations of the Rectangular Parallelepiped, *J Aco Soc America*, v. 48, #1, p. 290.

Gaines, J.H. and Volterra, E., Transverse Vibrations of Cantilevered Bars of Variable Cross-section, *J Aco Soc America*, v. 39, #4, p. 679.

Galerkin, B.G., Rods and Plates. Series Occurring in Various Questions Concerning the Elastic Equilibrium of Rods and Plates, Vestnik Inzhenerov, v. 19, 1915, p. 897.

Goland, M., Wickersham, P.D. and Dengler, M., Propagation of Elastic Impact in Beams in Bending, *J Appld Mech*, v. 22, #1, 1955, p. 1, See also discussion, Dec 1955, p. 608.

Golovin, G. Ya, The Vibrations of Bars in Presence of Frictional Forces, Nauch Zap Lvovsk Politekn, in-ta 27, 1955, p. 109.

Gonda, J., Bending Natural Frequencies of Short Simple Beams at Various Boundary Conditions, Stojnick Casopis, v. 17, #4, 1966, p. 364.

Hasan, S.A. and Barr, A.D.S., Cross-sectional Deformation Effects in Vibrations of Thin walled Beams of Arc Section, *J Mech Engg Sci.*, v. 7, #3, 1965, p. 292.

Hess, R., *Theoretical Vibrations of Beams*, Ph D Thesis, 1949, University of Michi- gan.

Hetenyi, M., A General Solution for the Bending of Beams of Arbitrary Discontinuity, *J Appld Phy*, v. 21, #1, 1950, p. 55.

Hoppmann, W.H., II: Forced Lateral Vibration of Beam Carrying a Concentrated Mass, *J Appld Mech*, v. 19, #3, 1952, p. 301.

Jacobsen, L.S., Natural Frequencies of Uniform Cantilever Beams of Symmetrical Cross-section, *J Appld Mech*, v. 5, #1, 1938, p. 1.

Jacobsen, L.S., Natural Frequencies of Uniform Cantilever Beams, *Proc ASCE*, v. 64, #3, 1938, p. 431 and 1536.

Jacobsen, L.S., Natural Periods of Uniform Cantilever Beams, *Trans ASCE*, v. 1014, 1939, p. 402.

Karnopp, B.H. and Fung, J.C., On the Approximate Determination of Natural Frequencies and Modes of Cantilever Beams, *Acta Mechanica*, v. 9, #12, 1970, p. 109.

Kelvin Lord, On Graphic Solution to Dynamical Problems, *Phil Mag*, v. 34, 1892.

Krishna Murthy, A.V., Vibrations of Short Beams, *AIAA Journal*, v. 8, #1, 1970, p. 34.

Lang, K.W. and Nemat-Nasser, S., An Approach for Estimating Vibration Characteristics of Non-uniform Rotor Blades, *AIAA J*, v. 17, 1979, p. 995.

Lee, W.F.Z. and Saibel, E., Free Vibrations of Constrained Beams, *J Appld Mech*, v. 19, #4, 1952, p. 471.

Macduff, J.N. and Felgar, R.P., Vibration Design Charts, *Trans ASME*, v. 79, #7, 1957, p. 1459.

Martin, A.I., Some Integrals Relating to Vibration of Cantilever Beams and Approximation for the Effect of Taper on Overtone Frequencies, *Aero Qly*, v. 7, 1956, p. 109.

Maryamova, F.A., New Method for Investigating the Appearance of Transverse Vibrations of a Rod, Sb Trudi Mosk Zoachn Poligr En-ta, v. 5, 1957, p.161.

Maryamova, F.A., Investigations of the Free Vibrations of Thin Elastic Rod, Visnik Akad Badwa i Arkhitekt Ukr SSR, v. 2, 1960, p. 38 and 74.

Meyer, F., Mathematische Theorie der Transversalen Schwingungen Eines Stabes von Veranderlichen Querchnitt, *Ann der Physik und Chemie*, v. 33, 1888, p. 661.

Mikhlowitz, J., Flexural Wave Solution of Coupled Equations representing the more Exact Theory of Bending, *J Appld Mech, Trans ASME*, v. 75, 1953, p. 511.

Mikhlowtiz, J., Recent Developments in Elastic Wave Propagation, *Appld Mech Rev.*, v. 13, 1960, p. 865.

Mindlin, R.D. and Goodman, L.E., Beam Vibrations with Time Dependent Boundary Conditions, *J Appld Mech*, v. 17, #4, 1950, p. 377.

Mok, G.H. and Murray, E.A., Free Vibrations of a Slender Bar with Nonuniform Characteristics, *J Aco Soc America*, v. 40, #2, 1966, p. 385.

Newman, M.K., Viscous Damping in Flexural Vibrations of Beams, *J Appld Mech*, v. 26, 1959, p. 367.

Nicholson, J.W., The Lateral Vibrations of Bars of Variable Cross Section, *Proc Roy Soc*, v. XCIIIA, 1917, p. 506.

Nobuo, I., A Technique for Computing Natural Frequencies of Trapezoidal Vibrators, *J Aero Sci.* v. 24, 1957, p. 239.

Nothmann, G.A., Vibration of a Cantilever Beam with Prescribed End Motion, *J Appld Mech*, v. 15, #4, 1948, p. 327.

Osinski, Z., Free Vibrations of a System with One Degree of Freedom in the Presence of Structural Damping, Izv Akad Nauk SSR, Mekh i Mashino Str. v. 5, 1963, p. 42.

Petrov, A.A., Stress Waves in Turbine Blades, Izv Uyssh Ucheba. Zaveden Energitica, v. 9, 1964, p. 95.

Pfeiffer, F., On the Differential Equations of Transverse Vibrations of a Rod, *ZAMM*, v. 27, 1947, p. 83.

Pickett, G., Equations for Computing Elastic Constants from Flexural and Torsional Resonant Frequencies of Vibrations of Prismatic Cylinders, *Proc American Soc of Testing Materials*, v. 45, 1945, p. 846.

Pickett, G., Flexural Motion of Unrestrained Cylinders and Disks, *J Appld Phy*, v. 16, #12, p. 820.

Prasad, C., Effect of Viscous Damping on Flexural Vibrations of a Rod, *Proc Natl Inst of Sci, of India*, v. 31A, #2, 1965, p. 152.

Rao, J.S., *Effect of Taper on Uncoupled Natural Frequencies of Cantilever Beams*, Ph D Thesis, I I T, Kharagpur, 1965.

Rao, J.S., The Fundamental Flexural Vibration of a Cantilever Beam of Rectangular Cross Section with Uniform Taper, *Aero Qly*, v. 16, 1965, p. 139.

Rao, J.S. and Carnegie, W., Determination of the Frequencies of Lateral Vibrations of Tapered Cantilever Beams by the Use of Ritz-Galerkin Process, *BMEE* v. 10, 1971, p. 239.

Ripperger, E.A. and Abramson, H.N., A Study of Wave Propagation of flexural Waves in Elastic Beams, *J Appld Mech*, v. 24, #3, 1957, p. 431.

Ritz, W., Uber Eine Neue Methods zur Losung Gewisser Variations Problem der Mathematischen Physik, *J Reine Angew Math*, v. 135, 1908, p. 1.

Robinson, A., Shock Transmission in Beams, *A R C Tech Report* 2265, 1950.

Scanlan, R.H., Note on Transverse Bending of Beams having both Translating and Rotating Mass Elements, *J Aero Sci*, v. 15, #7, 1948, p. 425, 426 and 434.

Snowdon, J.C., Response of a Simply Supported Beam to Vibratory Forces and Moments, *J Aco Soc of America*, v. 36, #3, 1964, p. 495.

Soler, A.I., Higher Order Effects in Thick Rectangular Elastic Beams, *Intl J Solids and Structures*, v. 4, 1968, p. 723.

Soler, A.I., Higher Order Theories for Structural Analysis using Legendre Polynomial Expansion, *J Appld Mech*, v. 36E, #4, 1969, p. 757.

Southwell, R., On a Graphical Method for Determining the Frequencies of Lateral Vibration for a Rod of Non Uniform Cross-section, *Phil Mag*, ser 6, v. 41, 1921, p. 419.

Suychiro, K., On the Damped Vibrations of Prismatic Bars, *Bull Earth Quake Res Inst*, Tokyo, v. 6, 1929, p. 63.

Tomar, J.S., Flexural Vibrations of Uniform Beams, *Bull Calcutta Math Soc.*, v. 54, #3, 1962, p. 151.

Uflyand, Ya S., The Propagation of Waves in the Transverse Vibrations of Bars and Plates, *P M M*, v. 12, 1948, p. 287.

Volterra, E., Some Recent Progress in Vibrations, *Proc 10th Midwestern Conf. on Mechanics*, 1967, p. 1.

Walker, P.B., Simple Formulae for Fundamental Natural Frequencies of Cantilevers, R & M 1831, 1938.

Wang, H.C., *A General Treatment of the Transverse Vibration of Tapered Beams*, Ph D Thesis, Univ of Illinois, 1966.

Ward, P.F., Transverse Vibrations of a Rod of Variable Cross Section, *Phil Mag*. v. 25, 1913, p. 85.

Warner, W.H., The Dynamical Equations of Beams, *Developments in Mechanics*, v. 3, #2, 1965, p. 119.

Watanabe, I., Natural Frequencies of Flexural Vibrations of Tapered Cantilevers with Uniform Thickness, *Proc Ist Jap Natl Cong Appl Mech*., 1951, p. 547.

White, W.T., An Integral Equation Approach to Problems of Vibrating Beams, *J Franklin Inst*, v. 245, 1948, p. 25 and 117.

Wrinch, D.M., Lateral Vibrations of Bars of Conical Type, *Proc Roy Soc* v. 101, 1922, p. 493.

Young, D., Vibration of a Beam with Concentrated Mass, Spring, and Dashpot, *J Appld Mech*, v. 15, #1, 1948, p. 65.

Young, D. and Felgar, R.P., Tables of Characteristic Functions Representing Normal Modes of Vibration of a Beam, Univ of Texas Rept 4913, 1949.

VIBRATION OF BEAMS WITH MOVING LOADS

Ayre, R.S., Jacobsen, L.S. and Hsu, C.S., Transverse Vibrations of One End of Two Span Beams under the Action of Moving Mass Load, *Proc 1st U.S. Natl Cong Appld Mech*, 1952, p. 81.

Filippov, A.P. and Kokhmanyuk, S.S., Dynamic Effect of Mobile Loads on Rods, *Izdatelstvo Naukov Dumko*, Kiev, 1967.

Gesund, H. and Young, D., Dynamic Response of Beams to Moving Loads, Intl Assn for Bridge and Str Engg Publication, 1961.

Kaczowski, Z., Vibrations of a Beam Under Moving Load, *Proc Vibration Problems*, v. 4, #4, 1963, p. 357.

Rao, D.K., *Solutions of Vibration Problems of Rods and Beams using Refined Theories*, Ph D Thesis, I I T, Kharagpur, 1972.

Saibel, E. and Lee, W., Vibrations of Continuous Beams under Constant Moving Forces, *J Franklin Inst*, v. 254, 1952, p. 499.

Schlack, A.L. and Raske, T.F., Resonances of Beams due to Cyclic Moving Loads, *Proc ASCE*, v. 92, EM6, 1966, p. 175.

Steele, C.R., The Timoshenko Beam with a Moving Load, *J Appld Mech*, v. 35E, #3, 1968, p. 481.

Valestsos, A.S.C., Analysis of Dynamic Response of Highway Bridges, *J Engng Mech, Proc ASCE*, v. 96, 1970, EM6, p. 593.

VIBRATION OF BEAMS WITH AXIAL FORCE

Szidarovszky, J., Natural Vibrations of a Bar under Axial Force, Taking into Consideration the Effect of Shearing Force and Rotary Inertia, *Acta Techn*, v. 39, 1/2, 1962, p. 29.

Woinowsky-Krieger, S., The Effect of an Axial Force on the Vibration of Hinged Bars, *J Appld Mech*, v. 17, #1, 1950, p. 35.

PRE-TWISTED BEAMS

Anliker, M. and Troesch, B.A., Lateral Vibrations of Pre-twisted Rods with Various Boundary Conditions, *ZAMP*, v. 14, 1963, p. 218.

Carnegie, W., Vibrations of Pre-twisted Cantilever Blading, *Proc Instn Mech Engrs*, v. 173, 1959, p. 343.

Carnegie, W., Dawson, B. and Thomas, J., Vibration Characteristics of Cantilever Blading, *Proc Inst Mech Engrs*, v. 180, #31, 1965-66, p. 71.

Dawson, B., *Vibration Characteristics of Cantilever Beams of Uniform Cross Section*, Ph D Thesis, Univ of London, 1967.

Dawson, B., Coupled Bending Vibrations of Pre-twisted Cantilever Blading Treated by Raleigh-Ritz Method, *JMES*, v. 10, 1968, p. 381.

Dawson, B. and Carnegie, W., Modal Curves of Pre-Twisted Beams of Rectangular Cross Section, *JMES* v. 11, 1969, p. 1.

Dawson, B., Ghosh, N.G. and Carnegie, W., Modal Curves of Pretwisted Beams Allowing for Shear Deformation and Rotary Inertia, *Proc 15th Cong Ind. Soc Theo and Appld Mech*, 1970.

Diprima, R.C. and Handelman, G.H., Vibration of Twisted Beams, *Qly Appl Math*, v. 12, 1954, p. 241.

Dunholter, R.J., Static Displacement and Coupled Frequencies of a Twisted Bar, *J Aero Sci*, v. 13, 1946, p. 214.

Martin, A.I., Approximation for the Effect of Twist on Vibration of a Cantilever Blade, *Aero Qly*, v. 8, 1957, p. 291.

Rao, J.S., Flexural Vibration of Pre-twisted Tapered Cantilever Beams Treated by Galerkin Method, *J Engng for Indus, ASME*, v. 94, 1972, p. 343.

Sato, Y., Vibrations of Twisted Rods and Effect of Position on Elastic Axis, *Trans JSME*, v. 24, 1958, p. 866.

Schimogo, T., Lateral Vibrations of Twisted Rods with Variable Cross Section, *Proc 7th Natl Cong Appl Mech*, 1957, p. 343.

Slyper, N.A., Coupled Bending Vibrations of Pre-twisted Cantilevers, *JMES* v. 4, 1962, p. 365.

Subrahmanyam, K.B., Kulkarni, S.V. and Rao, J.S., Coupled Bending-Bending Vibrations of Pre Twisted Cantilever Blading Allowing for Shear Deflection and Rotary Inertia by the Reissner Method, *Int J Mech Sci.*, v. 23, #9, 1981, p. 517.

Troesch, B.A., Anliker, M. and Zeigler, H., Lateral Vibrations of Twisted Rods, *Qly Appl Math*, v. 12, 1954, p. 163.

ASYMMETRIC BEAMS

Carnegie, W. and Dawson, B., Vibration Characteristics of Straight Blades of Asymmetric Airfoil Cross Section, *Aero Qly*, v. 20, 1969, p. 178.

Garland, C.F., The Normal Modes and Vibrations of Beams Having Noncollinear Elastic and Mass Axes, *J Appl Mech., Trans ASME*, 1948, p. A-97.

Houbolt, J.C. and Brooks, G., Differential Equations of Motion for Combined Flapwise Bending Chordwise Bending and Torsion of Twisted Nonuniform Rotor Blades, *NACA* 1346, 1958.

Kulkarni, S.V., Subrahmanyam, K.B., and Rao, J.S., Coupled Bending-Bending Torsion Vibrations of Turbomachine Blading Treated by Reissner Method, *Proc Ist Natl Conf Mechanisms and Machines*, Bombay, 1981, p. R-19.

Rao, J.S., Coupled Bending Bending Torsion Vibration of Cantilever Beams, *J Aero Soc India*, v. 24, 1972, p. 265.

Rao, J.S. and Carnegie, W., Solution of the Equations of Motion Coupled Bending-Bending Torsion Vibrations of Turbine Blades by the Method of Ritz-Galerkin, *Intl. J. Mech. Sci.* v. 12, 1970, p. 875.

Rao, J.S., Vibration of Turbine Blades, Aero. R and D. Board, Ministry of Defense, India, ARDB-STR-5002, 1977.

Subrahmanyam, K.B., Kulkarni, S.V. and Rao, J.S., Coupled Bending-Torsion Vibrations of Rotating Blades of Asymmetric Aerofoil Cross Section with Allowance for Shear Deflection and Rotary Inertia by Use of the Reissner Method, *J Sound Vib*, v. 75, #1, 1981, p. 17.

ROTATING BEAMS

Banerjee, S. and Rao, J.S., Relations for Natural Frequencies of Turbine Blading, *Proc 22nd Cong Ind Soc Theo & Appl Mech.*, 1977, p. 34.

Boyce, W.E., Effect of Hub Radius on the Vibrations of a Uniform Bar, *J Appl Mec., Trans ASME.* v. 23, 1956, p. 287.

Boyce, W.E., Diprima, R.C. and Handelman, G.R., Vibrations of Rotating Beams of Constant Section, *Proc 2nd U.S. Natl Cong Appld Mech*, 1954, p. 165.

Carnegie, W., Vibrations of Rotating Cantilever Blading: Theoretical Approaches to the Frequency Problem Based on Energy Methods, *JMES* v. 1, 1959, p. 235.

Carnegie, W., The Application of the Variational Method to Derive the Equations of Motion of Vibrating Cantilever Blading under Rotation, *BMEE*, v. 6, 1967, p. 29.

Hirsch, G., Investigation of Vibrations in Bending of Rotating Turbine Blades on Nonrigid Support, Jahrbuck Wissenschaft., Gessellsch Luftfahrt, 1958, p. 174.

Horway, G., "Chordwise and Beamwise Bending Frequencies of Hinged Rotor Blades", *J. Aero, Sci.* v. 15. p. 497 (1948).

Jadvani, H.M., *Vibration of Turbine Blades*, Ph D Thesis, IIT Delhi, 1982.

Kissel, W., The Lowest Natural Bending Frequency of a Rotating Blade of Uniform Cross Section, *Escher Wyss. News* v. 31, 1958, p. 28.

Krupka, R.M. and Baumanis, A.M., Bending-Bending Mode of a Rotating Tapered Twisted Turbomachine Blade including the Effect of Rotary Inertia and Shear Deformation, *J Engng for Indus, Trans ASME*, 1969, p. 1017.

Lo, H., A Nonlinear Problem in the Bending Vibration of a Rotating Beam, *J Appl Mech., Trans ASME*, v. 19, 1952, p. 461.

Lo, H. and Renbarger, J., Bending Vibrations of a Rotating Beam, *Proc Ist US Natl Cong Appl Mech.*, ASME, 1952, p. 75.

Niordson, F., Vibration of Turbine Blades with Loose Hinge Supports, *Engr Digest* v. 15, 1954, p. 359.

Rao, J.S., Flexural Vibration of Turbine Blades, Archiwum Budowy Maszyn, Tom XVII #3, 1970, p. 375.

Rao, J.S., ***Turbine Blade Vibration***, D Sc Thesis, IIT, Kharagpur, 1971.

Rao, J.S., Coupled Vibrations of Turbomachine Blades, Shock Vib. *Bull., U.S. Naval Res. Lab.*, Proc. No. 47, Pt. 2, 1977, p. 107.

Rao, J.S., "Turbine Blade Vibration", *23rd Cong. Ind. Soc. Theort Appl. Mech.*, p. 150 (1978).

Rao, J.S. and Banerji, S., Coupled Bending Torsion Vibrations of Rotating Blades, *ASME* 76 GT 43.

Rao, J.S. and Carnegie, W., Nonlinear Vibration of Rotating Cantilever Beams, *J Roy Aero Soc*, v. 74, 1970, p. 161.

Rao, J.S. and Carnegie, W., Nonlinear Vibrations of Rotating Cantilever Beams Treated by Ritz Averaging Procedure, *J Roy Aero Soc*, v. 76, 1972, p. 566.

Rao, J.S. and Jadvani, H.M., Free and Forced Vibration of Turbine Blades, Vibration of Bladed Disk Assemblies, *ASME publ*, 1983, p. 11.

Rao, J.S. and Rao, D.K., Equations of Motion of Rotating Pre-Twisted Blades in Bending-Bending-Torsion with Effects of Warping, Shear, Rotary Inertia, etc., *Proc Silver Jubilee Conf Aero Soc of India*, Bangalore, 1974, p. 4-13.

Schilhansl, M.J., Bending Frequency of Rotating Cantilever Beam, *J Appl Mech., ASME* v. 25, 1958, p. 28.

Subrahmanyam, K.B., Kulkarni, S.V. and Rao, J.S., Analysis of Lateral Vibrations of Rotating Cantilever Blades Allowing for Shear Deflection and Rotary Inertia by Reissner and Potential Energy Methods, *Mechanism and Machine Theory*, v. 17, #4, 1982, p. 235.

Swaminadham, M., Study of the Vibrations of Rotating Blades, *ASME* 77-DET-147.

Swaminatham, M. and Rao, J.S., Vibrations of Rotating Pretwisted and Tapered Blades, *Mech. Mach, Theory*, v. 12 #4, 1977, p. 331.

Taylor, J.L., Natural Vibrational Frequencies of Flexible Rotor Blades, *Aircraft Engr.* v. 30, 1958, p. 331.

Yntema, R.T., Simplified Procedures and Charts for the Rapid Estimation of Bending Frequencies of Rotating Beams, NACA TN 3459, 1955.

Zuladzinski, C., Effect of Centrifugal Forces on Lateral Deflection and Fundamental Frequency of Turbine Blades, *ASME* 77-Pet-12.

VISCOELASTIC BARS AND BEAMS

Achenbach, J.D. and Sun, C.T., Dynamic Response of Beam on Viscoelastic Subgrades, *Proc ASCE*, v. 91, EM5, 1965, p. 61.

Agbasiere, J.A. and Grootenhuis, P., Flexural Vibrations of Symmetrical Multilayer Beams with Viscoelastic Damping, *J Mech Engg Sci.*, v. 10, #3, 1968, p. 269.

Asnani, N.T. and Nakra, B.C., Vibration Analysis of Multilayered Beams with Alternating Elastic and Viscoelastic Layers, *J Instn Engrs (India)*, v. 50, #7, ME4, 1970, p. 187.

Chou, F.H. and Achenbach, J.D., Field Equations Governing the Mechanical Behaviour of Layered Circular Cylinder, *AIAA J*, v. 8, #8, 1970, p. 1445.

Ditaranto, R.A., Theory of Vibratory Bending for Elastic and Viscoelastic Layered Finite Length Beams, *J Appld Mech.*, v. 32E, 1965, p. 881.

Kerwin, E.M., Damping of Flexural Waves by a Constrained Viscoelastic Layer, *J Aco Soc America*, v. 31, # 7, 1959, p. 952.

Nakra, B.C., Vibration Response of a Sandwich Beam with Nonlinear Viscoelastic Core, *J Sci. & Engng Res*, v. 13, 1969, p. 137.

Lazaryeen, U.A., On the electrical Simulation of Transient Conditions of Motion of a Bar, Trudi Dne Propetroski Inta-Inzh Zh-a Transp No. 25, 1956, p. 84.

Oberst, H., Uber die Dampfung der Biegeschwingungen duner Bleiche durch fest haftenden Belage, Acoustica, v. 2, #4, 1952, p. 181.

Oberst, H. and Becker, G.W., Uber die Dampfung der Biegeschwingungen duner Bleiche durch fest haftenden Belage, *Acoustica*, v. 4, 1954, p. 433.

Plass, A.J., Damping of Vibrations in Elastic Rods of Sandwitch Structures by Incorporating Additional Viscoelastic Material, *Proc 3rd. Midwestern Conf. on Solid Mech*, 1957, p. 48.

Rao, D.K., *Solutions of Vibration Problems of Rods and Beams using Refined Theories*, Ph D Thesis, IIT, Kharagpur, 1972.

Rao, D.K. and Rao, J.S., A General Theory of Vibrations of Sandwich Beams, *Proc 17th Ind. Soc Theo & Appld Mech.*, 1972, p. 30.

Ripperger, E.A., Experimental Studies of Damping in Elastic Bars by Lateral Surface Treatment, Univ of Texas Def Res Lab, Report No. DRL-374.

Ruzicka, J.E., Damping Structural Resonances Using Viscoelastic Shear Damping Mechanisms, Pt I and II, *J Engng for Indus*, v. 83, 1961, p. 403.

Schwarz, L.F., Forced Bending and Extensional Vibrations of a Two Layer Compound Linear Viscoelastic Beam, *Acoustica*, v. 8, #3, 1958, p. 164.

BEAMS ON ELASTIC AND VISCOELASTIC FOUNDATION

Crandall, S.H., The Timoshenko Beam on Elastic Foundation, *Proc 3rd. Midwestern Conf. on Solid Mech*, 1957, p. 146.

Criner, H.E. and McCann, G.D., Rails on Elastic Foundation under the Influence of High Speed Travelling Loads, *J Appld Mech*, v. 75E, 1953, p. 13.

Essenberg, F., Shear Deformation in Beams on Elastic Foundation, *J Appld Mech*, v. 84E, #2, 1962, p. 313.

Filonenko-Borodich, M.M., Some Approximate Theories of the Elastic Foundation, Uchenyi Zapiski Ploskovskog Gosuderstvennogo Universiteta, Mechanica, v. 46, 1940, p. 3.

Flaherty, F.T., *Viscoelastic Timoshenko Beam Vibrations*, Ph D Thesis, 1963, New York University.

Hatenyi, M., Beams and Plates on Elastic Foundation and Related Problems, *Appld, Mech Reviews*, 1966, p. 95.

Kenney, J.T., Steady State Vibrations of a Beam on Elastic Foundation for Moving Load, *J Appld Mech*, v. 21, #2, p. 359.

Kerr, A.D., Elastic and Viscoelastic Foundation Models, *J Appld Mech*, v. 31, #3, 1964, p. 491.

Mathewa, P.M., Vibrations of a Beam on Elastic Foundation, *ZAMM*, v. 38, #3/4, 1958, p. 105.

Mathewa, P.M., Vibrations of a Beam on Elastic Foundation, *ZAMM*, v. 39, #1/2, 1959, p. 13.

Paria, G., Time Hardening and Time Softening Elastic Materials, Part I: Free Vibrations of a Finite Rod, *Ind. J. Mech and Math*, v. 4, #2, 1966, p. 117.

Rades, M., Steady State Response of Finite Beam on a Pasternak Type Foundation, *Intl J. Solids and Structures*, v. 6, 1970, p. 739.

Rao, D.K., *Solutions of Vibration Problems of Rods and Beams Using Refined Theories*, Ph D Thesis, IIT, Kharagpur, 1972.

Stadler, W. and Shreeves, R.W., The Transient and Steady State Response of an Infinte Bernoulli-Euler Beam with Damping and Elastic Foundation, *Qly J Mech Appld Math*, v. 23, #2, 1970, p. 192.

Tseitlin, A., On the Effect of Shear Deformation and Rotary Inertia in Vibrations of Beams on Elastic Foundation, *Appld Math Mech*, v. 25, #2, 1961, p. 531.

Volterra, E., Vibrations of Elastic Systems Having Hereditary Characteristics, *J Appld Mech*, v. 17, #4, 1950, p. 363.

Volterra, E., On Elastic Continuum with Hereditary Characteristics, *J Appld Mech*, v. 18, #3, p. 273.

Volterra, E., On the Response to External Load of Continuous Elastic Systems with Hereditary Damping Characteristics, *Proc 1st U.S. Natl Cong Appld Mech*, 1951, p. 57.

Winkler, E., Die Lehre von der Elasticitaet und Festig Keit, Prag, Domisicus, 1867.

Index